HOW TO USE THIS BOOK

- This book contains thousands of carefully constructed questions.

- The questions are organised into curriculum-aligned Question Sets that deal with individual learning objectives.

- The Question Sets are structured and categorised by topic and subtopic to help you plan using Learning by Questions.

- Each Question Set is set out in this book with all its questions, answers and diagrams or illustrations to allow you to check suitability for your class.

- Once you have found the perfect Question Set you'll be able to find it on our website at www.lbq.org using the Quick Search References.

Everything in this book is online at www.lbq.org

LbQ can be delivered in these ways:

Self-Paced Questioning
Set a task for your pupils – they each receive questions on their devices, respond at their own pace and receive instant feedback.

Ad Hoc Questioning
Pose questions on the spur of the moment – your pupils respond on their devices.

Teach
Turn any question into a slide, ideal for modelling concepts on your classroom display.

Self-Paced Questioning*

Lbq.org is built on tens of 1,000s of questions. Questions are grouped into carefully scaffolded Question Sets to provide structured support for learning and to help pinpoint problems.

When you click the 'Start' button, you're opening the door for your class to start a differentiated learning journey.

START BUTTON

Pupils connect with a simple code and start receiving questions straight away.

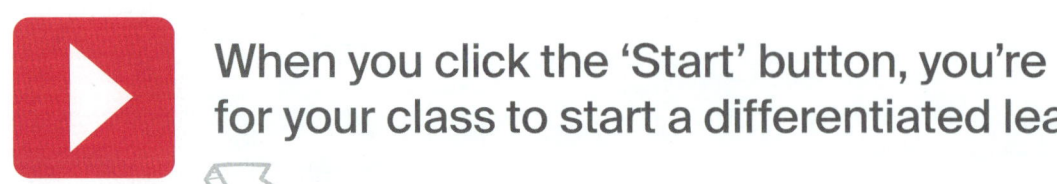

Enter your code

4 4 3

using LbQ Tasks

App Store | Google play | Get it from Microsoft

*After the initial free trial, a subscription is required for Self-Paced and Ad Hoc tasks. Teach mode remains free.

Register FREE at lbq.org

When your pupils hit a challenging question, **you'll know about it.**

Drill down to see every answer and be on top of every misconception.

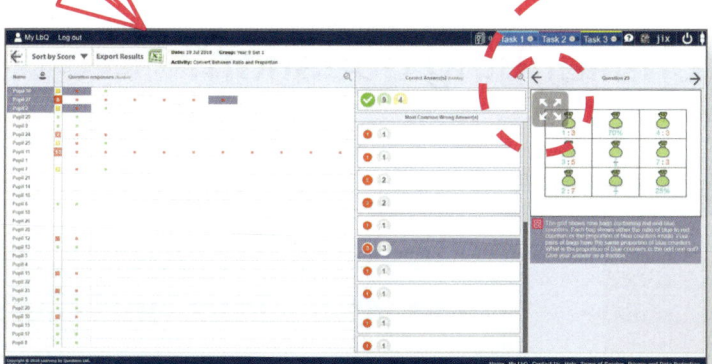

Pause, intervene, explore, explain and model using 'Teach' mode.

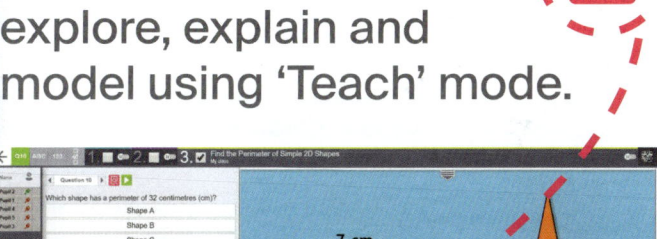

Try the question again with Ad Hoc questioning.

Everything in this book is online at www.lbq.org

USE THE BOOK TO PICK THE RIGHT TASKS FOR YOUR CLASS AND BE READY FOR INTERVENTION OPPORTUNITIES.

Ad Hoc Questioning

Have your pupils got tablets, laptops, Chromebooks, PCs? Lbq.org makes it fast, easy and super-productive to engage your whole class.

THE AD HOC QUESTION ICON

(always available top right on lbq.org) is your gateway to asking questions:

at any time,
of everyone,
about anything,
even during a Self-Paced task.

Are your class struggling with a challenging question?

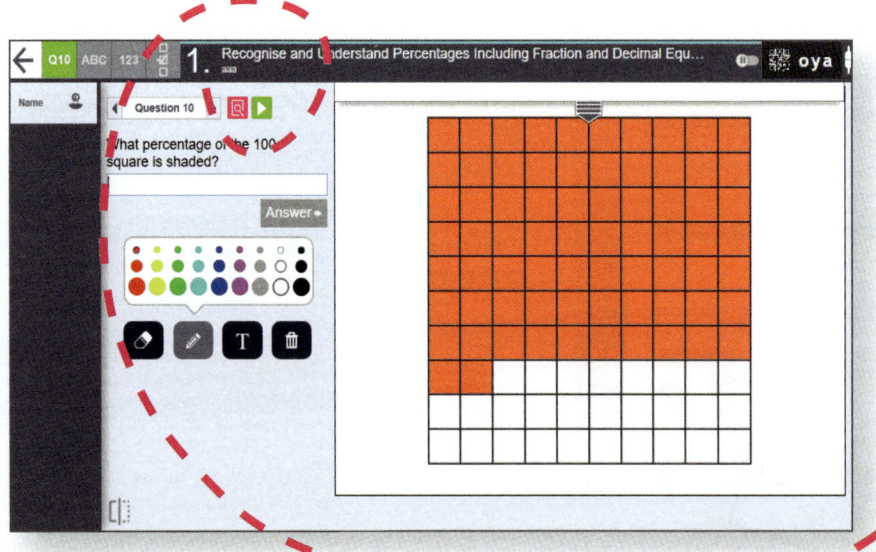

Use the green button to send the question to your class and give everyone another chance to answer.

TRY AD HOC QUESTIONING ON FREE TRIAL FOR *60 DAYS!

*Subject to change at LbQ discretion.

Register FREE at lbq.org

Want to build on an existing lbq.org question?

Annotate to explain, model, modify and extend...

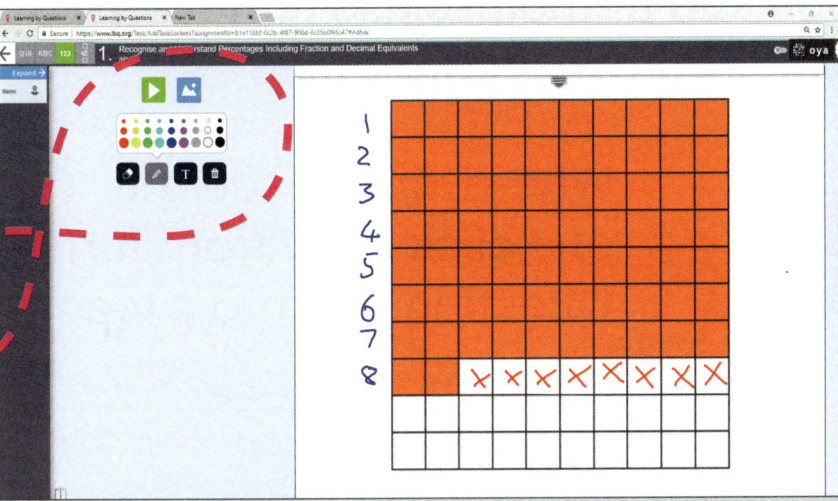

...and use Ad Hoc to send as a new question to your class.

Make your own questions on the fly.

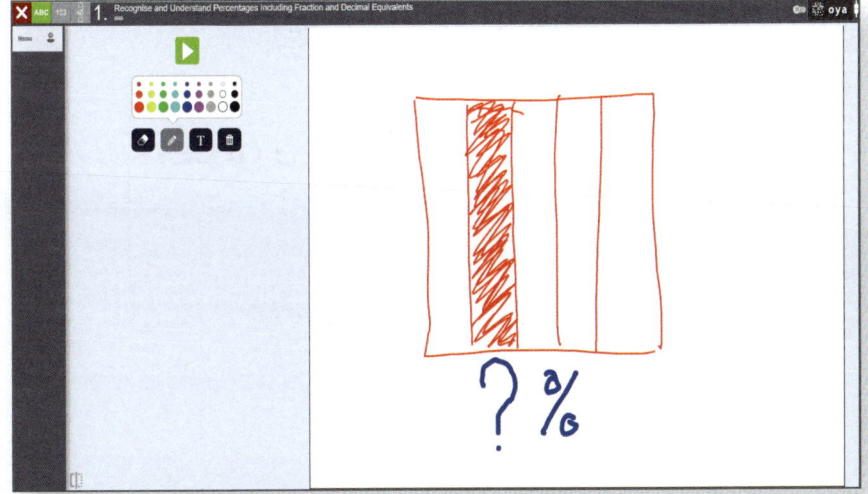

Use our Teach Tools to create the right question at the right time.

Write a question, draw a question or just ask a question.

Forget 'hands up'. With Ad Hoc questioning everyone answers, every question, every time!

USE THE BOOK TO CHOOSE, ORGANISE AND REHEARSE QUESTIONS TO TEACH IN YOUR CLASSES.

Teach

If you don't have pupil devices, you can still turn any question into a whole-class teaching resource on your classroom display.

Wherever you see the 'Teach' icon, you're one click away from turning a question or Question Set into a teaching resource.

TEACH ICON

Each question can be an ideal teaching point.

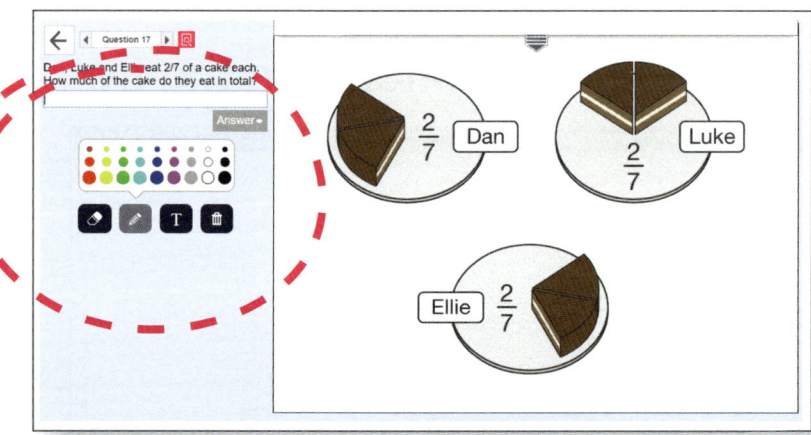

Annotate to explain, model and work as an example.

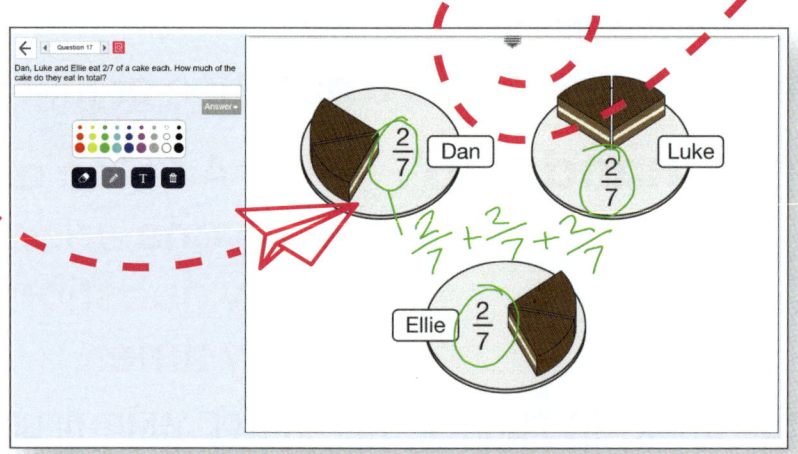

Use the pull-down pad to construct your own questions.

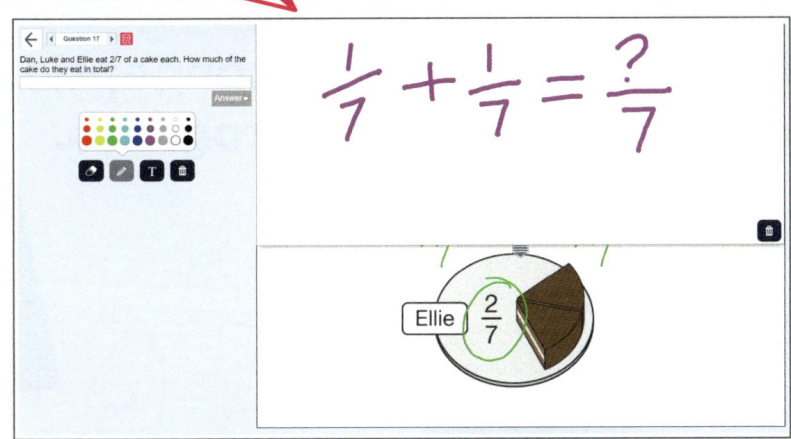

Work through multiple questions like a slide show.

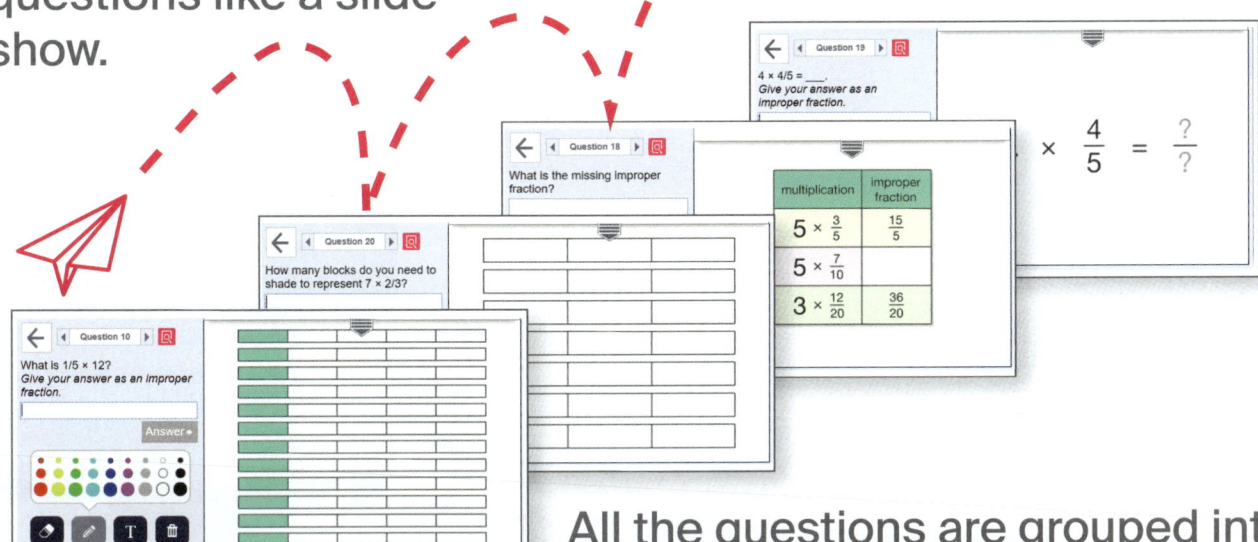

All the questions are grouped into carefully scaffolded Question Sets and provide great support for progression from understanding to problem solving.

TEACH MODE IS **FREE** TO ALL REGISTERED USERS

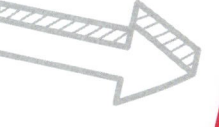

Register FREE at lbq.org

USE THE BOOK TO HELP FIND, ORGANISE AND REHEARSE QUESTIONS TO INCLUDE IN YOUR LESSONS.

Tailored Instant Feedback with Every Question

Feedback corrects misconceptions and reinforces learning to accelerate progress.

LbQ's digital platform provides pupils and teachers with instant feedback.

Our feedback is designed to reflect the kind of specific feedback a teacher would provide, given the time to do so.

Our feedback gives specific guidance and pupils are given the opportunity to retry the question.

Feedback is high quality and consistent across all Question Sets, relating directly to learning points taken from the National Curriculum.

Confidence and resilience building is developed through feedback. Encouraging students to try again takes away the 'fear of failure'.

STEPPED GUIDANCE

We don't just give the correct answer. We give feedback that provides short steps to reach the right answer. This is effective in its simplicity, providing guidance on the action needed to find the correct answer.

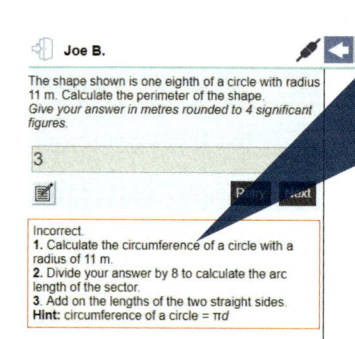

Incorrect.
1. Calculate the circumference of a circle with a radius of 11 m.
2. Divide your answer by 8 to calculate the arc length of the sector.
3. Add on the lengths of the two straight sides.
Hint: circumference of a circle $= \pi d$

Joe B.

The shape shown is one eighth of a circle with radius 11 m. Calculate the perimeter of the shape.
Give your answer in metres rounded to 4 significant figures.

3

Incorrect.
1. Calculate the circumference of a circle with a radius of 11 m.
2. Divide your answer by 8 to calculate the arc length of the sector.
3. Add on the lengths of the two straight sides.
Hint: circumference of a circle $= \pi d$

11 m

COMMON MISCONCEPTIONS

Common misconceptions are addressed in feedback to encourage pupils to find the correct solution themselves.

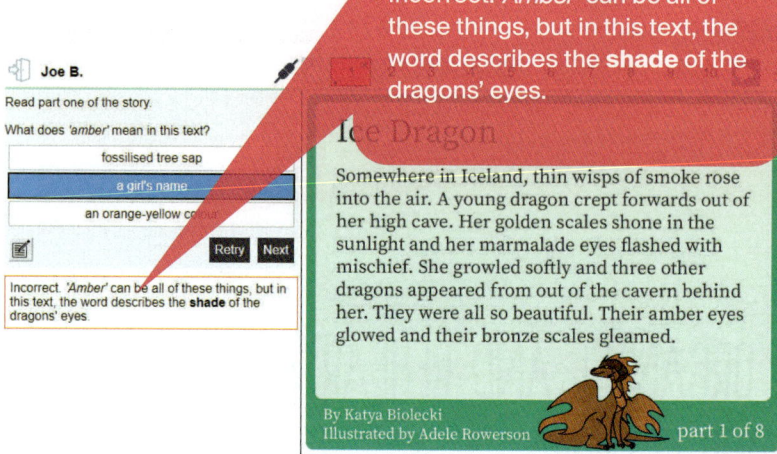

Incorrect. *'Amber'* can be all of these things, but in this text, the word describes the **shade** of the dragons' eyes.

Joe B.

Read part one of the story.

What does *'amber'* mean in this text?

fossilised tree sap

a girl's name

an orange-yellow colour

Incorrect. *'Amber'* can be all of these things, but in this text, the word describes the **shade** of the dragons' eyes.

Ice Dragon

Somewhere in Iceland, thin wisps of smoke rose into the air. A young dragon crept forwards out of her high cave. Her golden scales shone in the sunlight and her marmalade eyes flashed with mischief. She growled softly and three other dragons appeared from out of the cavern behind her. They were all so beautiful. Their amber eyes glowed and their bronze scales gleamed.

By Katya Biolecki
Illustrated by Adele Rowerson

part 1 of 8

REINFORCING DETAIL

Correct-answer feedback provides learners with reinforcing detail that boosts confidence and produces opportunities for deeper learning experiences.

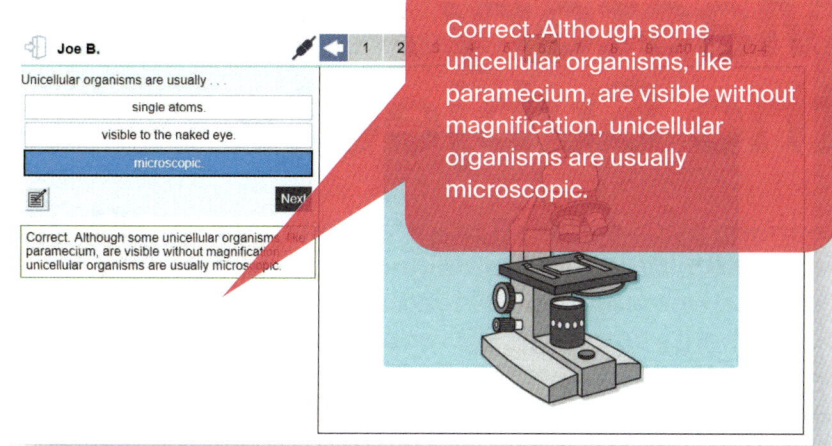

GROWTH MINDSET ENCOURAGEMENT

Feedback that provides information on how to correctly answer the question next time encourages a growth mindset by emboldening pupils to have another go.

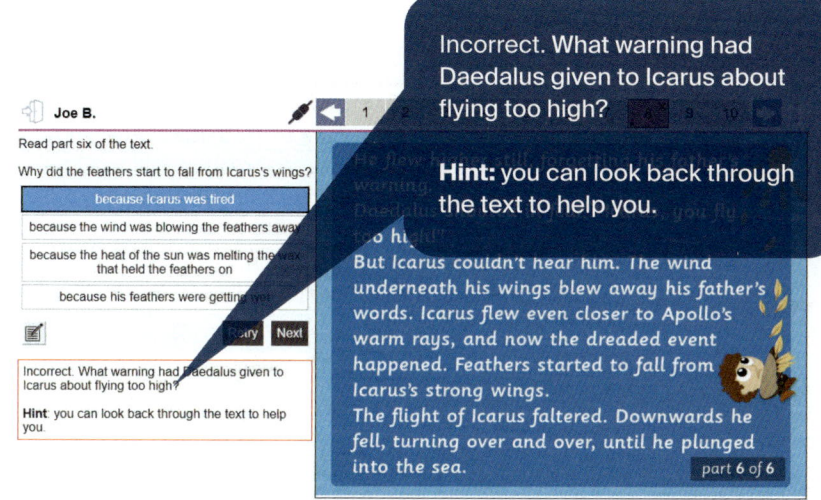

SUPPLEMENTARY INFORMATION

Supplementary information in feedback for correct answers develops understanding and nurtures curiosity.

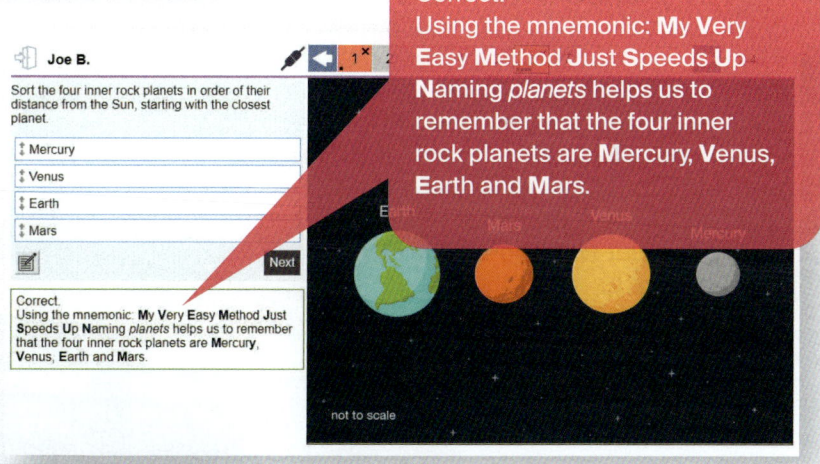

SUPPORTING IMAGERY

Feedback that explains how to use the image provided allows less-secure learners to use pictorial representations rather than abstract understanding.

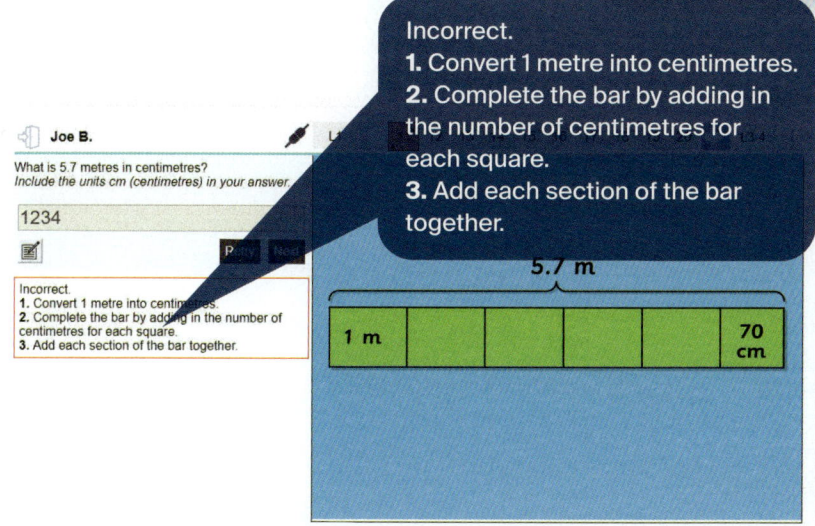

Mastery Across the Subjects

Learning by Questions (LbQ) is a classroom app filled with curriculum-aligned Question Sets and immediate pupil feedback to supercharge learning. LbQ reduces teacher workloads with automatic marking and real-time lesson reports to verify learning as it happens. LbQ covers KS2 maths, English and science, plus KS3 maths, English, chemistry, physics and biology.

Teachers use LbQ's live data to pinpoint where intervention is needed and why. Pupils use devices such as tablets, to progress through scaffolded Question Sets at their own pace and receive immediate feedback as they answer. Teachers receive live analysis and results are automatically recorded to support assessment and planning and to save time.

Mastery Approach

The mastery approach informs the creation of all of our Question Sets. At LbQ, we believe that mastering a topic or concept is vital. Each pupil's journey to mastery may be different to their peers' journeys, so it is important to us that scaffolding or guiding of learning is built into our Question Sets.

LbQ embraces digital tools to make the mastery approach to learning more achievable than it has ever been before.

Our mastery Question Sets are specifically designed to progress pupils only when they demonstrate a grasp of prerequisites, helping them to master concepts, build a growth mindset and take ownership of their learning. Equally, if a student demonstrates mastery level understanding, progression rules will allow the pupil to move on after a set amount of correct answers, pushing higher ability pupils at pace.

Mathematics

Mastery Question Sets

Maths mastery Question Sets take pupils on a learning journey through four different cognitive levels of a topic: understanding, fluency, reasoning and problem solving. To aid pupil progression, every question in every level provides contextualised feedback and addresses key misconceptions. Maths mastery Question Sets provide students with exposure to problem solving and reasoning questions, but not before previous skills and knowledge have been mastered.

WORKING TOWARDS MASTERY

Understanding
Define
Recognise
Identify

Fluency
Apply
Calculate
Solve

Reasoning
Compare
Describe
Explain

Problem Solving
Solve complex multi-stage problems

LOWER ORDER   THINKING SKILLS   HIGHER ORDER

Practice Question Sets

There are elements within the curriculum that require a different approach to learning. Practice Question Sets in maths drill one or two levels of cognition, where pupils recall and apply knowledge rapidly. Such Question Sets are perfect for practising times tables, number bonds and other basic recall of operations needed to progress in maths.

Reading Question Sets

This collection of Question Sets for both Key Stage 2 and 3 focuses on developing a wide range of reading skills. At Key Stage 2, skills in retrieval, inference and understanding vocabulary are covered by individual Question Sets on the same text. At Key Stage 3, students are prepared for the skills needed for examinations later in their education through the use of classic literature extracts and original texts.

WORKING TOWARDS MASTERY

Analysing
Language Question Sets
What do you think the author meant by using the phrase...?
Why do you think the author uses the word...?

Structure and Features Question Sets
What genre is this text?
Which paragraph explains...?
What is the purpose of...?

Inferring
Inference Question Sets
Why is/does/did...?
Which statement best describes the...?

Recalling
Retrieval Question Sets
How many...?
Which character...?
When did... take place?

Defining
Pre-Read Question Sets
What does...mean?
What do you expect this text to be about?

LOWER ORDER   THINKING SKILLS   HIGHER ORDER

Grammar, Spelling and Punctuation Question Sets

Key Stage 2 GPS Question Sets are aligned to the National Curriculum. These begin with questions that require pupils to define terminology and understand rules, before moving on to apply rules with contextualised examples.

Key Stage 3 GPS Question Sets build on the skills and knowledge students gained at Key Stage 2. GPS Question Sets at this level are created and labelled by ability to ensure each and every student in your classroom is catered for.

Reading Question Sets

Science mastery Question Sets enable pupils to progress through a carefully stepped sequence of learning. Blocks of questions are structured to guide learners as they move from knowledge, through understanding and application to analysis. Real-world applications are woven through the Question Sets, making learning relevant and relatable.

WORKING TOWARDS MASTERY

Analysing/ Evaluating
I can predict...
I can evaluate claims that...

Applying
I can apply my learning to...
I can suggest ways to...

Understanding
I can use a model to describe...
I understand how...

Knowing
I can recognise...
I can name...

LOWER ORDER   THINKING SKILLS   HIGHER ORDER

Practice Question Sets

LbQ's science practice Question Sets cover areas of the curriculum where repetition of a learning point is essential to ensure that it is thoroughly embedded. They often focus on a skill or area of the curriculum where a weaker grasp of knowledge may become a barrier to future learning, e.g. energy transfer terminology.

Investigation Question Sets

Question Sets support the skills of planning an investigation and analysing results. Pupils will be given opportunities to plan investigations and review data. In KS3, students are provided with further Question Sets that present a mix of exam-style questions and a wider application of the data.

Subscriptions

FULL ACCOUNT

Start a *60 day no-obligation trial of the FULL Account and view affordable pricing at www.lbq.org/trylbq

FULL Account subscription includes:
- connection to pupil devices
- instant pupil feedback
- automatic marking
- real-time lesson analysis
- lesson results saved

+

 **Self-Paced Questioning**

 **Ad Hoc Questioning**

 **Teach**

Use LbQ books to plan and quickly locate Question Sets at www.lbq.org/books

FREE ACCOUNT

Register at www.lbq.org/trylbq

FREE Account includes...
tens of 1,000s of questions with learning feedback for use by teachers on a classroom display.

+  **Teach**

*Subject to change at discretion of LbQ

Other Titles in the Series

Learning by Questions

PRIMARY KS2	SECONDARY KS3
Maths Year 3 Mathematics Primary Question Sets Year 4 Mathematics Primary Question Sets Year 5 Mathematics Primary Question Sets Year 6 Mathematics Primary Question Sets	**Maths** Year 7 Mathematics Secondary Question Sets Year 8 Mathematics Secondary Question Sets Year 9 Mathematics Secondary Question Sets
Science Years 3&4 Primary Science Question Sets Years 5&6 Primary Science Question Sets	**Biology** Years 7–9 Biology Secondary Question Sets
	Chemistry Years 7–9 Chemistry Secondary Question Sets
English Years 3&4 English Primary Question Sets Years 5&6 English Primary Question Sets	**Physics** Years 7–9 Physics Secondary Question Sets
	English Years 7–9 English Secondary Question Sets
Quick Search Reference Guide All Question Sets, all subjects, all years	

KS1 & KS4 Question Sets can also be found on our website www.lbq.org

See www.lbq.org/books for title availability

Understanding a Question Set

objective

Question Set title

topic

subtopic

Quick Search Reference number

level number & title

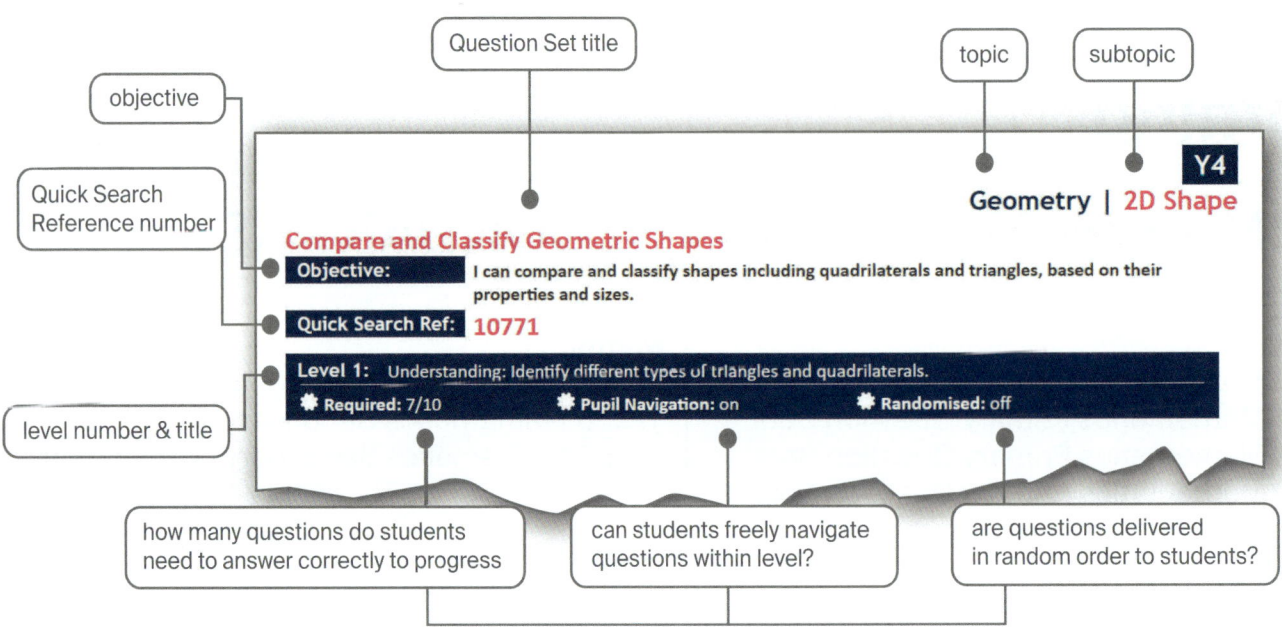

Y4

Geometry | 2D Shape

Compare and Classify Geometric Shapes

Objective: I can compare and classify shapes including quadrilaterals and triangles, based on their properties and sizes.

Quick Search Ref: 10771

Level 1: Understanding: Identify different types of triangles and quadrilaterals.

✴ Required: 7/10   ✴ Pupil Navigation: on   ✴ Randomised: off

how many questions do students need to answer correctly to progress

can students freely navigate questions within level?

are questions delivered in random order to students?

⚙ These settings can be adjusted by **adapting** the Question Set.

Understanding a question

question

answer – green for correct

29. Garth has five different-sized rectangles. He joins two rectangles together to make a square. Which two rectangles does Garth use?

2/5

- rectangle A ▪ **rectangle B** ▪ rectangle C
- **rectangle D** ▪ rectangle E

Question Type

▣ Multiple Choice

1 2 3 Numeric

2/5 (answers required)

a b c Text

↑ ↓ Sort

T F True or False

Y N Yes or No

question image or audio

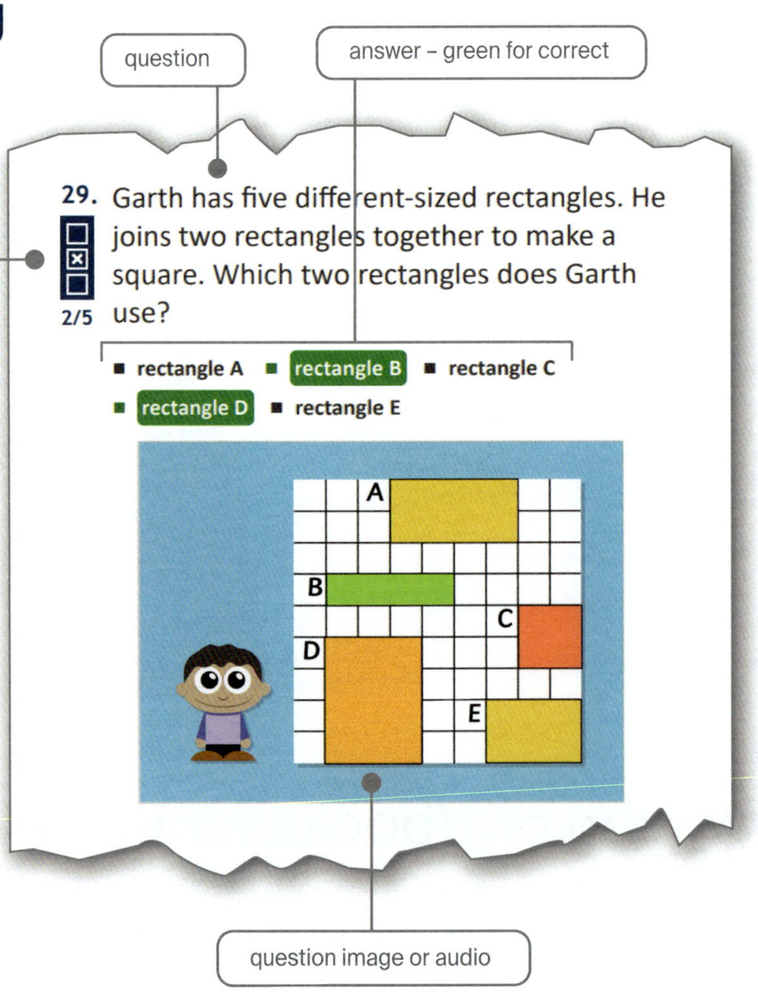

Finding a Question Set From This Book on the LbQ Platform

The **year**, **topic** and **subtopic** classifications used in this book relate directly to those used online.

The fastest way to find a specific Question Set on lbq.org is via the **Quick Search Reference Number** (e.g. 10771).

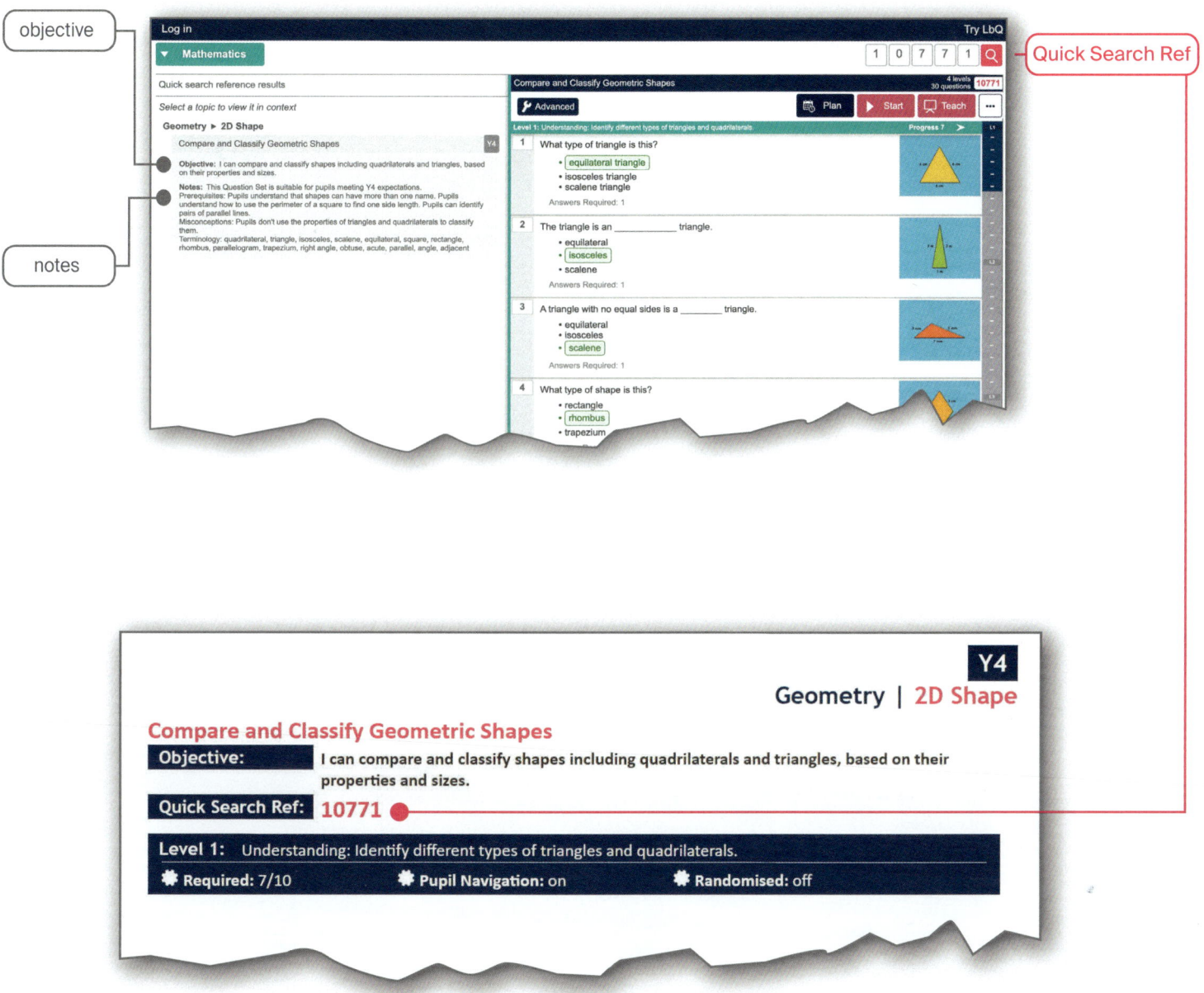

Note: The Question Sets detailed in this book are correct at time of compilation (October 2019) and correspond directly to the Question Sets as published on **www.lbq.org**.

Owing to the nature of **www.lbq.org**, we will from time to time extend, update or modify the Question Sets published, which will at times give rise to discrepancies between this book and the online resources.

Changes and additions to Question Sets will be notified on our website on a regular basis.

Topic Directory Y5

Mathematics Y5

Number

Counting

Count in Steps of Powers of 10

Objective: I can count forwards or backwards in steps of powers of 10 for any given number up to 1,000,000.

Quick Search Ref: 10104

Level 1: Understanding: Identify powers of 10 and count in powers of 10 with support.

⬡ **Required:** 7/10   ⬡ **Pupil Navigation:** on   ⬡ **Randomised:** off

1. Which **four** of these numbers are powers of 10?

4/6

▪ **10** ▪ 50 ▪ **100** ▪ **1,000** ▪ 1,010 ▪ **100,000**

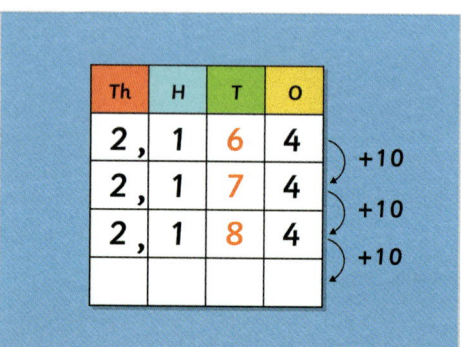

2. What number goes in the bottom row of the place value chart?

▪ **2194** ▪ 9

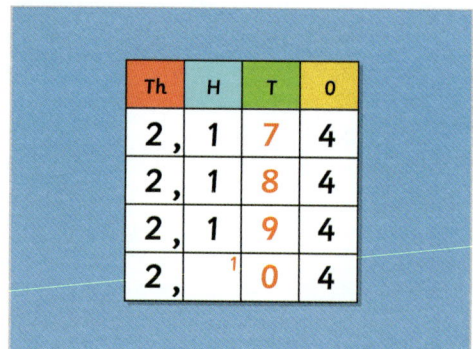

3. What is the next number in this sequence?
2,174; 2,184; 2,194; _____

▪ **2204** ▪ 2294 ▪ 2104

4. Count forwards in thousands from 2,204. What is the third number in the sequence?

▪ **4204** ▪ 3304

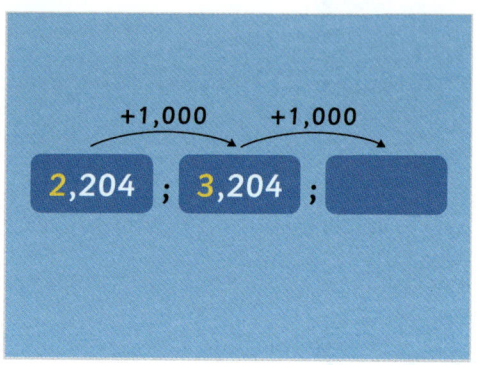

5. Count forwards in ten-thousands from 4,204. What is the fourth number in the sequence?

▪ 5204 ▪ **34204** ▪ 14204 ▪ 7204

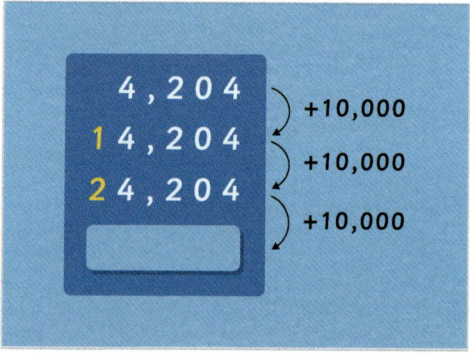

6. Count backwards in hundred-thousands from 874,693. What number goes in box C?

▪ 674693 ▪ **474693** ▪ 574693 ▪ 1074693

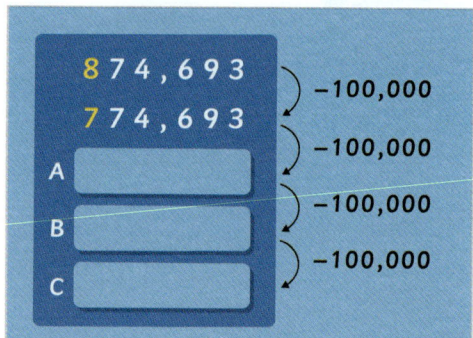

Level 1 continued

7. What is the next number in the sequence?

1 2 3

774,693; 874,693; 974,693; _____

▪ **1074693** ▪ 874693

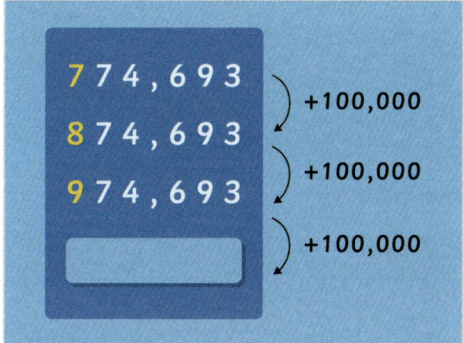

8. Count backwards in ten-thousands from 52,983. What number goes in box B?

a b c

▪ **22983** ▪ 62983 ▪ 32983 ▪ 82983

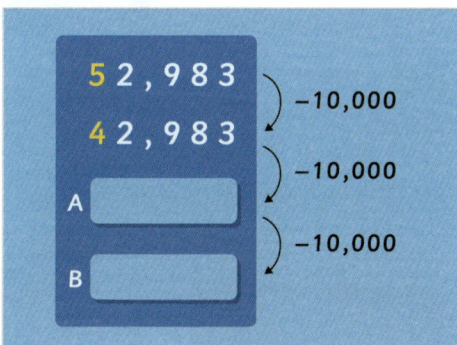

9. Count forwards in thousands from 5,602. What is the third number in the sequence?

1 2 3

▪ **7602** ▪ 6702 ▪ 5602

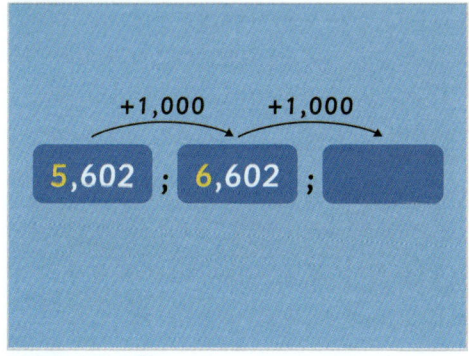

10. What number goes in the bottom row of the place value chart?

1 2 3

▪ **2381** ▪ 8

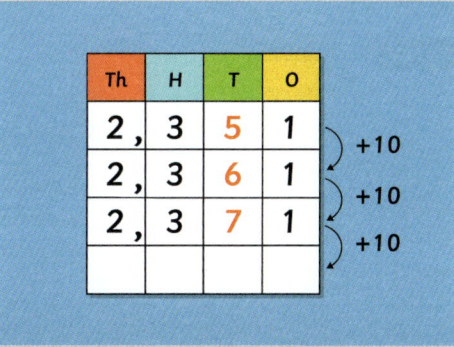

Level 2: Fluency: Count in powers of 10 and solve simple one-step calculations.

✻ **Required:** 7/10   ✻ **Pupil Navigation:** on
✻ **Randomised:** off

---

11. What number comes next in this sequence?

1 2 3

12,500; 13,500; 14,500; 15,500; _____

▪ **16500** ▪ 16000

12,500
13,500
14,500
15,500

12. Count forwards in ten-thousands from 185,038. What is the fifth number in the sequence?

1 2 3

▪ **225038** ▪ 215038 ▪ 205038

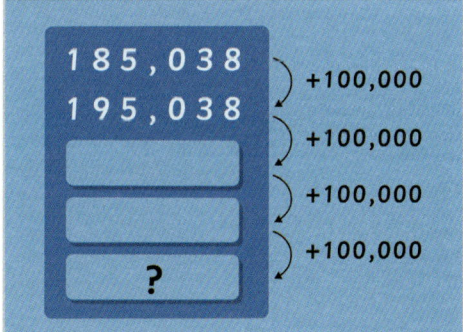

Level 2 continued

13. Count backwards in ten-thousands from 773,924. What is the fourth number in the sequence?

■ **743924** ■ 753924

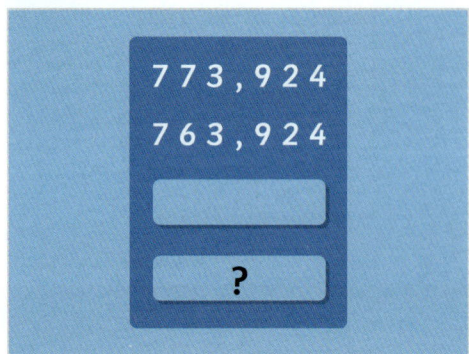

14. Charlie starts at 6,608,391 and subtracts 100,000 each time. What is the fifth number in his sequence?

■ **6208391** ■ 6108391

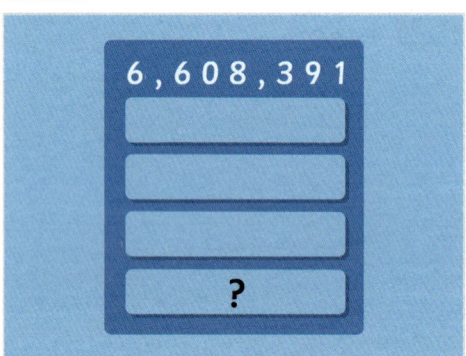

15. What number is missing from this sequence?

 ■ **74403**

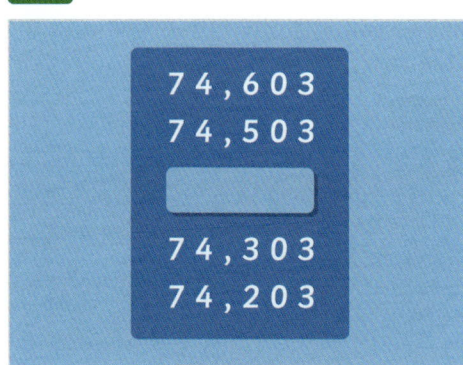

16. Jermaine makes a number using base-ten blocks. He counts forwards in ten-thousands 4 times from his number. What number does he end up with?

■ 10246 ■ **40246** ■ 246

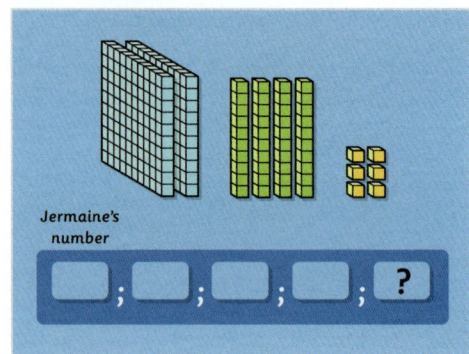

Jermaine's number

17. Wonkee's Chocolate have 42,201 bars of chocolate in stock. They sell 1,000 bars of chocolate each day. How many bars of chocolate do they have left after five days?

■ **37201** ■ 47201

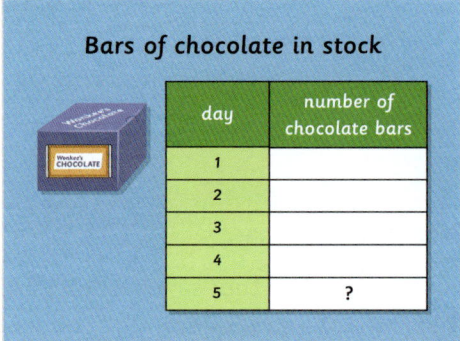

Bars of chocolate in stock

day	number of chocolate bars
1	
2	
3	
4	
5	?

18. Imogen makes a number using place value counters. She counts forwards in hundred-thousands 3 times from her number. What number does she end up with?

■ **510457** ■ 210457 ■ 310457

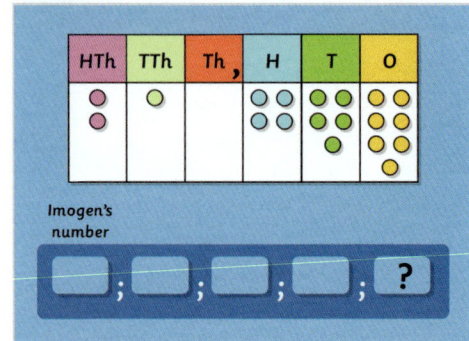

Imogen's number

Level 2 *continued*

19. What number is missing from this sequence?

 ■ 749082

20. What number comes next in this sequence?
212,300; 213,300; 214,300; _____

 ■ 215300

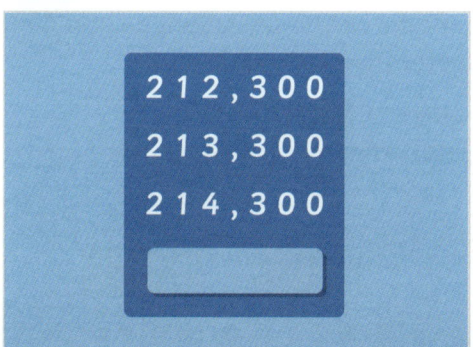

Level 3: Reasoning: Reason about counting in steps of powers of 10.

❋ **Required:** 5/5   ❋ **Pupil Navigation:** on
❋ **Randomised:** off

21. Tim is trying to count forwards in thousands. Explain what mistake Tim has made.

- Open question, no set answer

22. Jonah says, "To add 10 to a number, I only need to change one digit." Explain whether Jonah's statement is **always**, **sometimes** or **never** correct.

- Open question, no set answer

23. Maya is trying to write a number sequence, but she has made a mistake with one of the numbers. Find the mistake then enter the **correct number**.

■ 900429 ■ 909429 ■ 999429

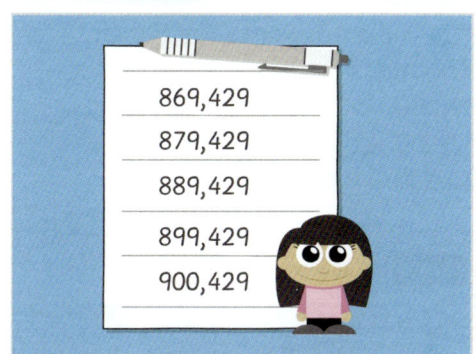

24. Naomi counts forwards in tens from 8,109. Which **three** of the following numbers will she say?

3/6 ■ 8,169 ■ 4,109 ■ 9,000.9 ■ 18,190 ■ 13,729
■ 89,119

Level 3 *continued*

25. Ashraf says, "If I count forwards in thousands
a
b
c
from 19, I will say the number 190,000."
Is Ashraf correct? Explain your answer.

- Open question, no set answer

Level 4: Problem solving with greater depth: Solve multi-step problems involving counting in powers of 10.

✹ **Required:** 5/5   ✹ **Pupil Navigation:** on
✹ **Randomised:** off

26. What is the **answer** to the fourth
1
2
3
calculation?
492,000 − 10,000 = 482,000
482,000 − 1,000 = 481,000
472,000 − 100 = 471,900
_____ − _____ = ?

■ 461990 ■ 462000 ■ 10

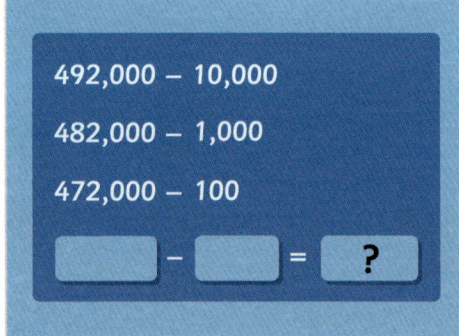

27. Isaac counts back five steps of 1,000 from
1
2
3
the number shown by the arrow. What
number does Isaac finish with?

■ 235000 ■ 240000 ■ 245000

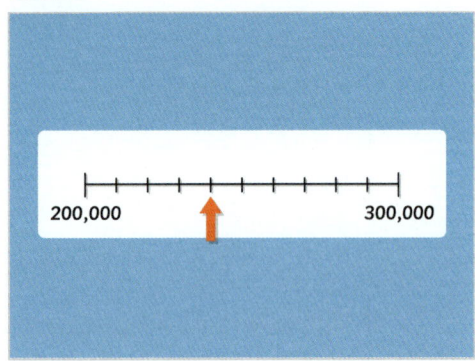

28. Liam uses place value counters to represent
1
2
3
a number. Starting with this number, he
counts forwards in thousands. How **many**
numbers will he say before he says a number
greater than 100,000?

■ 100348 ■ 79 ■ 21348 ■ 78

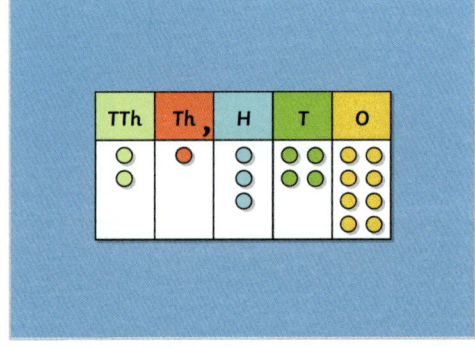

29. Ava is thinking of a number. She counts on
1
2
3
six steps of 10,000, then counts back four
steps of 10. Finally, she counts forwards
three steps of 100,000. Her answer is
894,129. What number did she start with?

■ 534169 ■ 594169 ■ 594129

Level 4 *continued*

30. Bling Motors have 12,200 cars in stock and make 100 cars each month. Jord Cars have 1,400 cars in stock and make 1,000 cars each month. How many **months** will it be until both companies have the same number of cars?

■ **12** ■ **13400** ■ **13**

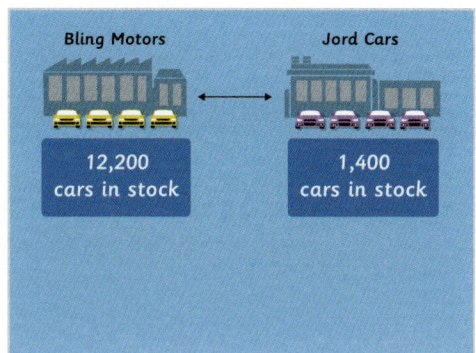

Solve Number Problems Involving Data from Tables

Objective: I can solve number and place value problems involving data from tables.

Quick Search Ref: 11722

Level 1: I can solve number and place value problems involving data from tables.

Required: 10/10   **Pupil Navigation:** on   **Randomised:** off

---

1. In degrees Celsius, what is the temperature in Bangkok rounded to the **nearest ten**? *Don't include the units in your answer.*

- **30** ▪ 20

Temperature in five cities

city	temperature (°C)
Bangkok	27
Churchill	–6
Dublin	10
Lisbon	17
Oulu	3

2. The diameters of which two planets **round down** to the nearest thousand?

2/7

- Mercury ▪ **Venus** ▪ Mars ▪ Jupiter ▪ Saturn
- Uranus ▪ **Neptune**

Diameter of planets in the solar system

planet	diameter (km)
Mercury	4,879
Venus	12,104
Mars	6,779
Jupiter	142,984
Saturn	120,536
Uranus	50,724
Neptune	49,244

3. Which planet's diameter is **closest** to 50,000 kilometres?

1/7

- Mercury ▪ Venus ▪ Mars ▪ Jupiter ▪ Saturn
- **Uranus** ▪ Neptune

Diameter of planets in the solar system

planet	diameter (km)
Mercury	4,879
Venus	12,104
Mars	6,779
Jupiter	142,984
Saturn	120,536
Uranus	50,724
Neptune	49,244

4. Which city's temperature is **closest** to 0°C?

1/5

- Bangkok ▪ Churchill ▪ Dublin ▪ Lisbon ▪ **Oulu**

Temperature in five cities

city	temperature (°C)
Bangkok	27
Churchill	–6
Dublin	10
Lisbon	17
Oulu	3

5. Sort the planets starting with the planet with the **smallest diameter** first.

- **Mercury** ▪ **Mars** ▪ **Venus** ▪ **Neptune** ▪ **Uranus**
- **Saturn** ▪ **Jupiter**

Diameter of planets in the solar system

planet	diameter (km)
Mercury	4,879
Venus	12,104
Mars	6,779
Jupiter	142,984
Saturn	120,536
Uranus	50,724
Neptune	49,244

6. What are the **two** true statements?

2/4

- **The temperature is higher in Lisbon than in Oulu.**
- The temperature is higher in Dublin than in Bangkok.
- The temperature in Churchill is higher than the temperature in Oulu.
- **The temperature in Oulu is greater than the temperature in Churchill.**

Temperature in five cities

city	temperature (°C)
Bangkok	27
Churchill	–6
Dublin	10
Lisbon	17
Oulu	3

Level 1 *continued*

7. The table shows the actual distances between different planets. To the **nearest ten thousand miles**, what is the distance between the planet Zeborg and the planet F81?

- ■ 112537   ■ **100000**   ■ 112530

Distance between planets in miles

Zeborg				
34,102	Thotov			
111,274	77,102	Akli		
112,537	78,709	574	F81	
224,932	188,084	116,157	102,411	Yuna

8. What is the total population of the **two** largest cities?

- ■ **1846662**

Populations of cities

city	population
Georgetown	348,661
Daleford	296,223
Bemford	1,225,122
Sunton	621,540
Ribham	270,463

9. Starting with the diameter of Mercury, count forwards in **steps of 100**. How many steps do you count before you reach the diameter of Mars?

- ■ **19**   ■ 1900

Diameter of planets in the solar system

planet	diameter (km)
Mercury	4,879
Venus	12,104
Mars	6,779
Jupiter	142,984
Saturn	120,536
Uranus	50,724
Neptune	49,244

10. In degrees, what is the **difference** between the temperatures in Churchill and Lisbon? *Don't include the units in your answer.*

- ■ **23**   ■ 11

Temperature in five cities

city	temperature (°C)
Bangkok	27
Churchill	−6
Dublin	10
Lisbon	17
Oulu	3

Solve Problems Involving Number and Place Value

Objective: I can solve problems involving place value, negative numbers, comparing, ordering and rounding numbers.

Quick Search Ref: **11721**

Level 1: I can solve multi-step problems involving number and place value.

✿ Required: 10/10 ✿ Pupil Navigation: on ✿ Randomised: off

1. Emily wants to buy a second-hand car that has been driven less than 55,000 miles. How many of the cars in the image could she buy?

■ **2**

2. Corey is thinking of a **two-digit number**. Use the following clues to find Corey's number:
• The number is less than 0.
• Both digits are odd.
• The sum of the digits is 16.
• The first digit is smaller than the second digit.

■ **-79** ■ 79 ■ -97

3. Jacob uses every digit card once to make the **smallest possible even number**.
Isla uses the same digit cards to make the **largest possible odd number**.
What is the difference between Jacob's number and Isla's number?

■ **752625** ■ 234798 ■ 234789 ■ 987423

4. Which **four** numbers round to 30,000 to the nearest ten-thousand?

4/7

■ **30,095** ■ 36,120 ■ **25,037** ■ **27,699** ■ 14,876
■ **34,999** ■ 22,374

nearest 10,000 → 30,000

Level 1 *continued*

5. An aircraft is 4,482 miles into a 10,000-mile journey from London to Australia. How many more miles does it need to travel so that it is closer to Australia than to London?
Don't include the units in your answer.

■ **518** ■ **5000**

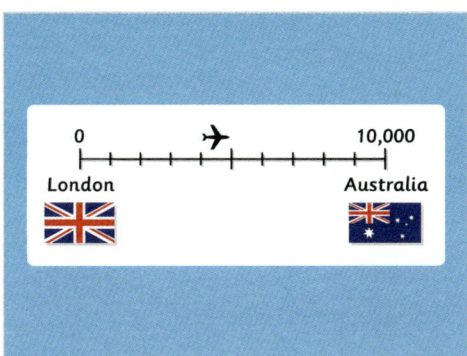

6. Anna represents a number using base-ten blocks. She counts backwards from this number in **steps of 10**. What is the first negative number that Anna says?

■ **-4** ■ **-6** ■ **18**

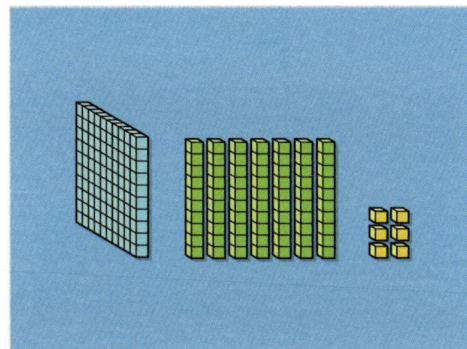

7. The Stirling family are moving house and their house is advertised at a price between £200,000 and £300,000. The Stirlings sell their house for **£10,000 less** than the advertised price, and then buy the **most expensive** house. How much more do they pay for their new house?
Include the £ sign in your answer.

■ **£233,500** ■ **233500** ■ **£243,500** ■ **£243500**
■ **243,500** ■ **£233500** ■ **243500** ■ **233,500**

8. Kiara finds the **largest** whole number that rounds to 32,900 to the nearest 100. Madison finds **smallest** whole number that rounds to 32,900 to the nearest 100. What is the difference between Kiara's number and Madison's number?

■ **99** ■ **32850** ■ **32949**

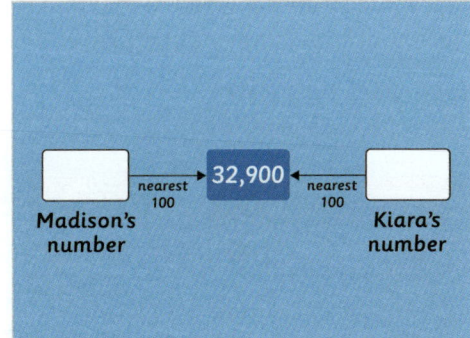

Level 1 continued

9. The table shows six children's scores on a computer game, but some of the values are missing. Use the following clues to complete the table, and then put the children's names in order, starting with the child with the **highest score**.

• Sam's score is 800,000 more than Poppy's score.

• Mia's score is 10 times smaller than Sam's score.

• Izzy's score is 131,000 less than Mia's score.

- Isaac - Sam - Poppy - Mia - Charlie - Izzy

Computer game scores

name	score
Sam	
Poppy	3,850,000
Charlie	420,000
Mia	
Izzy	
Izaac	5,970,000

10. Leo represents a number using place value counters. He can only use one counter in each place value column. Using exactly **five** place value counters, make a list of **all** of the different numbers he could make.

- Open question, no set answer

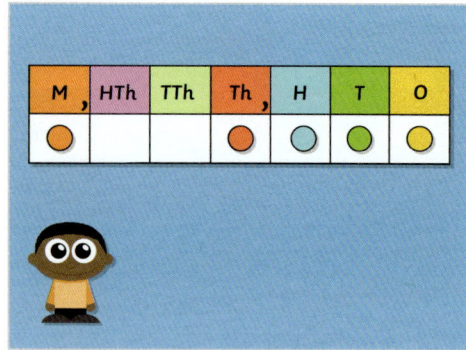

M	HTh	TTh	Th	H	T	O
●			●	●	●	●

Mathematics **Y5**

Place Value

Compare and Order
Read and Write
Negative Numbers
Rounding and Estimation

Compare and Order Numbers to at Least 1,000,000

Objective: I can order and compare whole numbers to at least 1,000,000.

Quick Search Ref: 11378

Level 1: Understanding: Compare and order numbers up to 1,000,000 with support.

✿ Required: 7/10   ✿ Pupil Navigation: on   ✿ Randomised: off

1. Which of these place value columns has the greatest value?

1/4

- thousands
- hundred-thousands
- hundreds
- ten-thousands

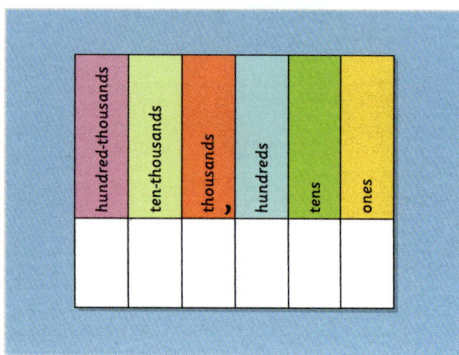

2. Which is the smallest number?

1/3

- five thousand, one hundred and seventeen
- 1,575
- 7,151

3. Which is the larger five-digit number?

- 36812

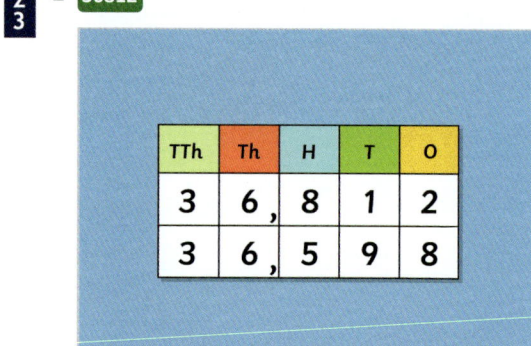

4. Which is the smallest number?

1/3

- 605,784
- 605,918
- 605,179

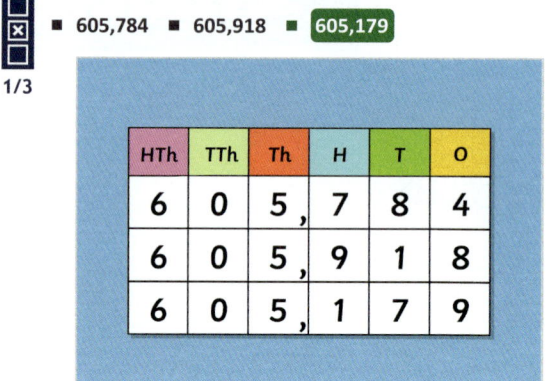

5. Sort the numbers starting with the smallest number first.

- 4,010
- 4,455
- 4,499
- 4,736

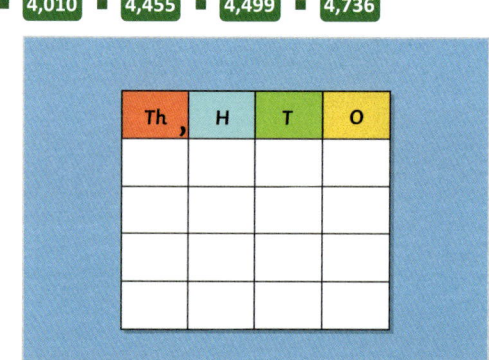

6. Sort the numbers in descending order (largest first).

- 731,764
- 727,355
- 726,010
- 720,736

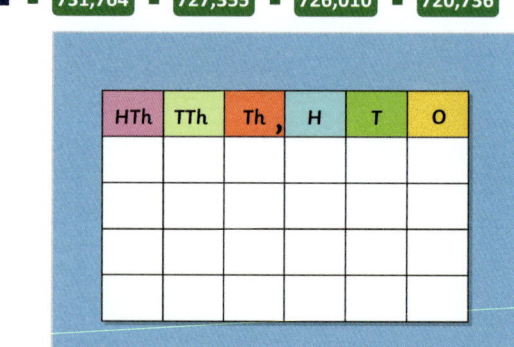

Level 1 continued

7. Sort the numbers from smallest to largest.

■ 22,604 ■ 202,600 ■ 220,000 ■ 2,200,601

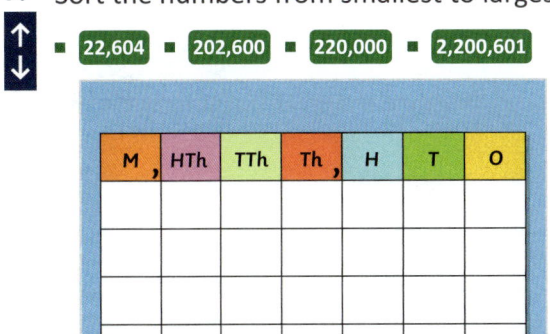

M ,	HTh	TTh	Th ,	H	T	O

8. Sort the numbers in ascending order (smallest first).

■ 29,650 ■ 30,270 ■ 30,560 ■ 31,080

TTh	Th ,	H	T	O

9. What is the largest amount of money in this list?

1/4

■ £77,950 ■ £705,900 ■ £17,500 ■ £175,095

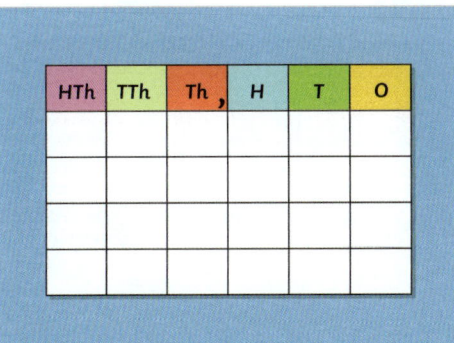

HTh	TTh	Th ,	H	T	O

10. Which is the smaller five-digit number?

1 2 3

■ 36598

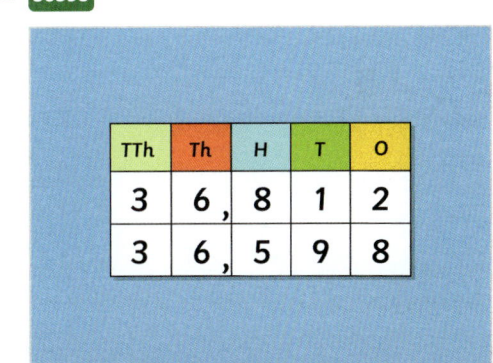

TTh	Th	H	T	O
3	6,	8	1	2
3	6,	5	9	8

Level 2: Fluency: Compare and order numbers to at least 1,000,000, including in simple contexts.

❋ **Required:** 7/10   ❋ **Pupil Navigation:** on
❋ **Randomised:** off

11. Select the longest distance.

1/4

■ 10,099 metres   ■ 9,999 metres   ■ 21,709 metres
■ 20,907 metres

> 10,099 metres
>
> 9,999 metres
>
> 21,709 metres
>
> 20,907 metres

12. Arrange these values in ascending order (smallest first).

■ 243,700 ■ 1,021,370 ■ 1,370,000 ■ 3,700,000

> 3,700,000
>
> 243,700
>
> 1,370,000
>
> 1,021,370

13. Which symbol makes the statement correct?

463,400 _____ 436,004

■ < ■ = ■ **>**

1/3

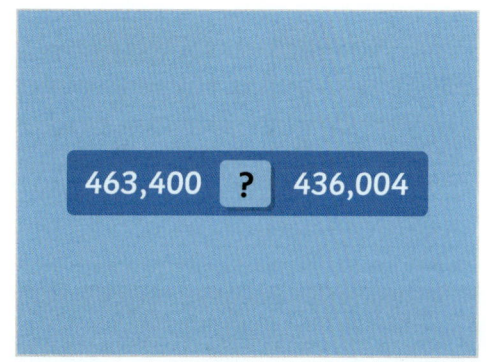

14. Which set of numbers is arranged from highest to lowest?

■ set A ■ set B ■ **set C** ■ set D

1/4

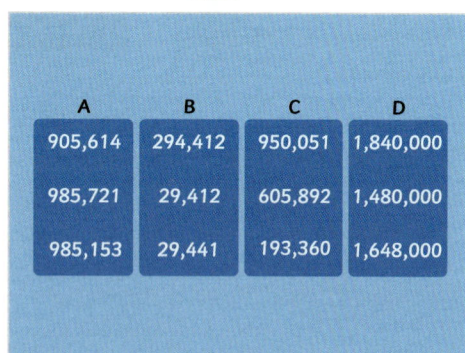

A	B	C	D
905,614	294,412	950,051	1,840,000
985,721	29,412	605,892	1,480,000
985,153	29,441	193,360	1,648,000

15. Five numbers have been sorted by size, from smallest to largest. Select the **two** missing numbers.

2/5

■ 99,800 ■ 24,000 ■ **155,000** ■ 13,900 ■ **135,300**

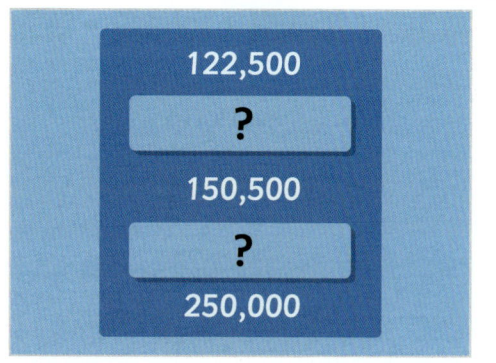

122,500

?

150,500

?

250,000

16. Which car has travelled the furthest?

■ car A ■ **car B** ■ car C ■ car D

1/4

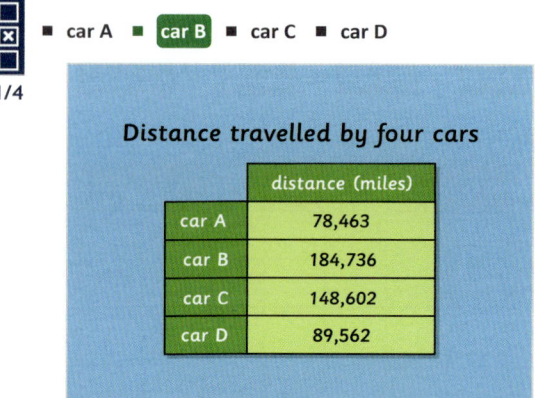

Distance travelled by four cars

	distance (miles)
car A	78,463
car B	184,736
car C	148,602
car D	89,562

17. Arrange the cities in descending order of population, starting with the city with the largest population.

■ **Bemford** ■ **Sunton** ■ **Georgetown** ■ **Daleford**
■ **Ribham**

Populations of cities

city	population
Georgetown	348,661
Daleford	296,223
Bemford	1,225,122
Sunton	621,540
Ribham	270,463

18. Which **two** houses cost less than £120,000?

■ house A ■ **house B** ■ house C ■ house D
■ **house E**

2/5

A
FOR SALE
£251,000

B
FOR SALE
£76,500

C
FOR SALE
£198,000

D
FOR SALE
£125,000

E
FOR SALE
£102,500

Level 2 continued

19. Select the largest volume of liquid.

■ **49,700 ml** ■ 745 ml ■ 1,150 ml ■ 17,820 ml

1/4

49,700 ml   745 ml

1,150 ml   17,820 ml

20. Which sign makes the statement correct?

1,499,000 ___ 1,501,000

■ **<** ■ = ■ >

1/3

1,499,000 **?** 1,501,000

Level 3: Reasoning: Reason about numbers to at least 1,000,000.

✱ **Required:** 5/5   ✱ **Pupil Navigation:** on

✱ **Randomised:** off

21. Some ink has been spilled on two of the digits in this number statement.
Which symbol (<, > or =) makes the statement true? Explain your answer.

- Open question, no set answer

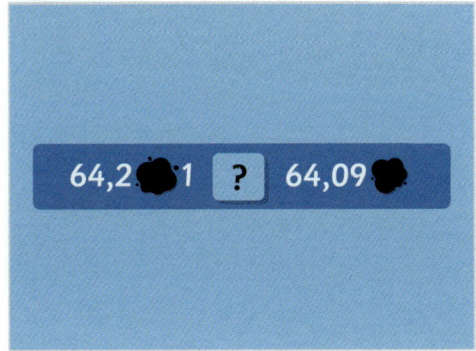

64,2●1 **?** 64,09●

22. Harriet says that she can order these numbers by looking at only the first three digits in each number. Is she correct? Explain your answer.

- Open question, no set answer

532,786
534,826
603,221
537,123
534,776
689,332

23. Abigail has written some numbers in a place value chart and arranged them from smallest to largest. The same digit has been covered by a star in each number. What digit is hidden behind the stars?

■ **2** ■ 1

TTh	Th	H	T	O
1	★ ,	3	4	2
★	2 ,	7	9	1
2	4 ,	★	0	8
★	4 ,	5	1	7
3	0 ,	★	5	1

24. Daniel says that when he orders these numbers from smallest to largest, the fourth number will be twenty-thousand more than the first number. Is he correct? Explain how you know.

- Open question, no set answer

55,015
35,015
53,013
35,033
55,555

Level 3 *continued*

25. If Layla uses each digit card once, what is the **second smallest number** that she could make?

- **123697** ■ 123679

Level 4: Problem solving with greater depth: Solve multi-step problems involving comparing and ordering numbers up to at least 1,000,000.

✴ **Required:** 5/5   ✴ **Pupil Navigation:** on
✴ **Randomised:** off

26. The Stirling family are moving house and their house is advertised at a price between £200,000 and £300,000. The Stirlings sell their house, and then buy the most expensive house. How much more do they pay for their new house?
Include the £ sign in your answer.

- **£233,500** ■ **£233500** ■ 233,500 ■ 233500

27. Emily wants to buy a second-hand car that has been driven less than 85,000 miles. How many of the cars in the image could she buy?

- **3**

28. Use the clues in the image to find a five-digit number.

- **48502**

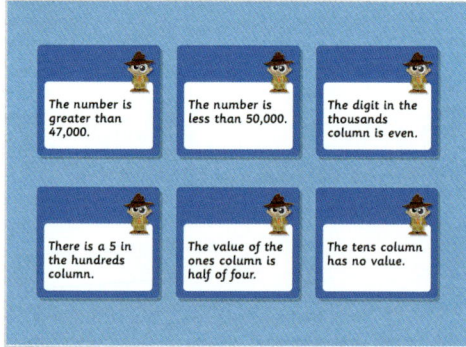

29. Luke has nine digit cards. He uses five of them to make the largest number he can. He makes the smallest number he can with the the remaining four cards.
Find the difference between the two numbers that Luke makes.

- **97531** ■ 1234 ■ 98765

Level 4 *continued*

30. Multiply the total of the digits of the largest number by the total of the digits of the smallest number.

- 540

Read, Write, Compare and Order Numbers to at Least 1,000,000

Objective: I can read, write, compare and order numbers to at least 1,000,000 and know the value of each digit.

Quick Search Ref: 10279

Level 1: Understanding: Read, write, compare and order numbers to at least 1,000,000 with support.

✱ Required: 7/10   ✱ Pupil Navigation: on   ✱ Randomised: off

---

1. What digit is in the ten-thousands column in the number 784,619?

- ■ **8**   ■ 80000

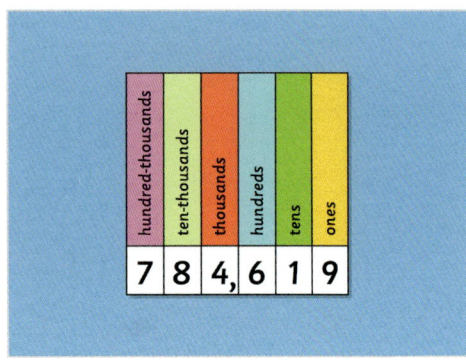

2. In digits, what number is represented by the place value counters?

- ■ **830947**   ■ 83947

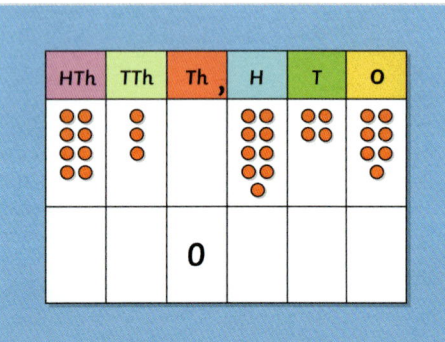

3. What is ninety-four thousand, one hundred and twenty-nine in digits?

- ■ **94129**   ■ 94000129

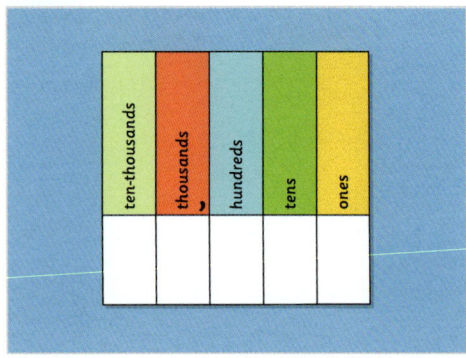

4. What is 3,452 written in words?

- ■ three thousand, five hundred and forty-two
- ■ thirty-four thousand and fifty-two
- 1/4 ■ **three thousand, four hundred and fifty-two**
- ■ three thousand, four hundred and twenty-five

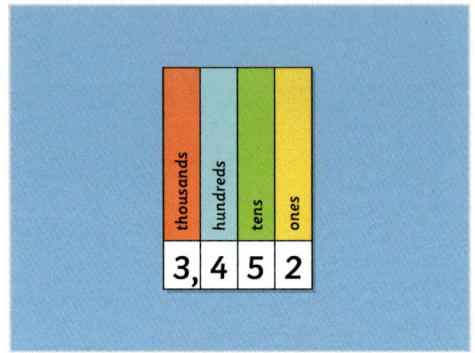

5. Which is the smaller five-digit number?

- ■ **36598**

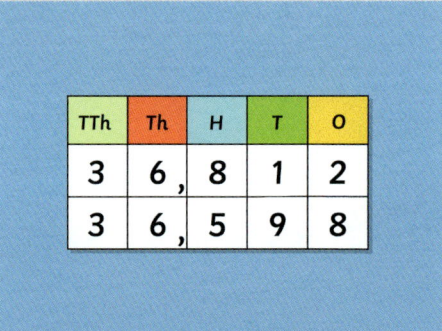

6. Which is the smallest number?

- ■ 605,784   ■ 605,918   ■ **605,179**

1/3

HTh	TTh	Th	H	T	O
6	0	5,	7	8	4
6	0	5,	9	1	8
6	0	5,	1	7	9

---

7. Sort the numbers from smallest to largest.

■ 22,604 ■ 202,600 ■ 220,000 ■ 2,200,601

M	HTh	TTh	Th	H	T	O

8. Sort the numbers in ascending order (smallest first).

■ 29,650 ■ 30,270 ■ 30,560 ■ 31,080

TTh	Th	H	T	O

9. Which is the larger five-digit number?

■ 36812

TTh	Th	H	T	O
3	6,	8	1	2
3	6,	5	9	8

10. What is the value of the 2 in the number 402,381?
Select two correct answers.

2/6

■ two hundred ■ two thousand
■ two thousand, three hundred and eighty-one ■ 200
■ 2,000 ■ 2,381

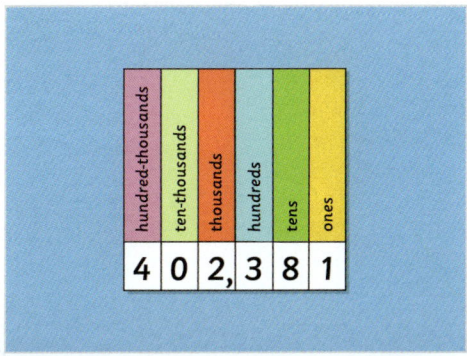

Level 2: Fluency: Read, write, compare and order numbers to at least 1,000,000, including in context.

❋ **Required:** 7/10 ❋ **Pupil Navigation:** on
❋ **Randomised:** off

11. What number is represented by the base-ten blocks in the diagram?

■ 1359 ■ 1000300509

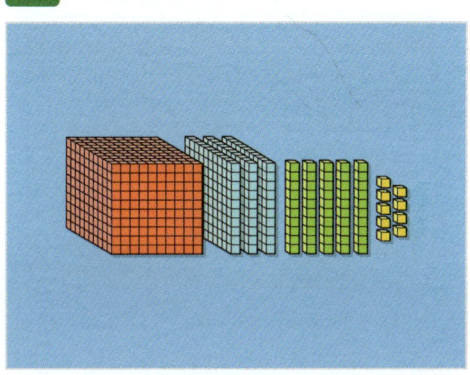

12. What is seven hundred and eighty thousand, four hundred and four in digits?

■ 780404

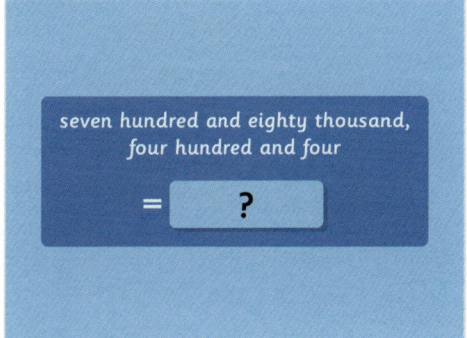

seven hundred and eighty thousand, four hundred and four

= [?]

Level 2 *continued*

13. What is 603,602 written in words?

1/4

- six hundred and three, six hundred and two
- six hundred thousand, three thousand, six hundred and two
- **six hundred and three thousand, six hundred and two**
- sixty-three thousand, six hundred and two

603,602

14. In words, what number is missing from the table?

1/3

- five hundred and ninety thousand, two hundred and fourteen
- **five hundred and nine thousand, two hundred and fourteen**
- five million and nine thousand, two hundred and fourteen

number in words	number in digits
seventeen thousand, six hundred and two	17,602
?	509,214
nine hundred and forty-three thousand, and twelve	943,012

15. Which car has travelled the furthest?

1/4

- car A
- **car B**
- car C
- car D

Distance travelled by four cars

	distance (miles)
car A	78,463
car B	184,736
car C	148,602
car D	89,562

16. Four hundred and fifty-five thousand people live in the city of Leeds. What is this number in digits?

- **455000**

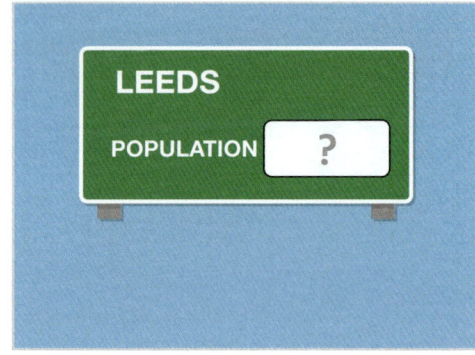

LEEDS

POPULATION ?

17. Arrange the cities in descending order of population, starting with the city with the largest population.

- Bemford
- Sunton
- Georgetown
- Daleford
- Ribham

Populations of cities

city	population
Georgetown	348,661
Daleford	296,223
Bemford	1,225,122
Sunton	621,540
Ribham	270,463

18. Which **two** houses cost less than £120,000?

2/5

- house A
- **house B**
- house C
- house D
- **house E**

A — FOR SALE — £251,000
B — FOR SALE — £76,500
C — FOR SALE — £198,000
D — FOR SALE — £125,000
E — FOR SALE — £102,500

Level 2 continued

19. Select the largest volume of liquid.

☒ 1/4

- **49,700 ml** ■ 745 ml ■ 1,150 ml ■ 17,820 ml

49,700 ml   745 ml

1,150 ml   17,820 ml

20. In digits, what number is missing from the table?

1 2 3

- **417511**

number in words	number in digits
sixty-five thousand, one hundred and two	65,102
four hundred and seventeen thousand, five hundred and eleven	?
eight hundred and three thousand, two hundred and twelve	803,212

Level 3: Reasoning: Reason about numbers to at least 1,000,000.

✱ Required: 5/5   ✱ Pupil Navigation: on
✱ Randomised: off

21. Sara has incorrectly written **3,900,466** in words. Explain what Sara has done wrong and write the number correctly.

a b c

- Open question, no set answer

thirty nine thousand, four hundred and sixty-six

22. What is the smallest 5-digit number you can make using all of the cards?

1 2 3

- **19539** ■ 39519

five   and   thirty
hundred   thousand
nineteen   nine

23. Owen has 5 counters. He wants to put the counters into the place value chart to make a 6-digit number, but he doesn't think he has enough counters.
Explain how Owen could make a 6-digit number using his 5 counters, giving an example of a number he could make.

a b c

- Open question, no set answer

hundred-thousands	ten-thousands	thousands	hundreds	tens	ones
		,			

24. Harriet says that she can order these numbers by looking at only the first three digits in each number. Is she correct? Explain your answer.

a b c

- Open question, no set answer

532,786
534,826
603,221
537,123
534,776
689,332

Level 3 *continued*

25. Abigail has written some numbers in a place value chart and arranged them from smallest to largest. The same digit has been covered by a star in each number. What digit is hidden behind the stars?

- **2** - **1**

TTh	Th	H	T	O
1	⭐,	3	4	2
⭐	2,	7	9	1
2	4,	⭐	0	8
⭐	4,	5	1	7
3	0,	⭐	5	1

Level 4: Problem solving with greater depth: Solve multi-step problems involving reading, writing, comparing and ordering numbers to at least 1,000,000.

✱ **Required:** 5/5   ✱ **Pupil Navigation:** on
✱ **Randomised:** off

26. Emily wants to buy a second-hand car that has been driven less than 85,000 miles. How many of the cars in the image could she buy?

- **3**

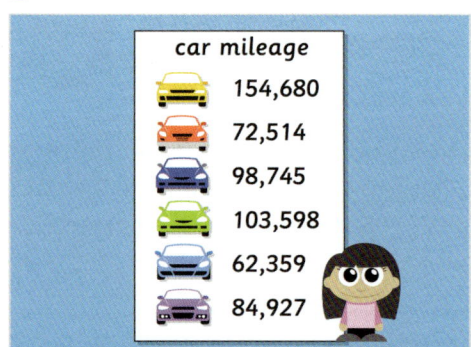

car mileage
🚗 154,680
🚗 72,514
🚗 98,745
🚗 103,598
🚗 62,359
🚗 84,927

27. Luke has nine digit cards. He uses five of them to make the largest number he can. He makes the smallest number he can with the the remaining four cards.
Find the difference between the two numbers that Luke makes.

- **97531** - **98765** - **1234**

28. Use the clues in the image to find a five-digit number.

- **48502**

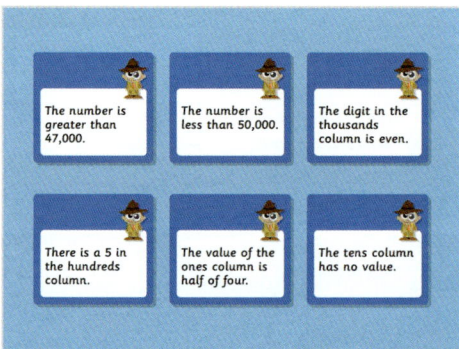

| The number is greater than 47,000. | The number is less than 50,000. | The digit in the thousands column is even. |
| There is a 5 in the hundreds column. | The value of the ones column is half of four. | The tens column has no value. |

29. Six children are playing a computer game. The table shows some of their scores.
Use the following clues to find **Izzy's score**:
• Sam's score is 800,000 more than Poppy's score.
• Mia's score is 10 times smaller than Sam's score.
• Izzy's score is 131,000 less than Mia's score.

- **465000** - **334000** - **4650000**

Computer game scores

name	score
Sam	
Poppy	3,850,000
Mia	
Charlie	420,000
Izzy	
Isaac	5,970,000

Level 4 *continued*

30. Elliot has written down four 5-digit numbers containing only the digits 2, 4, 6 and 8. He has represented each digit as a letter. Use the following clues to work out the value of **AABCA**:

- **DDBAC** is the smallest number.
- The sum of the digits in **BDBCD** equals 18.
- The sum of the digits in **CCCCC** equals 30.

■ 88468

DDBAC
BDBCD
CCCCC

AABCA

Read and Write Numbers to at Least 1,000,000

Objective: I can read and write numbers to at least 1,000,000 and know the value of each digit.

Quick Search Ref: 11367

Level 1: Understanding: Read and write numbers to at least 1,000,000 with image support.

❋ **Required:** 7/10 ❋ **Pupil Navigation:** on ❋ **Randomised:** off

---

1. What digit is in the ten-thousands column in the number 784,619?

- ■ **8** ■ 80000

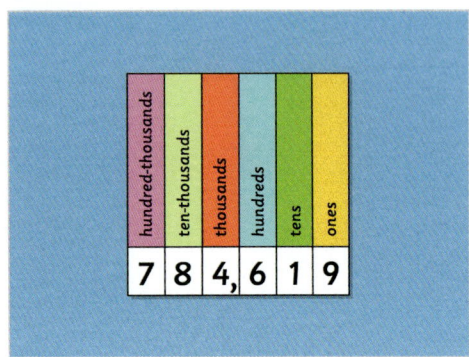

2. What is the value of the 2 in the number 402,381?

Select two correct answers.

2/6
- ■ two hundred ■ **two thousand**
- ■ two thousand, three hundred and eighty-one ■ 200
- ■ **2,000** ■ 2,381

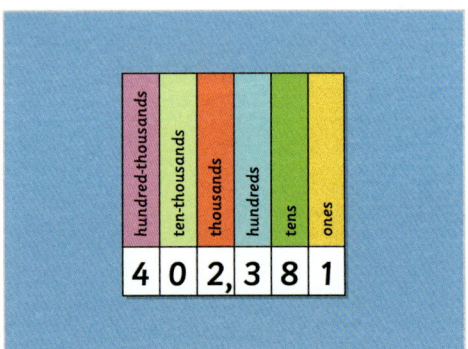

3. What is 3,452 written in words?

- ■ three thousand, five hundred and forty-two
- ■ thirty-four thousand and fifty-two

1/4
- ■ **three thousand, four hundred and fifty-two**
- ■ three thousand, four hundred and twenty-five

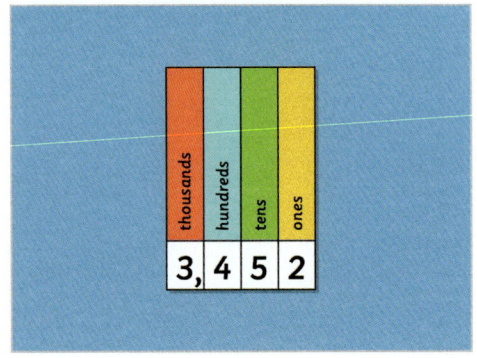

4. What is ninety-four thousand, one hundred and twenty-nine in digits?

- ■ **94129** ■ 94000129

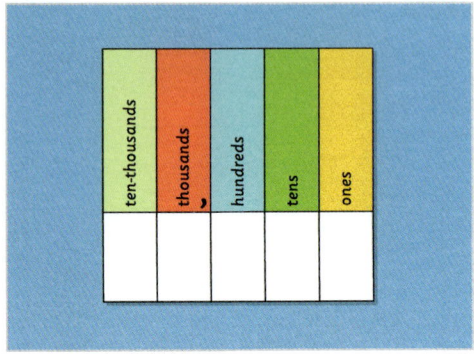

5. In digits, what number is represented by the place value counters?

- ■ **830947** ■ 83947

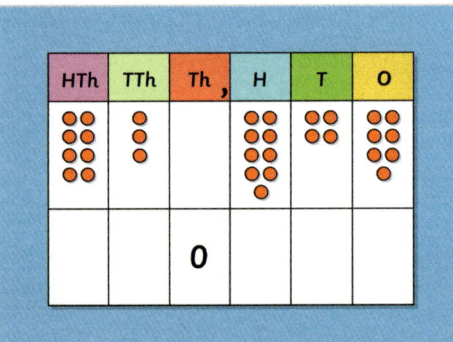

6. In the number seven hundred and two thousand, nine hundred and fifty-four, what digit is in the ten-thousands column?

- ■ **0** ■ 5 ■ 9 ■ 7 ■ 2 ■ 4

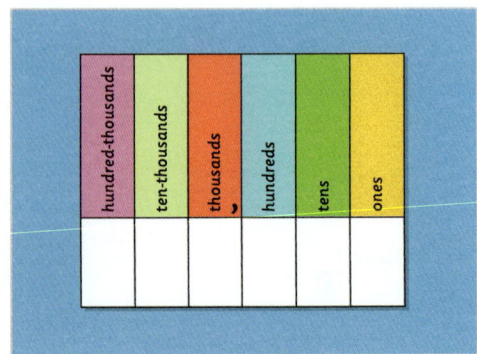

Level 1 *continued*

7. Freddie puts some counters onto a place value grid. What number does he make?

☐ ☒ ☐
1/3

- three hundred and twenty-four, seven hundred and nine
- **three hundred and twenty-four thousand, seven hundred and nine**
- three hundred-thousand, twenty thousand, four thousand, seven hundred and nine

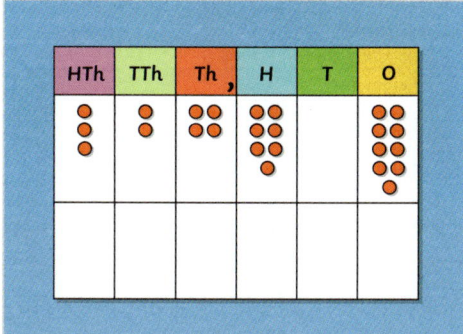

8. What number is represented by the place value counters?

☐ ☒ ☐
1/3

- nine hundred and ninety-nine thousand
- ninety-nine thousand, nine hundred and ninety-nine
- **nine hundred and ninety-nine thousand, nine hundred and ninety-nine**

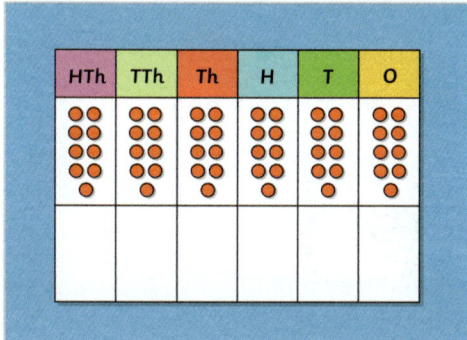

9. What is the value of the 6 in the number 846,513?

☐ ☒ ☐
1/5

- 60,000 ■ 6 ■ 1,000 ■ **6,000** ■ 600

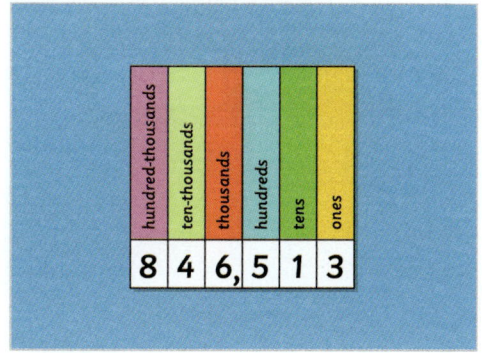

10. In the number 134,567 what digit is in the hundred-thousands column?

1
2
3

- 100000 ■ 4 ■ **1** ■ 3 ■ 5 ■ 6

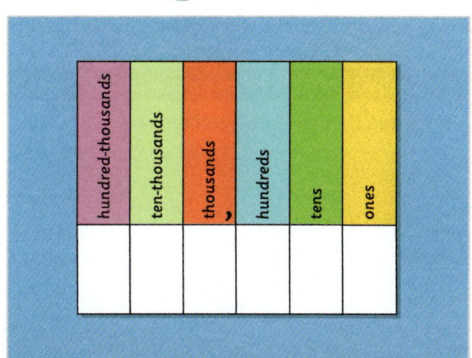

Level 2: Fluency: Read and write numbers to at least 1,000,000, including in simple contexts.

✻ **Required:** 7/10   ✻ **Pupil Navigation:** on
✻ **Randomised:** off

11. What number is represented by the base-ten blocks in the diagram?

1
2
3

- **1359** ■ 1000300509

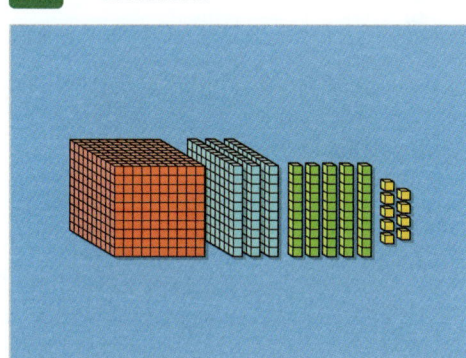

12. What is seven hundred and eighty thousand, four hundred and four in digits?

1
2
3

- **780404**

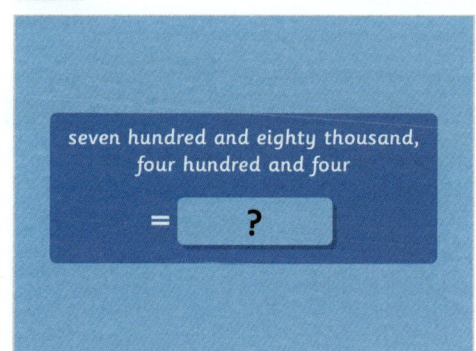

Level 2 *continued*

13. What is 603,602 written in words?

1/4

- six hundred and three, six hundred and two
- six hundred thousand, three thousand, six hundred and two
- six hundred and three thousand, six hundred and two
- sixty-three thousand, six hundred and two

603,602

14. One hundred and twenty-five thousand people attend a rock concert. What is this number in digits?

1 2 3

- 125000

one hundred and twenty-five thousand

15. Sort the numbers into the same order as they are represented in the diagrams, starting with the number that is represented by diagram A.

- nineteen thousand, nine hundred and ten
- four hundred and nine thousand, six hundred and forty-three
- four thousand, five hundred and twelve

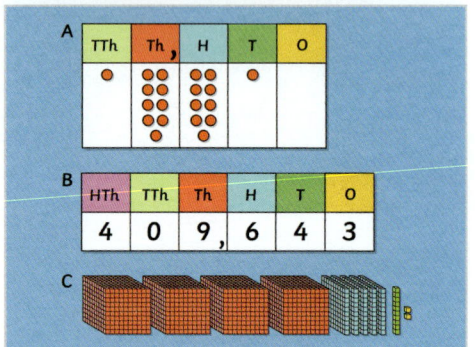

16. In digits, what number is missing from the table?

1 2 3

- 417511

number in words	number in digits
sixty-five thousand, one hundred and two	65,102
four hundred and seventeen thousand, five hundred and eleven	?
eight hundred and three thousand, two hundred and twelve	803,212

17. Mr Brown pays £215,000 for a house. What is this number in words?
Don't include the units in your answer.

a b c

- two hundred and fifteen thousand pounds
- two hundred and fifteen thousands
- two hundred fifteen thousand
- two hundred and fifteen thousand ■ 215 thousand

SOLD

£215,000

_____ _____ and _____ _____

18. Four hundred and fifty-five thousand people live in the city of Leeds. What is this number in digits?

1 2 3

- 455000

LEEDS

POPULATION ?

Level 2 *continued*

19. In words, what is Jemima's score?

1/5

- twenty-three thousand and four
- one thousand seven hundred and eighty
- seven hundred and ten thousand
- **ten thousand, seven hundred and eighty**
- four thousand, five hundred and ninety-two

Video game scores

name	latest score
Ella	4,592
Jemima	10,780
Grace	23,004

20. In words, what number is missing from the table?

1/3

- five hundred and ninety thousand, two hundred and fourteen
- **five hundred and nine thousand, two hundred and fourteen**
- five million and nine thousand, two hundred and fourteen

number in words	number in digits
seventeen thousand, six hundred and two	17,602
?	509,214
nine hundred and forty-three thousand, and twelve	943,012

Level 3: Reasoning: Reason about numbers up to 1,000,000.

✹ **Required:** 5/5   ✹ **Pupil Navigation:** on
✹ **Randomised:** off

21. How many **thousands** are the same as 40 hundreds?

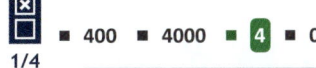

1/4

- 400
- 4000
- **4**
- 0

? thousands = 40 hundreds

22. What is the smallest 5-digit number you can make using all of the cards?

- **19539**
- 39519

| five | and | thirty |

| hundred | thousand |

| nineteen | nine |

23. Sara has incorrectly written **3,900,466** in words. Explain what Sara has done wrong and write the number correctly.

- Open question, no set answer

thirty nine thousand, four hundred and sixty-six

Level 4: Problem solving with greater depth: Solve multi-step problems involving numbers up to 1,000,000.

❋ **Required:** 5/5   ❋ **Pupil Navigation:** on
❋ **Randomised:** off

Level 3 *continued*

24. Using each digit card once, make the number that is closest to 200,000.

1 2 3

- ■ **201489** ■ 198420

25. Owen has 5 counters. He wants to put the counters into the place value chart to make a 6-digit number, but he doesn't think he has enough counters.

a b c

Explain how Owen could make a 6-digit number using his 5 counters, giving an example of a number he could make.

- Open question, no set answer

26. Use every digit card once to make the smallest possible **even number**.

1 2 3

Give your answer in digits.

- ■ **234798** ■ 234789

27. Six children are playing a computer game. The table shows some of their scores.

1 2 3

Use the following clues to find **Izzy's score**:
- Sam's score is 800,000 more than Poppy's score.
- Mia's score is 10 times smaller than Sam's score.
- Izzy's score is 131,000 less than Mia's score.

- ■ **334000** ■ 465000 ■ 4650000

Computer game scores

name	score
Sam	
Poppy	3,850,000
Mia	
Charlie	420,000
Izzy	
Isaac	5,970,000

Level 4 continued

28. Elliot has written down four 5-digit numbers containing only the digits 2, 4, 6 and 8. He has represented each digit as a letter. Use the following clues to work out the value of **AABCA**:

   • **DDBAC** is the smallest number.
   • The sum of the digits in **BDBCD** equals 18.
   • The sum of the digits in **CCCCC** equals 30.

   ■ 88468

29. The code is a 6-digit number. Use the clues to find the value of each digit and crack the code.

   ■ 154601

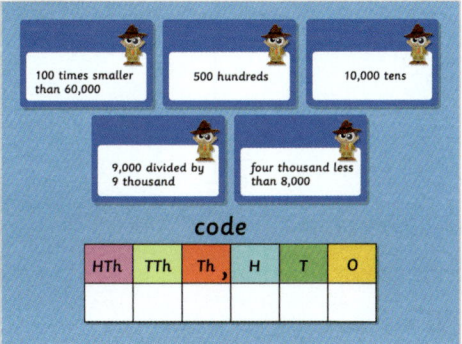

30. Ava thinks of a number. She adds twenty thousand to her number and then subtracts 500. Finally, she multiplies by 10 to get the number three million, nine hundred and seventy-eight thousand, two hundred and forty. What was Ava's original number? *Give your answer in digits.*

   ■ 378324   ■ 397824   ■ 398324

Interpret Negative Numbers

| Objective: | I can interpret negative numbers in context. |
| Quick Search Ref: | **10024** |

Level 1: Understanding: Identify negative numbers and count forwards and backwards with positive and negative numbers, including through zero, with support.

✿ **Required:** 7/10   ✿ **Pupil Navigation:** on   ✿ **Randomised:** off

1. Which **two** numbers are negative numbers?

☐
☒ ■ 20 ■ 0 ■ **–6** ■ **–12** ■ 91
☐

2/5

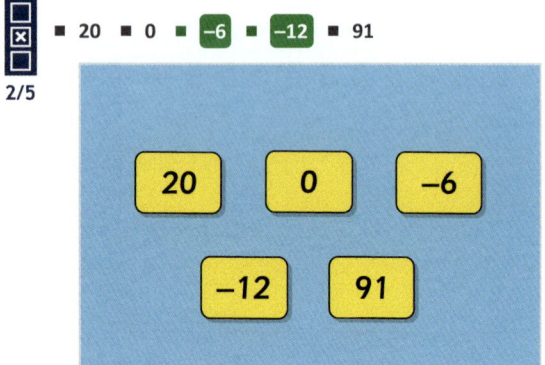

2. What is the missing **negative number** from the number line?

1
2
3 ■ **-3** ■ 3

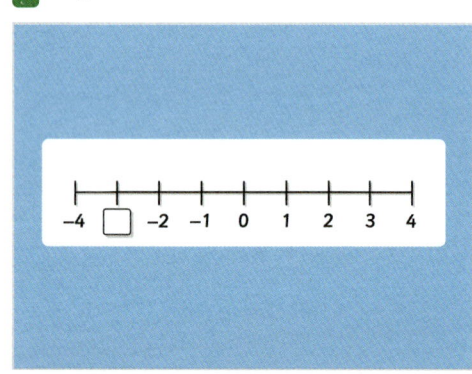

3. If you count back five from zero, what number do you reach?

1
2
3 ■ **-5** ■ -4 ■ 5

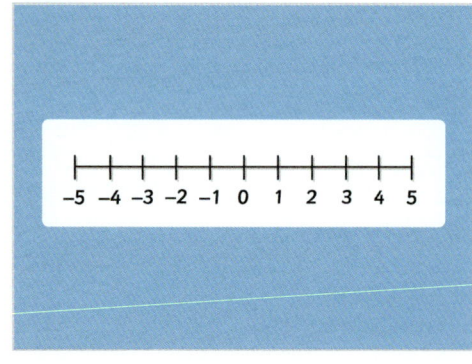

4. What number goes in box B?

1
2
3 ■ **-15** ■ 15

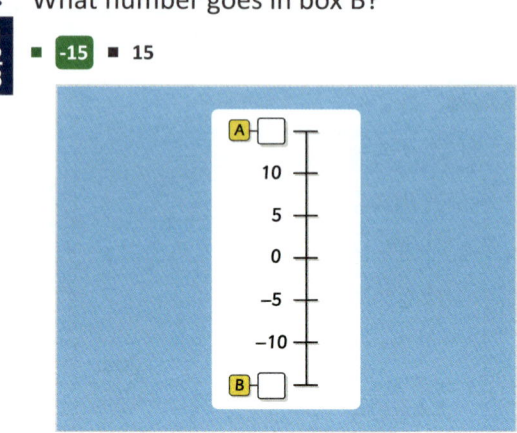

5. What is the missing number?

1
2
3 ■ 8 ■ **-8** ■ -7

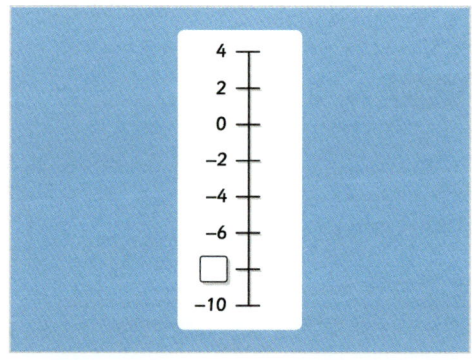

6. What is five less than 2?

1
2
3 ■ **-3** ■ 7 ■ 3 ■ -7

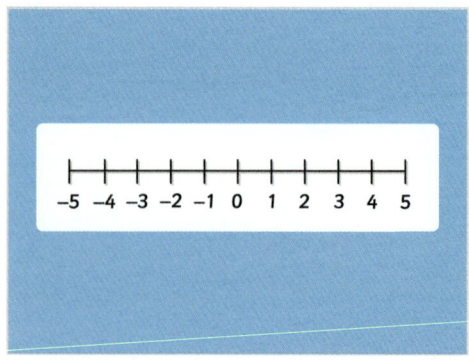

Level 1 *continued*

7. What is ten more than −6?

1
2
3
■ 6 ■ 4 ■ 16 ■ -16

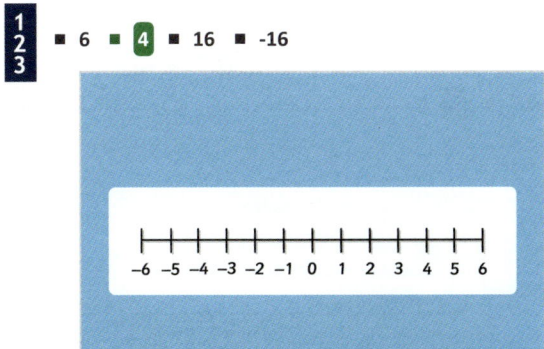

8. What is the missing number?

1
2
3
■ -1 ■ -2 ■ 2

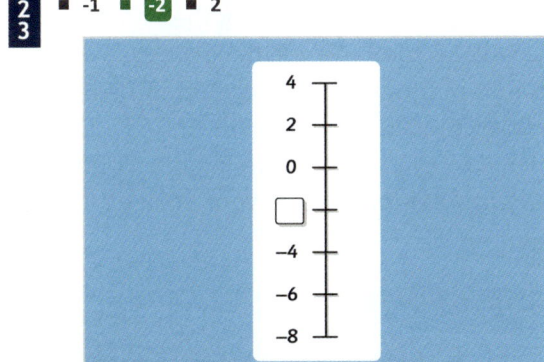

9. What is three less than −1?

1
2
3
■ -4 ■ 2 ■ -2

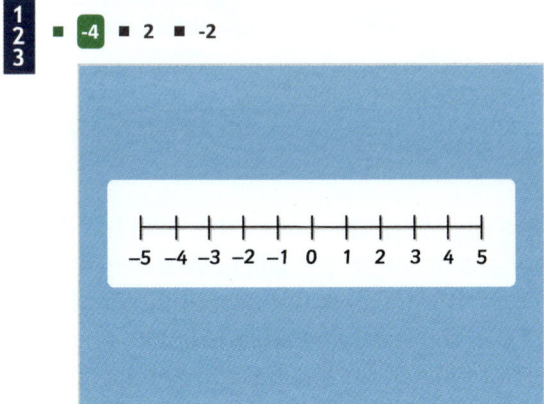

10. What is the missing **negative number** from the number line?

1
2
3
■ -5 ■ 5

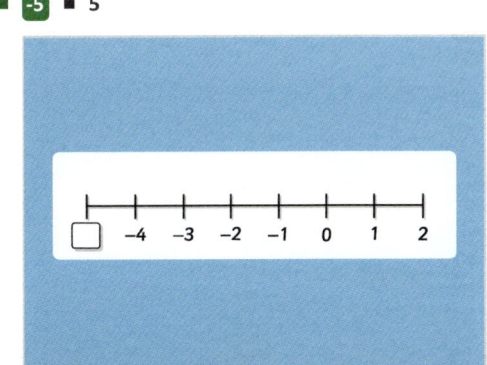

Level 2: Fluency: Interpret negative numbers in context and count forwards and backwards with positive and negative numbers, including through zero.

✱ **Required:** 7/10   ✱ **Pupil Navigation:** on
✱ **Randomised:** off

11. What number is two less than −4?

1
2
3
■ -6 ■ -2

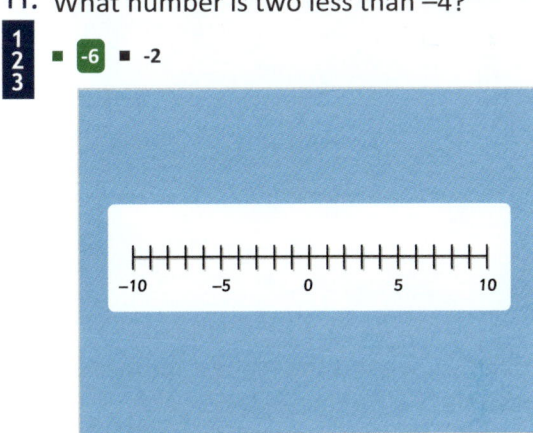

12. What number goes in box A?

1
2
3
■ 20 ■ -20 ■ -4

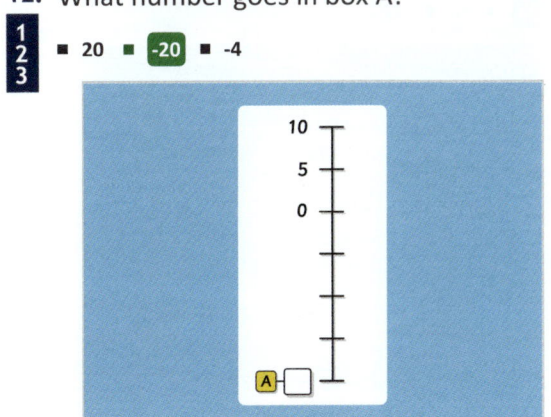

Level 2 *continued*

13. The temperature shown on the
thermometer is _____°C.

- **-6** ■ -3

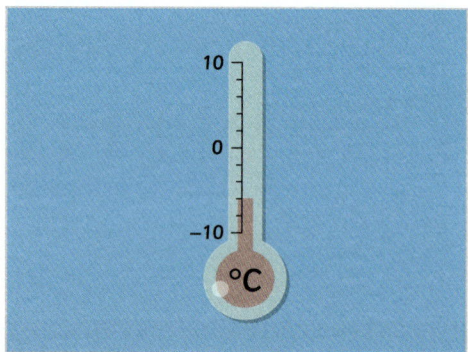

14. What temperature is six degrees less than −6°C?
Don't include the units in your answer.

- -9 ■ **-12** ■ 0

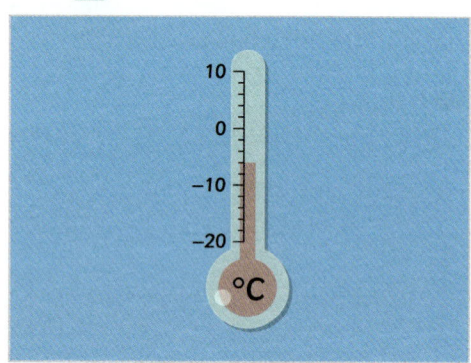

15. In degrees, what is the highest temperature shown on the thermometers?
Don't include the units in your answer.

- **-2** ■ -6 ■ -12

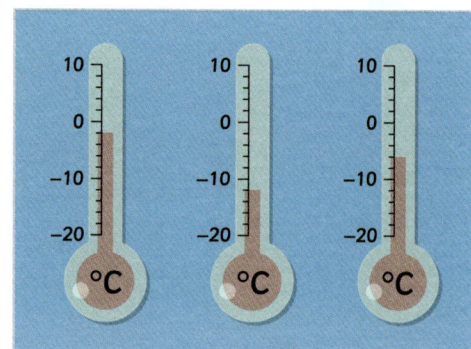

16. The temperature in Kiev is 6°C warmer than the temperature in Moscow. What is the temperature in Kiev?
Don't include the units in your answer.

- -16 ■ **-4** ■ -10

Temperatures in different cities

city	temperature (°C)
Kiev	?
Sydney	19
New York	−3
Moscow	−10

17. If David spends more money than he has in his bank account, he will have a negative amount of money in his account. If David has £5.00 in the bank and spends £10.00, how much money does he have in his account?
Don't include the units in your answer.

- **-5** ■ 15

David's bank account

money before shopping	money spent	money after shopping
£20.00	£15.00	£5.00
£5.00	£10.00	?

18. If Eva spends more money than she has in her bank account, she will have a negative amount of money in her account. If Eva has £5.00 in the bank and spends £8.00, how much money does she have in her account?
Don't include the units in your answer.

- **-3** ■ 3

Eva's bank account

money before shopping	money spent	money after shopping
£30.00	£25.00	£5.00
£5.00	£8.00	?

Level 2 *continued*

19. What temperature is five degrees less than −10°C?
Don't include the units in your answer.

■ -5   ■ **-15**   ■ -35

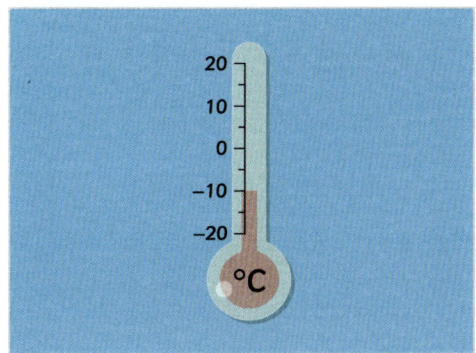

20. What number is twelve more than −5?

■ **7**   ■ -17

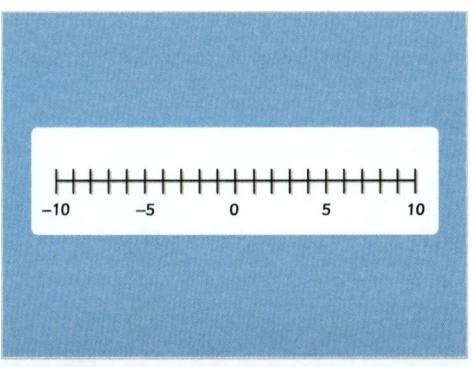

Level 3: Reasoning: Reason about negative numbers.

✿ **Required:** 5/5   ✿ **Pupil Navigation:** on
✿ **Randomised:** off

21. Billy has written a sequence of numbers, but one of his numbers has been covered in ink. What is the hidden number?

■ **-3**   ■ -9

22. Alice says that 12 is greater than 6, so −12 is greater than −6. Is Alice correct? Explain your answer.

- Open question, no set answer

23. Scott counts backwards in twos from 4. What is the sixth number he says?

■ **-6**   ■ 14

24. Steve has tried to write a sequence of numbers that increases in fives. Explain what mistake he has made.

- Open question, no set answer

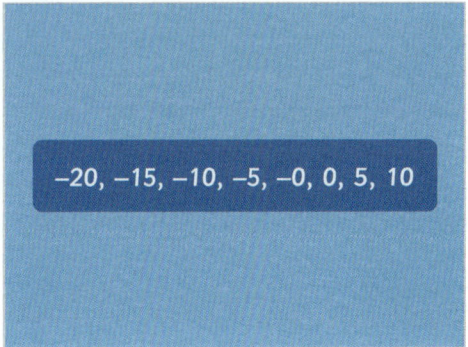

Level 3 continued

25. Select the statement about the temperatures in the table that is **not true**.

☐
☒
☐

1/4

- The temperature was 9°C colder on Friday than on Tuesday.
- The temperature was 5°C warmer on Saturday than on Friday.
- **The temperature was 5°C warmer on Tuesday than on Friday.**
- The temperature was 5°C warmer on Wednesday than on Friday.

Daily temperatures

day	temperature (°C)
Monday	4
Tuesday	7
Wednesday	3
Thursday	1
Friday	−2
Saturday	3
Sunday	11

Level 4: Problem solving with greater depth: Solve multi-step problems involving negative numbers.

❉ **Required:** 5/5   ❉ **Pupil Navigation:** on
❉ **Randomised:** off

26. In degrees, what is the difference in temperature between May and October?

1
2
3

- **-13** ▪ **13** ▪ 5

Average monthly temperatures

month	temperature (°C)
January	−22
February	−18
March	−10
April	0
May	9
June	15
July	17
August	14
September	8
October	−4
November	−17
December	−25

27. If the temperature on Monday is 9°C and every day the temperature gets 2°C colder, what is the temperature on Saturday? *Don't include the units in your answer.*

1
2
3

- **-1**

28. The bar chart shows the temperatures in different cities around the world. What is the difference between the hottest and the coldest temperatures? *Don't include the units in your answer.*

1
2
3

- **22** ▪ **-22**

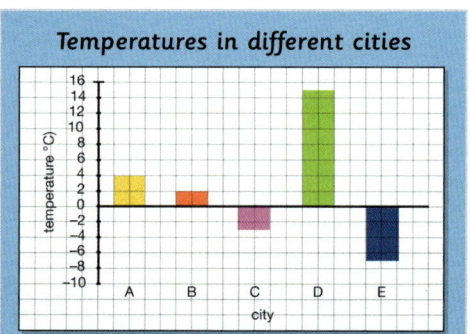

29. In the County Schools competition, teams are awarded two points for each correct answer and one point is deducted for each incorrect answer.

Potterton School has −7 points at half time. If they don't get any more questions wrong, **how many questions** do they need to answer correctly to reach the winning score of 10 points?

■ **9** ■ 17

30. Use the following clues to find the temperature in Cardiff:
- London is 7°C warmer than Manchester.
- Newcastle is 10°C colder than London.
- Cardiff is 4°C warmer than Newcastle.

■ **-3** ■ 3 ■ -7

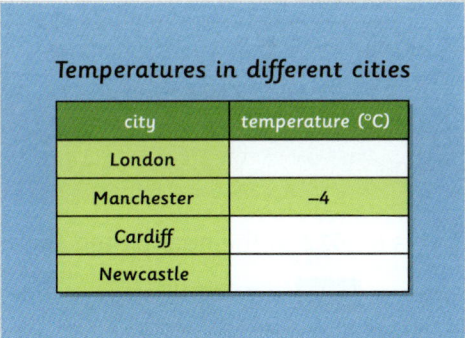

Temperatures in different cities

city	temperature (°C)
London	
Manchester	−4
Cardiff	
Newcastle	

Round Numbers up to 1,000,000

Objective: I can round any number up to 1,000,000 to the nearest 10, 100, 1,000, 10,000 and 100,000.

Quick Search Ref: 10124

Level 1: Understanding: Round numbers up to 1,000,000 to the nearest 10, 100, 1,000, 10,000 and 100,000 with support.

❄ **Required:** 7/10 ❄ **Pupil Navigation:** on ❄ **Randomised:** off

1. What is the multiple of 100 that is immediately before 272?

▪ 300 ▪ **200** ▪ 100

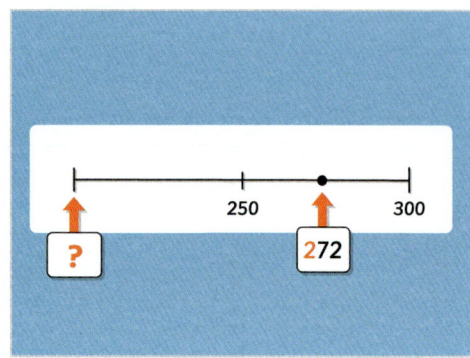

2. What is the next multiple of 100 after 272?

▪ **300** ▪ 200

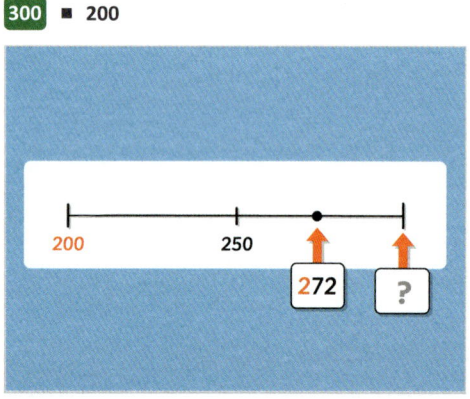

3. What is 272 to the nearest hundred?

▪ **300** ▪ 270 ▪ 200 ▪ 372

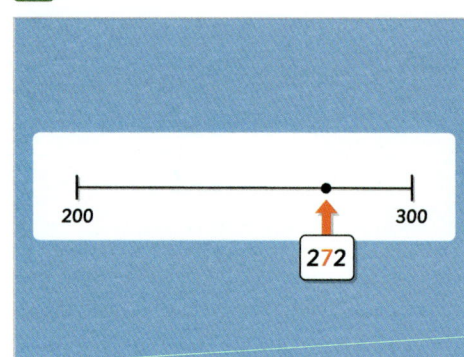

4. What is 1,581 rounded to the nearest thousand?

▪ 1000 ▪ **2000** ▪ 2581 ▪ 1600

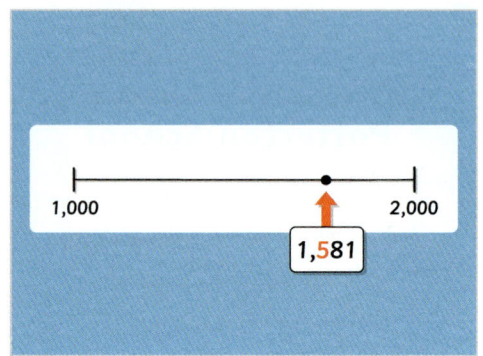

5. What is the next multiple of 10,000 after 63,912?

1/4

▪ 61,000 ▪ 700 ▪ 7,000 ▪ **70,000**

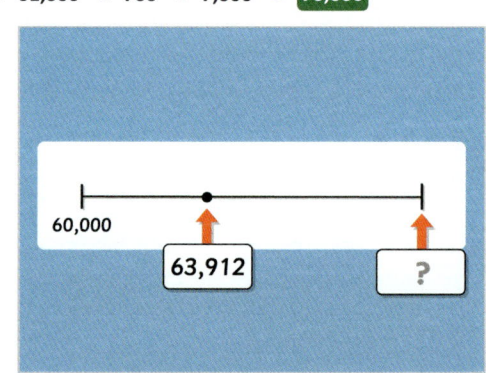

6. What is 63,912 to the nearest ten-thousand?

▪ 60912 ▪ **60000** ▪ 64000 ▪ 50000 ▪ 70000

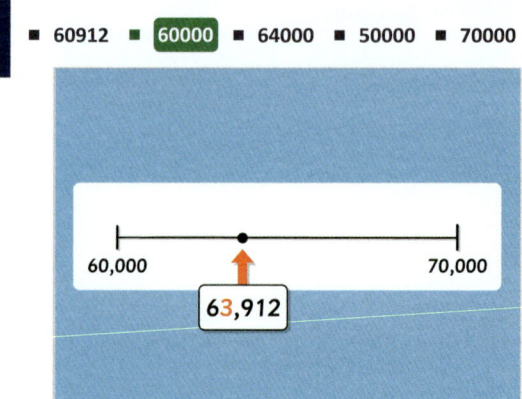

Level 1 continued

7. What is 298,147 to the nearest hundred-thousand?

1 2 3

- **300000** ▪ 298000 ▪ 200000

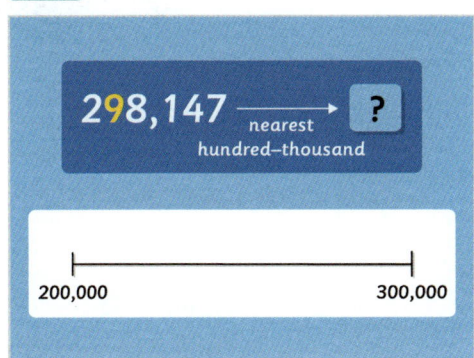

8. What is 24,571 to the nearest ten-thousand?

1 2 3

- **20000** ▪ 25000 ▪ 30000

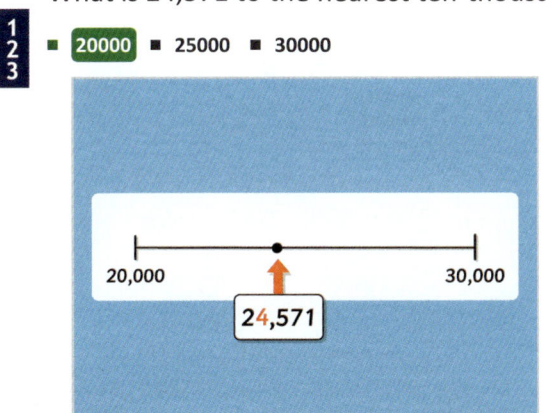

9. What is the next multiple of 10,000 after 128,312?

▢ ▢
▢ ✕ ▢
▢ ▢
1/4

▪ 110,000 ▪ 129,000 ▪ **130,000** ▪ 120,000

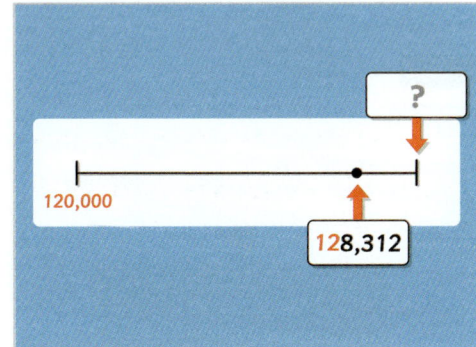

10. What is 865 rounded to the nearest hundred?

1 2 3

▪ **900** ▪ 870 ▪ 800

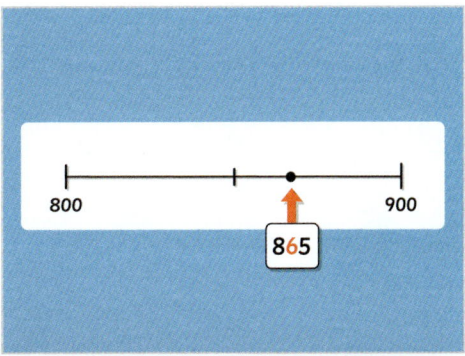

Level 2: Fluency: Round numbers up to 1,000,000 to the nearest 10, 100, 1,000, 10,000 and 100,000, including in context.

✱ **Required:** 7/10   ✱ **Pupil Navigation:** on
✱ **Randomised:** off

11. What is 21,871 rounded to the nearest hundred?

1 2 3

▪ 21800 ▪ **21900** ▪ 20000 ▪ 21870

12. What is 21,871 rounded to the nearest thousand?

1 2 3

▪ **22000** ▪ 21000 ▪ 20000

Level 2 *continued*

13. What is 49,997 rounded to the nearest ten?

■ 49000 ■ 50000 ■ 49900 ■ 49990

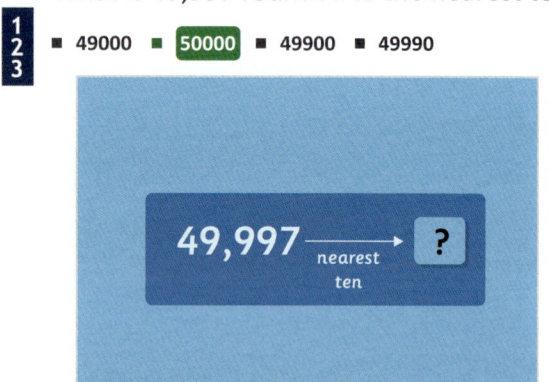

14. What is 999,950 rounded to the nearest hundred-thousand?

■ 1000000 ■ 900000

15. Which two numbers **do not round** to 50,000 to the nearest ten-thousand?

2/5

■ 51,200 ■ 55,000 ■ 45,800 ■ 44,999 ■ 54,000

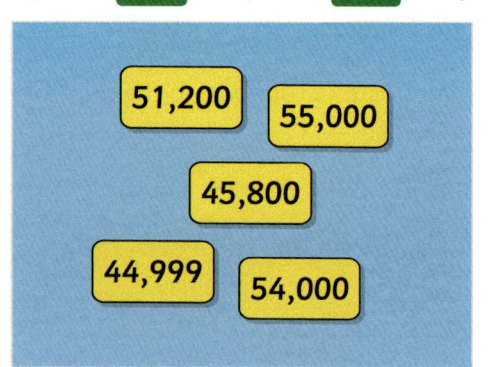

16. There are 12,105 spectators at a volleyball game. How many spectators are there to the nearest ten-thousand?

■ 10000 ■ 20000 ■ 12000

attendance: 12,105

17. The population of Luxembourg is 576,243. What is the population to the nearest hundred-thousand?

■ 500000 ■ 600000 ■ 576000

Luxembourg

POPULATION 576,243

18. The population of Newcastle is 268,064. What is the population to the nearest thousand?

■ 268000 ■ 269000

Newcastle

POPULATION 268,064

Level 2 *continued*

19. Which **three** numbers round to 300,000 to the nearest hundred-thousand?

3/5
- 249,100 ▪ 255,000 ▪ 355,800 ▪ 345,999
- 301,999

20. What is 17,921 rounded to the nearest ten?

1 2 3
- 17920 ▪ 17900 ▪ 17930 ▪ 18000

Level 3: Reasoning: Reason about rounding numbers up to 1,000,000.

✿ **Required:** 5/5   ✿ **Pupil Navigation:** on
✿ **Randomised:** off

21. What is the largest possible whole number that rounds to 23,400 to the nearest hundred?

1 2 3
- 23449 ▪ 23350

22. Amber has completed her maths homework. Check her homework carefully. Explain any mistakes that she has made, and find the correct answer(s).

a b c

- Open question, no set answer

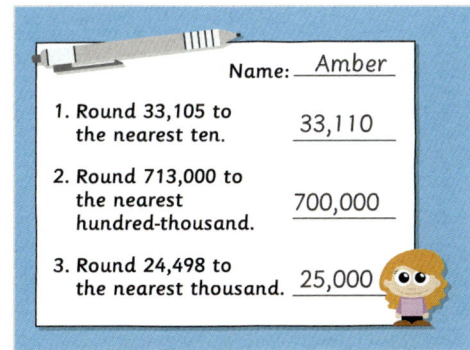

23. Which symbol makes the statement true?
153,646 to the nearest hundred-thousand ____ 196,999 to the nearest ten-thousand

1/3
- > ▪ = ▪ <

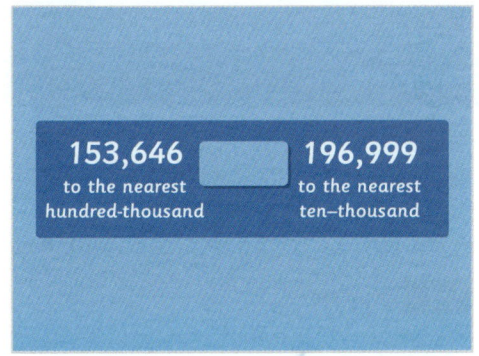

24. A train breaks down 598 miles into a 1,000-mile journey from Barcelona to Naples. Is it closer to Barcelona or Naples? Explain your answer.

a b c

- Open question, no set answer

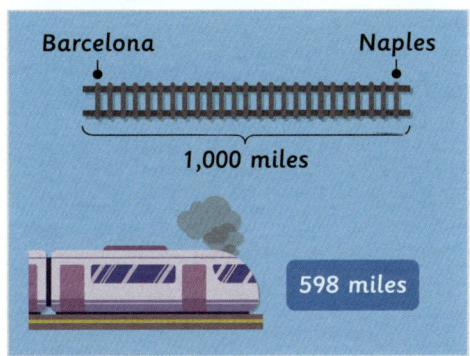

Level 3 *continued*

25. Ben is trying to find which planets have

a b c populations that **round down** to the nearest ten-thousand. He says that he only needs to look at one digit of each number. Do you agree with Ben? Explain your answer.

- Open question, no set answer

Population of planets

planet	population
Troyton	13,233
Zen	76,022
Xorbie	90,277
Balec	544,987
Mundane	776,086

Level 4: Problem solving with greater depth: Solve multi-step problems involving rounding numbers to a required degree of accuracy.

✿ **Required:** 5/5   ✿ **Pupil Navigation:** on
✿ **Randomised:** off

26. The table shows how many miles Becky

1 2 3 travels in her car each year. Use rounding to estimate how many miles Becky travels in her car in total to the **nearest 1,000 miles**.

▪ 42000 ▪ 42326

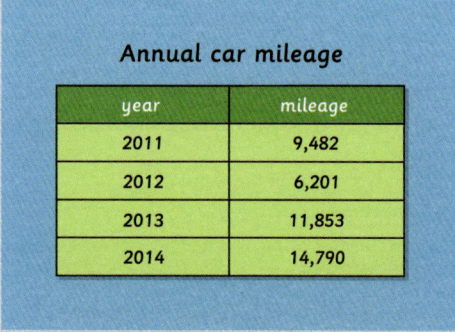

Annual car mileage

year	mileage
2011	9,482
2012	6,201
2013	11,853
2014	14,790

27. In a year, a shop sells 138,211 pairs of men's

1 2 3 shoes, 52,694 pairs of women's shoes and 92,042 pairs of children's shoes. To the **nearest 10,000**, how many pairs of shoes does the shop sell in total?

▪ 280000 ▪ 282947

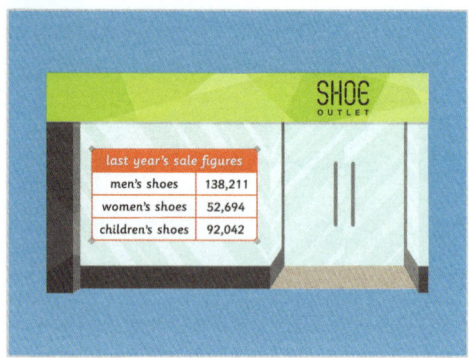

28. William is thinking of a 5-digit number. His

1 2 3 number rounds to 50,000 to the nearest 10,000. All of the digits are different, and they are all odd numbers. What is the **largest** possible number William could be thinking of?

▪ 53971 ▪ 97531 ▪ 59731 ▪ 51379

Level 4 *continued*

29. Ali is thinking of a whole number. Use the following clues to find the number that Ali is thinking of:

• The number rounds to 21,200 to the nearest 100.

• The number rounds to 21,150 to the nearest 10.

• The digits in the ten-thousands and ones columns are the same.

■ 21152 ■ 21154

30. Using every digit card once, how many numbers can you make that round to 2,000 to the **nearest 1,000**?

■ 6

Place Value Topic Review

Objective:	I can answer questions about place value.
Quick Search Ref:	**11428**

Level 1: Understanding

❁ **Required:** 7/10   ❁ **Pupil Navigation:** on   ❁ **Randomised:** off

1. What number is represented by the base-ten blocks in the diagram?

▪ **1359**   ▪ 1000300509

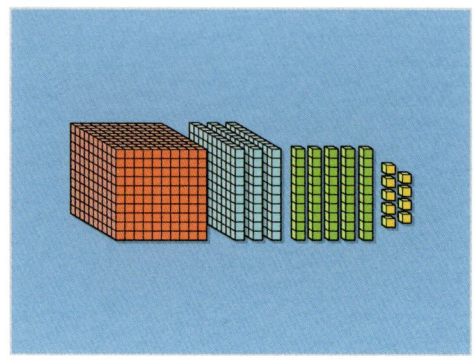

2. What is 680 rounded to the nearest **100**?

▪ **700**   ▪ 600

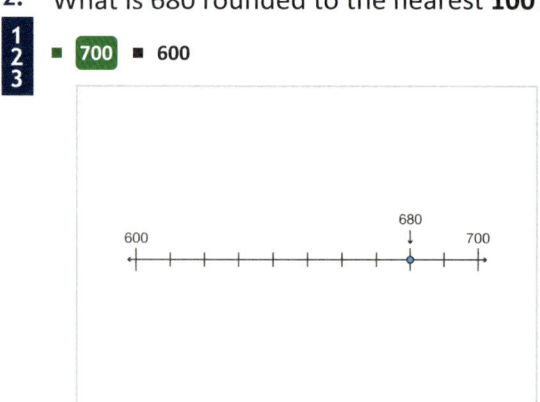

3. What is seven hundred and eighty thousand, four hundred and four in digits?

▪ **780404**   ▪ **780,404**   ▪ 7844   ▪ 70080404

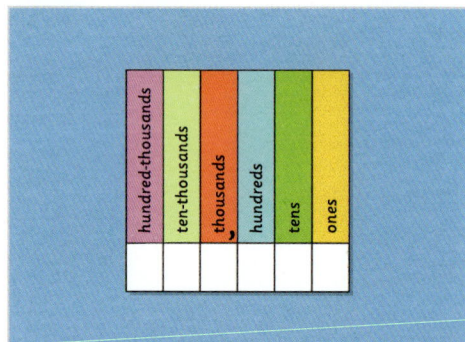

4. How is 800 represented in Roman numerals?

▪ **DCCC**   ▪ CCM

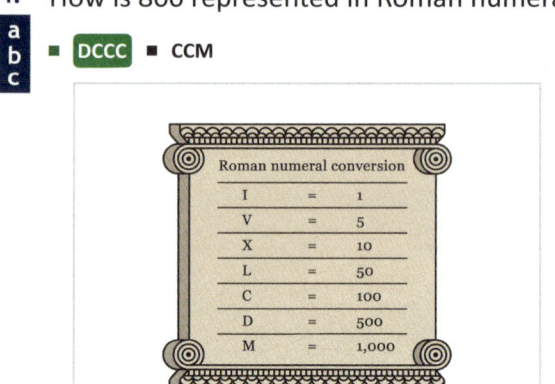

5. Which number has the **lowest** value?

1/6

▪ 0   ▪ 7   ▪ 101   ▪ **-872**   ▪ -2   ▪ 10,000

6. In the number 328.974 which digit represents the thousandths?

▪ **4**   ▪ 8   ▪ 3   ▪ 2   ▪ 9   ▪ 7

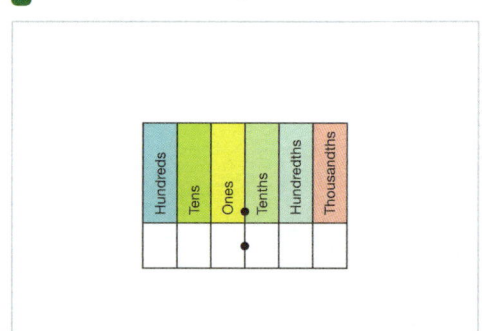

7. Which car has travelled the furthest?

1/4

▪ car A   ▪ **car B**   ▪ car C   ▪ car D

Distance travelled by four cars

	distance (miles)
car A	78,463
car B	184,736
car C	148,602
car D	89,562

Level 1 *continued*

8. Which **two** houses cost less than £120,000?

☐☒☐ ■ house A ■ house B ■ house C ■ house D
2/5 ■ house E

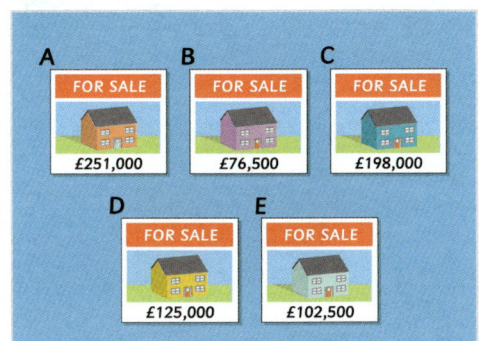

9. Which is the smallest number?

☐☒☐ ■ 605,784 ■ 605,918 ■ 605,179
1/3

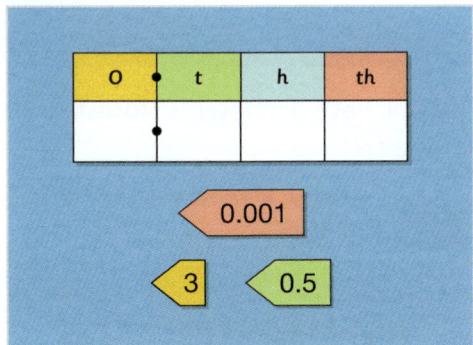

10. What decimal number is made using these arrow cards?

1 2 3 ■ 3.501 ■ 3.51 ■ 3501

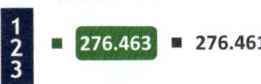

Level 2: Fluency

✹ **Required:** 7/10 ✹ **Pupil Navigation:** on
✹ **Randomised:** off

11. In words, what is Jemima's score?

☐☒☐ ■ twenty-three thousand and four
 ■ one thousand seven hundred and eighty
1/5 ■ seven hundred and ten thousand
 ■ ten thousand, seven hundred and eighty
 ■ four thousand, five hundred and ninety-two

12. What is the value of Y shown on the number
1 2 3 line?
 ■ -3

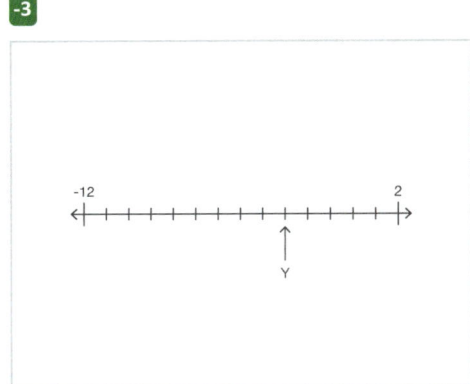

13. What number is missing from the sequence?
1 2 3 -15, -12, -9, -6, ___, 0, 3, 6, 9
 ■ -3

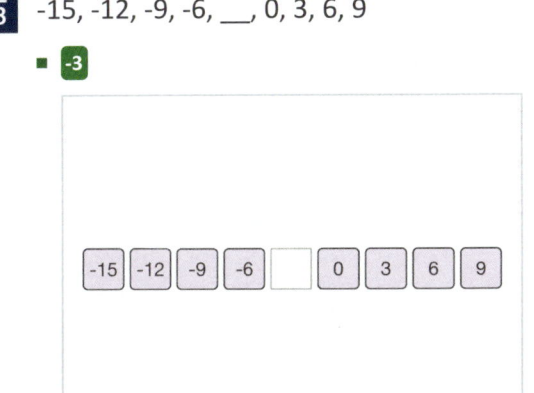

14. What is a **thousandth** more than 276.462?
1 2 3 ■ 276.463 ■ 276.461

Level 2 *continued*

15. What number does **XCVI** represent?

a b c
- 96 - 1,010,056 - 906 - 116 - 1010056

16. There are 8,276 seashells on a beach. What
is this rounded to the nearest **1,000?**

a b c
- 8,000 - 9000 - 8000 - 8,280 - 8280 - 8,300
- 7000 - 9,000 - 7,000 - 8300

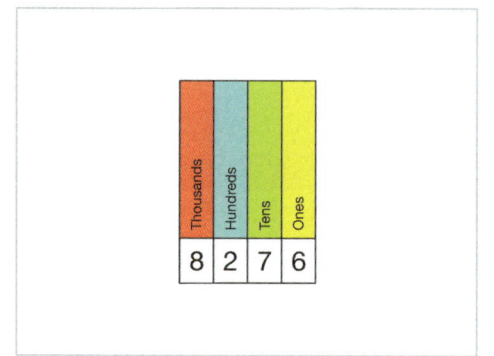

Thousands	Hundreds	Tens	Ones
8	2	7	6

17. Arrange the cities in descending order of
population, starting with the city with the
largest population.

- Bemford - Sunton - Georgetown - Daleford
- Ribham

Populations of cities

city	population
Georgetown	348,661
Daleford	296,223
Bemford	1,225,122
Sunton	621,540
Ribham	270,463

18. What is 99 in Roman numerals?

a b c
- IC - XCIX - XXXXXXXXXIX - XCIIIIIIIII
- XXXXXXXXXIIIIIIIII

19. Sort the numbers from smallest to largest.

- 22,604 - 202,600 - 220,000 - 2,200,601

M	HTh	TTh	Th	H	T	O

20. What is 168,342 rounded to the nearest
1,000?

a b c
- 168000 - 200,000 - 167,000 - 168,000
- 168,340 - 168340 - 170000 - 169000
- 168,300 - 167000 - 200000 - 168300
- 169,000 - 170,000

Hundred-thousands	Ten-thousands	Thousands	Hundreds	Tens	Ones
1	6	8	3	4	2

Level 3: Reasoning

✿ **Required:** 5/5 ✿ **Pupil Navigation:** on
✿ **Randomised:** off

21. Scott counts backwards in twos from 4. What
is the sixth number he says?

1 2 3
- -6 - 14

4, 2, ...

22. Which planet's diameter is **closest** to 50,000
kilometres?

1/7
- Mercury - Venus - Mars - Jupiter - Saturn
- Uranus - Neptune

Diameter of planets in the solar system

planet	diameter (km)
Mercury	4,879
Venus	12,104
Mars	6,779
Jupiter	142,984
Saturn	120,536
Uranus	50,724
Neptune	49,244

Level 3 *continued*

23. Using every digit card once, make the
smallest possible seven-digit whole number.

a
b
c

- `1023456` - `1,023,456` - `6543210` - `1234560`
- `0123456`

24. Abigail has written some numbers in a place
value chart and arranged them from smallest
to largest. The same digit has been covered
by a star in each number. What digit is
hidden behind the stars?

1
2
3

- `2` - `1`

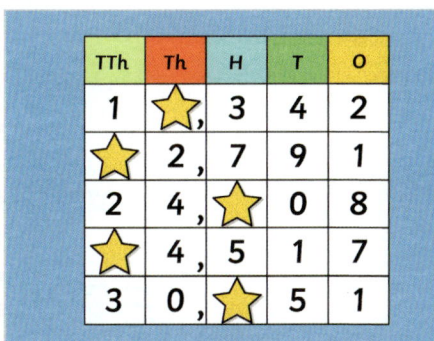

25. What is the largest possible whole number
that rounds to 23,400 to the nearest
hundred?

1
2
3

- `23449` - `23350`

Level 4: Problem solving with greater depth

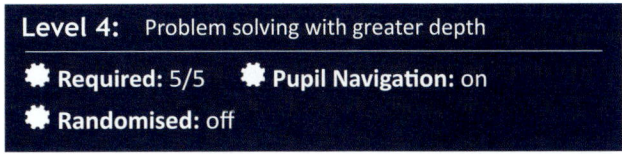

Required: 5/5 **Pupil Navigation:** on
Randomised: off

26. Using every digit card once, make the closest
possible number to 65.

1
2
3

- `64.951` - `65.149`

27. Colin works in a skyscraper. His office is on
the 14th floor and the car park is on the -2nd
floor. How many floors does he travel to get
from his car to his office?

1
2
3

- `16` - `12`

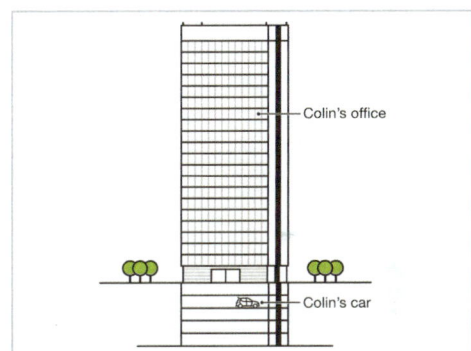

28. Use the clues in the image to find a five-digit
number.

1
2
3

- `48502`

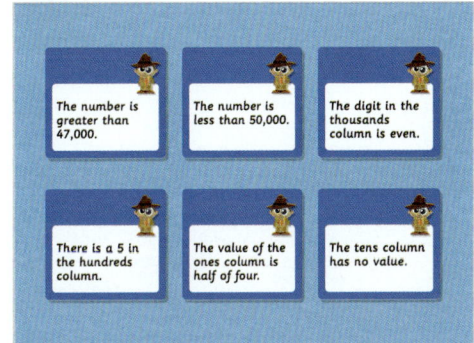

29. The gravestone shows the dates of the Roman Emperor Nero's birth and death. How old was he when he **died**?

■ **30**   ■ 68   ■ 31

30. Mary is thinking of a **negative** number.
- The number is greater than −162.
- All of the digits in the number are odd.
- The number ends in a 5.

How many possible numbers can Mary be thinking of?

■ **9**   ■ 16

Mathematics  Y5

Addition and Subtraction

Mental Calculation
Written Methods Addition
Written Methods Subtraction
Decimals

Add and Subtract Numbers Mentally – Understanding

Objective: I can use mental strategies to add and subtract numbers mentally.

Quick Search Ref: 10027

Level 1: Understanding: Use mental strategies for addition, including counting on, number bonds, partitioning, compensation and near doubles.

✿ **Required:** 6/8   ✿ **Pupil Navigation:** on   ✿ **Randomised:** off

---

1. What is 194 + 40?

■ **234** ■ 134

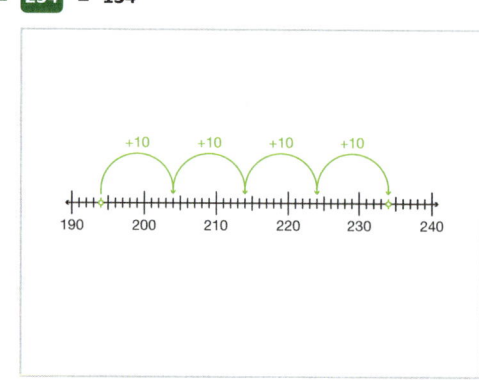

2. What is 25 + 13 + 75?

■ **113**

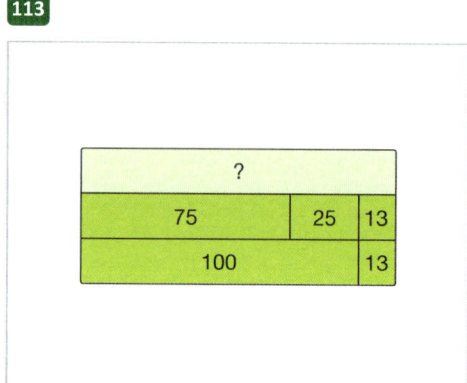

3. What is the total of 382 and 130?

■ **512**

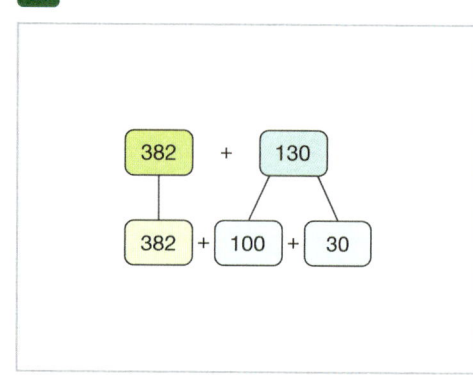

4. Add together 436 and 55.

■ **491**

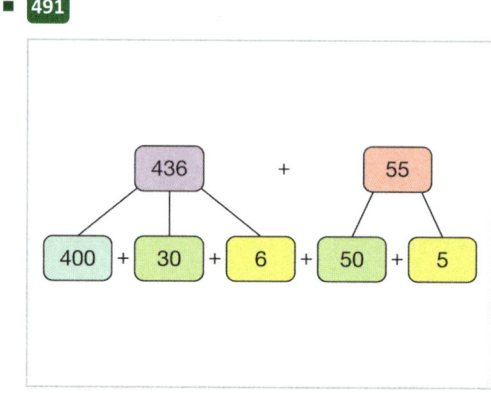

5. Add together 199 and 105.

■ **304**

6. Add together 3.6 and 3.5.

■ **7.1**

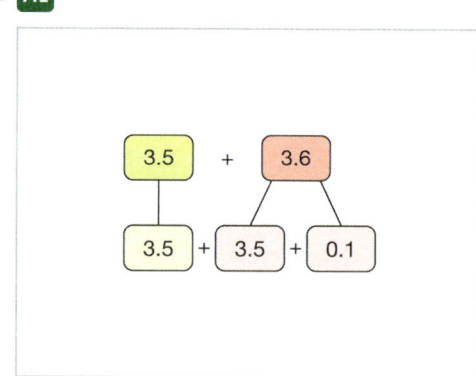

7. What is 469 + 329?

■ **798** ■ 800

---

8. What is the sum of 301 and 298?

1 2 3 ▪ 601 ▪ **599** ▪ 598

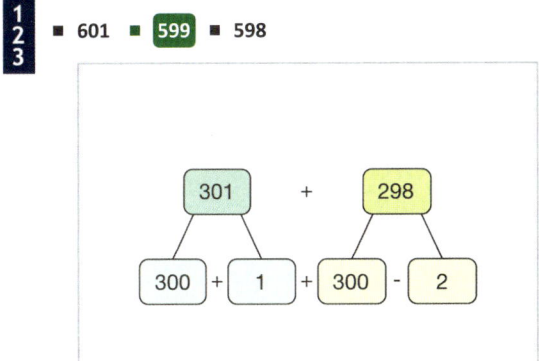

Level 2: Understanding: Mental strategies for subtraction: count back, count on, partitioning, compensation

❋ **Required:** 6/8   ❋ **Pupil Navigation:** on
❋ **Randomised:** off

9. What is 50 subtracted from 328?

1 2 3 ▪ **278**

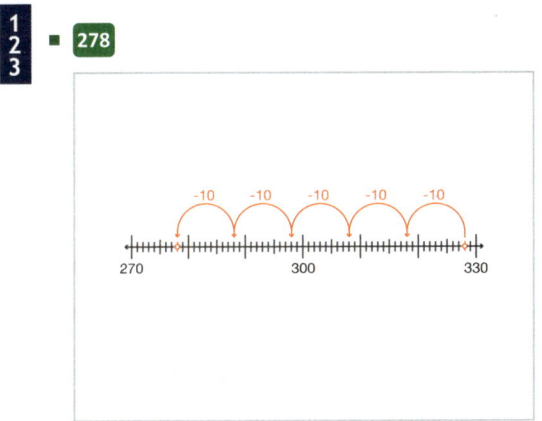

10. What is 27 less than 74?

1 2 3 ▪ **47** ▪ 101

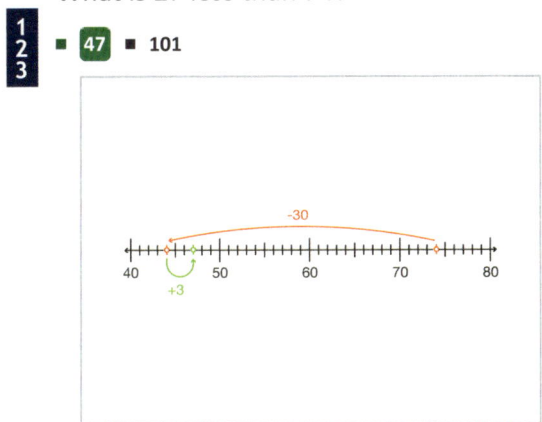

11. How many fewer is 214 than 534?

1 2 3 ▪ **320**

12. How many more is 7,003 than 3,996?

1 2 3 ▪ **3007** ▪ 10999

13. What is 856 - 735?

1 2 3 ▪ **121**

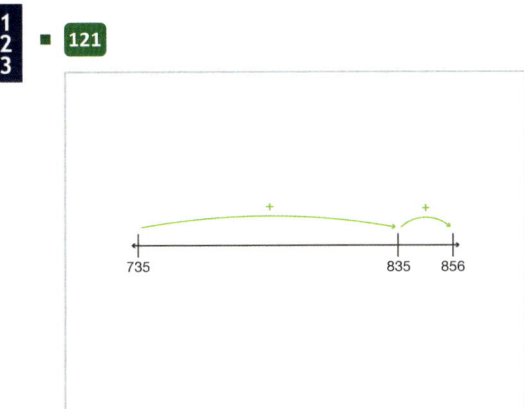

14. What is 160 less than 540?

1 2 3 ▪ **380** ▪ 420

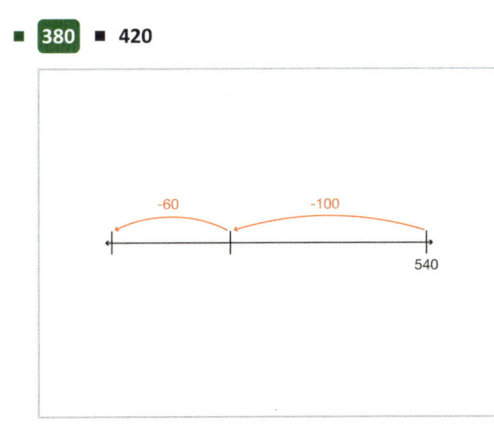

15. How many fewer is 14.6 than 15.3?

1 2 3 ▪ **0.7**

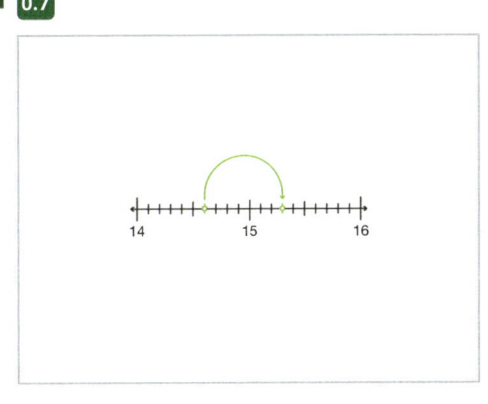

16. What is the difference between 109 and
 1,008?

■ 899

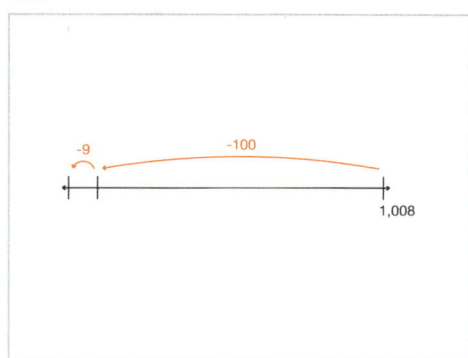

Add Numbers Mentally

Objective: I can use mental strategies for addition, including counting on, number bonds, partitioning and compensation.

Quick Search Ref: 11265

Level 1: Understanding: Use mental strategies for addition including counting on, number bonds, partitioning, compensation and near doubles.

✿ **Required:** 7/10   ✿ **Pupil Navigation:** on   ✿ **Randomised:** off

---

1. What is 194 + 40?

1 2 3 ■ **234** ■ 198 ■ 594 ■ 134

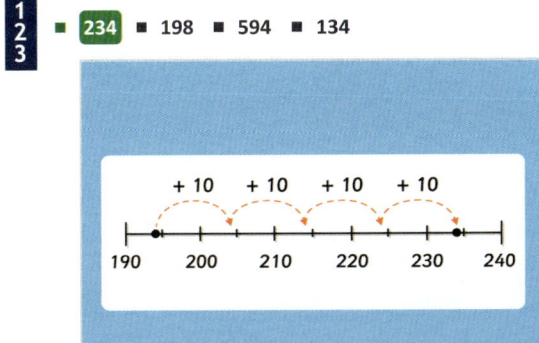

2. What is 25 + 13 + 75?

1 2 3 ■ **113**

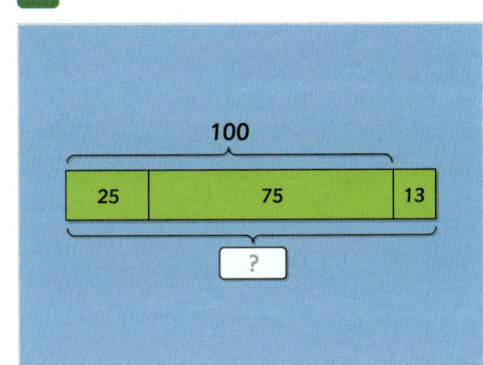

3. What is the total of 382 and 130?

1 2 3 ■ **512**

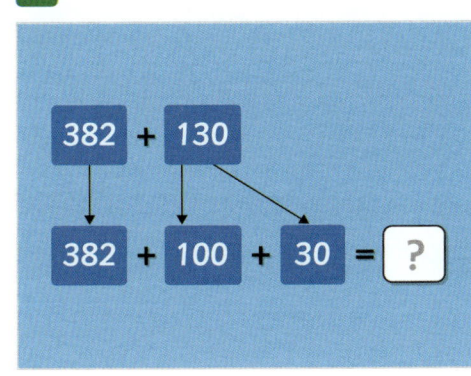

4. Add together 436 and 55.

1 2 3 ■ **491**

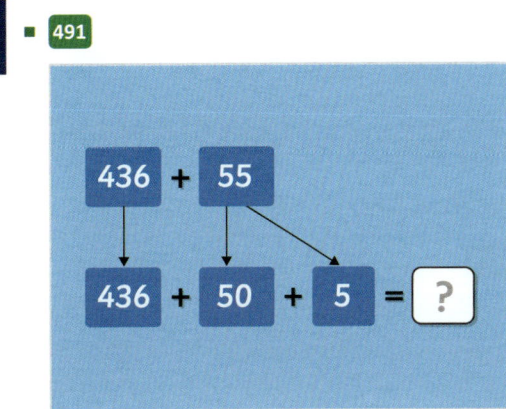

5. Add together 199 and 105.

1 2 3 ■ **304** ■ 305

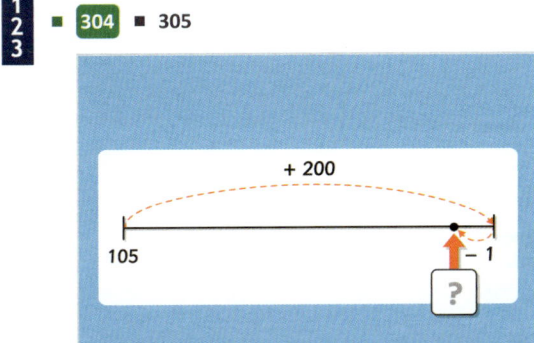

6. Add 250 and 253.

1 2 3 ■ **503** ■ 500

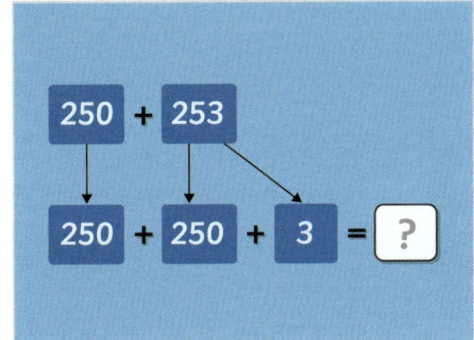

Level 1 *continued*

7. What is the sum of 301 and 298?

1 2 3 ▪ 599 ▪ 600

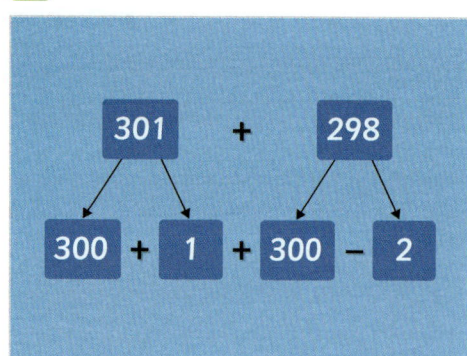

8. Add 70 to 673.

1 2 3 ▪ 743

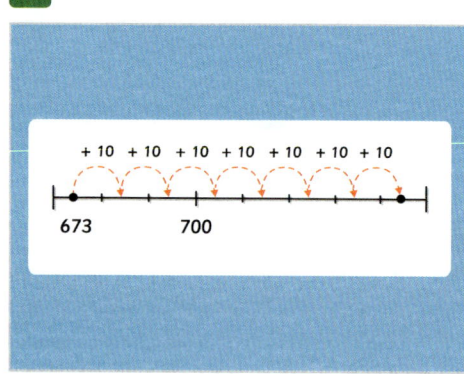

9. What is 228 + 65?

1 2 3 ▪ 293 ▪ 295

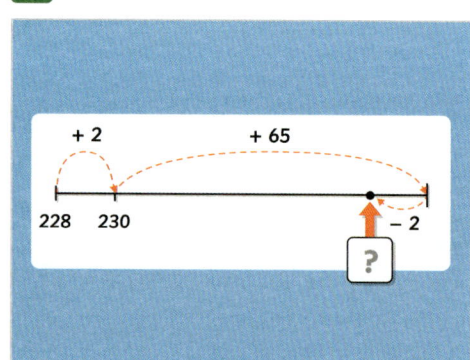

10. Find the total of 357 and 240.

1 2 3 ▪ 597

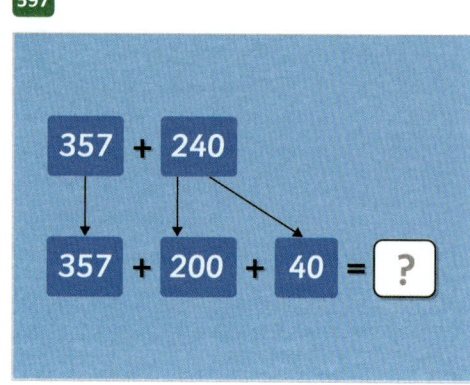

Level 2: Fluency: Add numbers (including decimals) mentally by counting on or using number bonds, partitioning and compensation.

✹ **Required:** 7/10   ✹ **Pupil Navigation:** on
✹ **Randomised:** off

11. Add together 3.5 and 3.6.

1 2 3 ▪ 7.1 ▪ 7

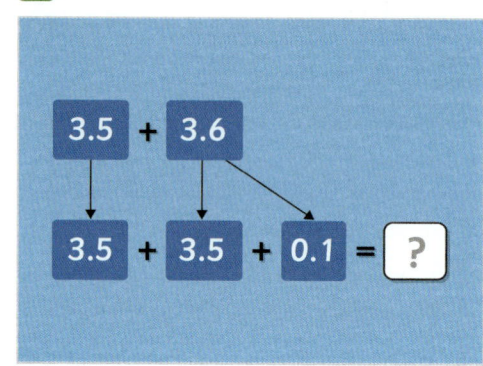

12. What is 469 + 329?

1 2 3 ▪ 798 ▪ 800

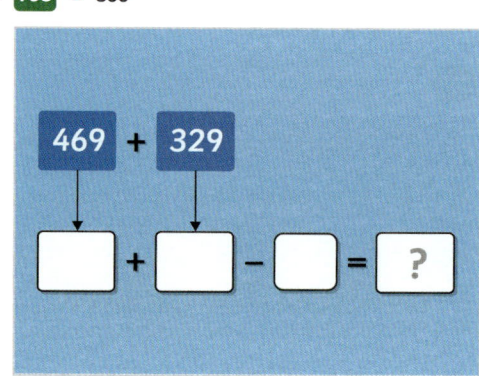

13. What is 1,049 + 2,999?

1 2 3 ▪ 4048 ▪ 4050

Level 2 *continued*

14. Joe runs 478 metres (m) and takes a break.
a He then runs a further 800 m. How far does
b Joe run altogether?
c
Include the units m (metres) in your answer.

- 1.278 km ▪ **1,278 metres** ▪ **1278 metres**
- 1.278 kilometres ▪ **1,278 m** ▪ **1278 m** ▪ 1,278
- 1278 ▪ 1.278

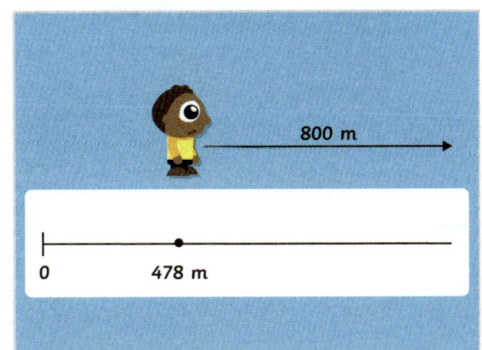

15. There are 252 boys and 296 girls at
1 Whiteside School. How many children are
2 there at the school in total?
3

- **548**

16. What is the total cost of the playground
a equipment?
b
c *Include the £ sign in your answer.*

- **£1,360.00** ▪ **£1,360** ▪ **£1360** ▪ **£1360.00**
- 1,360 ▪ 1360

17. Daisy buys a pair of trousers and a jacket.
a How much does she spend in total?
b
c *Include the £ sign in your answer.*

- 40.00 ▪ **£40.03** ▪ 40.03 ▪ £40.03p ▪ 40 ▪ £40
- £40.00

18. Find the total of 3,540 and 2,190.
1
2 ▪ **5730**
3

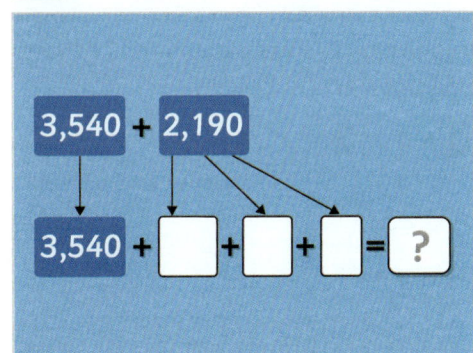

19. What is the sum of 250 + 550 + 450?
1
2 ▪ **1250**
3

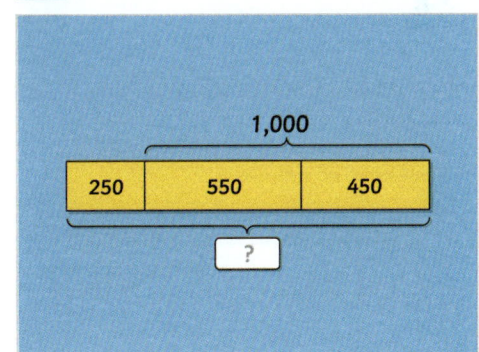

Level 2 continued

20. What is $17.5 + 21.2$?

 ▪ 38.7

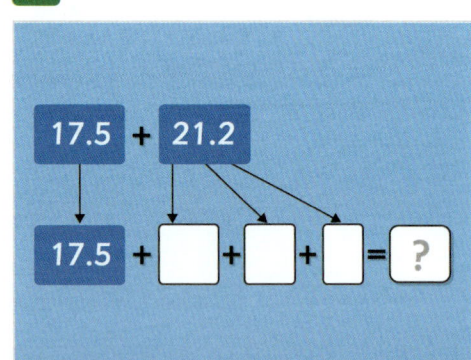

Level 3: Reasoning: Reason about addition using mental methods.

❋ **Required:** 5/5 ❋ **Pupil Navigation:** on
❋ **Randomised:** off

21. Which **two** of these calculations have the same answer?

2/3 ▪ A ▪ B ▪ C

22. Explain what Ellie has done wrong in her calculation.

- Open question, no set answer

$39 + 52$
$= 40 + 50 + 1 - 2$
$= 89$

23. Explain which mental method you would use to solve this calculation. Give your reasons for choosing this method and complete the calculation.

- Open question, no set answer

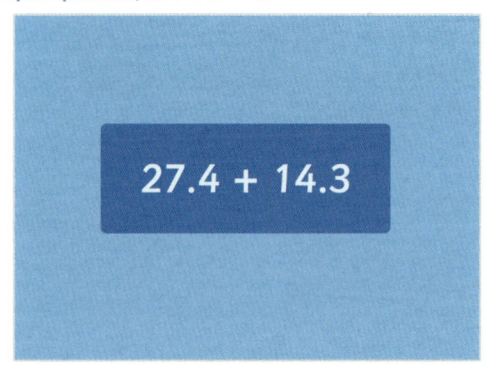

27.4 + 14.3

24. Use partitioning to help you to identify which of these calculations is **incorrect**.

▪ A ▪ B ▪ C ▪ D

1/4

A $10,002 + 697 = 10,699$

B $9,997 + 702 = 10,699$

C $5,001 + 5,697 = 10,699$

D $4,998 + 5,701 = 10,699$

25. The children are discussing the best method for adding £40.05 and £29.97. They decide to round the numbers to £40 and £30 to make a total of £70, but they have different ideas about what to do next. Who is correct?

1/3

▪ Noah ▪ Poppy ▪ Liam

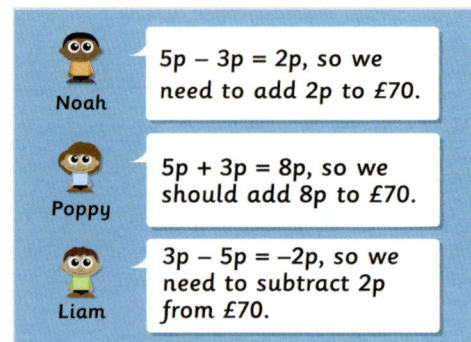

Noah: 5p − 3p = 2p, so we need to add 2p to £70.

Poppy: 5p + 3p = 8p, so we should add 8p to £70.

Liam: 3p − 5p = −2p, so we need to subtract 2p from £70.

Level 4: Problem solving with greater depth: Solve multi-step questions involving mental addition.

✸ **Required:** 5/5   ✸ **Pupil Navigation:** on
✸ **Randomised:** off

26. The score board shows Ava's scores for each level of a computer game. What is Ava's total score?

■ 1001 ■ **21412** ■ 20411

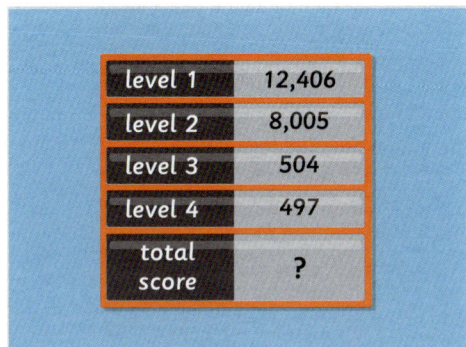

level 1	12,406
level 2	8,005
level 3	504
level 4	497
total score	?

27. What is the answer to the calculation with the highest total?

■ **4634** ■ 4324 ■ 4332

A 1,632 + 1,499 + 1,503

B 2,299 + 1,029 + 1,004

C 2,101 + 2,099 + 124

28. The answer to two of these calculations is 682. What is the total of the other calculation?

■ **681**

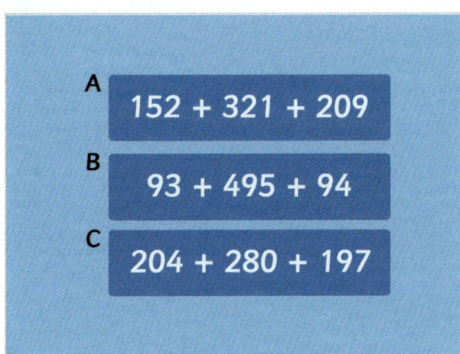

A 152 + 321 + 209

B 93 + 495 + 94

C 204 + 280 + 197

29. The children at Greenbank Primary have been raising money for charity. The poster shows how much money the children have raised so far. The local football club have offered to donate £1,250 if the children raised more than £5,000 on their own. How much **more money** do the children need to raise before the football club will make their donation?
Include the £ sign in your answer.

■ 4,095 ■ £4,095 ■ **£905.00** ■ **£905** ■ 905.00

■ 905 ■ 4095 ■ £4095

Greenbank Primary
so far we have raised:

cake sale £345   donations £750

sponsored run £2,005   toy sales £995

30. Ben, Alice and Chris each throw a javelin three times. What is the total distance for the child who throws the javelin the greatest total distance?
Include the units m (metres) in your answer.

■ **14.8 m** ■ **14.8 metres** ■ 14.8 ■ 12.9 m

■ 13.8 metres ■ 13.8 ■ 12.9 ■ 13.8 m

■ 12.9 metres

Javelin throws

	throw 1 (m)	throw 2 (m)	throw 3 (m)	total distance (m)
Ben	3.1	7.2	4.5	
Alice	2.8	6.9	3.2	
Chris	3.0	7.1	3.7	

Subtract Numbers Mentally

Objective: I can use mental strategies for subtraction including counting back, counting on, partitioning and compensation.

Quick Search Ref: 11266

Level 1: Understanding: Use mental strategies for subtraction including: counting back, counting on, partitioning and compensation.

✸ **Required:** 7/10   ✸ **Pupil Navigation:** on   ✸ **Randomised:** off

---

1. What is 378 − 50?

1
2
3

■ **328**   ■ 428

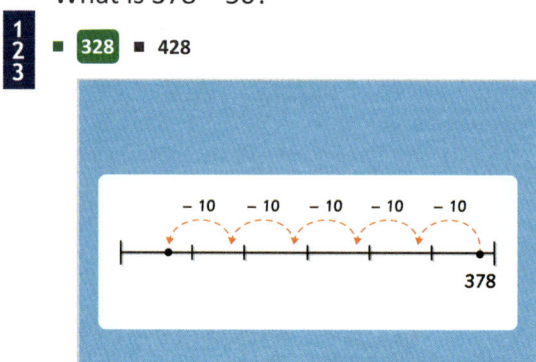

2. What is the difference between 163 and 19?

1
2
3

■ **144**   ■ 143

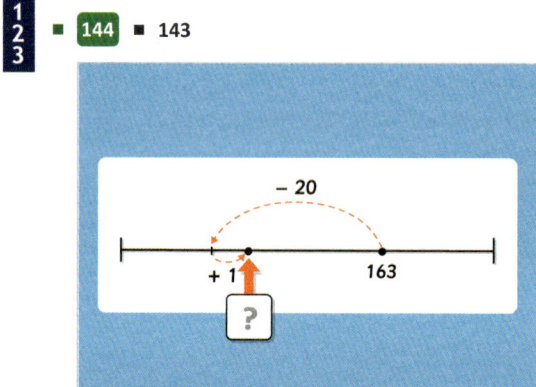

3. Subtract 70 from 824.

1
2
3

■ **754**   ■ 894

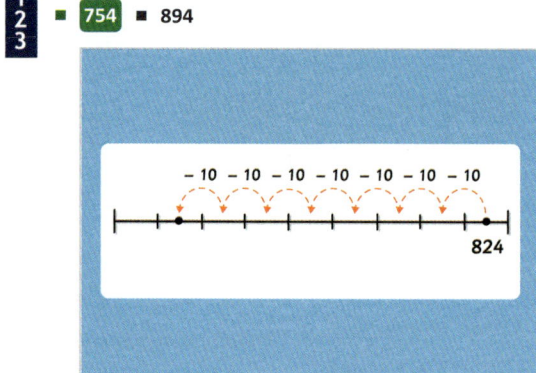

4. Find the answer to 534 − 214.

1
2
3

■ **320**

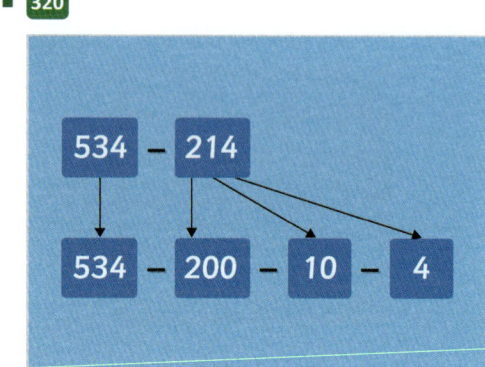

5. What is 524 − 131?

1
2
3

■ **393**   ■ 394

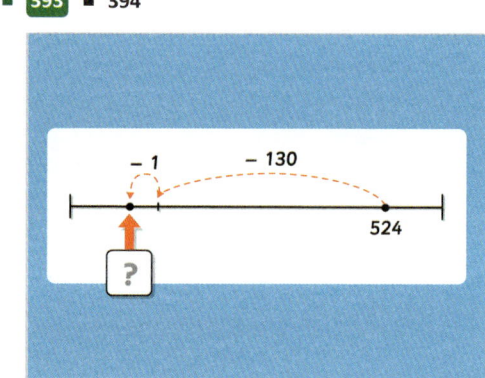

6. How many more than 735 is 856?

1
2
3

■ **121**

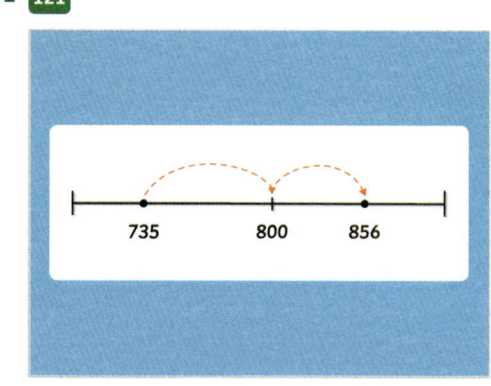

Level 1 *continued*

7. What is 160 less than 540?

1 2 3 ▪ **380**

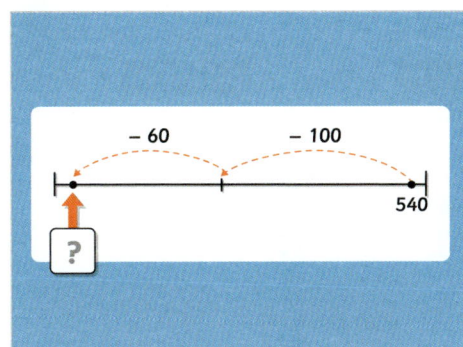

8. What is 27 less than 74?

1 2 3 ▪ 101 ▪ **47** ▪ 44

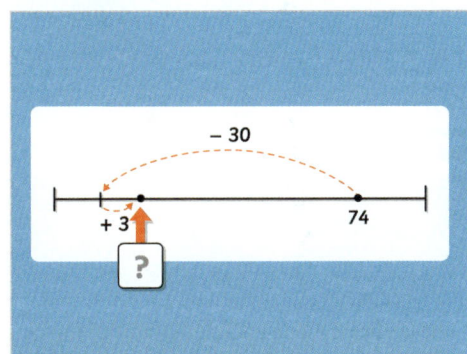

9. What is 328 subtract 50?

1 2 3 ▪ **278** ▪ 378

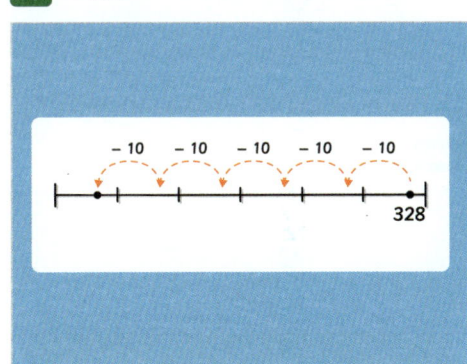

10. How much more than 374 is 452?

1 2 3 ▪ **78**

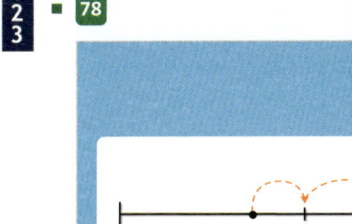

Level 2: Fluency: Use mental strategies for subtracting, including with decimals and in context.

✱ **Required:** 7/10   ✱ **Pupil Navigation:** on
✱ **Randomised:** off

11. The difference between 5.6 and 4.2 is _____.

1 2 3 ▪ **1.4** ▪ 9.8

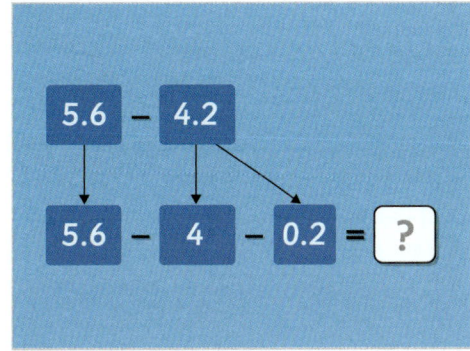

12. What is 6.3 minus 4.9?

1 2 3 ▪ **1.4** ▪ 1.3

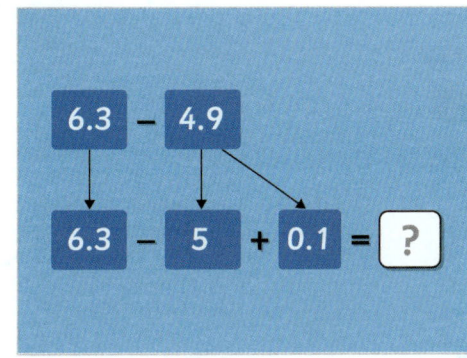

13. 14.6 + _____ = 15.3

1 2 3 ▪ **0.7** ▪ 29.9

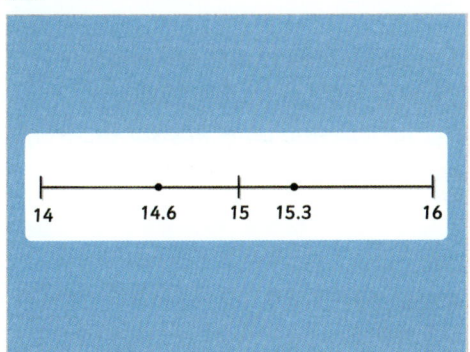

Level 2 continued

14. Ellie runs 1,265 metres (m). If Rose runs 400
a b c m less than Ellie, what distance does Rose
run?
Include the units m (metres) in your answer.

- 1665 metres ■ **865 m** ■ 1,665 ■ 865
- **865 metres** ■ 1,665 m ■ 1,665 metres ■ 1665 m
- 1665

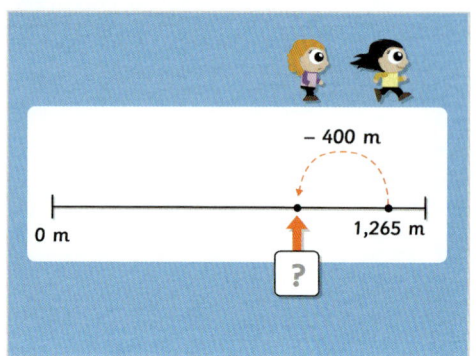

15. There are 1,602 chocolate drops in a jar and
1 2 3 799 jelly sweets in another jar. How many
more chocolate drops are there than jelly
sweets?

- **803**

16. If Megan spends £9.36, she will get _____
1 2 3 pence (p) change from £10.00.
Don't include the units in your answer.

- **64**

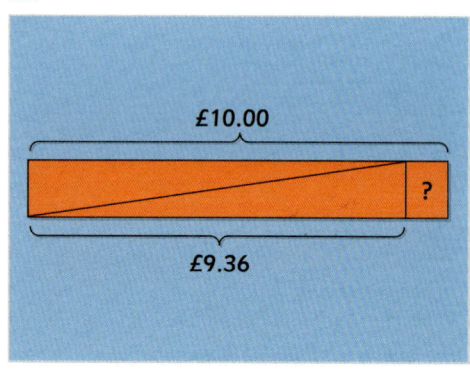

17. If Billy drinks 439 millilitres (ml) of water
a b c from a bottle containing 750 ml, how much
water is left in the bottle?
*Include the units ml (millilitres) in your
answer.*

- **311 ml** ■ 310 ml ■ **311 millilitres** ■ 310 millilitres
- 311 ■ 310

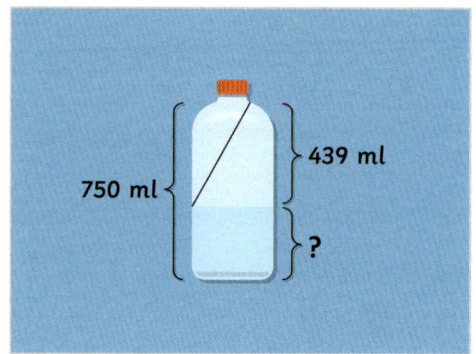

18. What is the difference between 109 and
1 2 3 1,008?

- **899** ■ 1117

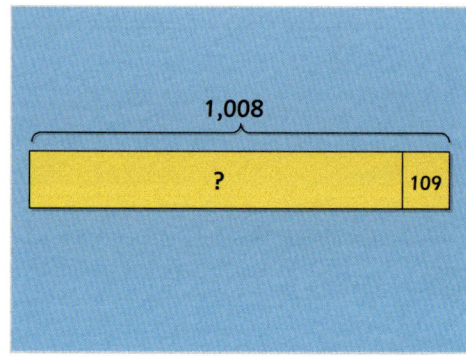

19. How many larger is 7,003 than 3,996?
1 2 3

- **3007** ■ 3003

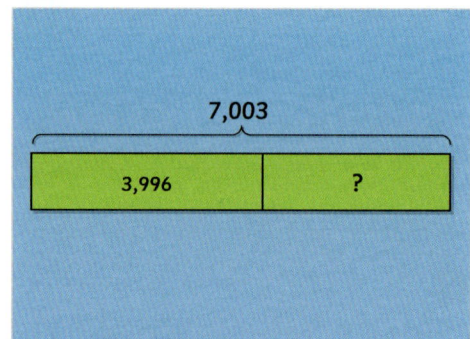

Level 2 *continued*

20. What is 12,047 − 8,034?

 ▪ **4013**

Level 3: Reasoning: Reason about subtraction using mental methods.

✱ **Required:** 5/5   ✱ **Pupil Navigation:** on
✱ **Randomised:** off

21. Which **two** of these calculations have the same answer?

 ▪ **calculation A** ▪ calculation B ▪ **calculation C**
2/3

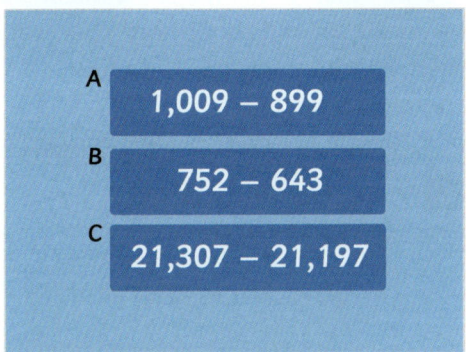

22. Explain what Leo has done wrong in his calculation.

- Open question, no set answer

23. Explain which mental method you would use to solve the calculation, and find the answer.

- Open question, no set answer

24. One of the calculations has an incorrect answer. Identify which calculation is incorrect and **give the correct answer**.

▪ **212** ▪ **208**

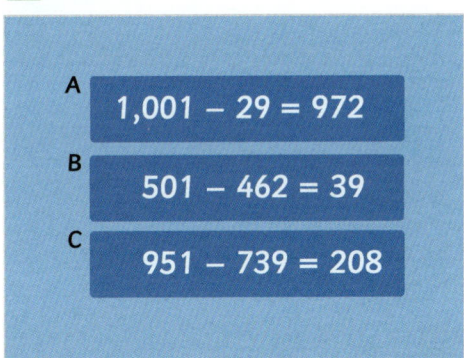

25. The children are discussing different mental methods for subtraction calculations. Who has the **wrong idea** about how to use their method?

1/3

▪ Daisy ▪ Ben ▪ **Alice**

Level 4: Problem solving with greater depth: Solve multi-step questions involving mental subtraction.

✹ **Required:** 5/5   ✹ **Pupil Navigation:** on
✹ **Randomised:** off

26. Miss Brown cuts three shelves from a 2-metre long plank of wood. If the shelves measure 47 cm, 34 cm and 65 cm, what length of **wood will be left in centimetres**?
Include the units cm (centimetres) in your answer.

- 0.54 m   ▪ 54 cm   ▪ 54 centimetres   ▪ 0.54 metres
- 54

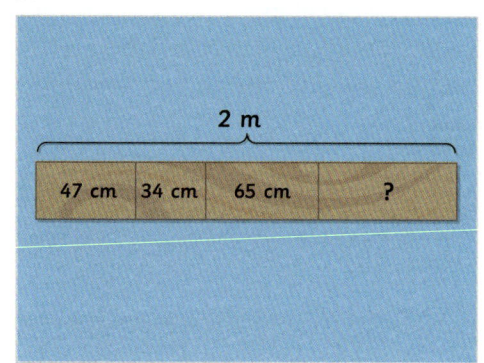

27. Work out the answers to all three calculations and give the smallest answer.

▪ 497   ▪ 503   ▪ 498

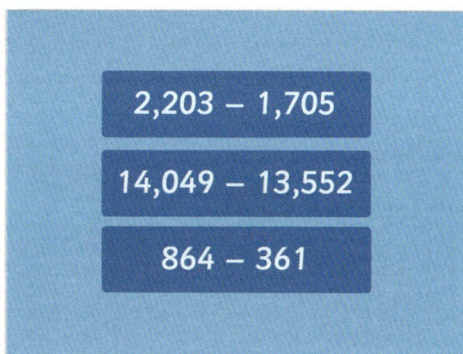

2,203 – 1,705

14,049 – 13,552

864 – 361

28. What is the difference between the answers to the two calculations shown?

▪ 0   ▪ 0.1   ▪ 7.4   ▪ 7.3

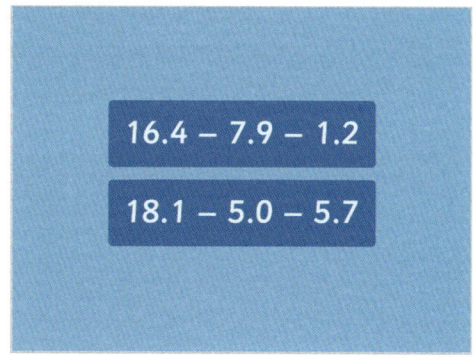

16.4 – 7.9 – 1.2

18.1 – 5.0 – 5.7

29. The table shows Hannah's level scores and total scores for three levels of a video game. In level 2, she scored 5,000 fewer points than she scored in level 3. How many points did Hannah score in level 1?

- 6394

Hannah's game scores

	level score	total score
level 1	?	
level 2		12,994
level 3	11,600	24,594

30. Rory's toy parachute takes 3.6 seconds to reach the ground. When he hangs a paperclip from the parachute, the parachute takes 0.3 seconds less to reach the ground. Each **extra** paperclip takes 0.2 seconds off the time taken to fall the same distance. How many seconds does Rory's parachute take to reach the ground with **four paperclips** attached?
Don't include the units in your answer.

▪ 2.7   ▪ 3.3   ▪ 2.5   ▪ 3.1   ▪ 2.9

Rory's parachute times

number of paperclips	time to reach ground (seconds)
0	3.6
1	
2	
3	
4	

Add Numbers up to 2 Decimal Places Using the Column Method

Objective:	I can add numbers with up to 2 decimal places using the written column method.
Quick Search Ref:	11380

Level 1: Understanding: Add decimals with no carry.

⚙ **Required:** 7/10 ⚙ **Pupil Navigation:** on ⚙ **Randomised:** off

1. What is 48 + 121?

■ **169**

2. Which image shows the calculation 13.2 + 4.7 lined up correctly?

■ **A** ■ B ■ C

1/3

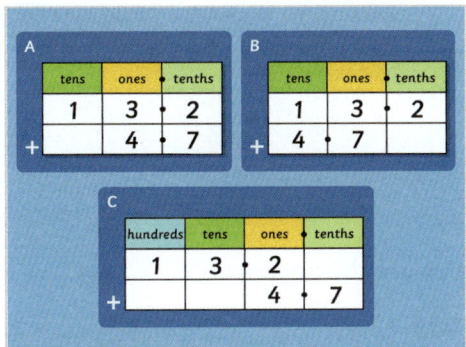

3. What is 2.4 + 5.3?

■ 77 ■ **7.7** ■ 0.77

1/3

4. Calculate the answer to 6.1 + 13.4.

■ **19.5** ■ 1.95 ■ 195

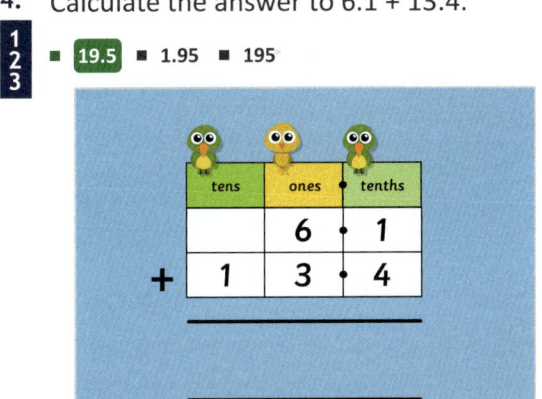

5. What is 31.4 + 8.5?

■ **39.9** ■ 3.99 ■ 399

6. Find the sum of 4.73 and 5.21.

■ 994 ■ **9.94** ■ 99.4

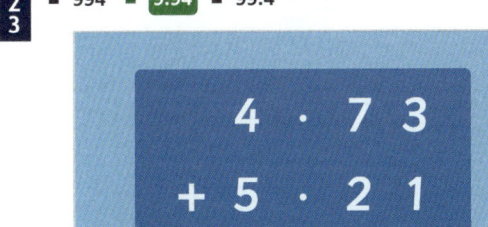

7. Calculate the answer to 1.5 + 2.37.

■ 387 ■ **3.87** ■ 38.7

8. Add 6.4 and 2.5.

■ **8.9** ■ 89

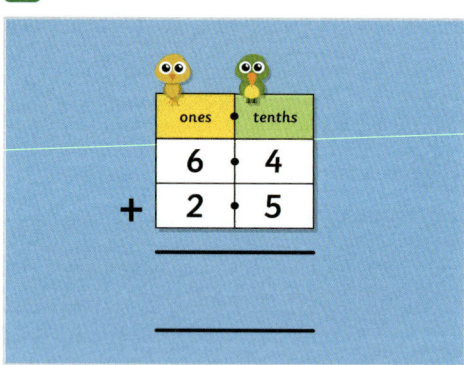

9. Which image shows the calculation 34.7 + 4.12 lined up correctly?

1/3

■ A ■ **B** ■ C

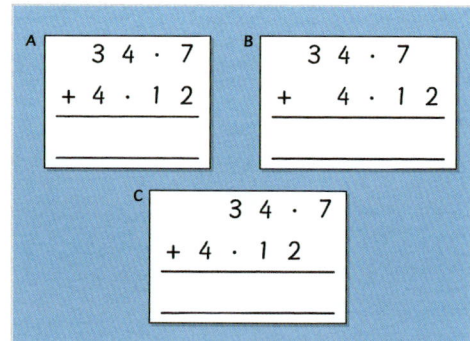

10. Calculate the answer to 1.2 + 18.4.

■ **19.6** ■ 1.96 ■ 196

Level 2: Fluency: Add decimals with carrying.

❋ Required: 7/10   ❋ Pupil Navigation: on
❋ Randomised: off

11. What is 6.3 + 2.9?

■ 8.12 ■ **9.2** ■ 8.2

12. Calculate the answer to 8.7 + 14.1.

■ **22.8** ■ 112.8 ■ 12.8 ■ 228

Level 2 *continued*

13. Add 1.5 to 32.6.

■ 341 ■ **34.1** ■ 33.11 ■ 33.1

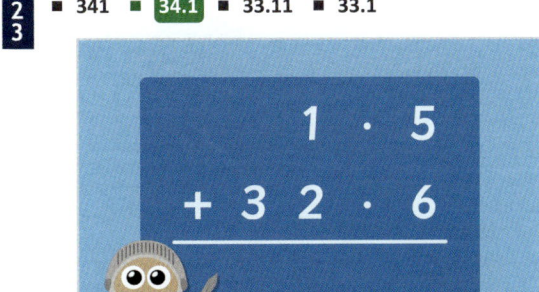

14. Add 1.47 to 6.35.

■ 7.712 ■ **7.82** ■ 782 ■ 7.72

15. What is 2.78 + 9.4?

■ 1218 ■ **12.18** ■ 11.18 ■ 11.118

16. Jon has 18 grams of sand and Bella has 85.4 grams of sand. How many grams of sand do they have altogether?
Don't include the units in your answer.

■ **103.4** ■ 913.4 ■ 1034 ■ 93.4

17. Courtney builds a tower that's 49.6 centimetres (cm) tall. If she adds more bricks to make it 3.7 centimetres taller, how many centimetres tall would it be?
Don't include the units in your answer.

■ 42.3 ■ **53.3** ■ 52.3 ■ 43.3 ■ 533

18. Calculate the answer to 1.47 + 6.75.

■ 7.22 ■ **8.22** ■ 8.12 ■ 822 ■ 7.12

Level 2 *continued*

19. What is 9.87 + 1.3?

1 2 3 ■ 10.117 ■ 11.17 ■ 1117 ■ 10.17

20. Chelsea adds 25.6 millilitres of cordial to 134.8 millilitres of water. What volume of liquid does Chelsea have altogether? *Don't include the units in your answer.*

■ 150.4 ■ 160.4 ■ 159.4 ■ 1604

Level 3: Reasoning: Add decimals with up to 2 decimal places using the column method.

✱ **Required:** 5/5 ✱ **Pupil Navigation:** on
✱ **Randomised:** off

21. What is the missing digit?

1 2 3 ■ 2

22. Erico adds 12.49 and 7.83 to get the answer 20.22. What mistake has Erico made?

a b c

- Open question, no set answer

23. What number do you add to 12.21 to get 27.48?

1 2 3 ■ 15.27

24. Delilah wants to use the column method to add 87.4 and 12.03. What mistake has she made?

a b c

- Open question, no set answer

Level 3 *continued*

25. Jonah is calculating the total weight of three
a
b bags. He uses the column method to add
c together the first two weights, then to add
the total to the third weight. Explain how
Jonah could have found the answer more
quickly.

- Open question, no set answer

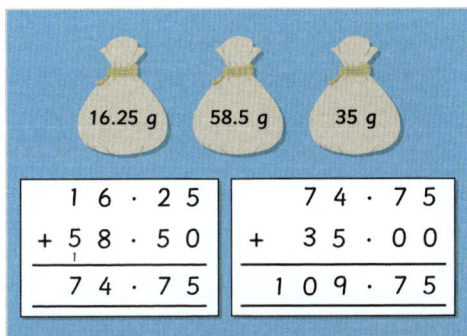

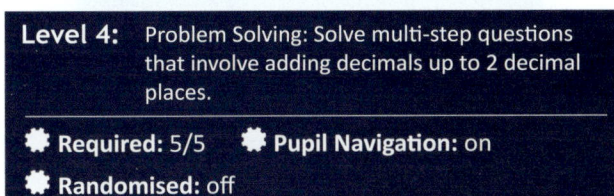

Level 4: Problem Solving: Solve multi-step questions
that involve adding decimals up to 2 decimal
places.

❋ **Required:** 5/5   ❋ **Pupil Navigation:** on
❋ **Randomised:** off

26. Lexi spends £39.33 on two items from the
list. Which two items did she buy?

2/5

■ soft toy ■ motorised car ■ slide ■ board game
■ doll's house

27. Rose has spilled ink on her homework. What
1 is the answer to Rose's calculation?
2
3
■ 60.4 ■ 50.4

28. What is the missing number in the
1
2 sequence?
3
■ 11.1 ■ 10.1

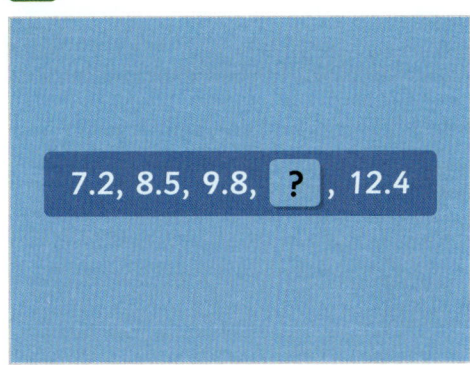

29. What is the smallest total that can be made
1
2 by filling in the addition with the digit cards
3 shown?
Use each card once.

■ 3.86

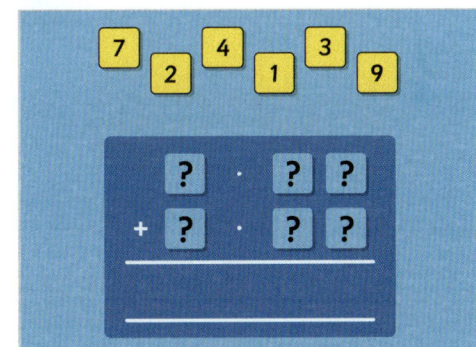

30. The code is a four digit number.
1
2 Work out the values of **A**, **B**, **C** and **D** and
3 crack the code.

■ 4127 ■ 4128 ■ 4228 ■ 4227

Add Numbers with More Than 4 Digits Using the Column Method

Objective: I can add numbers with more than 4 digits using the formal written method.

Quick Search Ref: 11769

Level 1: Understanding: Add numbers with more than 4 digits with support.

✱ **Required:** 7/10 ✱ **Pupil Navigation:** on ✱ **Randomised:** off

---

1. What is the total of the hundreds?

▪ 800 ▪ 8

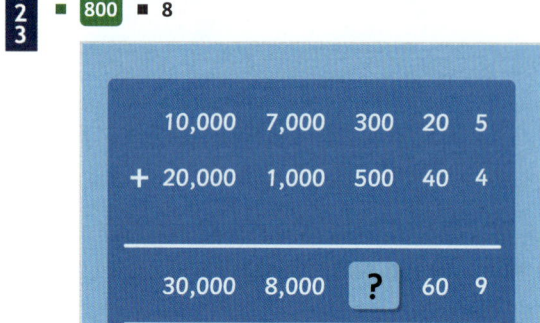

2. Calculate the answer to 49,270 + 10,628.

▪ 59898

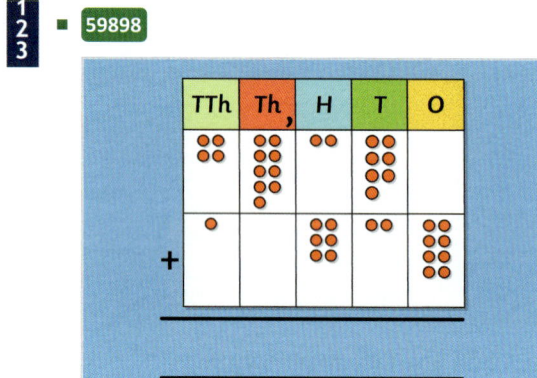

3. Calculate the answer to 12,935 + 11,427.

▪ 24362 ▪ 4362

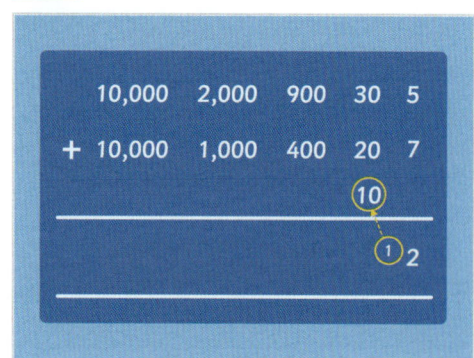

4. What is 30,004 + 42,005?

▪ 72009 ▪ 729

5. What is 89,154 added to 10,638?

▪ 99792 ▪ 99782

6. What is the total of 26,903 and 12,488?

▪ 39381 ▪ 39391 ▪ 38391

Level 1 continued

7. What is 99,138 + 6,249?

■ **105387** ■ 95387 ■ 105377

8. Calculate 34,521 + 12,498.

■ **47019** ■ 46919

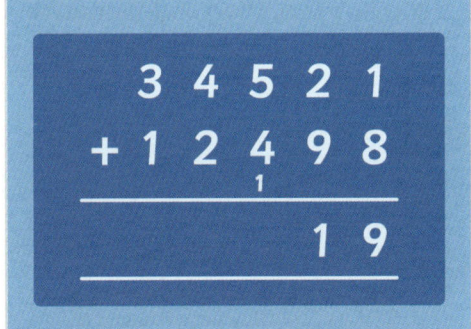

9. What is the sum of 71,709 and 24,162?

■ **95871** ■ 95861

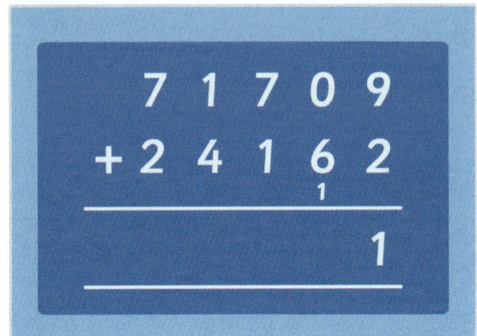

10. Calculate the answer to 20,043 + 40,056.

■ **60099** ■ 699

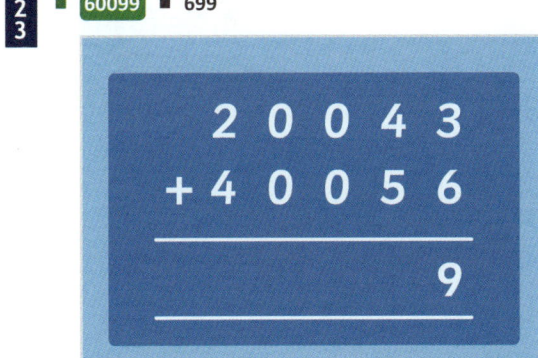

Level 2: Fluency: Add numbers with more than 4 digits, including questions in context.

❋ **Required:** 7/10   ❋ **Pupil Navigation:** on
❋ **Randomised:** off

11. What is 45,195 + 22,103?

■ **67298**

12. Calculate 52,947 + 14,216.

■ **67163** ■ 66153

13. 130,575 + 317,573 = _____

■ **448148** ■ 447048

14. What is the total of 1,208,391 and 1,341,768?

- **2550159** ■ 2549059

15. Calculate 51,142 + 23,108 + 1,790.

- **76040** ■ 92150

16. A company buys a new machine that costs £64,529. They also have to pay £12,906 in taxes and fees. What is the total cost of the new machine?
Include the £ sign in your answer.

- **£77435** ■ **£77,435** ■ £51,623 ■ 77,435
- £51623 ■ 77435

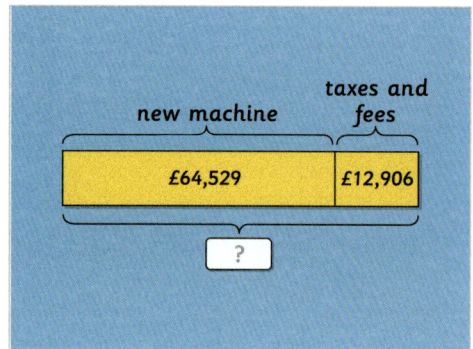

17. A football club sells 17,536 adult tickets, 4,893 family tickets and 3,206 children's tickets for a match. How many tickets does the club sell in total?

- **25635**

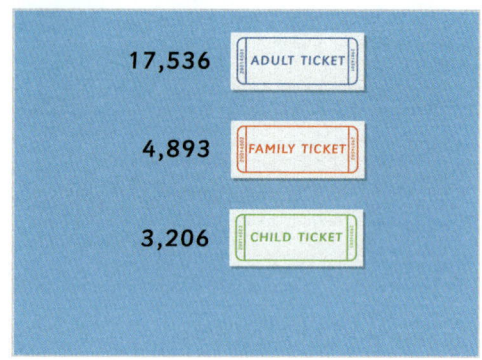

18. In 2017, there were 798,307 visitors at Green Park Zoo. In 2018, there is an **increase** of 16,858 visitors. How many visitors are there in **2018**?

- **815165** ■ 781449

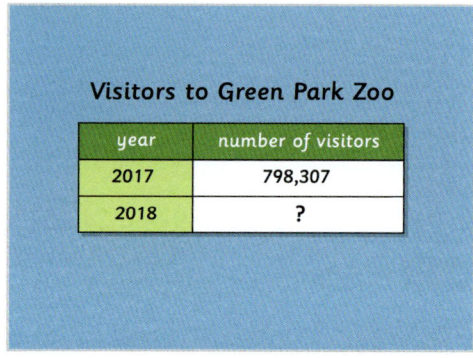

19. What is the total of 1,257,145 and 1,213,864?

- **2471009** ■ 2460909

Level 2 *continued*

20. Calculate 15,721 + 62,240.

1
2
3
■ 77961

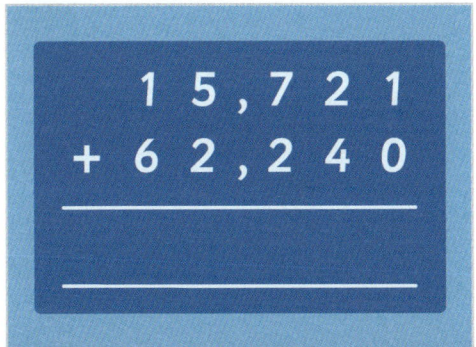

Level 3: Reasoning: Reason about adding numbers with more than 4 digits.

✱ **Required:** 5/5   ✱ **Pupil Navigation:** on
✱ **Randomised:** off

21. What is the missing number in the following

1
2
3
calculation?

_____ − 19,349 = 34,061

■ 53410   ■ 14712

22. Aria and Matt are trying to find the value of

a
b
c
the missing digit in the calculation.
Aria says, "There is only one possible missing digit."
Matt says, "The question must be wrong because no digit adds to 4 to make 3."
Who do you agree with? Explain your answer.

- Open question, no set answer

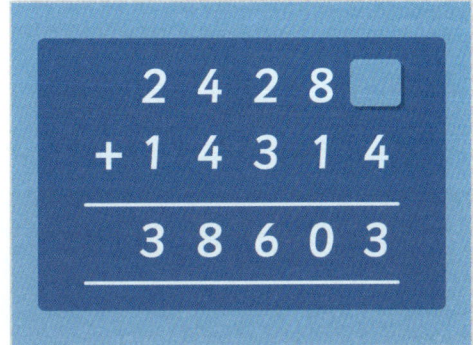

23. Lorna says, "If you add a 4-digit number and

a
b
c
5-digit number, your answer will be a 9-digit number."
Is Lorna correct? Explain your answer.

- Open question, no set answer

24. Mohammed uses place value counters to

1
2
3
represent an addition calculation. What is the **second number** in his calculation?
Give your answer in digits.

■ 20182   ■ 20122

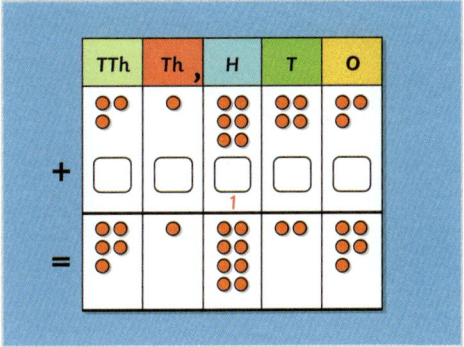

Level 3 *continued*

25. Which calculation is the odd one out?
Explain your answer.

a
b
c - Open question, no set answer

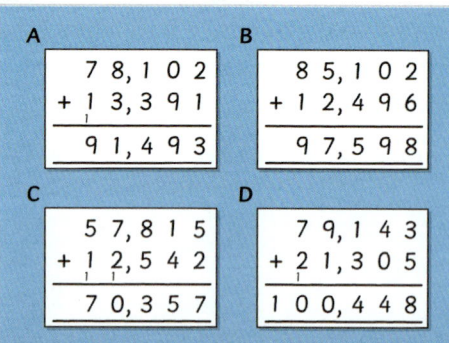

A
```
  7 8, 1 0 2
+ 1 3, 3 9 1
  9 1, 4 9 3
```

B
```
  8 5, 1 0 2
+ 1 2, 4 9 6
  9 7, 5 9 8
```

C
```
  5 7, 8 1 5
+ 1 2, 5 4 2
  7 0, 3 5 7
```

D
```
  7 9, 1 4 3
+ 2 1, 3 0 5
1 0 0, 4 4 8
```

Level 4: Problem solving with greater depth: Solve multi-step problems involving adding numbers with more than 4 digits.

✸ **Required:** 5/5   ✸ **Pupil Navigation:** on
✸ **Randomised:** off

26. The bar chart shows the number of people
living in different towns. In total, how many
people live in Ristown or Catville?

1
2
3

■ 95000   ■ **135000**   ■ 130000   ■ 40000

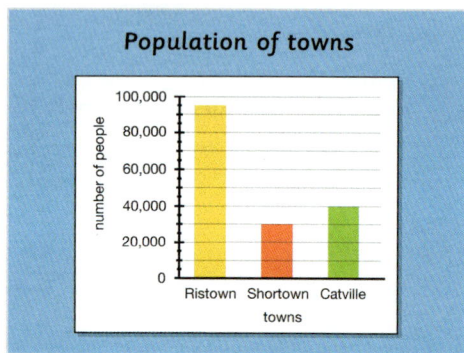

Population of towns

27. The four digit cards complete the addition
calculation. What is the value of *a*?

1
2
3

■ 6059   ■ **9**   ■ 1

```
  0   5   4   a

    7 4 1 3
+   6 ▢ ▢ ▢
  5 3 4 7 2
```

28. Which digit goes in position B?

1
2
3

■ **4**   ■ 5

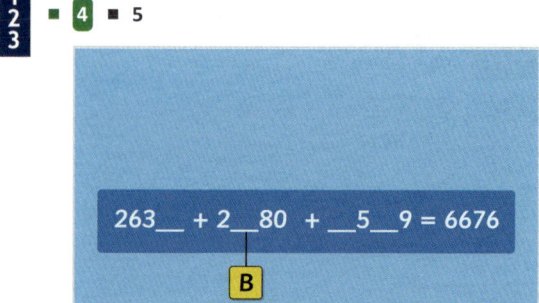

$263\_\_ + 2\_80 + \_5\_9 = 6676$

B

29. Arrange the number cards so that each line
has the same total. What is the total of each
line?
The same card can be used more than once.

1
2
3

■ **59056**   ■ 118112   ■ 86007

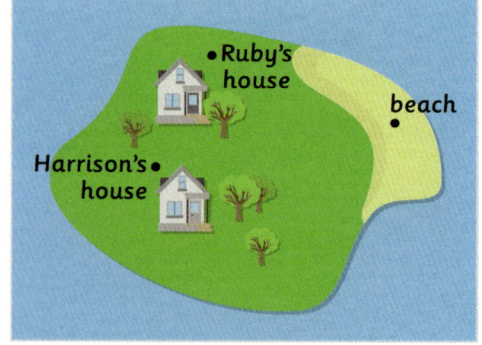

```
 12,580   16,819   10,132
    32,105   14,371
```

30. Harrison drives 11,312 m (metres) to Ruby's
house. He then drives them both 3,718 m to
the beach. Later in the day, he drops Ruby
back off at her house, and then he drives
home. How many metres does Harrison
drive in total during the day?
Include the units m (metres) in your answer.

a
b
c

■ 15,030 metres   ■ **30,060 metres**   ■ **30060 metres**
■ **30,060 m**   ■ 15030 metres   ■ **30060 m**   ■ 30,060
■ 30060   ■ 15,030 m   ■ 15030 m

Solve Multi-Step Addition and Subtraction Problems

Objective: I can solve addition and subtraction multi-step problems and decide which operations and methods to use.

Quick Search Ref: 10088

Level 1: Problem solving with greater depth: Solve multi-step addition and subtraction multi-step problems in context.

❖ Required: 7/10   ❖ Pupil Navigation: on   ❖ Randomised: off

---

1. **a b c** James goes to a restaurant and buys one portion of mushrooms, a lasagne and a fudge cake. He pays with a £20.00 note. How much change does he get?
Include the £ sign in your answer.

- **£4.44** ▪ **4.44** ▪ **£15.56**

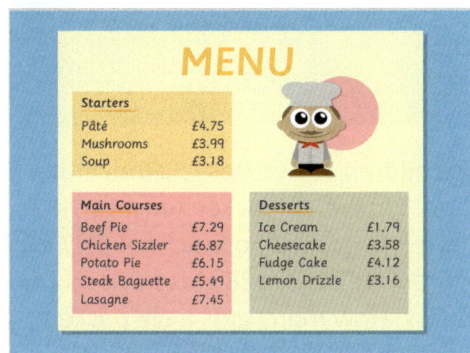

2. **a b c** Mick pays for a football with a £20.00 note and gets £4.82 change.
Charlie pays for a rugby ball with a £10.00 note and gets £1.05 change.
What is the total cost of a rugby ball and a football?
Include the £ sign in your answer.

- **£15.18** ▪ **£24.13** ▪ **£8.95** ▪ **24.13**

3. **1 2 3** A cricket club sell 246 adult tickets, 117 children's tickets and 359 over-60s tickets. At the next match, they sell 84 fewer tickets in total. How many tickets do they sell for the two cricket matches **in total**?

- **722** ▪ **1360** ▪ **638**

4. **1 2 3** Some children are asked what their favourite sports are. The results are shown in the bar chart.
How many more children chose basketball or netball than chose hockey?

- **30000** ▪ **45000** ▪ **15000**

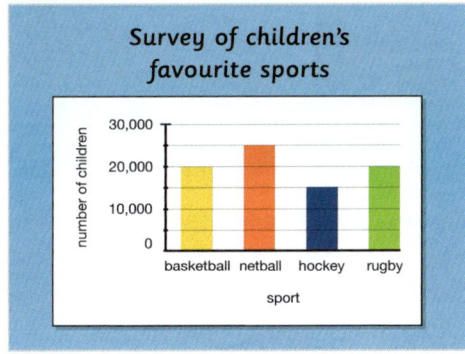

Level 1 continued

5. A supermarket sells three types of juice. Use the following clues to find the **total number of bottles of juice** the supermarket sells in a week:

• The supermarket sells 846 more bottles of orange juice than apple juice.

• It sells 118 more bottles of apple juice than peach juice.

• It sells 656 bottles of apple juice.

■ 1502 ■ 2696 ■ 538 ■ 2902

6. 348 children go to Quiverton Primary School. Use the table to work out **how many girls** go to the school.

■ 145 ■ 203 ■ 68

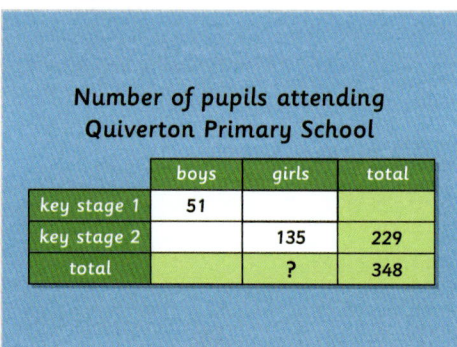

Number of pupils attending
Quiverton Primary School

	boys	girls	total
key stage 1	51		
key stage 2		135	229
total		?	348

7. Sally and Dan have a total of £8.42. Sally has £1.96 more than Dan. How much money does **Sally** have?
Include the £ sign in your answer.

■ £5.19 ■ 5.19 ■ £3.23 ■ £6.46

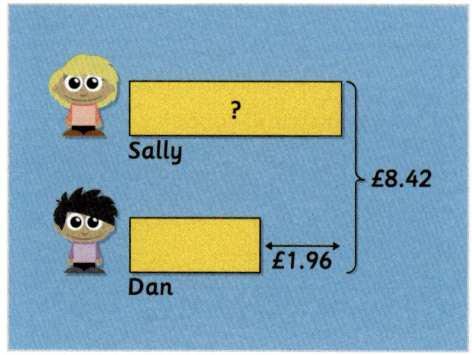

8. Joel has £7,061 in his bank account. He buys a new car for £5,892 and a new computer for £427. He is paid into his bank account £2,764 in wages from his job. How much money does Joel have left in his bank account?
Include the £ sign in your answer.

■ £3,506 ■ £3506 ■ 3506 ■ £6319 ■ £6,319
■ £742 ■ 3,506

9. The children at Red Rose School either have school dinners or packed lunches. 463 children go to Red Rose School. 257 of the children are boys. 128 girls have school dinners. How many girls have packed lunches?

■ 78 ■ 335 ■ 129 ■ 206

Level 1 *continued*

10. Pat buys 1 kilogram (kg) of mangoes, 1 kg of
grapes and 4 kg of cherries. He gives the
shopkeeper £25.00 and gets £3.26 change.
How much does 1 kilogram of **cherries** cost?
Include the £ sign in your answer.

 ▪ £3.14 ▪ £12.56 ▪ £9.18 ▪ 3.14 ▪ £21.74

Fruit prices

fruit	cost per kilogram
mangoes	£5.27
grapes	£3.91
melons	£1.16
oranges	£1.84
apples	£2.37
cherries	?

Solve Problems Involving Addition, Subtraction, Multiplication and Division

Objective: I can solve problems involving addition, subtraction, multiplication and division and understand the meaning of the equals sign.

Quick Search Ref: 10234

Level 1: I can solve multi-step problems involving addition, subtraction, multiplication and division.

❋ Required: 10/10 ❋ Pupil Navigation: on ❋ Randomised: off

1. Use the totals at the end of each row and column to find the value of **one star**.

■ **52** ■ 87 ■ 99

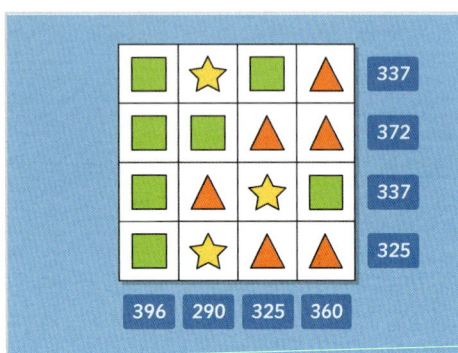

2. There are 1,028 passengers on board a ferry. There are 451 women and 392 men. How many **children** are on board?

■ **185** ■ 843

3. Three friends hire bikes on holiday. Use the following clues to find the **total** amount the children pay to hire the bikes:
• Kira and Alex hire TMX bikes for four days.
• Jordan hires a triathlon bike for one day and a mountain bike for three days.
Include the £ sign in your answer.

■ **£600.00** ■ £170 ■ £400.00 ■ **£600** ■ £600p
■ 600 ■ £400

Bike hire price list

bike	price per day
triathlon	£80.00
TMX	£50.00
mountain	£40.00

4. Three consecutive numbers (numbers that follow each other in order) have a total of 75. What is the **largest** of these three numbers?

■ **26** ■ 24 ■ 25

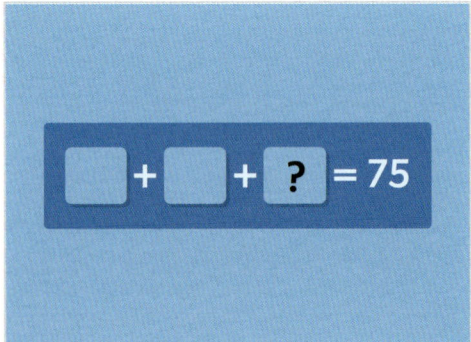

Level 1 *continued*

5. Oranges are sold in boxes of 24, and apples are sold in boxes of 15. A shopkeeper buys 42 boxes of oranges and 76 boxes of apples. How many pieces of fruit does the shopkeeper buy in total?

■ 1008 ■ **2148** ■ 1140

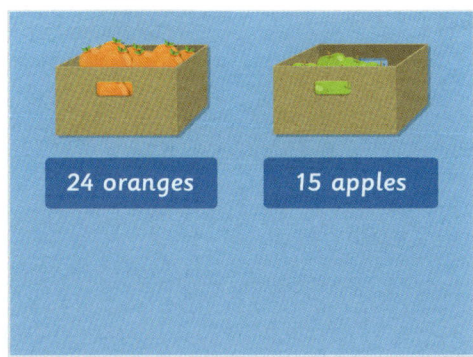

6. The numbers on each line have the **same** total. What is the value of the missing number?

■ 20683 ■ **8807** ■ 11876

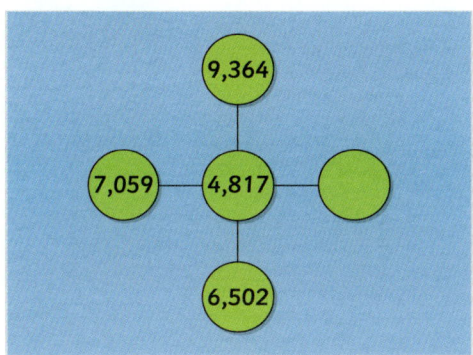

7. A car park has room for 325 cars. It costs £7.00 to park each car for a concert, and the car park company collects £1,869 in total. How many **empty spaces** are there?

■ **58** ■ 267

8. The owner of a clothes shop buys 1,215 red jackets and 936 blue jackets to sell. They pay £9.00 for each jacket. How much does it cost the shop owner to buy all the jackets? *Include the £ sign in your answer.*

■ £19359p ■ £19,359p ■ **£19,359** ■ **£19359**
■ 2,151 ■ 19359 ■ 2151 ■ 19,359

9. Hannah's tablet screen is 123 millimetres (mm) long and 61 mm wide. Kai's tablet screen is 147 mm long and 65 mm wide. In square millimetres (mm²), how much **larger** is the area of Kai's screen than Hannah's screen? *Don't include the units in your answer.*

■ **2052** ■ 9555 ■ 7503

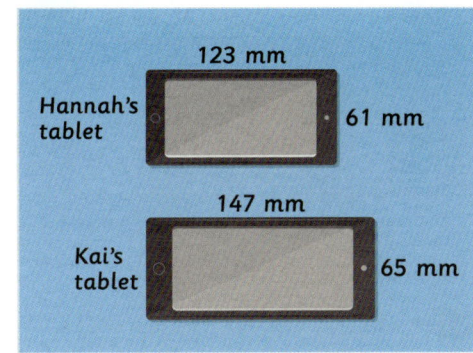

10. How much does it cost to buy ten apples and eight bananas? *Include the £ sign in your answer.*

■ **£2.84** ■ £2.84p ■ £34.08 ■ 2.84 ■ 34.08

Subtract Numbers up to 2 Decimal Places Using the Column Method

Objective: I can subtract numbers with up to 2 decimal places using the written column method.

Quick Search Ref: 11382

Level 1: Understanding: Subtract decimals with no exchanging.

⚙ **Required:** 8/10 ⚙ **Pupil Navigation:** on ⚙ **Randomised:** off

---

1. What is 147 subtract 12?

■ **135**

2. Which image shows the calculation 15.81 − 2.3 correctly lined up in the column method?

1/2 ■ A ■ **B**

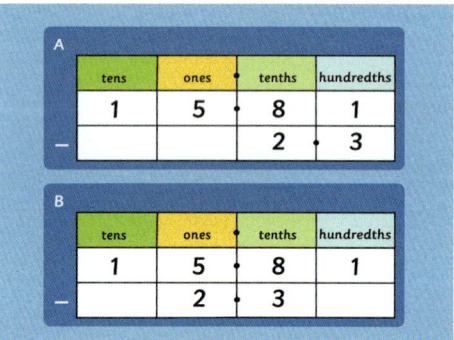

3. Calculate the answer to 4.2 − 3.1.

■ **1.1** ■ 11

4. Subtract 14.7 from 78.9.

■ **64.2** ■ 642

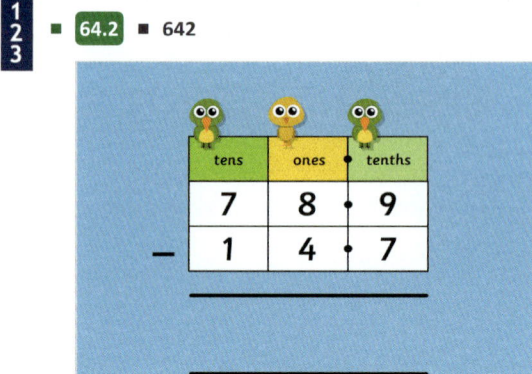

5. What is 8.64 subtract 2.01?

■ **6.63** ■ 663

6. Find the difference between 1.8 and 5.92.

■ **4.12** ■ 412

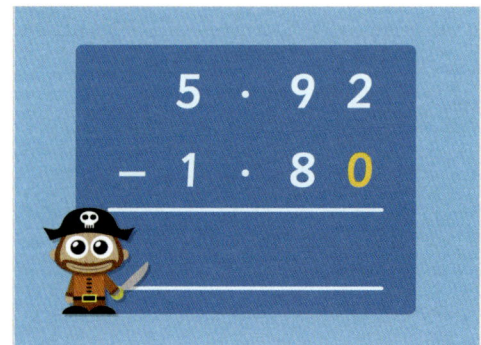

Level 1 *continued*

7. Calculate the answer to 27.49 − 6.27.

 ▪ **21.22** ▪ **2122**

8. Subtract 6.2 from 9.5.

 ▪ **3.3** ▪ **33**

9. Which image shows the calculation 27.8 − 4.31 correctly lined up in the column method?

1/3 ▪ A ▪ **B** ▪ C

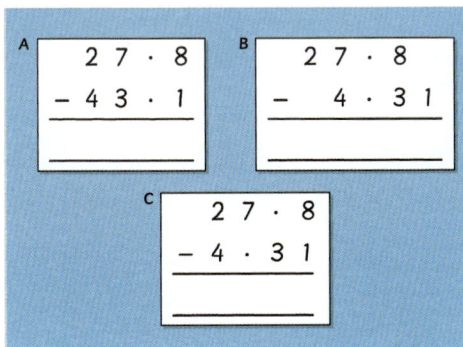

10. What is 42.4 subtract 31.2?

▪ **11.2** ▪ **112**

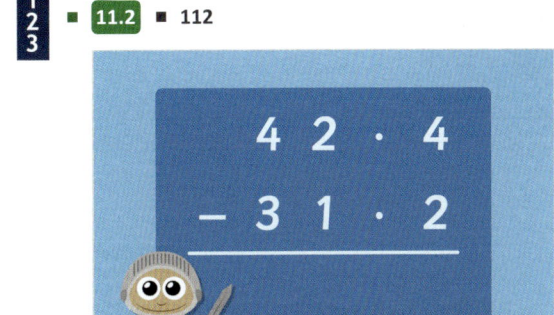

Level 2: Fluency: Subtract decimals with exchanging.

✿ **Required:** 7/10   ✿ **Pupil Navigation:** on
✿ **Randomised:** off

11. What is 29.4 − 7.6?

▪ **21.8** ▪ **218**

12. Calculate the answer to 14.7 − 8.3.

▪ **6.4** ▪ **16.4** ▪ **64** ▪ **14.4**

Level 2 *continued*

13. Subtract 19.2 from 27.3.

▪ **8.1** ▪ 81 ▪ 12.1 ▪ 18.1

$$\begin{array}{r} 2\ 7\ \cdot\ 3 \\ -\ 1\ 9\ \cdot\ 2 \\ \hline \end{array}$$

14. Find the difference between 83.44 and 47.82.

▪ 44.42 ▪ **35.62** ▪ 3562

15. What is 152.76 subtract 27.8?

▪ **124.96** ▪ 135.16 ▪ 12496

$$\begin{array}{r} 1\ 5\ 2\ \cdot\ 7\ 6 \\ -\ \ \ 2\ 7\ \cdot\ 8\ 0 \\ \hline \end{array}$$

16. Annie has 752.25 millilitres of lemonade. If she drinks 238.4 millilitres, how much lemonade does she have left?
Don't include the units in your answer.

▪ **513.85** ▪ 51385

17. Katie is 127.67 centimetres tall. Ben is 124.8 centimetres tall. How much taller is Katie than Ben?
Don't include the units in your answer.

▪ **2.87** ▪ 287

18. What is 45.6 − 9.8?

▪ **35.8** ▪ 358

$$\begin{array}{r} 4\ 5\ \cdot\ 6 \\ -\ \ \ 9\ \cdot\ 8 \\ \hline \end{array}$$

Level 2 *continued*

19. Calculate the answer to 37.18 − 4.87.

 ▪ `32.31`

20. Sam has 247.4 grams of grapes. If he gives 89.3 grams of grapes to Jax, how many grams of grapes does he have left?
Don't include the units in your answer.

▪ `158.1`

Level 3: Reasoning: Subtract numbers up to 2 decimal places using the column method.

✱ **Required:** 5/5   ✱ **Pupil Navigation:** on
✱ **Randomised:** off

21. What is the missing digit in this calculation?
4_.17 − 26.29 = 21.88

▪ `8` ▪ `7`

22. Eve says the missing number in this subtraction must be 7, because 7 − 6 = 1. Explain why she's wrong.

- Open question, no set answer

23. The three missing digits all have the same value. What is the value of the missing digit?

▪ `8`

24. Joshua subtracts 24.79 from 82.31. What mistake has he made?
Explain your answer.

- Open question, no set answer

Level 3 *continued*

25. What number do you need to subtract from 73.34 to get 27.9?

■ 45.44

Level 4: Problem Solving: Subtract numbers up to 2 decimal places using the column method.

❋ **Required:** 5/5   ❋ **Pupil Navigation:** on
❋ **Randomised:** off

26. The yellow boxes contain the totals of the two green circles on either side of them. What is the value of the bottom right circle?

■ 7.8   ■ 3.5

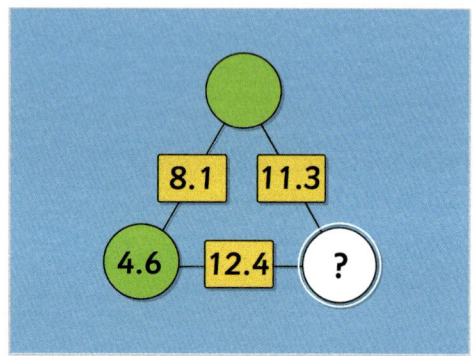

27. What is the missing number in the sequence?

■ 38.56   ■ 15.12

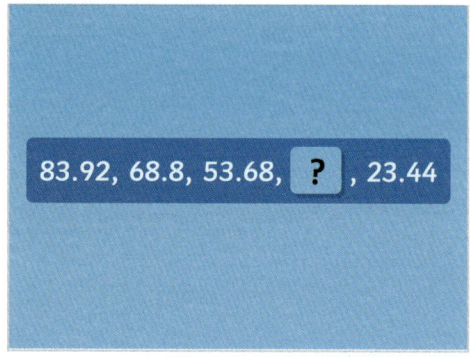

83.92, 68.8, 53.68, ? , 23.44

28. Using every digit card once, what is the **largest answer** you can make?

■ 85.2   ■ 1.24   ■ 97.6

29. The code is a four-digit number. Work out the values of **A**, **B**, **C** and **D** and crack the code.

■ 2441   ■ 2241   ■ 24.41   ■ 1442   ■ 2451

30. Jude buys two items from the list with a £20 note and gets £8.12 change. Which two items does he buy?

2/4

■ car   ■ doll   ■ crayons   ■ board game

price list

car...................£6.42
doll.................£7.81
crayons............£4.96
board game....£5.46

Subtract Numbers with More Than 4 Digits Using the Column Method

Objective: I can subtract numbers with more than 4 digits using the formal written method.

Quick Search Ref: 11767

Level 1: Understanding: Subtract numbers with more than 4 digits with support.

❈ **Required:** 7/10   ❈ **Pupil Navigation:** on   ❈ **Randomised:** off

---

1. What is the missing value in the answer?

■ **1000**   ■ 1   ■ 11302

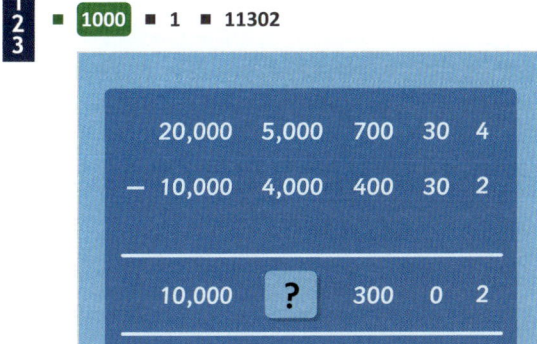

2. Calculate 43,173 − 31,142.

■ **12031**

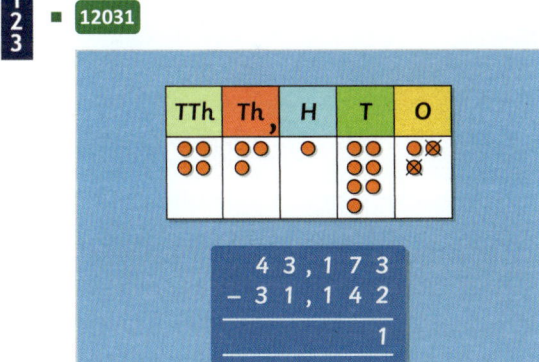

3. What is 36,271 subtract 22,433?

■ **13838**   ■ 13848

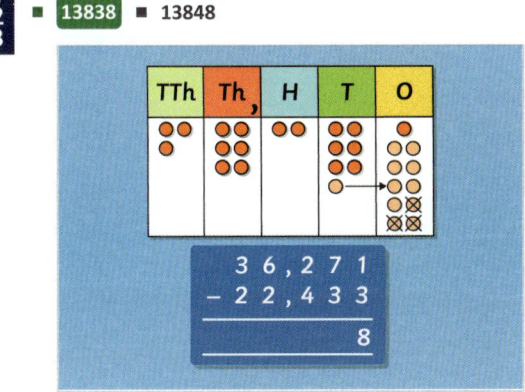

4. What is 46,819 − 23,608?

■ **23211**   ■ 70427

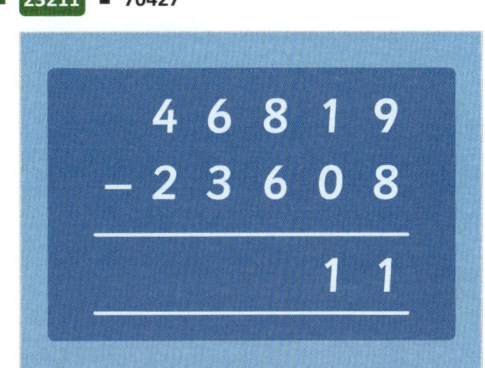

5. What is 22,354 − 11,615?

■ **10739**   ■ 33969   ■ 11341

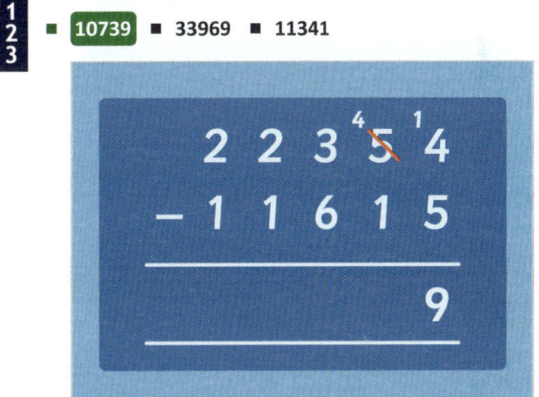

6. What is 58,201 − 45,102?

■ **13099**   ■ 13101

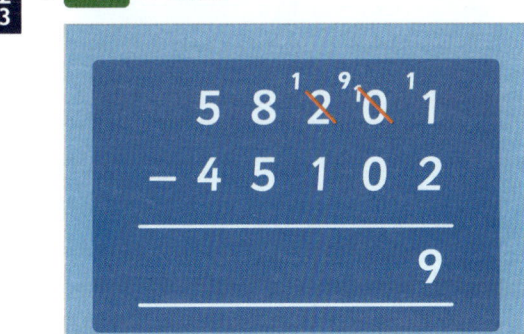

Level 1 *continued*

7. Calculate 658,002 − 174,891.

📋 ■ **483111** ■ 524891

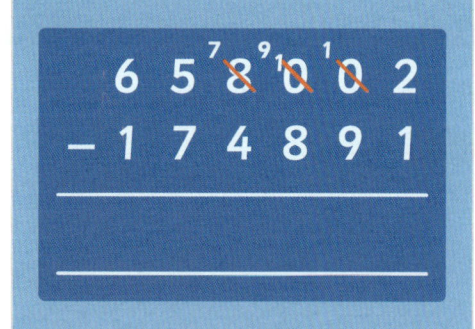

8. Subtract 31,194 from 73,405.

📋 ■ **42211** ■ 42391

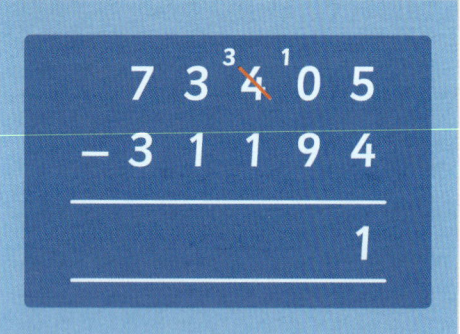

9. What is 28,176 − 15,085?

📋 ■ **13091** ■ 13111

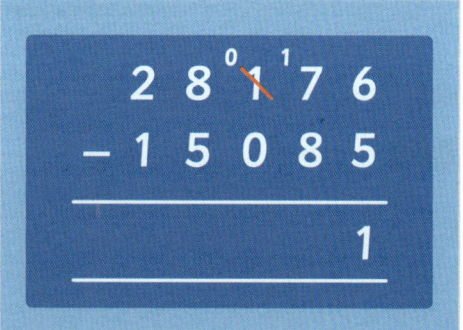

10. What is 49,734 − 13,424?

📋 ■ **36310** ■ 63158

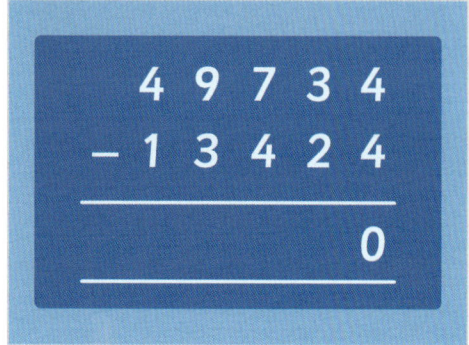

Level 2: Fluency: Subtract numbers with more than 4 digits, including in context.

✳ **Required:** 7/10   ✳ **Pupil Navigation:** on
✳ **Randomised:** off

11. What is 67,193 − 35,072?

📋 ■ **32121**

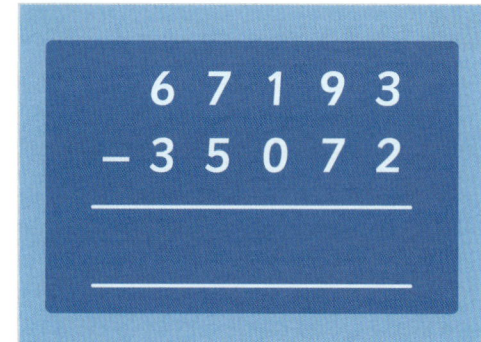

12. Calculate 97,284 − 35,672.

📋 ■ **61612** ■ 62412

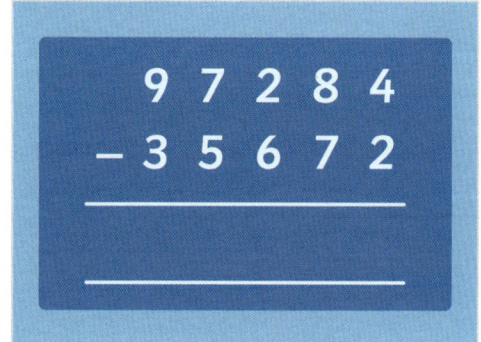

13. Subtract 878,210 from 935,168.

📋 ■ **56958** ■ 143042 ■ 143158

Level 2 *continued*

14. 2,292,571 − 1,181,650 = _____

- **1110921** ∎ 1111121

15. What is 452,539 less than 1,280,812?

- **828273** ∎ 1232327

16. John borrowed £63,084 from the bank to buy a house. He has paid back £20,275 to the bank. How much money does he still owe?
Include the £ sign in your answer.

- 42809 ∎ £83359 ∎ **£42809** ∎ **£42,809** ∎ £83,359
- £43211 ∎ 42,809 ∎ £43,211

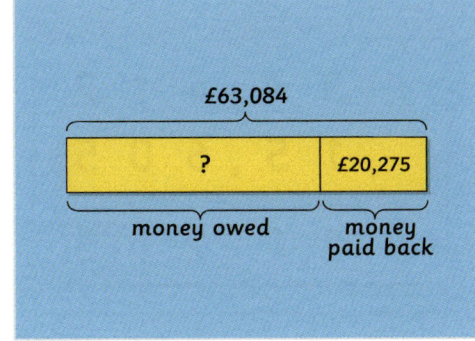

17. Dale and Nick buy one car each and spend £21,274 in total. Dale buys a Tafi for £9,436. Which model of car does Nick buy?

1/4 ∎ Kasod ∎ Dorf ∎ **Sinsan** ∎ Tafi

18. MegaBeanz has 92,745 tins of beans in stock. It delivers 4,317 tins to different stores. How many tins of beans does MegaBeanz have left?

∎ 97062 ∎ **88428** ∎ 92432

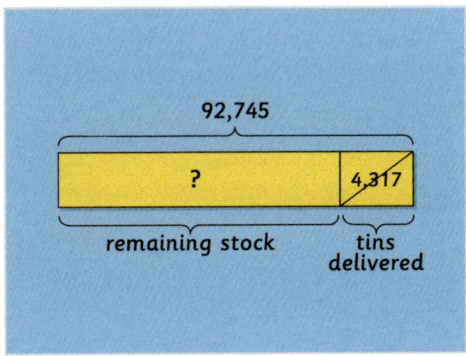

19. 192,571 − 181,650 = _____

- **10921** ∎ 11121

Level 2 continued

20. Calculate 65,128 − 24,161.

■ **40967** ■ 41047

Level 3: Reasoning: Reason about subtracting numbers with more than 4 digits.

✸ **Required:** 5/5   ✸ **Pupil Navigation:** on
✸ **Randomised:** off

21. Zoe says that you can add more than two
    a
    b numbers using the column method, so you
    c can subtract more than two numbers using
    the column method. Do you agree with Zoe?
    Explain your answer.

   - Open question, no set answer

22. Is it possible to subtract a 5-digit whole
    a
    b number from a 7-digit whole number and
    c get a 2-digit answer? Explain your answer,
    and use an example to prove that you are
    correct.

   - Open question, no set answer

23. What is the missing number?
    1
    2 34,061 = _____ + 19,349
    3

   ■ **14712**

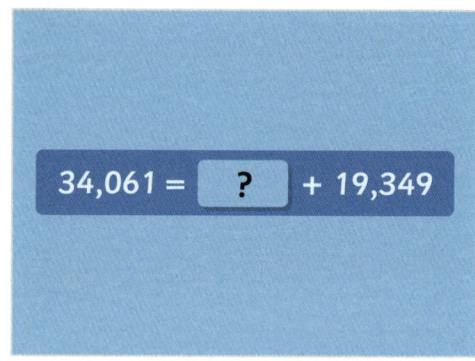

$$34,061 = \boxed{?} + 19,349$$

24. Millie and Callum are calculating the answer
    a
    b to 20,001 − 19,999. Millie uses column
    c subtraction, but Callum works out the
    answer in his head. Which method do you
    think is best? Explain your answer.

   - Open question, no set answer

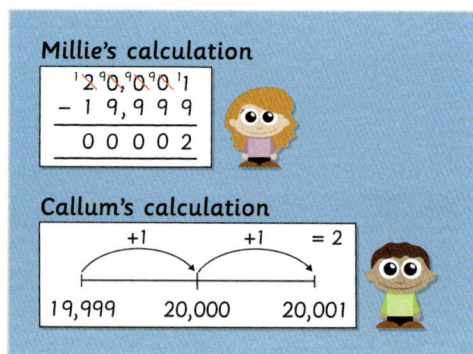

25. Jacob tries to solve 34,412 − 15,605, but he
    a
    b gets the answer wrong. Explain Jacob's
    c mistake and find the correct answer.

   - Open question, no set answer

$$\begin{array}{r} 3\,4\,,\,4\,1\,2 \\ -\ 1\,5\,,\,6\,0\,5 \\ \hline 2\,1\,,\,2\,1\,3 \\ \hline \end{array}$$

Level 4: Problem solving with greater depth: Solve multi-step problems involving subtracting numbers with more 4 digits.

✻ **Required:** 5/5   ✻ **Pupil Navigation:** on

✻ **Randomised:** off

26. The five digit cards complete the subtraction calculation. What is the value of *a*?

■ 6059 ■ 9 ■ 1

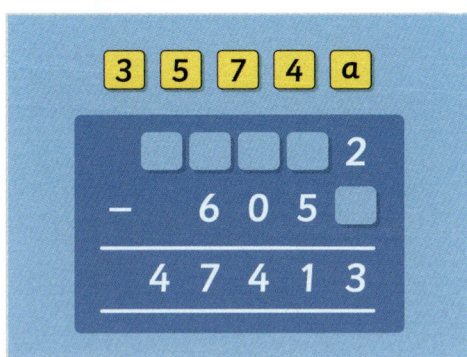

27. The pictogram shows the favourite sports of children in Warwickshire. How many more children choose football than choose snowboarding?

■ 20000 ■ 12500 ■ 32500

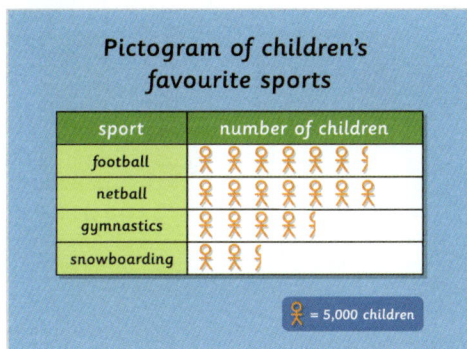

28. Tim wins £540,283 on the lottery. He buys a house and a car. How much of the prize money does Tim have left?
Include the £ sign in your answer.

■ 287548 ■ £252735 ■ £287,548 ■ £287548

■ £252,735 ■ 287,548

29. Four friends each borrow money from the bank to buy their houses. They each record how much they have already paid back. Sort the names by how much each person **still owes to the bank,** starting with the person who owes the most.

■ Paul ■ Cath ■ Phil ■ Carol

name	amount borowed	amount paid back	amount owed
Carol	£102,940	£23,075	
Paul	£98,735	£9,818	
Phil	£95,992	£14,557	
Cath	£88,729	£6,943	

Money borrowed and paid back to the bank

30. All the numbers in a straight line have the same total. What is the value of the missing number?

■ 22750 ■ 63000 ■ 40250

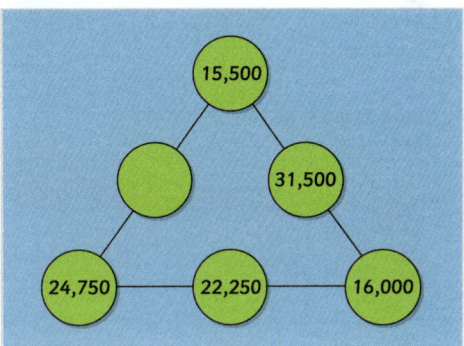

Addition and Subtraction Topic Review

Objective: I can answer questions involving addition and subtraction.

Quick Search Ref: **11429**

1. What is 228 + 65?

▪ **293** ▪ 295

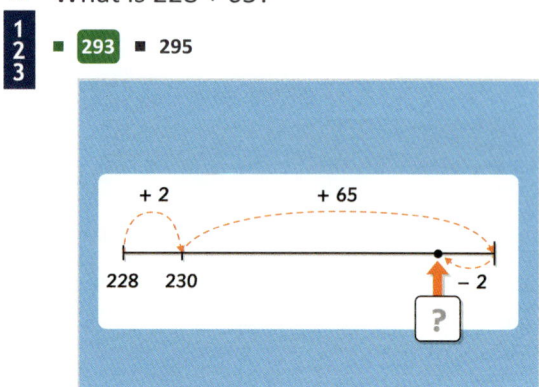

2. Subtract 200 from 741.

▪ 739 ▪ **541** ▪ 721 ▪ 941

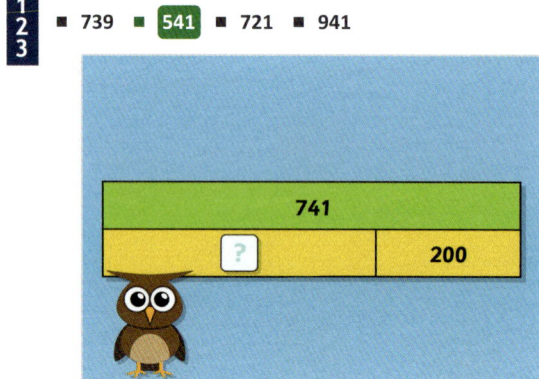

3. What is the difference between 109 and 1,008?

▪ **899**

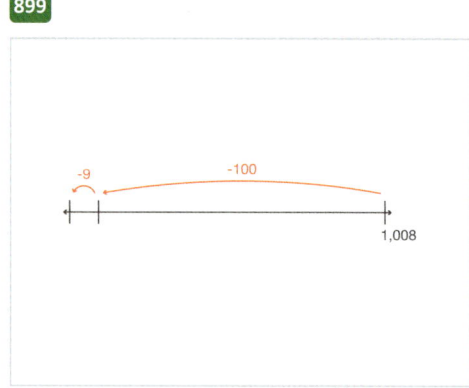

4. What is 30,004 + 42,005?

▪ **72009** ▪ 729

5. Estimate the answer to 48,684 − 15,563 by rounding each number to the **nearest 1,000.**

▪ 33121 ▪ 33120 ▪ **33000** ▪ **33,000** ▪ 33100
▪ 33,120 ▪ 33,100 ▪ 33,121

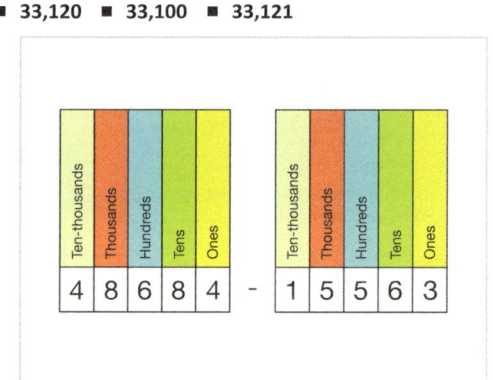

6. Add 6.4 and 2.5.

▪ **8.9** ▪ 89

Level 1 *continued*

7. What is 42.4 subtract 31.2?

 ▪ **11.2**   ▪ 112

8. Calculate the answer to 20,043 + 40,056.

123 ▪ **60099**   ▪ 699

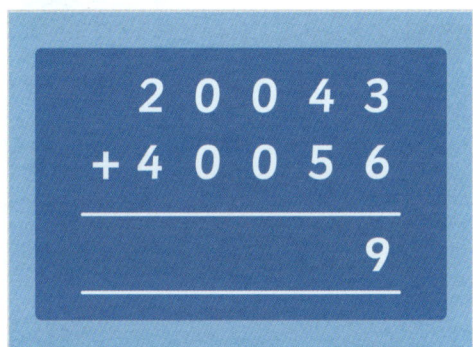

9. What is 58,201 − 45,102?

123 ▪ **13099**   ▪ 13101

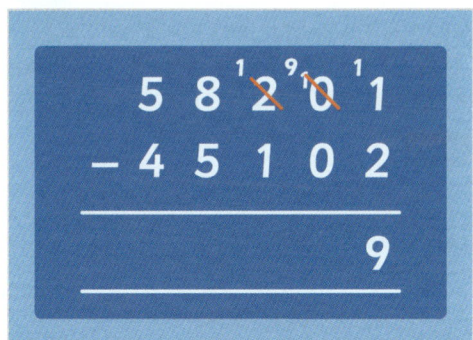

10. What is 328 subtract 50?

123 ▪ **278**   ▪ 378

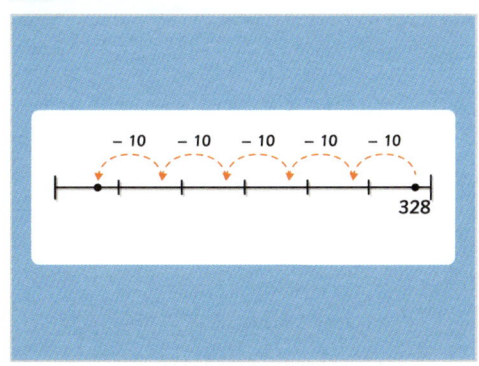

Level 2: Fluency

✱ **Required:** 7/10   ✱ **Pupil Navigation:** on
✱ **Randomised:** off

11. What is 469 + 329?

123 ▪ **798**   ▪ 800

12. Find the total of 3,540 and 2,190.

123 ▪ **5730**

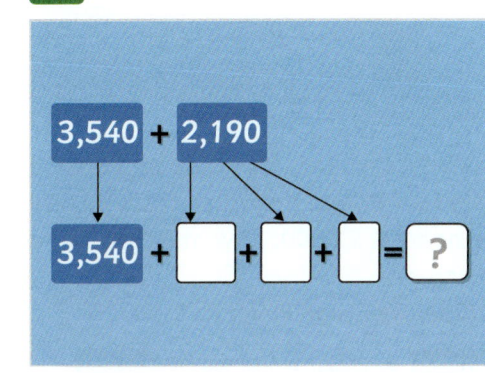

13. How many larger is 7,003 than 3,996?

123 ▪ **3007**   ▪ 3003

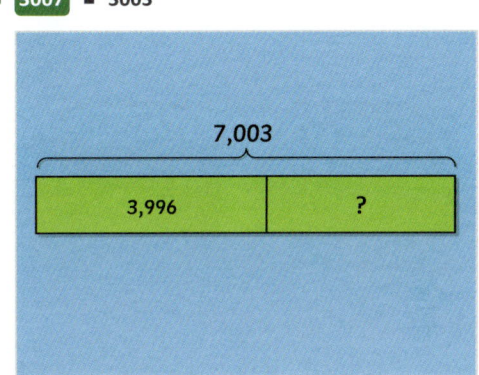

14. Use the inverse operation to find the missing number in the equation **41,470 - _____ = 10,595**.

abc ▪ **30875**   ▪ **30,875**   ▪ 52065   ▪ 31125   ▪ 31,125
▪ 52,065

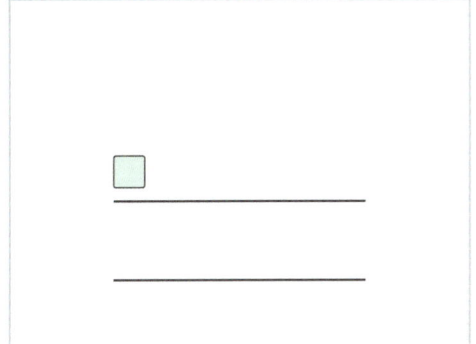

Level 2 *continued*

15. Add 1.47 to 6.35.

`1 2 3` ■ 7.82 ■ 7.72 ■ 7.712 ■ 782

16. Alex buys two cars for his daughters. One
`a b c` costs £9,450 and the other costs £7,699. **To
the nearest £1,000**, how much do the two
cars cost?
Include the £ sign in your answer.

■ £17000 ■ 17000 ■ £18,000 ■ £17,000
■ £17,149 ■ 17,000 ■ £16,000 ■ £16000 ■ £18000
■ £17149

17. Sam has 247.4 grams of grapes. If he gives
`1 2 3` 89.3 grams of grapes to Jax, how many
grams of grapes does he have left?
Don't include the units in your answer.

■ 158.1

18. 130,575 + 317,573 = _____

`1 2 3` ■ 448148 ■ 447048

19. MegaBeanz has 92,745 tins of beans in stock.
`1 2 3` It delivers 4,317 tins to different stores. How
many tins of beans does MegaBeanz have
left?

■ 88428 ■ 97062 ■ 92432

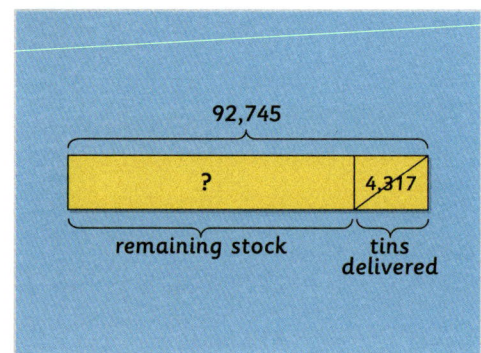

20. A football club sells 17,536 adult tickets,
`1 2 3` 4,893 family tickets and 3,206 children's
tickets for a match. How many tickets does
the club sell in total?

■ 25635

Level 3: Reasoning

❋ **Required:** 5/5   ❋ **Pupil Navigation:** on

❋ **Randomised:** off

21. What is the missing digit?

[1 2 3] ■ **2**

22. Explain what Ellie has done wrong in her calculation.

[a b c]

- Open question, no set answer

23. One of the calculations has an incorrect answer. Identify which calculation is incorrect and **give the correct answer**.

[1 2 3]

■ **212**   ■ **208**

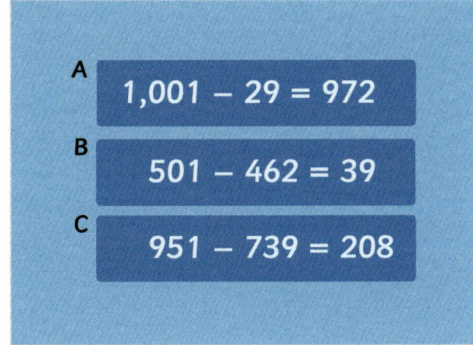

24. Lorna says, "If you add a 4-digit number and 5-digit number, your answer will be a 9-digit number."

[a b c]

Is Lorna correct? Explain your answer.

- Open question, no set answer

25. The four digit cards complete the addition calculation. What is the value of *a*?

[1 2 3]

■ **9**   ■ **6059**   ■ **1**

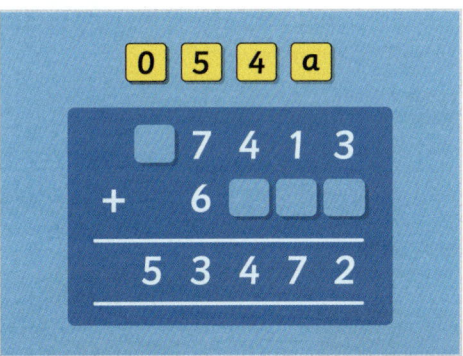

Level 4: Problem solving with greater depth

❋ **Required:** 5/5   ❋ **Pupil Navigation:** on

❋ **Randomised:** off

26. There are 1,028 passengers on board a ferry. There are 451 women and 392 men. How many **children** are on board?

[1 2 3]

■ **185**   ■ **843**

27. The numbers on each line have the **same**
total. What is the value of the missing
number?

■ **8807** ■ 20683 ■ 11876

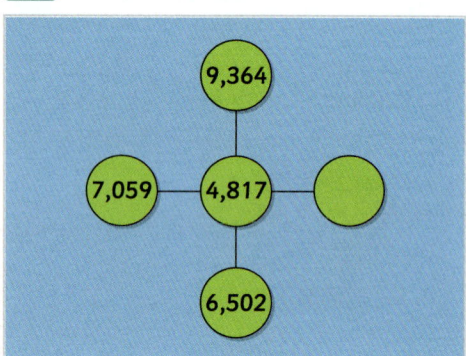

28. The table shows Hannah's level scores and
total scores for three levels of a video game.
In level 2, she scored 5,000 fewer points than
she scored in level 3. How many points did
Hannah score in level 1?

■ **6394**

Hannah's game scores

	level score	total score
level 1	?	
level 2		12,994
level 3	11,600	24,594

29. Chris buys one kilogram each of grapes,
apples and oranges. He pays with a £20 note.
How much **change** does Chris get back?
Include the £ sign in your answer.

■ **£11.88** ■ 11.88 ■ £8.12

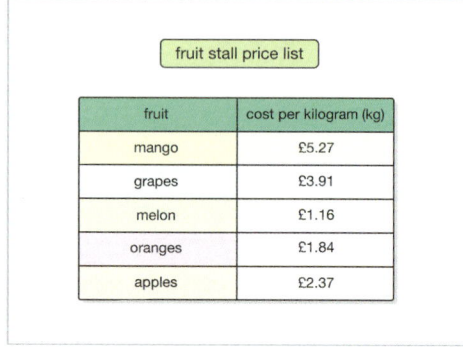

fruit stall price list

fruit	cost per kilogram (kg)
mango	£5.27
grapes	£3.91
melon	£1.16
oranges	£1.84
apples	£2.37

30. 348 children go to Quiverton Primary School.
Use the table to work out **how many girls** go
to the school.

■ **203** ■ 145 ■ 68

Number of pupils attending Quiverton Primary School

	boys	girls	total
key stage 1	51		
key stage 2		135	229
total		?	348

Mathematics  Y5

Multiplication and Division

Mental Multiplication
Mental Division
Written Methods Multiplication
Written Methods Division
Decimals
Mental Tables Practice

Multiply and Divide Numbers Including Decimals by 10, 100 and 1,000

Objective: I can multiply and divide decimal numbers by 10, 100 and 1,000 and give answers up to three decimal places.

Quick Search Ref: 11648

Level 1: Understanding: Multiply and divide decimals by 10,100 and 1,000 with image support.

❖ **Required:** 7/10   ❖ **Pupil Navigation:** on   ❖ **Randomised:** off

---

1. What is 7/10 as a decimal?

abc   ■ **0.7**   ■ 0.70   ■ 7.10   ■ 7.1

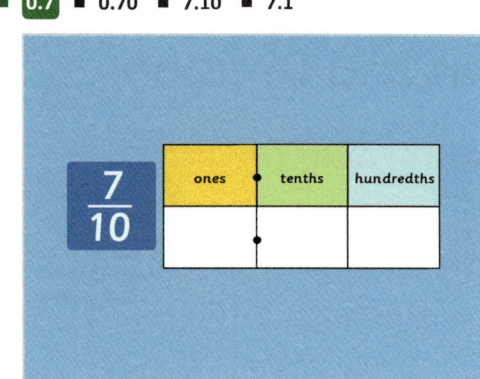

2. What is 5.6×100?

abc   ■ **560**   ■ 56   ■ 5.600   ■ 0.056

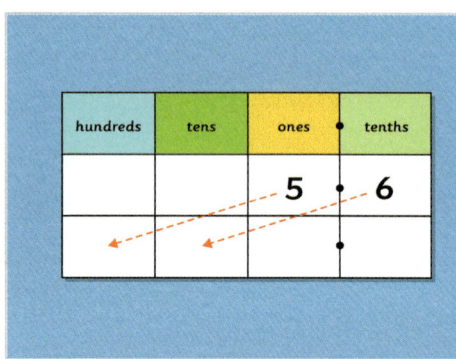

3. What is $5.6 \times 1{,}000$?

abc   ■ 5.6000   ■ **5,600**   ■ **5600**   ■ 56   ■ 0.0056

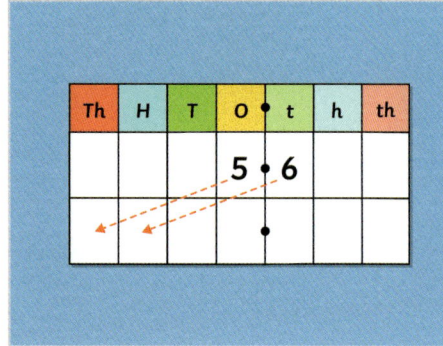

4. $342 \div 10 =$ _____

123   ■ **34.2**   ■ 34   ■ 3420

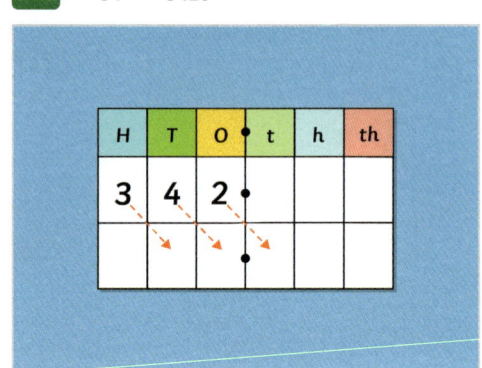

5. $342 \div 100 =$ _____

123   ■ 34.2   ■ **3.42**   ■ 34200   ■ 34

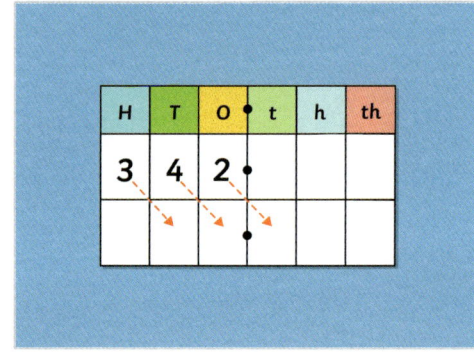

6. $1.06 \times 1{,}000 =$ _____

123   ■ **1060**

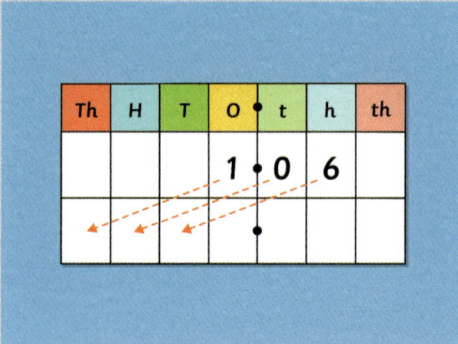

---

7. What is 0.08 divided by 10?

 ▪ 0.008 ▪ 0.8

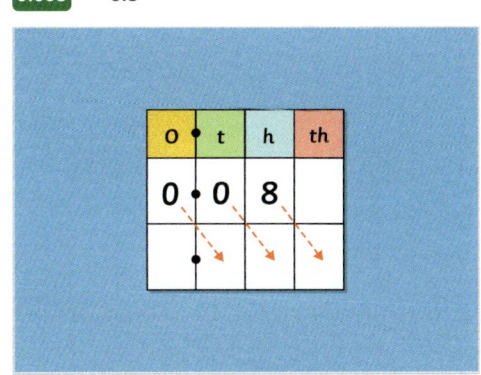

8. 4,250 divided by 1,000 = _____

▪ 425 ▪ 4.25 ▪ 4.250 ▪ 42.5 ▪ 4250000
▪ 4,250,000 ▪ 4

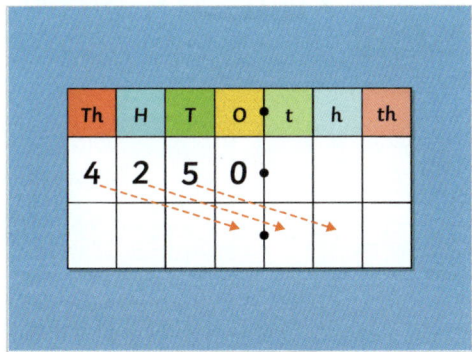

9. What is 1.2 divided by 100?

▪ 0.12 ▪ 0.012 ▪ 120

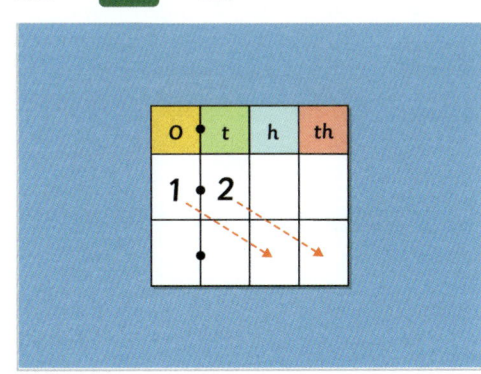

10. What is 4/100 as a decimal?

▪ 4.01 ▪ 0.04 ▪ 4.100 ▪ 4.1 ▪ 0.4

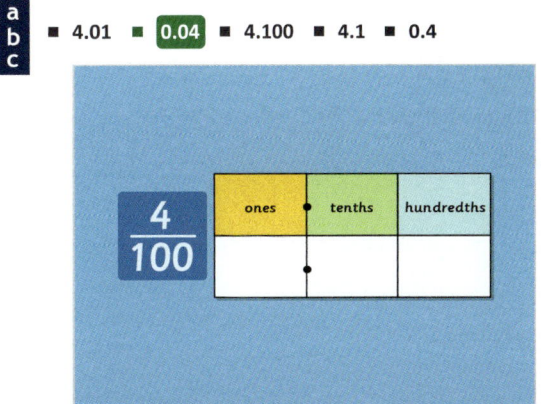

Level 2: Fluency: Multiply and divide decimals by 10, 100 and 1,000.

✱ **Required:** 7/10   ✱ **Pupil Navigation:** on
✱ **Randomised:** off

11. What is 3.2 × 100?

▪ 320 ▪ 3.200

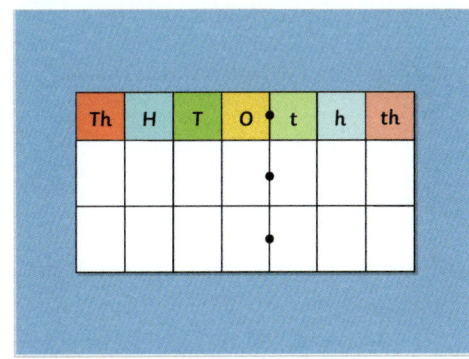

12. What is 9.274 × 10?

▪ 92.74

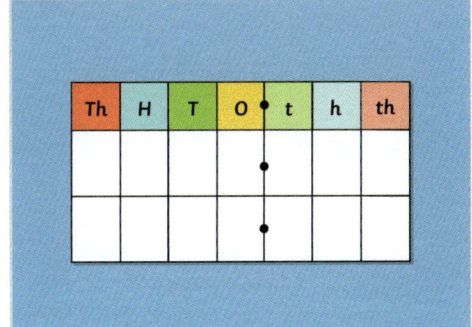

13. What is 87 ÷ 100?

■ 0.87

87 ÷ 100 = ?

14. What is 3,983 ÷ 1,000?

■ 3.983

3,983 ÷ 1,000 = ?

15. What number is missing from the following calculation?

_____ × 1,000 = 4,761

■ 4.761   ■ 4761000

? × 1,000 = 4,761

16. What number is missing from the following calculation?

_____ × 63.807 = 6,380.7

■ 100   ■ 10

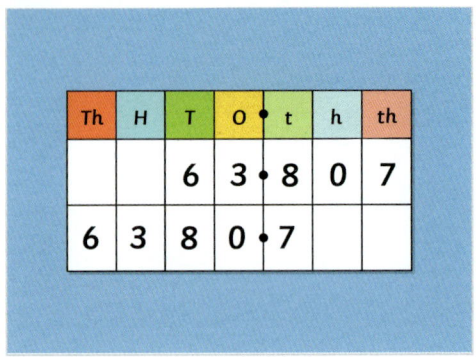

Th	H	T	O	t	h	th
		6	3	8	0	7
6	3	8	0	7		

17. Select the **two** calculations that have the same answer.

2/4

■ 54,823 ÷ 100   ■ 100 × 54.823   ■ 54.823 ÷ 100

■ 10 × 548.23

54,823 ÷ 100	100 × 54.823
54.823 ÷ 100	10 × 548.23

18. What is 521.854 × 100?

■ 52185.4

521.854 × 100 = ?

Level 2 *continued*

19. What is 7.3 divided by 100?

■ 0.073

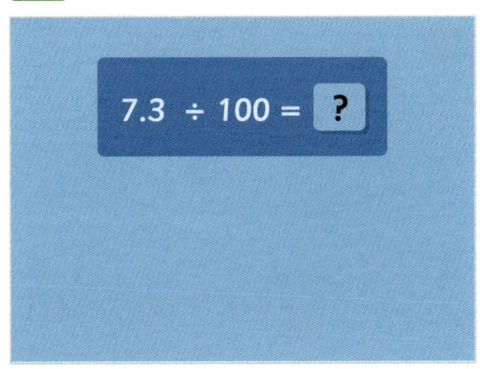

$$7.3 \div 100 = \boxed{?}$$

20. 4.281 multiplied by 100 = _____

■ 428.1

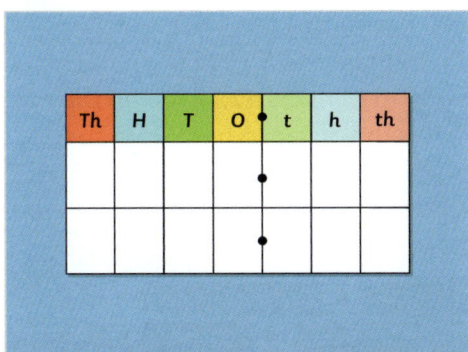

Th	H	T	O	t	h	th

Level 3: Reasoning: Reason about multiplying and dividing by 10, 100 and 1,000.

✱ **Required:** 5/5 ✱ **Pupil Navigation:** on
✱ **Randomised:** off

21. Which **two** problems can you solve by multiplying by 1,000?

2/3

■ 1,000 packets of strawberries have a total weight of 16.7 kilograms (kg). What is the average weight of one packet?

■ 100 art students each have to make 10 clay models. Every model uses 164.25 grams (g) of clay. How much clay is needed altogether?

■ A tower is made up of 1,000 blocks on top of each other. Each block is 0.47 centimetres (cm) high. How tall is the tower?

22. Tobey says, "To multiply a number by 100, I can just add two zeros on the end of the number." Is Tobey correct? Explain your answer.

a
b
c

- Open question, no set answer

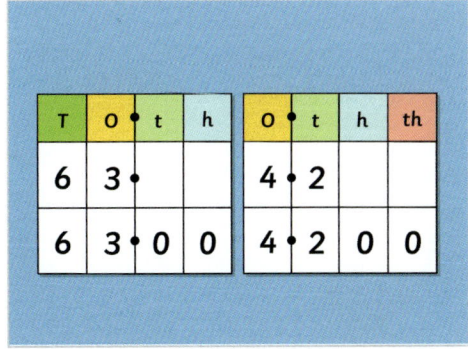

T	O	t	h		O	t	h	th
6	3·				4·	2		
6	3·	0	0		4·	2	0	0

23. Which symbol makes the following statement true?

3.923 × 100 _____ 3,918.8 ÷ 10

1/3 ■ < ■ = ■ >

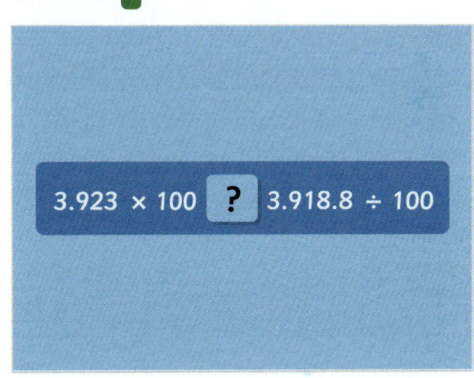

$$3.923 \times 100 \boxed{?} 3.918.8 \div 100$$

24. Which calculation is the odd one out? Prove that you are correct.

a
b
c

- Open question, no set answer

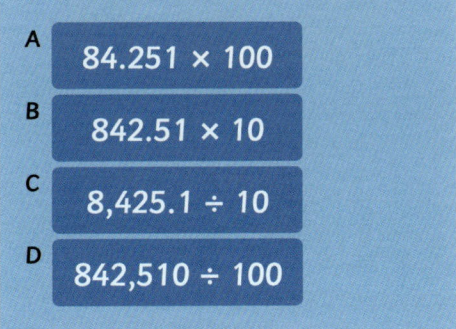

A 84.251 × 100

B 842.51 × 10

C 8,425.1 ÷ 10

D 842,510 ÷ 100

Level 3 *continued*

25. Class 6 have 403 pencils and Class 5 have 43
a b c pencils. Ben says, "Class 6 have exactly 10 times as many pencils as Class 5." Do you agree with Ben? Explain your answer.

- Open question, no set answer

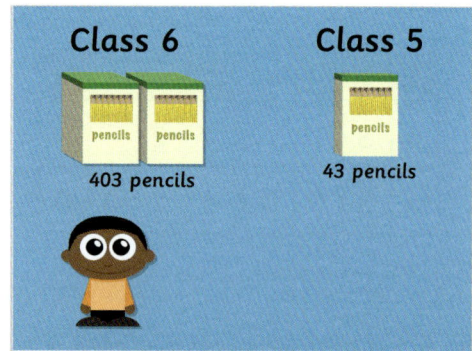

Level 4: Problem solving with greater depth: Solve multi-step problems involving decimal numbers.

✱ **Required:** 5/5 ✱ **Pupil Navigation:** on
✱ **Randomised:** off

26. Using every digit card and the decimal point
1 2 3 once, make the number closest to 65. What number do you get if you multiply your answer by 1,000?

- ■ 64951 ■ 64.951

27. Ali thinks of a number. He multiplies his
1 2 3 number by 10, subtracts 50 and then divides his answer by 1,000. He gets 0.062. What number did he start with?

- ■ 11.2

28. The scale on a map is 10 centimetres (cm) =
a b c 1,000 kilometres (km)
Bern to Prague is 8.124 cm on the map. Prague to Minsk is 12.24 cm on the map. What is the distance from Bern to Minsk via Prague in kilometres?
Include the units km (kilometres) in your answer.

- ■ 2036.4 km ■ 2,036.4 kilometres
- ■ 2036.4 kilometres ■ 2,036.4 km ■ 20.364 cm
- ■ 20.364 ■ 20.364 centimetres ■ 2,036.4 ■ 2036.4

Level 4 *continued*

29. At an athletics track, Henry completes 10
a laps of track 2 and 100 laps of track 4. How
b many **kilometres** (km) does he run in total?
c *Include the units km (kilometres) in your
answer.*

- ■  7.7 km ■ 7,700 m ■ 7.7 kilometres ■ 7,700
- ■ 7.7 ■ 7,700 metres

Length of running tracks

track	distance (metres)
track 1	400
track 2	352
track 3	75.4
track 4	41.8

30. Pedro has four number cards: **a**, **b**, **c** and **d**.
a Use the clues in the image to calculate the
b answer to **a ÷ d**.
c **Hint:** You can use any value for **a** and get the
same answer.

- ■ 10

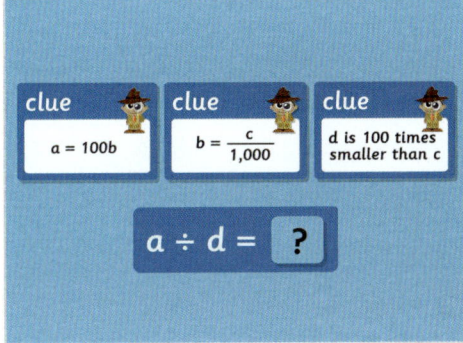

clue: $a = 100b$

clue: $b = \dfrac{c}{1,000}$

clue: d is 100 times smaller than c

$a \div d = \boxed{?}$

Multiply and Divide Numbers Mentally

Objective: I can multiply and divide numbers mentally using times tables facts.

Quick Search Ref: 10106

Level 1: Understanding: Mental strategies, up to 3-digit numbers by 1-digit numbers.

✱ **Required:** 7/10   ✱ **Pupil Navigation:** on   ✱ **Randomised:** off

1. What is 3 x 40?

1 2 3 ▪ 120 ▪ 43

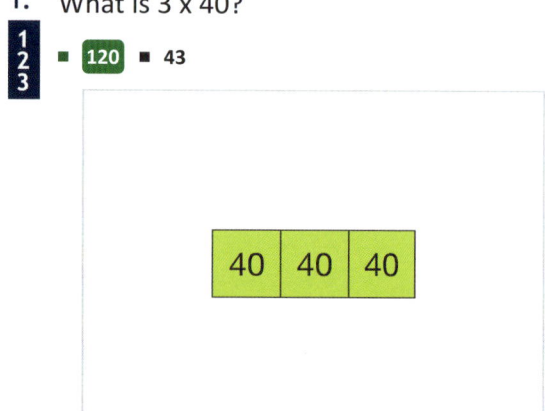

2. What is 32 ÷ 8?

1 2 3 ▪ 4 ▪ 256

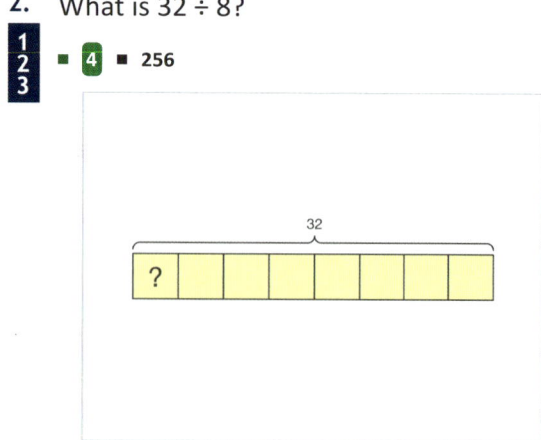

3. What is 16 multiplied by 5?

1 2 3 ▪ 80

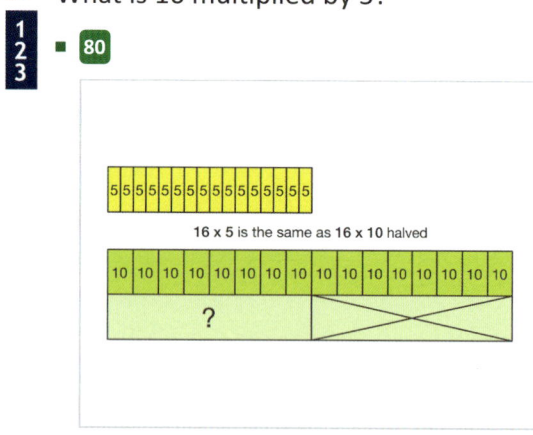

4. Find the product of 7 and 50.

1 2 3 ▪ 350 ▪ 57

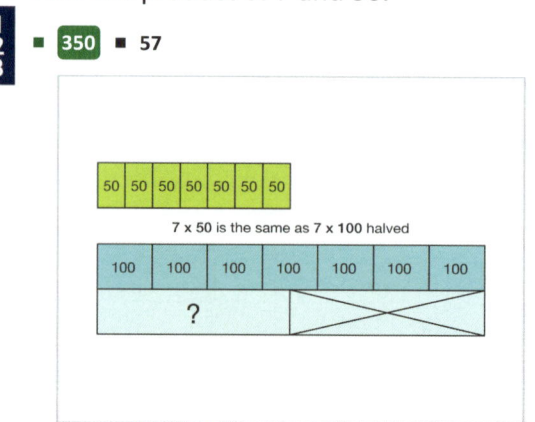

5. What is 100 doubled?

▪ 200 ▪ 50 ▪ 10,000

1/3

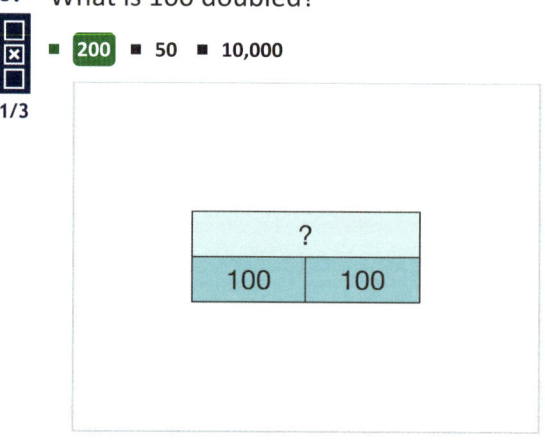

6. What is half of 38?

1 2 3 ▪ 19 ▪ 76

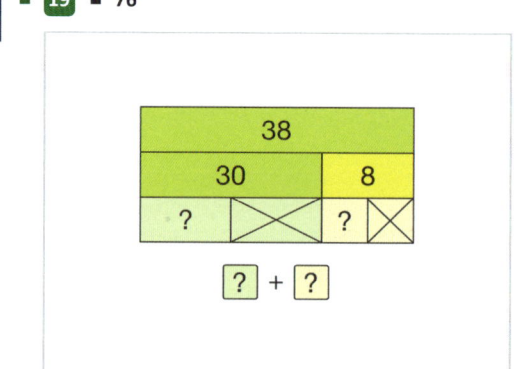

Level 1 continued

7. Find the product of 16 and 3.

 ▪ **48** ▪ **19**

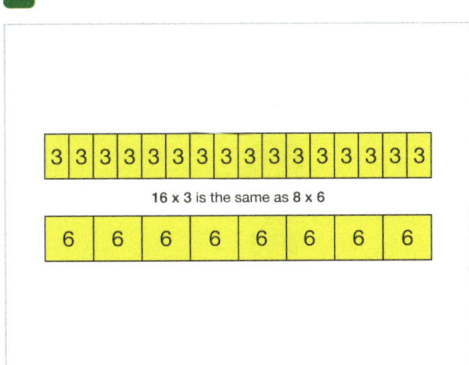

8. What is 320 divided by 5?

 ▪ **64** ▪ **1600**

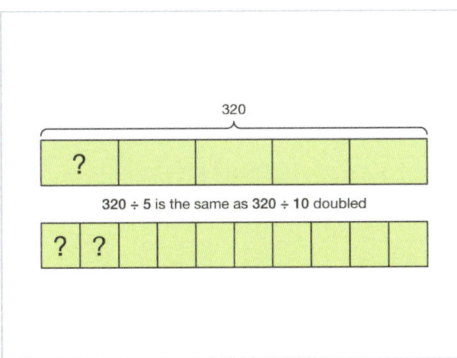

9. What is 248 shared into 4 equal parts.

 ▪ **62** ▪ **992**

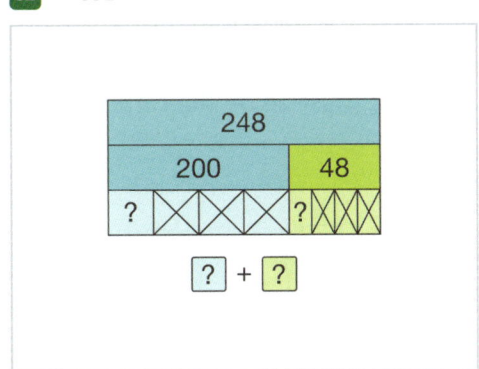

10. What is 610 divided into 5 equal parts?

 ▪ **122**

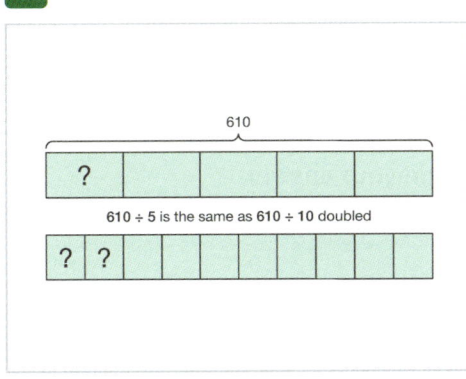

Level 2: Fluency: Mental strategies, up to 4-digit numbers by 2-digit numbers.

✿ **Required:** 7/10   ✿ **Pupil Navigation:** on
✿ **Randomised:** off

11. What is 17 multiplied by 9?

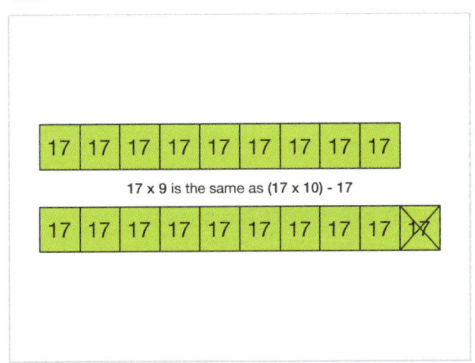 ▪ **153** ▪ **170**

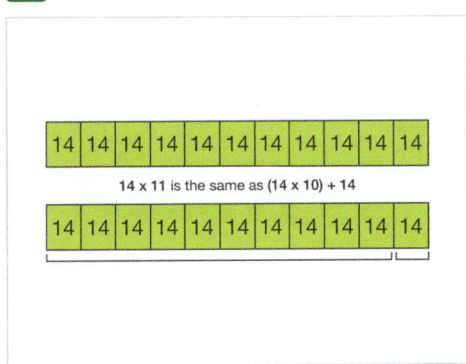

12. Find the product of 14 and 11.

▪ **154** ▪ **25**

13. Find the product of 25 and 13.

▪ **325** ▪ **38**

14. What is 143 x 21?

▪ **3003** ▪ **2860**

15. What is 800 divided between 50?

▪ **16** ▪ **40000**

16. What is £16 multiplied by 25?
Give your answer in pounds.
Include the £ sign in your answer.

▪ **£400** ▪ **£400.00** ▪ **400** ▪ **£41.00** ▪ **£41**

Level 2 *continued*

17. What is 115 kg x 18?
a b c *Give your answer in kg (kilograms).*
- **2070 kg** - **2,070 kilograms** - **2070 kilograms**
- **133 kilograms** - **2070** - **2,070 kg** - **2,070** - **133**
- **133 kg**

18. What is 4,530 divided by 15?
1 2 3 - **302** - **32**

19. 1,500 blueberries are shared equally
1 2 3 between 50 children.
How many blueberries does each child get?
- **30**

20. 3,960 paperclips are divided equally into
1 2 3 30 boxes. How many paperclips are in each box?
- **132** - **1188** - **15**

Level 3: Reasoning: To use and apply multiplication and division facts.
✹ **Required:** 6/8   ✹ **Pupil Navigation:** on
✹ **Randomised:** off

21. What is the missing number?
1 2 3 ____ ÷ 12 = 12
- **144** - **1**

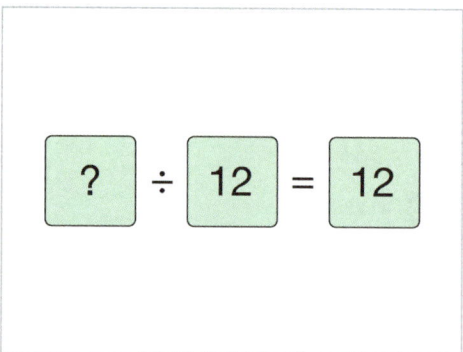

22. If 2 x 133 = 266, what is 4 times
1 2 3 133?
- **532** - **1064** - **399**

23. Joel says, "Every number that ends in 4 is a
a b c multiple of 4".
Is he correct? Explain your answer.

- *Open question, no set answer*

Every number that ends in 4 is a multiple of 4.

24. Which **two missing numbers** balance the
calculation?
4 x ____ = ____ ÷ 2
2/7 - **3** - **9** - **12** - **36** - **64** - **72** - **84**

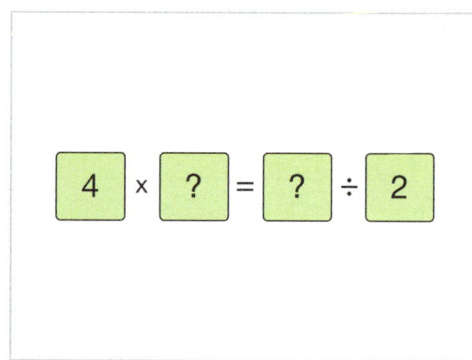

4 x ? = ? ÷ 2

25. Three of the cards complete the calculation,
1 2 3 but one of the cards has been turned over.
What digit must be on the turned over card?
- **3**

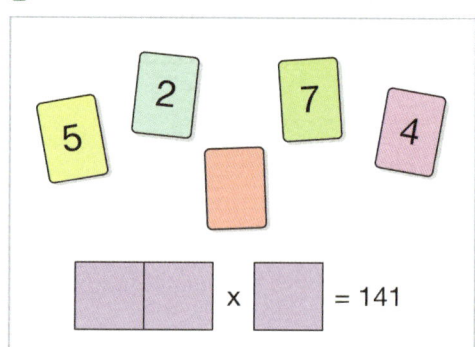

5   2   7   4

☐☐ x ☐ = 141

26. 10 times a number is 4,220, what is
a b c 9 times the same number?
Explain your answer.

- *Open question, no set answer*

27. Which **two missing numbers** balance the calculation?

180 ÷ _____ = 12 x _____

2/7 ▪ 2 ▪ **3** ▪ 4 ▪ **5** ▪ 6 ▪ 7 ▪ 8

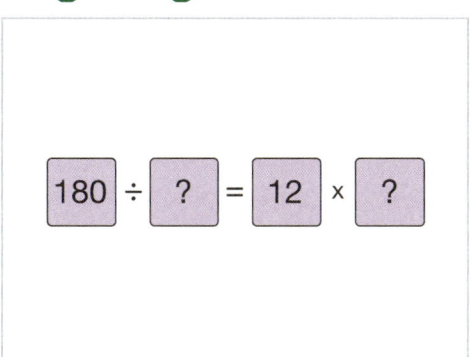

28. Sarah says, 'At least one of the missing numbers must be an even number because the answer is an even number'.
Is Sarah correct? Explain your answer.

- Open question, no set answer

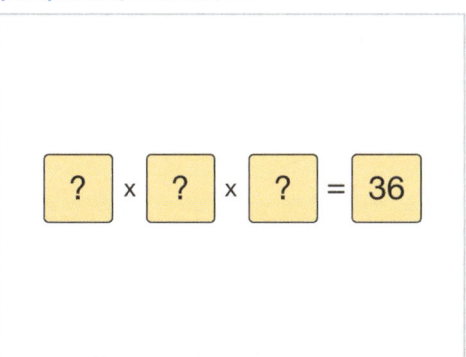

Level 4: Problem Solving: Using multiplication and division facts.

❋ **Required:** 8/8   ❋ **Pupil Navigation:** on
❋ **Randomised:** off

29. Bedding plants are sold in trays of 6.
Jacob needs 70 bedding plants to fill his containers.
How many trays must he buy?

▪ **12** ▪ 11

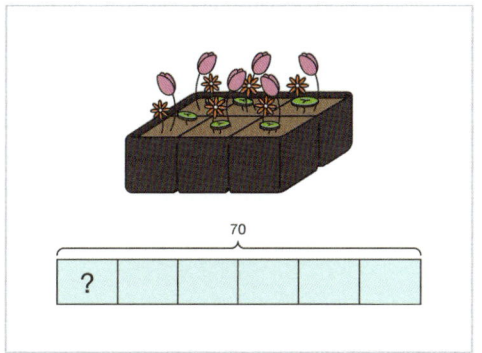

30. The numbers in the outer circles are all factors of the missing number, which is greater than 20 but less than 50.
What is the missing number?

▪ **36** ▪ 54 ▪ 18

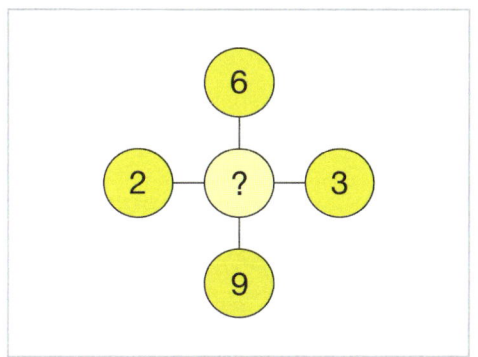

31. In the multiplication table what are the **two missing numbers**?

2/6 ▪ 2 ▪ **3** ▪ 4 ▪ 6 ▪ **8** ▪ 9

x	2	?	5
4	8	12	20
6	12	18	30
?	16	24	40

32. The same number is missing from each of the boxes.
What is the missing number?

▪ 13 ▪ **11** ▪ 12

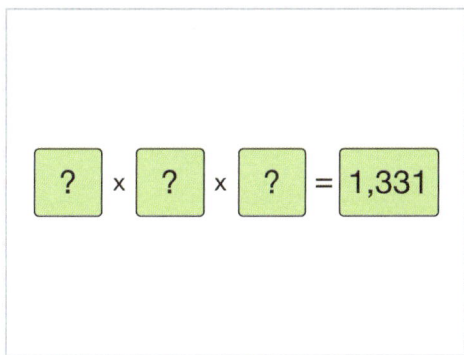

33. Kyle is doing a sponsored 164 mile walk for charity. If he walks 12 miles each day, how many days will it take him to complete the walk?

▪ **14** ▪ 13

34. One pack of football stickers costs 50 pence.
Xavier spends £7.50 on stickers and then stacks the packs on top of each other. Each pack is 1.8 millimetres thick.
In **centimetres**, how high is the stack of stickers?

1
2
3

- 27   - **2.7**   - 15

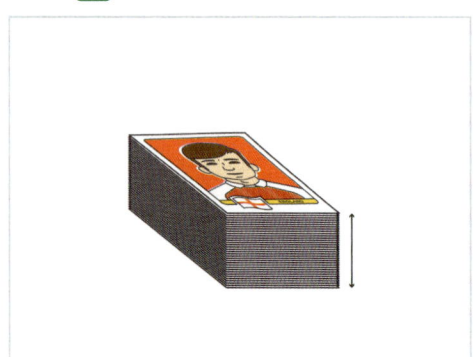

35. Grace's grandma is knitting her a new cardigan and she needs three balls of wool and six buttons.

a
b
c

The buttons cost 75 pence each.
The total cost of the cardigan is £15.00.
In **pounds**, what is the cost of **one ball of wool**?

- **£3.50**   - **3.50**   - £4.50   - £10.50

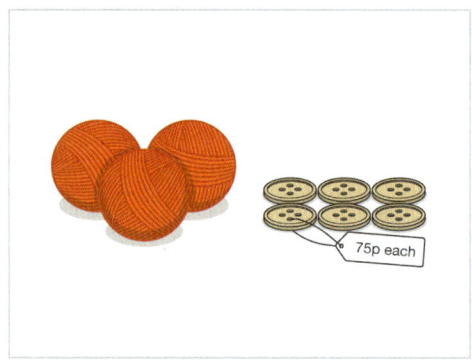

36. Adam is making a garden gate that needs to be 96 centimetres wide.

a
b
c

Lengths of wood are sold in packs of 6 for £36.00 and each length of wood is 12 centimetres wide.
How much will the new garden gate cost in **pounds**?

- £2.00   - **£72.00**   - **72**   - £8   - **£72**   - £288
- £2   - £8.00

Multiply Numbers Mentally

Objective:	I can use mental strategies to multiply numbers using known facts.
Quick Search Ref:	**11267**

Level 1: Understanding: Use mental strategies to multiply numbers up to 3 digits by 1-digit numbers.

✿ **Required:** 7/10   ✿ **Pupil Navigation:** on   ✿ **Randomised:** off

1. What is 3 × 40?

■ **120**   ■ 12

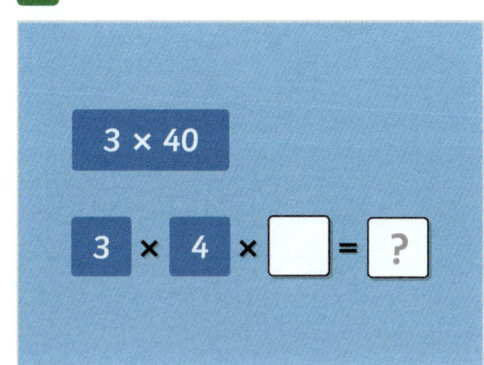

2. Calculate 16 multiplied by 5.

■ **80**   ■ 160

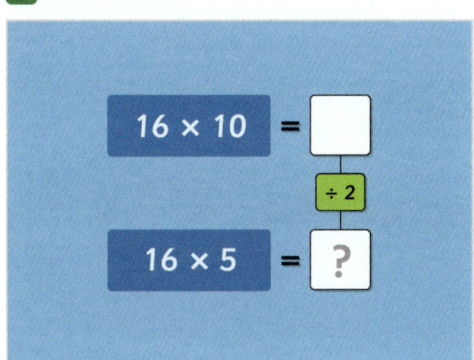

3. Find the product of 3 and 27.

■ **81**

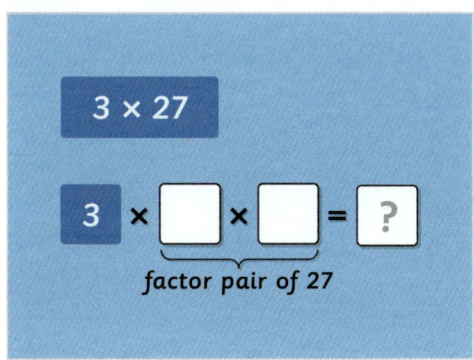

4. What is 32 × 4?

■ **128**   ■ 64

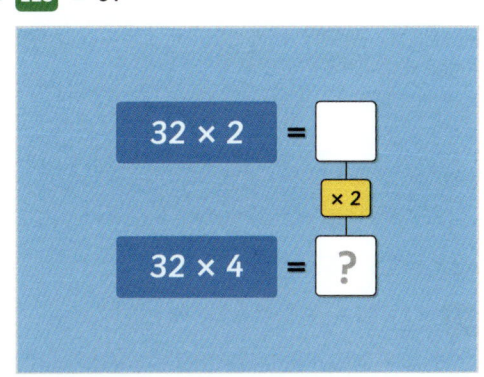

5. Calculate 64 × 7.

■ **448**   ■ 420

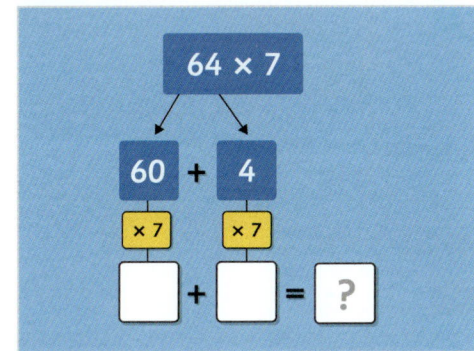

6. What is the product of 9 and 17?

■ **153**   ■ 170

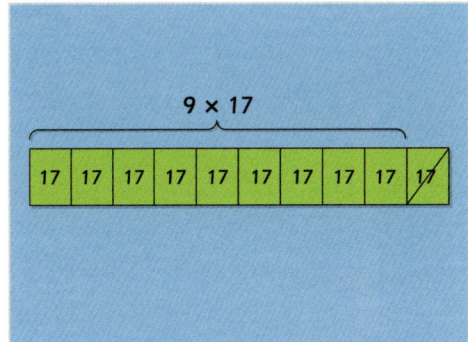

Level 1 *continued*

7. Complete this number sentence.

[1 2 3] 24 × 6 = _____

■ **144**

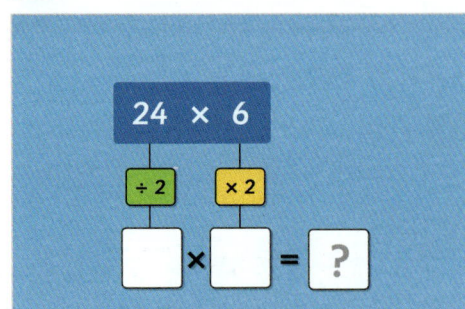

8. What is 14 × 6?

[1 2 3] ■ **84**

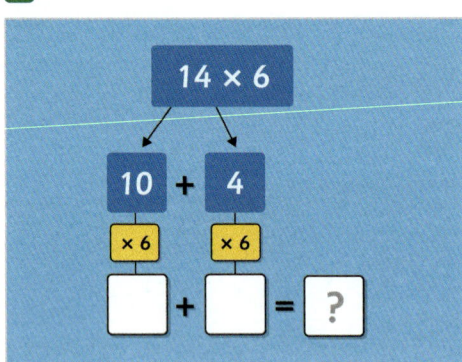

9. Calculate 5 × 80.

[1 2 3] ■ **400**

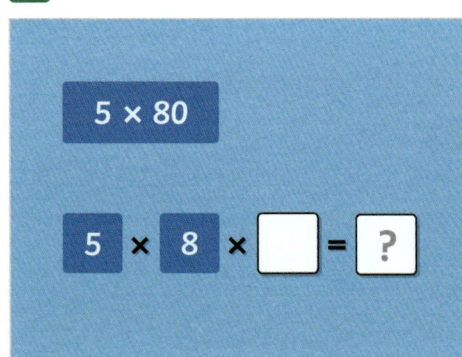

10. What is double 150?

[1 2 3] ■ **300**

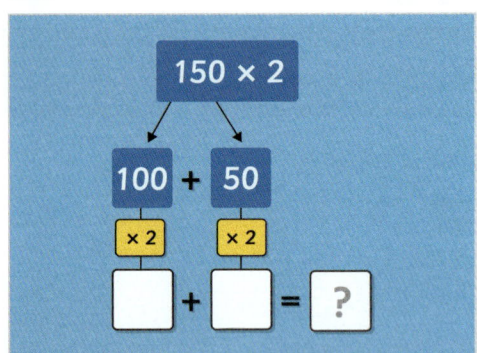

Level 2: Fluency: Use mental strategies to multiply numbers up to 3 digits by 2-digit numbers, including decimals.

✱ **Required:** 7/10 ✱ **Pupil Navigation:** on
✱ **Randomised:** off

11. Multiply 60 by 30.

[1 2 3] ■ **1800** ■ 180

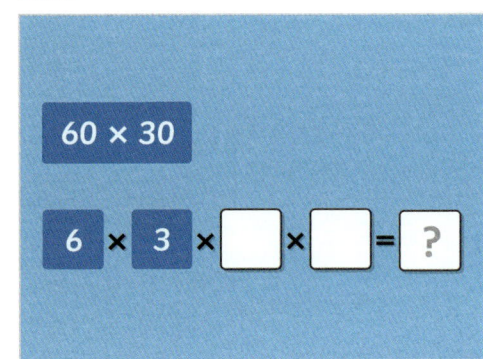

12. What is double 176?

[1 2 3] ■ **352**

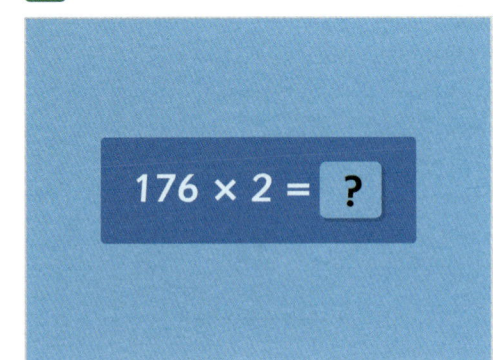

13. What is 2.4 × 20?

 ▪ 48

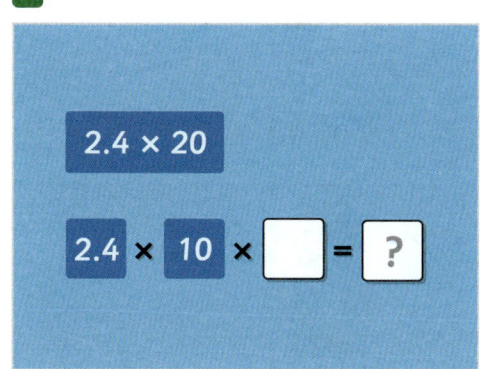

14. Find the product of 68 and 50.

▪ 3400 ▪ 6800

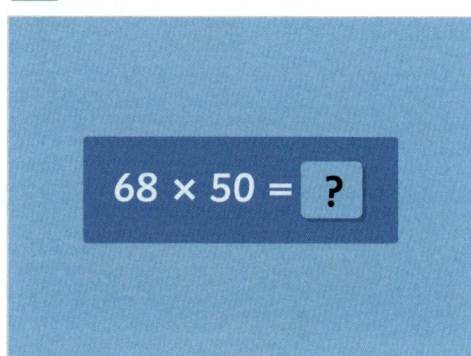

68 × 50 = ?

15. Calculate 36 × 25.

▪ 900 ▪ 3600

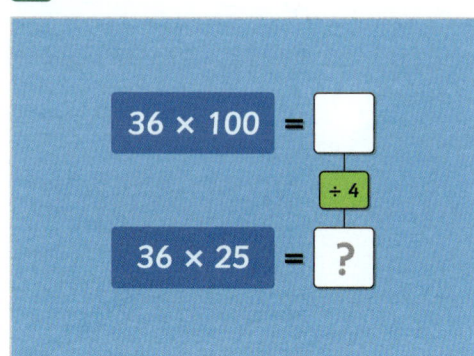

16. Mo runs 14 laps of a 25-metre (m) running track. How far does Mo run in total?
Include the units m (metres) in your answer.

▪ 0.35 kilometres ▪ 350 metres ▪ 350 m ▪ 350

▪ 0.35 ▪ 0.35 km

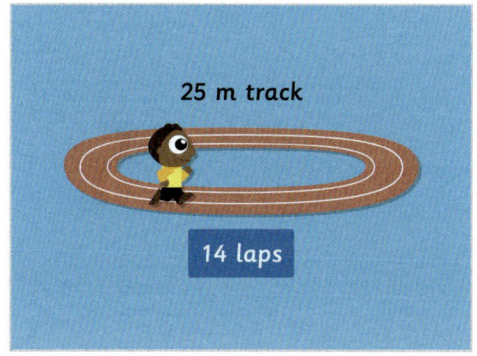

25 m track

14 laps

17. One bag of flour weighs 27 kg. What is the total weight of 19 bags of flour?
Include the units kg (kilograms) in your answer.

▪ 513 kg ▪ 513 kilograms ▪ 513 ▪ 540 kg

▪ 540 kilograms ▪ 540

flour 27 kg   flour 27 kg   flour 27 kg

20 bags − 1 bag = 19 bags

18. Find 8 lots of 45.

▪ 360 ▪ 320 ▪ 180

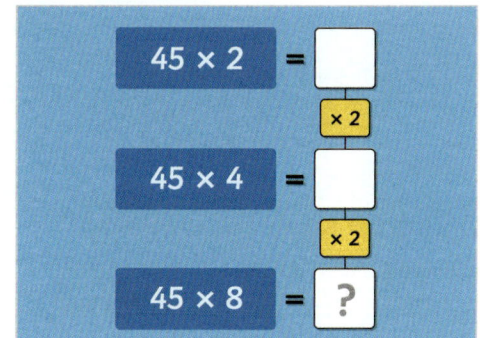

45 × 2 =

× 2

45 × 4 =

× 2

45 × 8 = ?

Level 2 continued

19. Find the product of 28 and 3.

1
2
3

- ■ 84 ■ 60

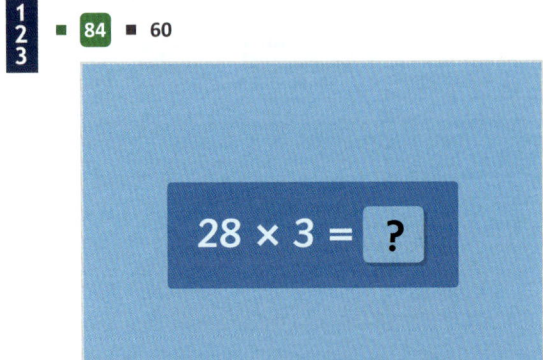

$$28 \times 3 = \boxed{?}$$

20. $32 \times 20 = $ _____

1
2
3

- ■ 64 ■ 640 ■ 320

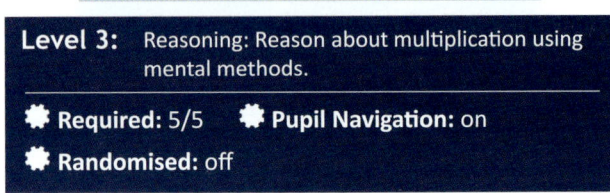

$$32 \times 20$$

$$32 \times \boxed{} \times \boxed{} = \boxed{?}$$

Level 3:   Reasoning: Reason about multiplication using mental methods.

✸ **Required:** 5/5   ✸ **Pupil Navigation:** on
✸ **Randomised:** off

21. Which of these calculations has a different answer to 90×60?

1/4

- ■ 180×30 ■ $20 \times 3 \times 90$ ■ $9 \times 600 \times 10$ ■ $3 \times 60 \times 30$

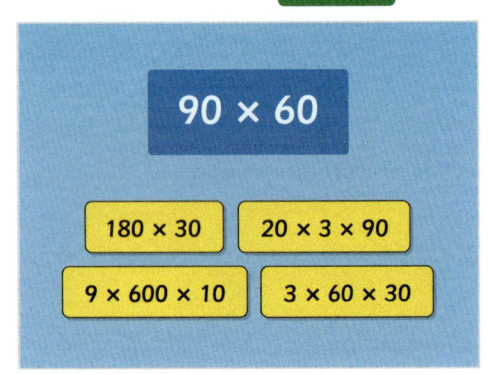

$$90 \times 60$$

180×30 $20 \times 3 \times 90$
$9 \times 600 \times 10$ $3 \times 60 \times 30$

22. Zara says that to calculate 16×45, she can
a
b
c
multiply 8 by 90. Is Zara correct? Explain your answer.

- Open question, no set answer

$$16 \times 45 = 8 \times 90$$

23. Which calculation gives the same answer as 7.3×800?

1/3

- ■ 73×0.8 ■ 8×0.73 ■ 80×73

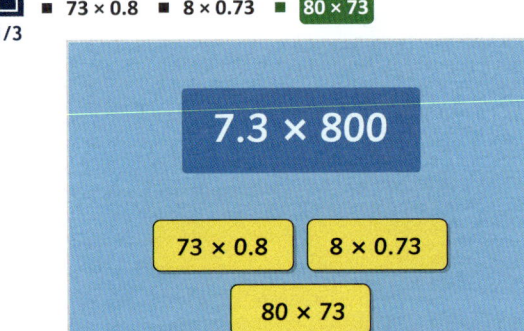

$$7.3 \times 800$$

73×0.8 8×0.73
80×73

24. Tom knows that $15 \times 12 = 180$ and $15 \times 4 = $
a
b
c
60. Explain how Tom can use these calculations to find 15×160, and calculate the answer.

- Open question, no set answer

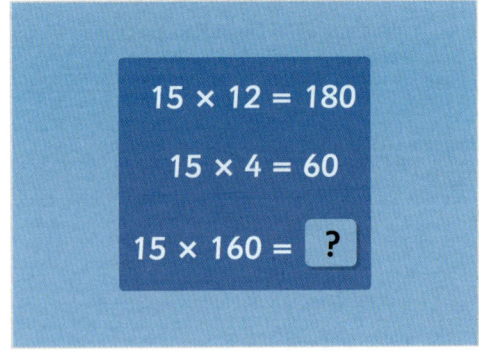

$$15 \times 12 = 180$$
$$15 \times 4 = 60$$
$$15 \times 160 = \boxed{?}$$

Level 3 *continued*

25. William has multiplied 70 by 600 and got the
a answer 4,200. Explain what William has
b done wrong and what he needs to do to
c calculate the correct answer.

- Open question, no set answer

Level 4: Problem solving with greater depth: Use mental multiplication methods to solve multi-step problems.

❋ **Required:** 5/5   ❋ **Pupil Navigation:** on
❋ **Randomised:** off

26. Animal cards are sold in packs of 8. There are
1 50 packs in a box. How many cards are there
2 in 28 boxes?
3

▪ **11200** ▪ 224 ▪ 400

27. In the number pyramid, each number is the
1 product of the two numbers in the boxes
2 below it. What is the value of box C?
3

▪ **13** ▪ 11.7 ▪ 6.5 ▪ 2

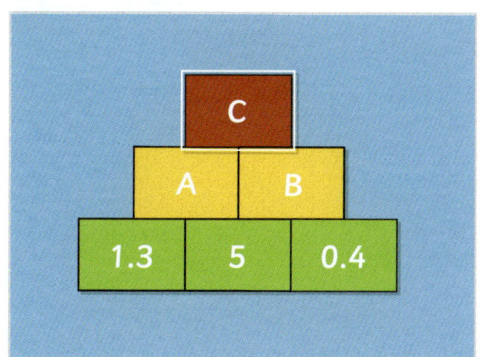

28. Aidan calculates the area of Riverford
1 swimming pool by multiplying the length of
2 the pool by its width. If Watersmeet
3 swimming pool is the same length as
Riverford swimming pool, but twice as wide,
what is the area of Watersmeet swimming
pool in square metres?
Don't include the units in your answer.

▪ **800** ▪ 400

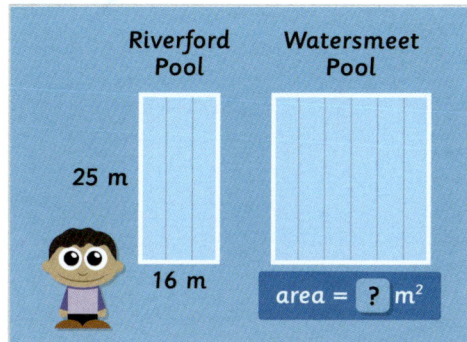

29. If Olivia can swim 19 lengths of a 25-metre
a (m) pool every 15 minutes, how many
b metres can she swim in one hour?
c *Include the units m (metres) in your answer.*

▪ **1,900 metres** ▪ 76 m ▪ **1,900 m** ▪ **1900 m** ▪ 76
▪ 475 ▪ **1900 metres** ▪ 1900 ▪ 475 m ▪ 1,900
▪ 76 metres ▪ 475 metres

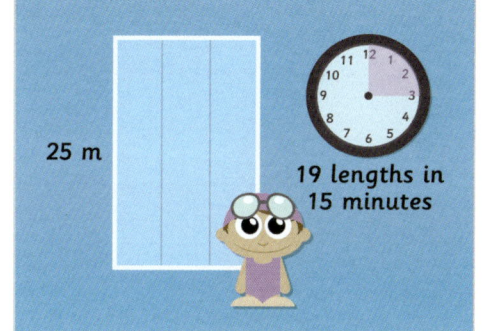

30.
a
b
c
George jumps 1.4 metres (m) forward 19 times. What is the total distance he jumps?
Include the units m (metres) in your answer.

- **26.6 m** ■ **28** ■ **26.6** ■ **26.6 metres** ■ **28 m**
- **28 metres**

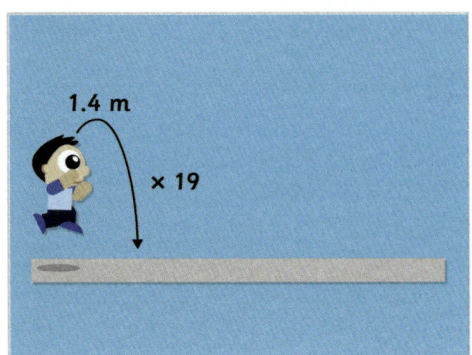

1.4 m

× 19

Practise 10 Times Table Decimal Multiplication

Objective:	I can quickly recall decimal multiplication facts related to the 10 times table with numbers to 1 decimal place.

Quick Search Ref: 10390

Level 1: I can quickly recall decimal multiplication facts related to the 10 times table with numbers to 1 decimal place.

⚙ **Required:** 11/11   ⚙ **Pupil Navigation:** off   ⚙ **Randomised:** on

1. 10 × 0.1

2. 10 × 0.2

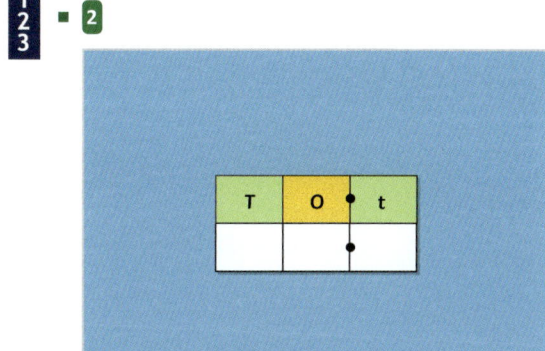

3. 10 × 0.3

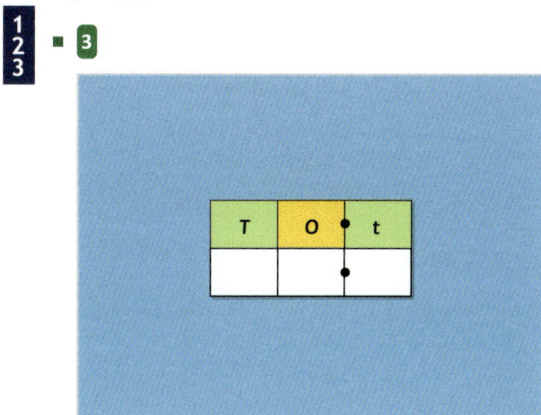

4. 10 × 0.4

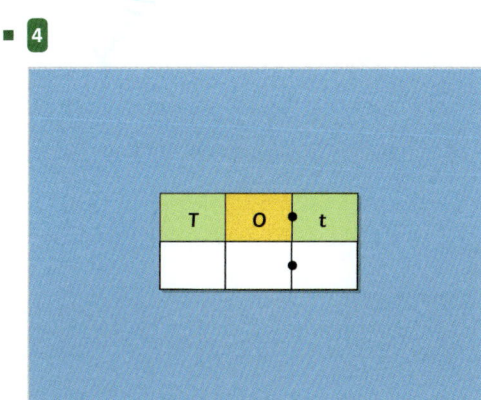

5. 10 × 0.5

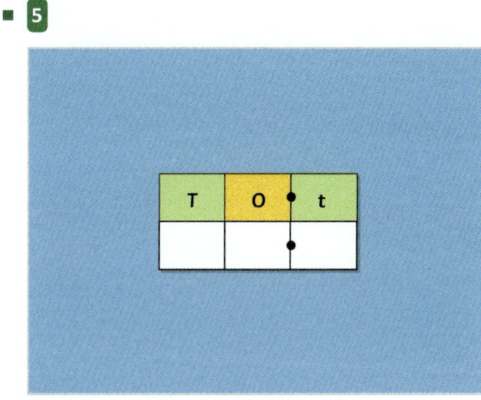

6. 10 × 0.6

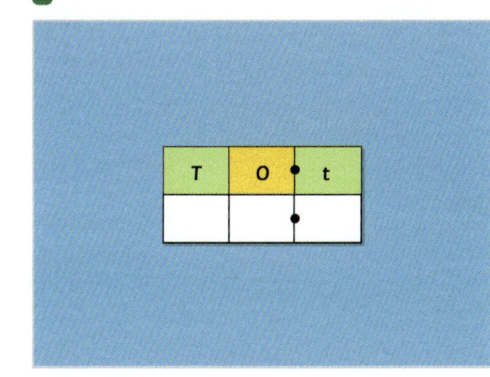

Objective: I can quickly recall decimal multiplication facts related to the 10 times table with numbers to 1 decimal place.

Level 1 *continued*

7. 10 × 0.7

 ▪ 7

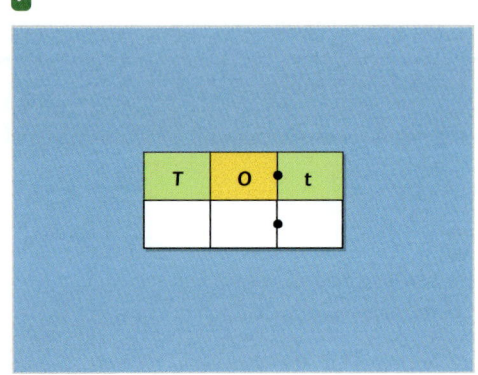

8. 0.8 × 10

 ▪ 8

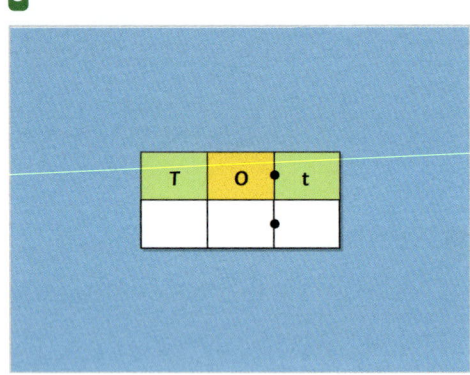

9. 0.9 × 10

 ▪ 9

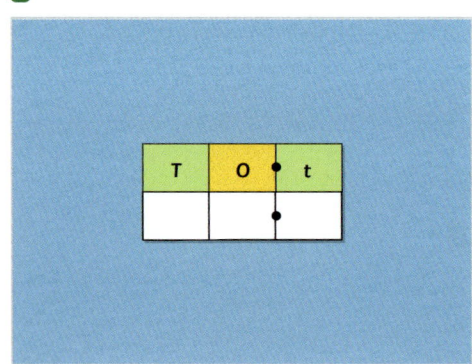

10. 1.1 × 10

 ▪ 11

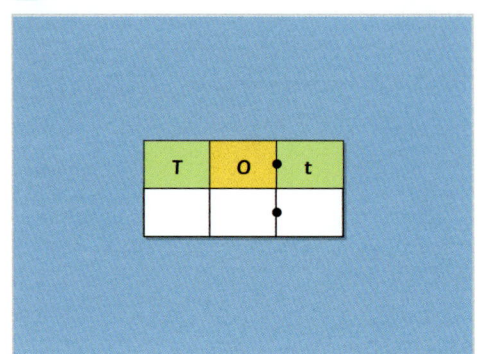

11. 10 × 1.2

 ▪ 12

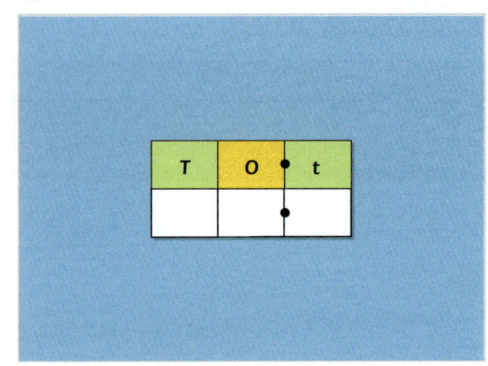

Practise 11 Times Table Decimal Multiplication

Objective:	I can quickly recall decimal multiplication facts related to the 11 times table with numbers to 1 decimal place.

Quick Search Ref: 10391

1. 0.1×11

 ▪ 1.1

2. 0.2×11

 ▪ 2.2

3. 1.1×3

 ▪ 3.3

4. 11×0.4

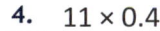 ▪ 4.4

5. 11×0.5

 ▪ 5.5

6. 11×0.6

 ▪ 6.6

Level 1 *continued*

7. 0.7 × 11

 ▪ 7.7

8. 0.8 × 11

 ▪ 8.8

9. 11 × 0.9

 ▪ 9.9

10. 1.1 × 10

 ▪ 11

11. 11 × 1.1

 ▪ 12.1

12. 11 × 1.2

 ▪ 13.2

13. 1.1 × 0

 ▪ 0

Practise 12 Times Table Decimal Multiplication

| Objective: | I can quickly recall decimal multiplication facts related to the 12 times table with numbers to 1 decimal place. |

Quick Search Ref: 10392

Level 1: I can quickly recall decimal multiplication facts related to the 12 times table with numbers to 1 decimal place.

❋ Required: 13/13　　❋ Pupil Navigation: off　　❋ Randomised: on

1. 0.1 × 12

■ 1.2

2. 0.2 × 12

■ 2.4

3. 12 × 0.3

■ 3.6

4. 12 × 0.4

■ 4.8

5. 12 × 0.5

■ 6

6. 12 × 0.6

■ 7.2

Level 1 *continued*

7. 0.7 × 12

■ 8.4

8. 12 × 0.8

■ 9.6

9. 12 × 0.9

■ 10.8

10. 10 × 1.2

■ 12

11. 11 × 1.2

■ 13.2

12. 1.2 × 12

■ 14.4

13. 1.2 × 0

■ 0

Practise 2 Times Table Decimal Multiplication

Objective:	I can quickly recall decimal multiplication facts related to the 2 times table with numbers to 1 decimal place.

Quick Search Ref:	**10393**

Level 1: I can quickly recall decimal multiplication facts related to the 2 times table with numbers with 1 decimal place.

✿ Required: 13/13   ✿ Pupil Navigation: off   ✿ Randomised: on

1. 0×0.2

 ▪ 0

2. 0.1×2

 ▪ 0.2

3. 0.2×2

 ▪ 0.4

4. 0.2×3

 ▪ 0.6

5. 0.2×4

 ▪ 0.8

6. 0.2×5

 ▪ 1

Level 1 *continued*

7. 0.2×6

1 2 3 ▪ 1.2

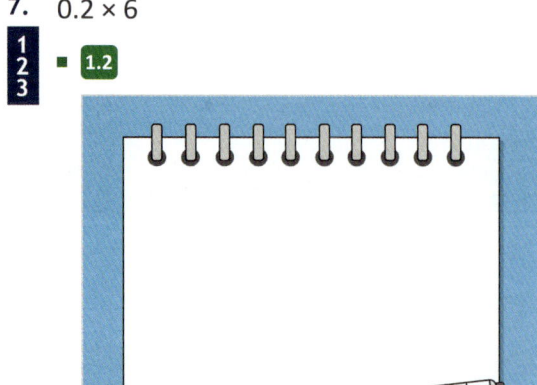

8. 7×0.2

1 2 3 ▪ 1.4

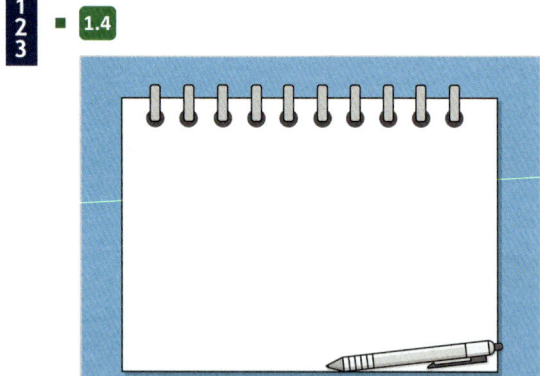

9. 8×0.2

1 2 3 ▪ 1.6

10. 9×0.2

1 2 3 ▪ 1.8

11. 10×0.2

1 2 3 ▪ 2

12. 0.2×11

1 2 3 ▪ 2.2

13. 0.2×12

1 2 3 ▪ 2.4

Practise 3 Times Table Decimal Multiplication

Objective:	I can quickly recall decimal multiplication facts related to the 3 times table with numbers to 1 decimal place.

Quick Search Ref:	**10394**

Level 1: I can quickly recall decimal multiplication facts related to the 3 times table with numbers to 1 decimal place.

✿ Required: 13/13   ✿ Pupil Navigation: off   ✿ Randomised: on

1. 0.3×0

123 ▪ **0**

2. 0.1×3

123 ▪ **0.3**

3. 0.2×3

123 ▪ **0.6**

4. 0.3×3

123 ▪ **0.9**

5. 0.3×4

123 ▪ **1.2**

6. 0.3×5

123 ▪ **1.5**

Level 1 *continued*

7. 0.3 × 6

123 ▪ 1.8

8. 0.3 × 7

123 ▪ 2.1

9. 8 × 0.3

123 ▪ 2.4

10. 9 × 0.3

123 ▪ 2.7

11. 10 × 0.3

123 ▪ 3

12. 11 × 0.3

123 ▪ 3.3

13. 12 × 0.3

123 ▪ 3.6

Practise 4 Times Table Decimal Multiplication

Objective: I can quickly recall decimal multiplication facts related to the 4 times table with numbers to 1 decimal place.

Quick Search Ref: 10395

Level 1: I can quickly recall decimal multiplication facts related to the 4 times table with numbers to 1 decimal place.

✿ **Required:** 13/13　　✿ **Pupil Navigation:** off　　✿ **Randomised:** on

1. 0.4 × 0

123 ▪ 0

2. 0.1 × 4

123 ▪ 0.4

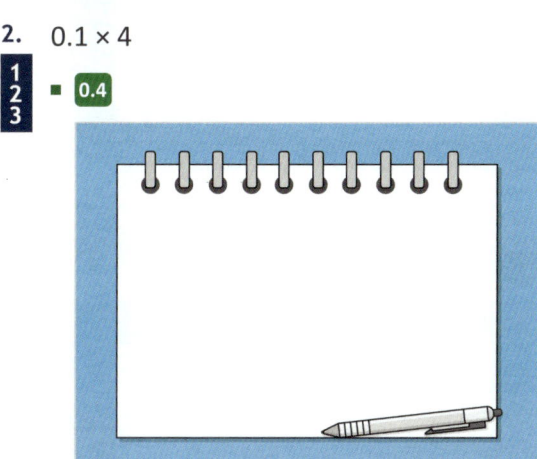

3. 0.2 × 4

123 ▪ 0.8

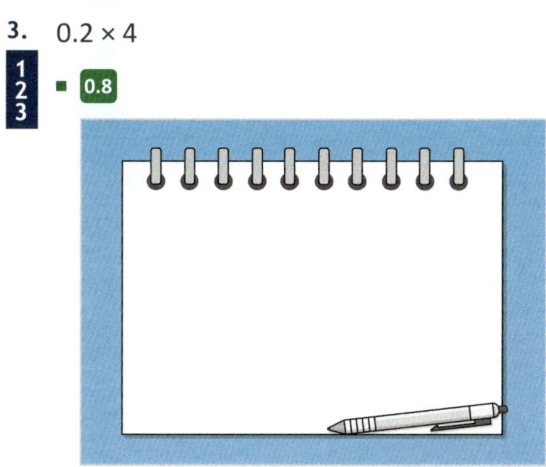

4. 0.3 × 4

123 ▪ 1.2

5. 0.4 × 4

123 ▪ 1.6

6. 0.4 × 5

123 ▪ 2

Level 1 *continued*

7. 0.4 × 6

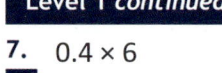 ▪ 2.4

8. 0.4 × 7

123 ▪ 2.8

9. 0.4 × 8

123 ▪ 3.2

10. 9 × 0.4

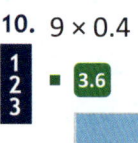 ▪ 3.6

11. 10 × 0.4

123 ▪ 4

12. 11 × 0.4

123 ▪ 4.4

13. 12 × 0.4

123 ▪ 4.8

Practise 5 Times Table Decimal Multiplication

Objective:	I can quickly recall decimal multiplication facts related to the 5 times table with numbers to 1 decimal place.

Quick Search Ref:	**10396**

Level 1: I can quickly recall decimal multiplication facts related to the 5 times table with numbers to 1 decimal place.

❋ **Required:** 13/13   ❋ **Pupil Navigation:** off   ❋ **Randomised:** on

1. 0.5 × 0

123 ▪ 0

2. 0.1 × 5

123 ▪ 0.5

3. 0.2 × 5

123 ▪ 1

4. 0.3 × 5

123 ▪ 1.5

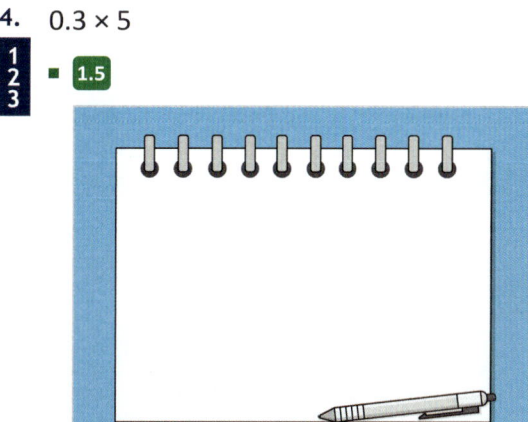

5. 0.4 × 5

123 ▪ 2

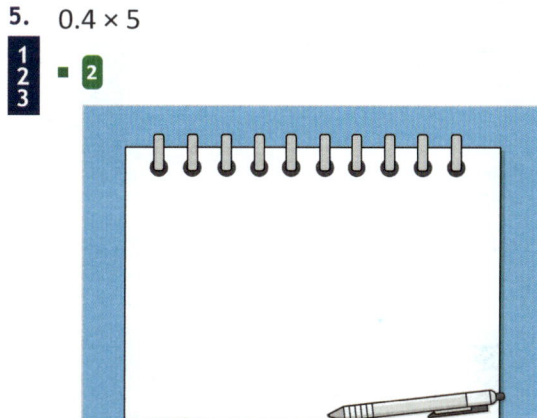

6. 0.5 × 5

123 ▪ 2.5

Level 1 *continued*

7. 0.5 × 6

123 ▪ **3**

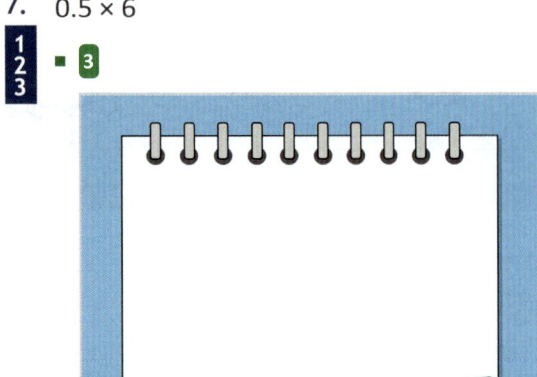

8. 0.5 × 7

123 ▪ **3.5**

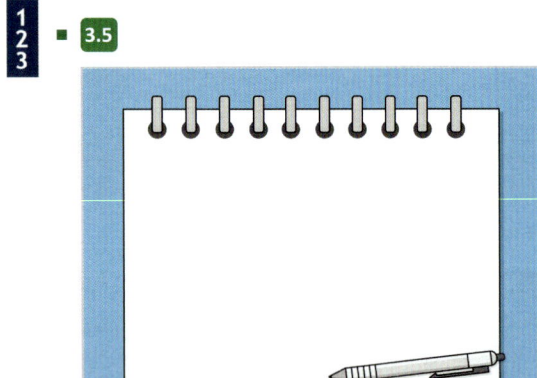

9. 0.5 × 8

123 ▪ **4**

10. 9 × 0.5

123 ▪ **4.5**

11. 10 × 0.5

123 ▪ **5**

12. 11 × 0.5

123 ▪ **5.5**

13. 12 × 0.5

123 ▪ **6**

Practise 6 Times Table Decimal Multiplication

Objective: I can quickly recall decimal multiplication facts related to the 6 times table with numbers to 1 decimal place.

Quick Search Ref: 10397

Level 1: I can quickly recall decimal multiplication facts related to the 6 times table with numbers to 1 decimal place.

⚙ **Required:** 13/13   ⚙ **Pupil Navigation:** off   ⚙ **Randomised:** on

1. 0.6×0

1 2 3 ▪ 0

4. 0.3×6

1 2 3 ▪ 1.8

2. 6×0.1

1 2 3 ▪ 0.6

5. 0.4×6

1 2 3 ▪ 2.4

3. 0.2×6

1 2 3 ▪ 1.2

6. 0.5×6

1 2 3 ▪ 3

Level 1 *continued*

7. 0.6 × 6

 ▪ 3.6

8. 0.6 × 7

123 ▪ 4.2

9. 0.6 × 8

123 ▪ 4.8

10. 0.6 × 9

123 ▪ 5.4

11. 10 × 0.6

123 ▪ 6

12. 11 × 0.6

123 ▪ 6.6

13. 12 × 0.6

123 ▪ 7.2

Practise 7 Times Table Decimal Multiplication

Objective: I can quickly recall decimal multiplication facts related to the 7 times table with numbers with 1 decimal place.

Quick Search Ref: 10398

Level 1: I can quickly recall decimal multiplication facts related to the 7 times table with numbers to 1 decimal place.

✿ **Required:** 13/13   ✿ **Pupil Navigation:** off   ✿ **Randomised:** on

1. 0×0.7
   ▪ 0

2. 7×0.1
   ▪ 0.7

3. 7×0.2
   ▪ 1.4

4. 0.3×7
   ▪ 2.1

5. 0.4×7
   ▪ 2.8

6. 0.5×7
   ▪ 3.5

7. 0.6×7
   ▪ 4.2

8. 0.7×7
   ▪ 4.9

9. 0.7×8
   ▪ 5.6

10. 9×0.7
    ▪ 6.3

11. 10×0.7
    ▪ 7

12. 0.7×11
    ▪ 7.7

13. 0.7×12
    ▪ 8.4

Practise 8 Times Table Decimal Multiplication

Objective:	I can quickly recall decimal multiplication facts related to the 8 times table with numbers to 1 decimal place.

Quick Search Ref: 10399

Level 1: I can quickly recall decimal multiplication facts related to the 8 times table with numbers to 1 decimal place.

❖ **Required:** 13/13 ❖ **Pupil Navigation:** off ❖ **Randomised:** on

1. 0 × 0.8

123 ▪ **0**

2. 8 × 0.1

123 ▪ **0.8**

3. 8 × 0.2

123 ▪ **1.6**

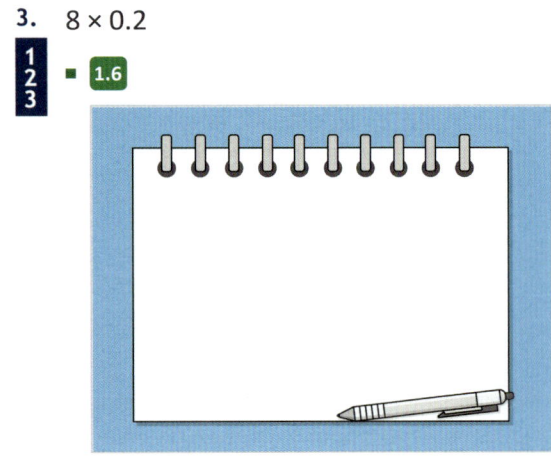

4. 8 × 0.3

123 ▪ **2.4**

5. 0.4 × 8

123 ▪ **3.2**

6. 0.5 × 8

123 ▪ **4**

7. 0.6 × 8

1 2 3 ▪ 4.8

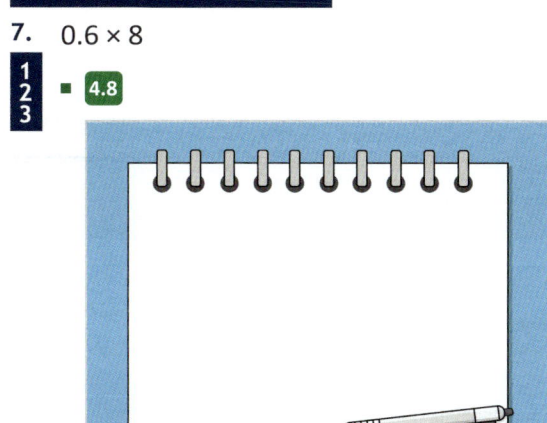

8. 0.7 × 8

1 2 3 ▪ 5.6

9. 0.8 × 8

1 2 3 ▪ 6.4

10. 9 × 0.8

1 2 3 ▪ 7.2

11. 0.8 × 10

1 2 3 ▪ 8

12. 0.8 × 11

1 2 3 ▪ 8.8

13. 12 × 0.8

1 2 3 ▪ 9.6

Practise 9 Times Table Decimal Multiplication

| **Objective:** | I can quickly recall decimal multiplication facts related to the 9 times table with numbers to 1 decimal place. |

Quick Search Ref: 10400

Level 1: I can quickly recall decimal multiplication facts related to the 9 times table with numbers to 1 decimal place.

✿ **Required:** 13/13　　✿ **Pupil Navigation:** off　　✿ **Randomised:** on

1. 0×0.9

123 ▪ 0

2. 9×0.1

123 ▪ 0.9

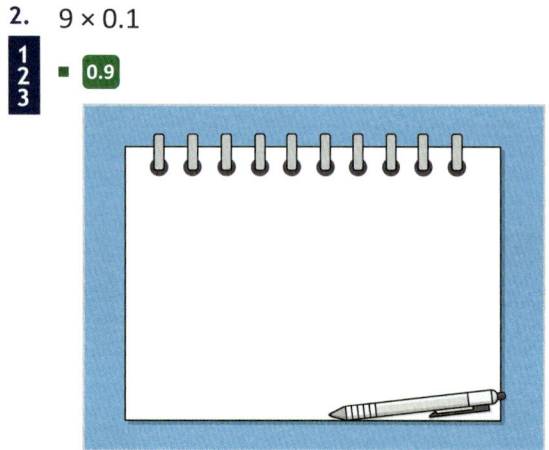

3. 0.9×2

123 ▪ 1.8

4. 9×0.3

123 ▪ 2.7

5. 9×0.4

123 ▪ 3.6

6. 9×0.5

123 ▪ 4.5

Level 1 *continued*

7. 0.6 × 9

1
2
3 ▪ 5.4

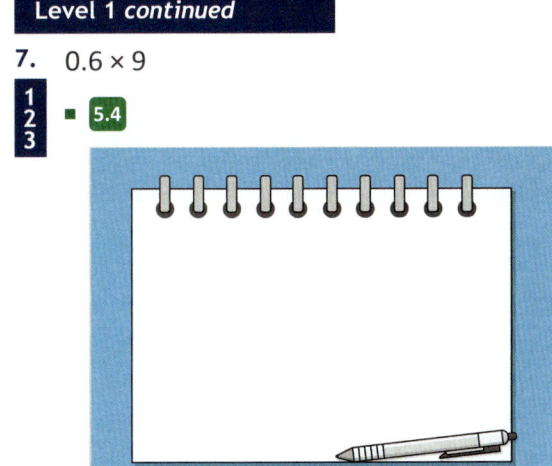

8. 9 × 0.7

1
2
3 ▪ 6.3

9. 9 × 0.8

1
2
3 ▪ 7.2

10. 0.9 × 9

1
2
3 ▪ 8.1

11. 0.9 × 10

1
2
3 ▪ 9

12. 11 × 0.9

1
2
3 ▪ 9.9

13. 12 × 0.9

1
2
3 ▪ 10.8

Multiplication and Division Topic Review

Objective: I can answer questions involving multiplication and division from the Year 5 curriculum.

Quick Search Ref: **10737**

Level 1: Understanding

❋ **Required:** 7/10   ❋ **Pupil Navigation:** on   ❋ **Randomised:** off

1. What is 320 divided by 5?

▪ **64** ▪ 1600

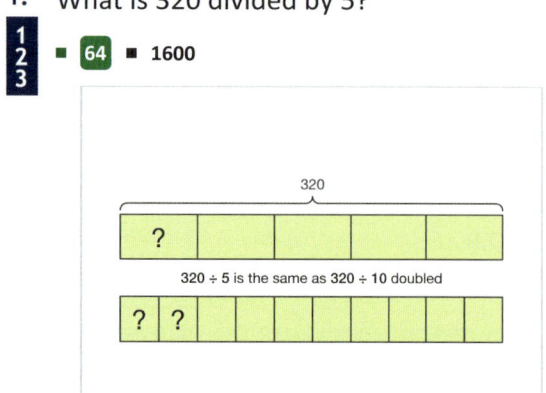

2. Find the product of 16 and 3.

▪ **48** ▪ 19

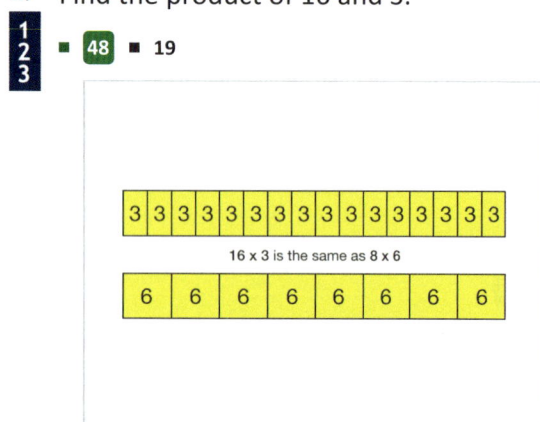

3. 8.9 × 1,000 =

▪ **8900** ▪ 0.0089 ▪ 890 ▪ 89

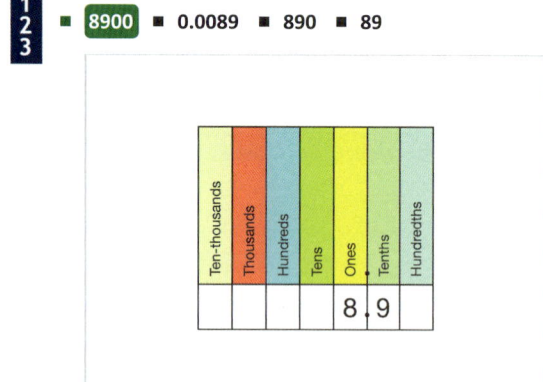

4. Select **three** multiples of 4.

3/6

▪ **12** ▪ 14 ▪ **16** ▪ 26 ▪ 34 ▪ **48**

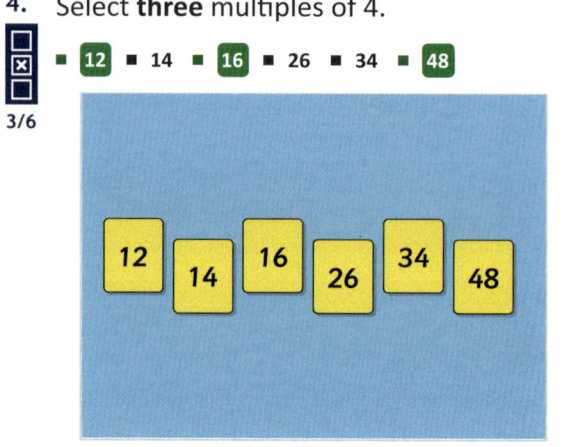

5. Select the **factor pair** (2 factors) of 15 which will make the following equation complete:

2/5   __ × __ = 15

▪ **3** ▪ 4 ▪ **5** ▪ 6 ▪ 7

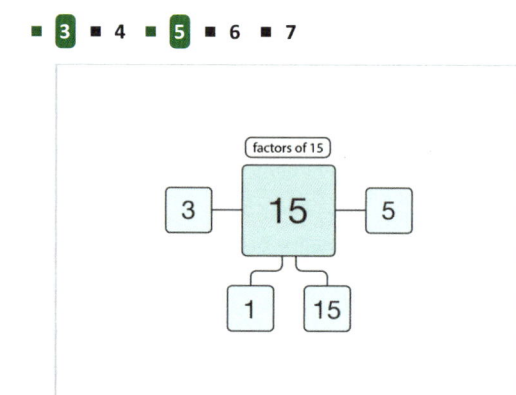

6. Find the answer to 8,523 × 14.

▪ **119322** ▪ 85230 ▪ 42615

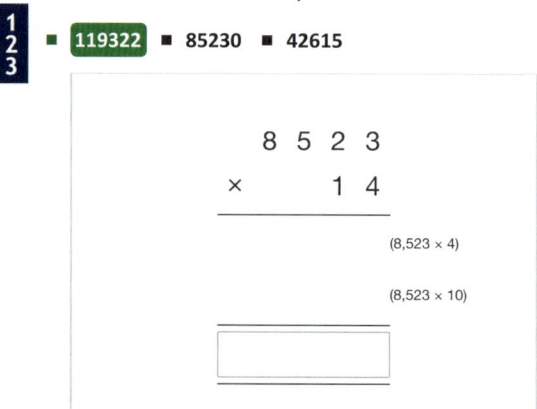

Level 1 *continued*

7. What are the first **three** prime numbers?

■ 1 ■ 2 ■ 3 ■ 4 ■ 5

3/5

number	factors
1	
2	
3	
4	
5	

8. Find the answer to 1,432 × 12.

■ 17184 ■ 4296 ■ 2864 ■ 14320

9. Calculate 4,215 × 3.

■ 12645 ■ 12635 ■ 12665

$$4\ 2\ 1\ 5$$
$$\times\qquad 3$$
$${}_{1}$$
$$4\ 5$$

10. Calculate the answer to 735 ÷ 5.

■ 147 ■ 3675

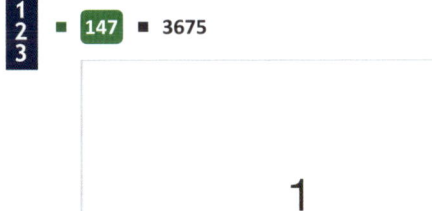

$$5 \overline{)7\ {}^{2}3\ 5}\ \ ^{1}$$

11. What is 194 **divided** by 6?

Use an **r** to show any remainders. For example, 7 ÷ 2 = 3r1.

■ 32 ■ 32r2 ■ 1,164 ■ 1164

$$6 \overline{)1\ 9\ 4}\ \ ^{r}$$

12. What is the product of 2,016 × 6?

■ 12096 ■ 12066 ■ 2066

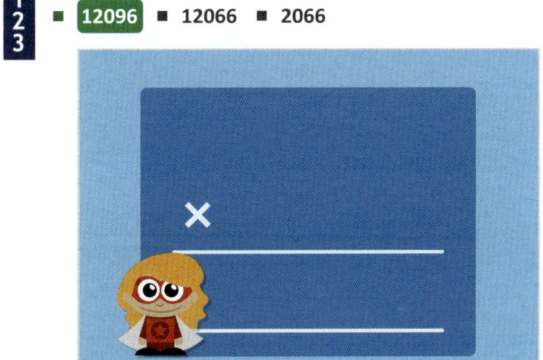

13. What is the missing number from the calculation?

_____ ÷ 1,000 = 79.4

■ 79400 ■ 0.0794

14. What is 4,321 × 12?

■ 51852 ■ 43210 ■ 12963 ■ 8642

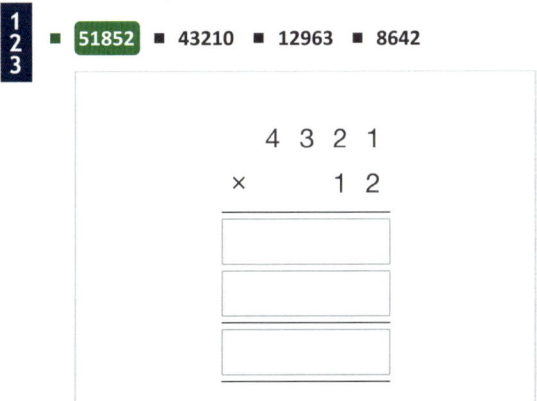

$$4\ 3\ 2\ 1$$
$$\times\qquad 1\ 2$$

Level 2 *continued*

15. Select the pair of square numbers which **equal 20** when added together.

▪ 1² ▪ **2²** ▪ 3² ▪ **4²**

2/4

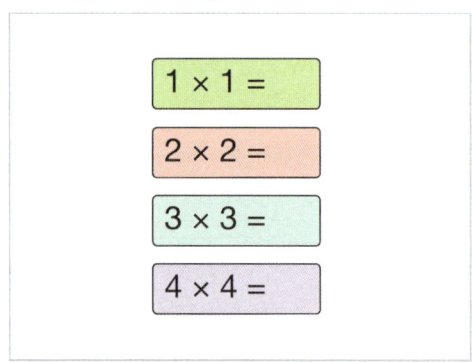

16. Katie is paid £1,764 every month. How much does she earn in one year?
Include the £ sign in your answer.

▪ **£21,168.00** ▪ **£21168** ▪ 21168.00 ▪ **£21168.00**
▪ **£21,168** ▪ 21,168 ▪ 21,168.00 ▪ 21168

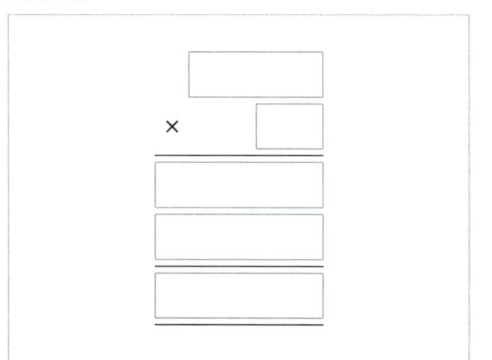

17. Asifa has 294 t-shirts and splits them into packs of 4. How many spare t-shirts does she have left?

▪ **2** ▪ 73

18. Select the symbol that makes the following statement true.
7² _____ 4³

1/3

▪ **<** ▪ > ▪ =

19. What number is one hundred times larger than 2.39?

▪ 23.9 ▪ **239** ▪ 0.0239

20. Calculate 885 **divided** by 8.

▪ 11r5 ▪ 110 ▪ **110r5** ▪ 115

1/4

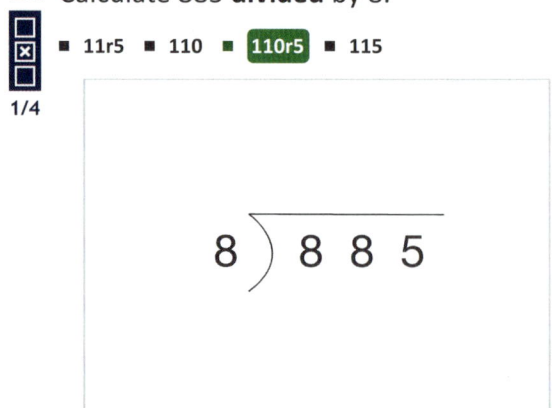

Level 3:   Reasoning

❋ **Required:** 5/5   ❋ **Pupil Navigation:** on
❋ **Randomised:** off

21. What is the missing number?
3,440 ÷ 8 = _____ × 2

▪ **215** ▪ 430 ▪ 862

22. Franky says, "Before calculating, I know that there will be a remainder in the calculation 1,876 ÷ 5." Explain how Franky can know this without doing any calculations.

- Open question, no set answer

23. Kane says, "When multiplying whole or decimal numbers by 10, 100 or 1,000, you can just add zeros to the end of the number". Is Kane correct? Explain your answer.

- Open question, no set answer

24. 8 × 9 = 72. How can this multiplication fact be used to solve the following calculation?

800 × 900 = _____

Explain your method.

- Open question, no set answer

Level 3 *continued*

25. Henry has completed his homework, but he has made a mistake. Explain what mistake he has made, and then calculate the correct answer.

a
b
c

- Open question, no set answer

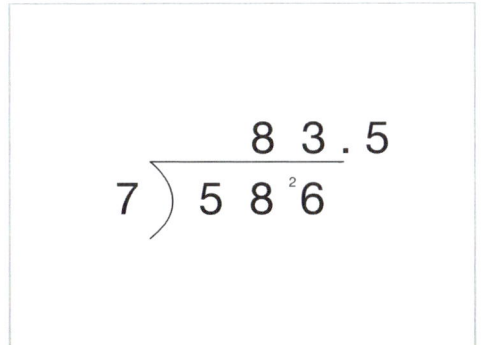

Level 4: Problem Solving

✿ **Required:** 5/5   ✿ **Pupil Navigation:** on
✿ **Randomised:** off

26. Hannah is thinking of 3 numbers x, y and z.
She says, 'y is 10 times smaller than x, and z is 100 times larger than x'.
What is the value of $z \div y$?

1
2
3

▪ 1000

27. Light bulbs are sold in boxes of 5 and there are 20 boxes in each carton. How many light bulbs are there in 250 cartons?

1
2
3

▪ 25000 ▪ 1250 ▪ 5000 ▪ 100

28. Which digit goes in box C?

a
b
c

▪ 5 ▪ 905 ▪ 3 ▪ 129 r 2

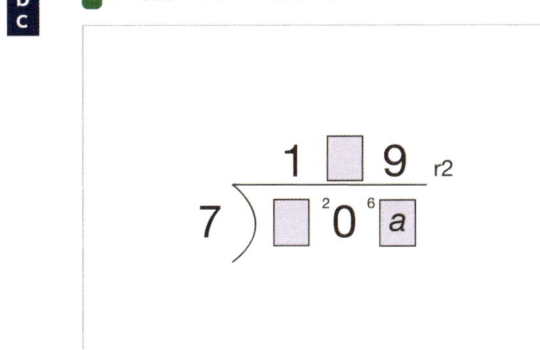

29. If the number in each box is the product of the two boxes underneath it, what number goes in box C?

1
2
3

▪ 136 ▪ 272 ▪ 544

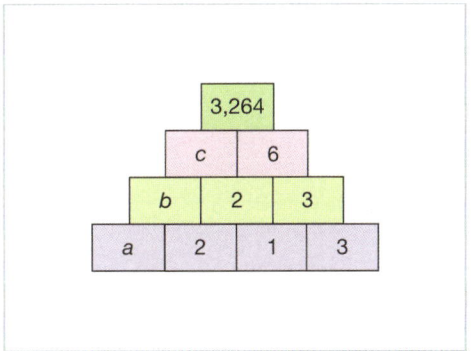

30. Fill in the factors missing from the multiplication grid to find the value of y.

1
2
3

▪ 8 ▪ 30 ▪ 2 ▪ 5 ▪ 6

Mathematics Y5

Fractions

Addition
Subtraction
Multiplication
Division
Compare and Order
Counting
Sequences
Equivalent Fractions
Improper Fractions and Mixed Numbers
Decimals and Percentages

Add and Subtract Fractions with Related Denominators (Proper, Improper and Mixed Number Fractions)

Objective: I can add and subtract proper, improper and mixed number fractions with the same denominator and denominators that are multiples of the same number.

Quick Search Ref: 10181

Level 1: Understanding: Add and subtract fractions with image support.

✿ **Required:** 7/10   ✿ **Pupil Navigation:** on   ✿ **Randomised:** off

---

1. What is 1/12 + 5/6?

a
b
c
Put a forward slash (/) between the numerator and denominator. For example, one-half is 1/2.

- ■ 66/72 ■ 11/12 ■ 11/24 ■ 6/18

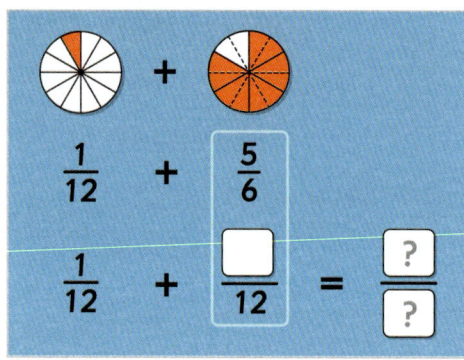

2. 1/3 − 1/9 = ___

a
b
c
Put a forward slash (/) between the numerator and denominator. For example, one-half is 1/2.

- ■ 4/18 ■ 6/27 ■ 2/9 ■ 1/6 ■ 4/9

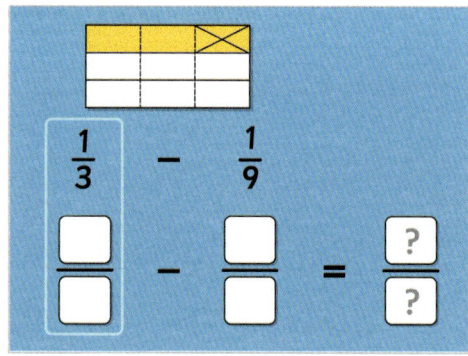

3. What is the total of 7/3 + 11/9?

a
b
c
Give your answer as an improper fraction. For example, 3/2.

- ■ 18/12 ■ 32/9 ■ 32/18

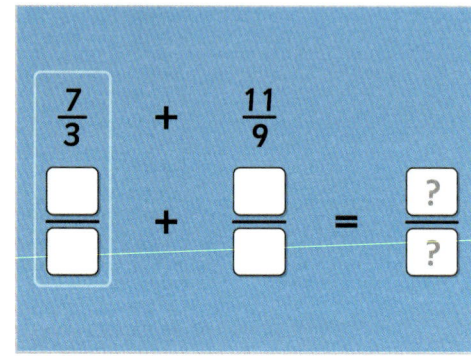

4. What is 2 − 5/4?

a
b
c
Give your answer as an improper fraction. For example, 3/2.

- ■ 3/4 ■ 13/4

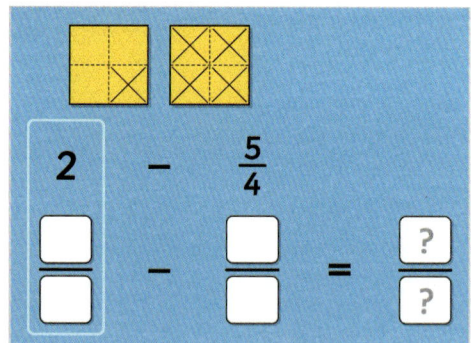

Level 1 *continued*

5. What is the total of 1 3/7 + 1 5/7?

a b c *Give your answer as a mixed number fraction. For example, five and one-half is 5 1/2.*

- 2 8/7 ▪ **3 1/7** ▪ 1 1/7 ▪ 22/7

6. Calculate 4 3/5 − 7/10.

a b c *Give your answer as a mixed number fraction. For example, five and one-half is 5 1/2.*

- **3 9/10** ▪ **3 45/50** ▪ 39/10 ▪ 195/50

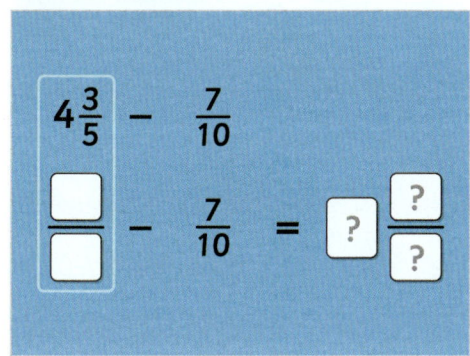

7. What is 5 1/8 − 1 3/8?

a b c *Give your answer as a mixed number fraction. For example, five and one-half is 5 1/2.*

- **3 3/4** ▪ 30/8 ▪ **3 6/8** ▪ 15/4 ▪ 4 2/8 ▪ 4 1/4

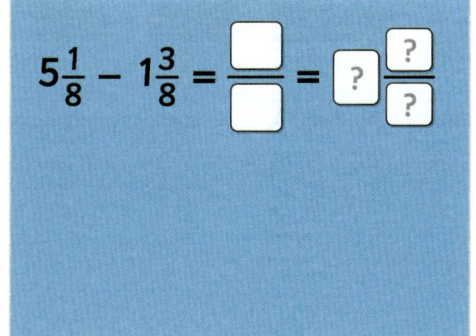

8. What is 14/2 + 7/4?

a b c *Give your answer as an improper fraction. For example, 3/2.*

- **35/4** ▪ 35/8 ▪ 21/6

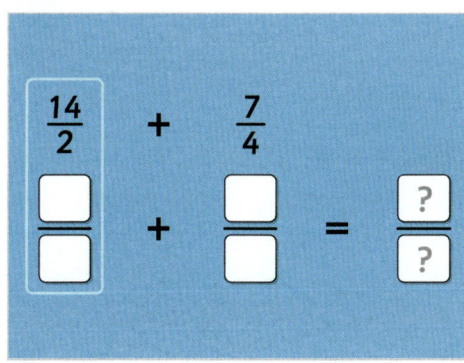

9. Calculate 1 3/5 − 7/10.

a b c *Give your answer as a fraction. For example, one-half is 1/2.*

- **9/10** ▪ 1 1/10

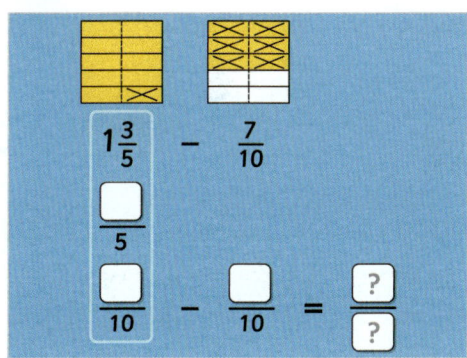

10. What is 1/10 + 1/30?

a b c *Put a forward slash (/) between the numerator and denominator. For example, one-half is 1/2.*

- **2/15** ▪ 8/60 ▪ **4/30** ▪ **40/300** ▪ 2/40 ▪ 4/60

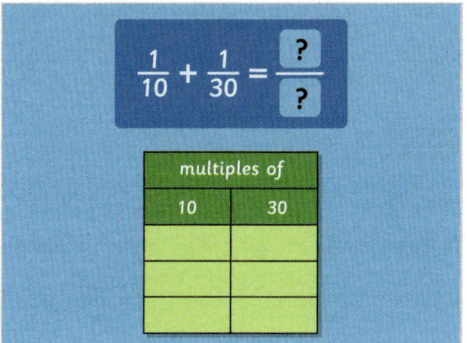

Level 2: Fluency: Add and subtract two or three fractions including 1-step word problems.

❋ **Required:** 7/10   ❋ **Pupil Navigation:** on

❋ **Randomised:** off

11. What is 3/4 + 1/8 + 1/16?

a
b
c
Give your answer as improper fraction. For example 3/2.

■ **15/16** ■ **15/48**

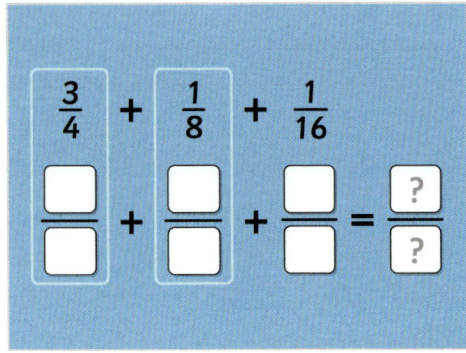

12. What is 13/4 – 3/2 – 9/8?

a
b
c
Give your answer as a fraction. For example, 1/2.

■ **5/8**

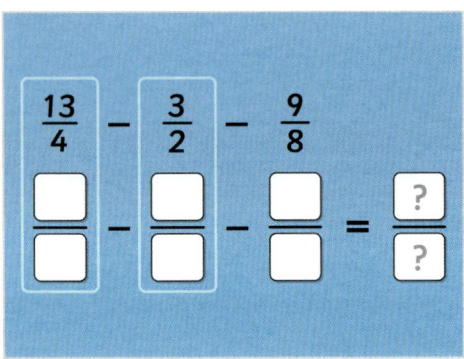

13. What is 2 1/4 + 5/8 + 1 1/16?

a
b
c
Give your answer as a mixed number fraction. For example, five and one-half is 5 1/2.

■ **3 15/16** ■ **63/16**

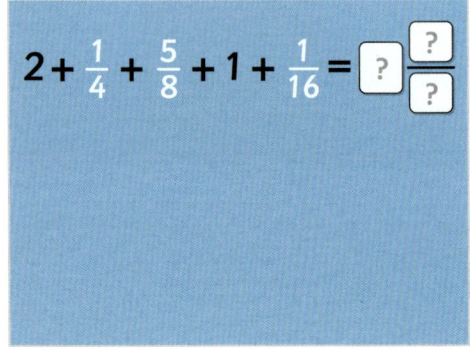

14. What is 3 1/3 – 1 1/6 – 1/3?

a
b
c
Give your answer as a mixed number fraction.

■ **1 5/6** ■ **11/6**

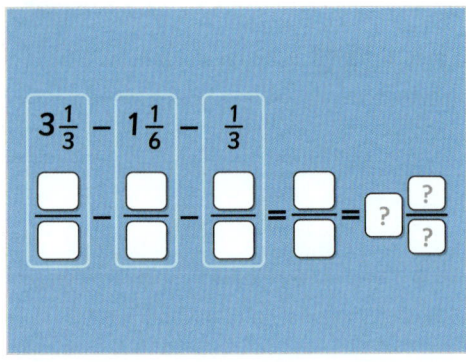

15. 14/25 of a class have school dinners. If 2/5 of the class change to packed lunches, what fraction of the class still has school dinners?

a
b
c
Put a forward slash (/) between the numerator and denominator. For example, one-half is 1/2.

■ **4/25** ■ **20/125** ■ **24/25**

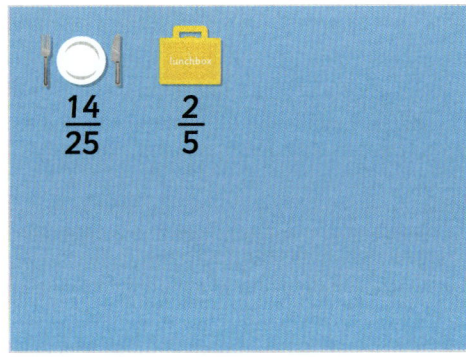

16. A jug contains 55/50 litres of water. If another 12/10 litres of water are poured into the jug, the total amount of water in the jug will be _____ litres.

a
b
c
Enter the missing improper fraction. Don't include the units in your answer.

■ **2.3** ■ **115/50 litres** ■ **115/50 l** ■ **115/50**

■ **2.3 litres** ■ **2.3 l**

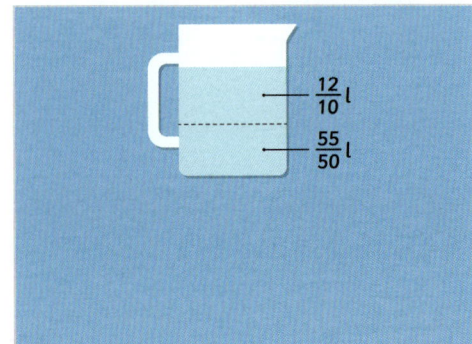

Level 2 *continued*

17. The shorter flower in the diagram is _____ cm
_a_b_c tall.

Enter the missing mixed number fraction.
Don't include the units in your answer.

- **4 3/4 cm** - **4 3/4 centimetres** - **4 3/4** - 4.75
- 4.75 centimetres - 4.75 cm

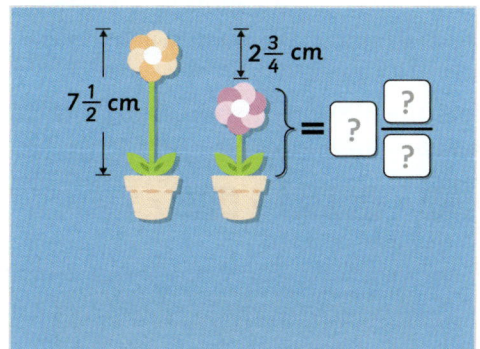

18. Mr Brown has 3 1/10 bags of corn. If Mr
₁₂₃ Brown's chickens eat 3/5 of a bag of corn,
how many **full bags** of corn does Mr Brown
have left?

- **2**

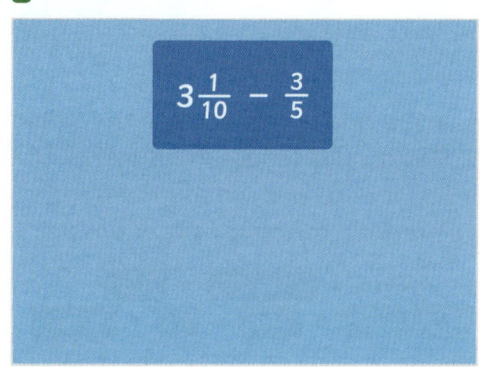

19. What is 9/4 + 5/8 + 17/16?
_a_b_c *Give your answer as an improper fraction.*
For example, 3/2.

- **63/16**

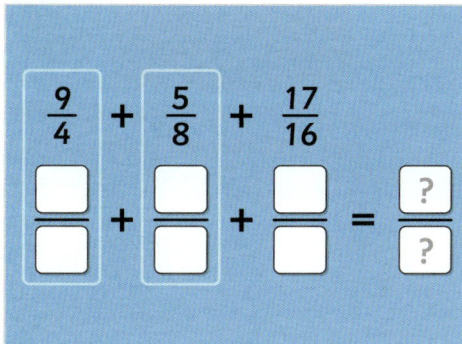

20. Calculate 1 1/8 + 1 1/4.
_a_b_c *Give your answer as a mixed number*
fraction. For example, five and one-half is 5
1/2.

- **2 3/8** - 19/8

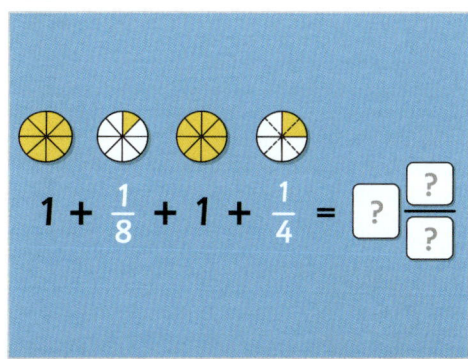

Level 3: Reasoning: Reason about adding and subtracting fractions with related denominators.

❋ **Required:** 5/5   ❋ **Pupil Navigation:** on
❋ **Randomised:** off

21. Dylan says he knows that 4/6 − 1/2 is less
_a_b_c than 2/3 − 5/12 without having to complete
the calculations. Explain how he might know
this.

- Open question, no set answer

22. Which **three** fractions give a total of 15/18
when added together?

3/5

- 1/6 - **5/18** - 7/9 - **1/3** - **2/9**

$$\frac{?}{?} + \frac{?}{?} + \frac{?}{?} = \frac{15}{18}$$

Level 3 *continued*

23.
a b c
Abbie says, "When you subtract two smaller mixed number fractions from a larger mixed number fraction, your answer will **never** be a whole number." Is Abbie correct? Explain your answer and give an example.

- Open question, no set answer

$$5\frac{1}{4} - 1\frac{1}{2} - 1\frac{1}{8} = 2\frac{5}{8}$$

24.
a b c
Oscar is finding the total of 17/12 and 19/6. He converts both fractions to have a denominator of 72 before adding them. Explain how Oscar can make his calculation easier.

- Open question, no set answer

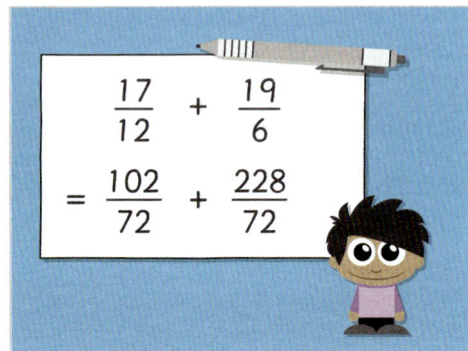

$$\frac{17}{12} + \frac{19}{6}$$
$$= \frac{102}{72} + \frac{228}{72}$$

25.
a b c
There are enough children in Year 5 to make 3 2/3 netball teams. If 5/6 of a team are ill, will there still be enough children to make three full teams? Explain your answer and use a calculation to prove how many full teams can be made.

- Open question, no set answer

$$3\frac{2}{3} -$$

Level 4: Problem solving with greater depth: Solve multi-step problems involving adding and subtracting fractions with related denominators.

❋ **Required:** 5/5 ❋ **Pupil Navigation:** on
❋ **Randomised:** off

26.
a b c
Jane pours 2 2/5 litres of paint into a 5 litre bucket. If she pours another 2 11/15 litres into the bucket, the paint will overflow by _____ of a litre.
Enter the missing fraction. Don't include the units in your answer.

- 2/15 - 2/15 litres - 2/15 l - 5 2/15 l
- 5 2/15 litres - 5 2/15

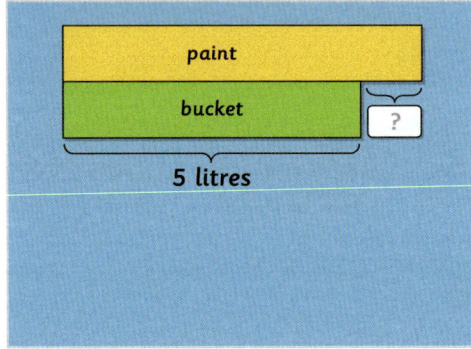

27.
a b c
A rectangle has a perimeter of 31/5 metres (m) and a side length of 17/10 m. What does the shorter side length measure?
Give your answer as an improper fraction. Don't include the units in your answer.

- 14/5 metres - 7/5 metres - 14/10
- 14/10 metres - 7/5 m - 7/5 - 14/10 m
- 28/10 m - 28/10 - 28/10 metres - 14/5 m
- 14/5

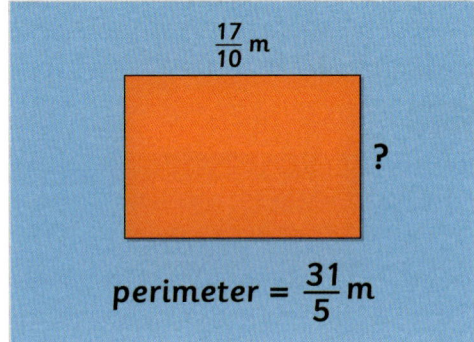

Level 4 *continued*

28. Dom and his friends complete a 100-piece jigsaw.

a
b
c

Aaron completes 1/5 of the jigsaw.
Beth completes 3/25 of the jigsaw.
Cara completes 25 pieces.
Dom finishes the rest of the jigsaw.

What **fraction** of the jigsaw does Dom complete?

▪ 43/100 ▪ 57/100

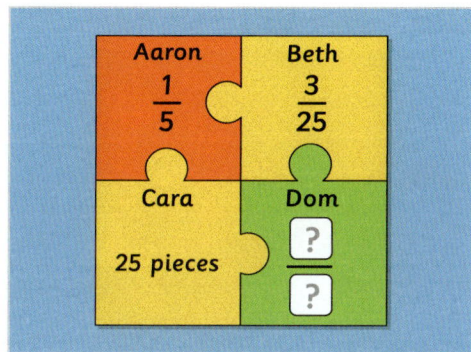

29. Max is exactly 10 2/3 years old. Ben is 7 months younger than Max.

a
b
c

What fraction of a year is it until Ben's next birthday?
Give your answer as a fraction. Don't include the units in your answer.

▪ 11/12 ▪ 11 ▪ 10 1/12

30. On Thursday, Beth records how long she spends on different activities during the day. The next day Beth spends ½ an hour less playing and she goes to bed ¾ of an hour later. On **Friday**, Beth spends _____ more hours sleeping than playing.
Enter the missing mixed number fraction. Don't include the units in your answer.

a
b
c

▪ 9 1/8 ▪ 9 3/8 ▪ 9 1/8 hours ▪ 9 1/8 hrs
▪ 10 3/4 ▪ 1 5/8

Time Beth spends on different activities on Thursday	
activity	time spent (hours)
sleeping	$11\frac{1}{2}$
eating	$1\frac{1}{4}$
lessons	$5\frac{1}{2}$
playing	$2\frac{1}{8}$
homework	$\frac{3}{8}$

Add Fractions with Related Denominators (Proper, Improper and Mixed Number Fractions)

Objective: I can add proper, improper and mixed number fractions with the same denominator and denominators that are multiples of the same number.

Quick Search Ref: 10103

Level 1: Understanding: Add fractions with image support.

✿ **Required:** 7/10   ✿ **Pupil Navigation:** on   ✿ **Randomised:** off

---

1. What is 2/3 + 1/6?

a b c *Put a forward slash (/) between the numerator and denominator. For example, one-half is 1/2.*

■ 15/18 ■ 5/6 ■ 3/9 ■ 5/12

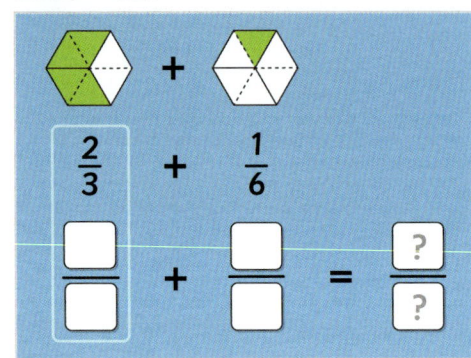

2. Calculate 7/5 + 13/10.

a b c *Give your answer as an improper fraction. For example, 3/2.*

■ 27/10 ■ 27/20 ■ 20/15

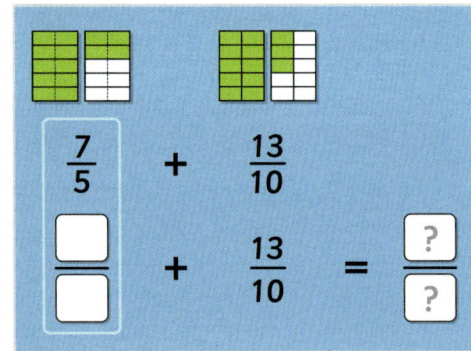

3. What is the total of 1 1/6 + 1 2/6?

a b c *Give your answer as a mixed number fraction. For example, five and one-half is 5 1/2.*

■ 2 1/2 ■ 3/6 ■ 2 3/6 ■ 1/2 ■ 15/6

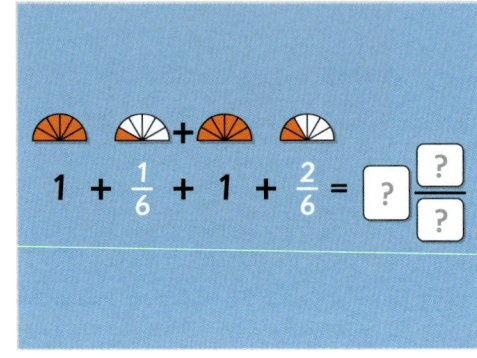

4. What is 1 5/8 + 3/4?

a b c *Give your answer as a mixed number fraction. For example, five and one-half is 5 1/2.*

■ 2 12/32 ■ 1 3/8 ■ 2 6/16 ■ 2 3/8 ■ 1 11/8
■ 38/16 ■ 76/32 ■ 19/8

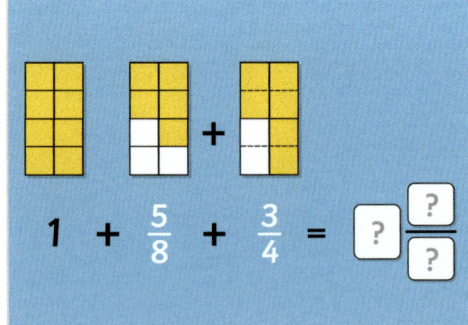

Level 1 continued

5. What is 3/16 + 1/4?

a b c *Put a forward slash (/) between the numerator and denominator. For example, one-half is 1/2.*

■ 28/64 ■ 14/32 ■ 7/16 ■ 7/32 ■ 4/20

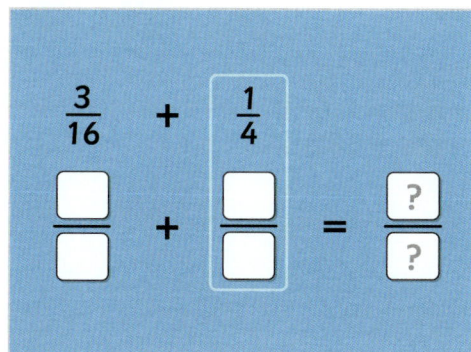

6. What is 11/8 + 9/4?

a b c *Give your answer as an improper fraction. For example, 3/2.*

■ 29/8 ■ 20/12 ■ 29/16

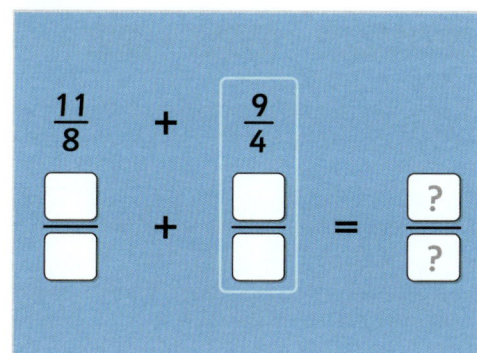

7. What is 2 5/8 + 1 7/8?

a b c *Give your answer as a mixed number fraction. For example, five and one-half is 5 1/2.*

■ 36/8 ■ 3 12/8 ■ 4 2/4 ■ 1 2/4 ■ 4 1/2
■ 4 4/8 ■ 3 6/4 ■ 1 1/2 ■ 1 4/8 ■ 3 3/2

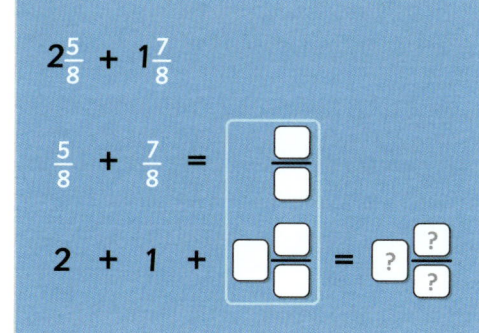

8. What is the total of 19/8 + 7/2?

a b c *Give your answer as an improper fraction. For example, 3/2.*

■ 47/8 ■ 26/10 ■ 47/16

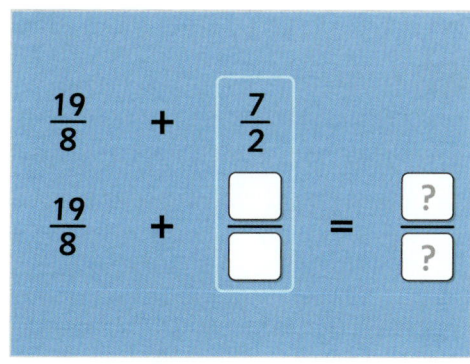

9. Calculate 1 1/3 + 1/6.

a b c *Give your answer as a mixed number fraction. For example, five and one-half is 5 1/2.*

■ 9/6 ■ 1 1/2 ■ 1 3/6 ■ 1 9/18 ■ 1/2 ■ 3/6
■ 27/18

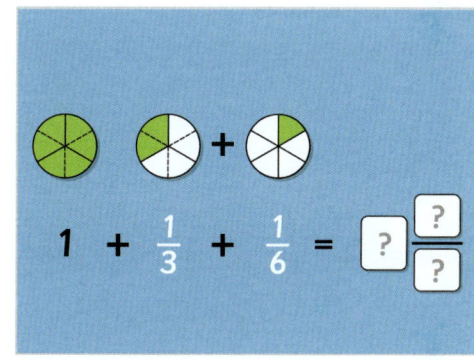

10. What is 1/12 + 5/6?

a b c *Put a forward slash (/) between the numerator and denominator. For example, one-half is 1/2.*

■ 66/72 ■ 11/12 ■ 6/18 ■ 11/24

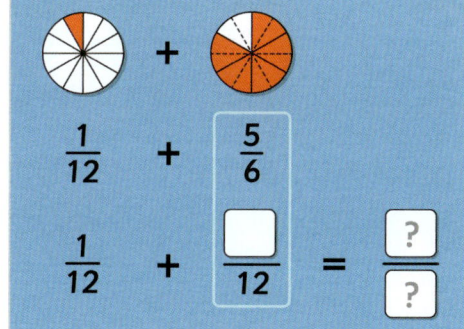

Level 2: Fluency: Add fractions including adding more than two fractions and 1-step word problems.

✱ **Required:** 7/10   ✱ **Pupil Navigation:** on
✱ **Randomised:** off

11. What is 1/6 + 1/12 + 1/24?

a
b
c
■ **7/24** ■ **7/42**

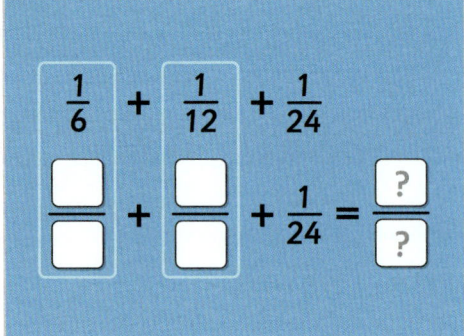

12. Find the total of 7/3 + 1/6 + 7/6.

a
b
c
Give your answer as an improper fraction. For example, 3/2.

■ **22/6** ■ **22/18**

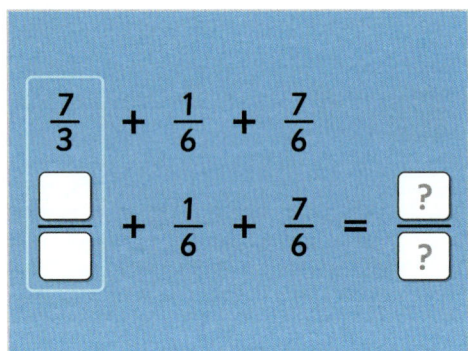

13. What is 1 1/4 + 2 3/4 + 2 1/8?

a
b
c
Give your answer as a mixed number fraction. For example, five and one-half is 5 1/2.

■ **6 1/8** ■ **5 9/8** ■ **49/8**

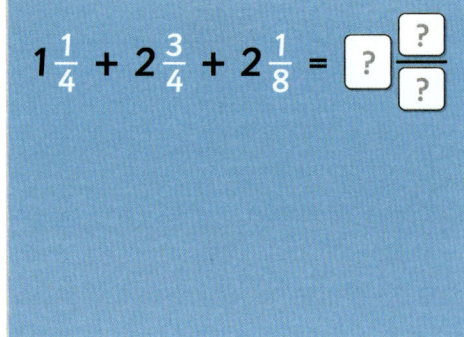

14. Two children have identical bottles of juice. Oliver drinks half of his bottle and Lily drinks three-quarters of her bottle of juice. The children drink _____ bottles of juice in total.

a
b
c
Enter the missing improper fraction. Don't include the units in your answer.

■ **5/4**

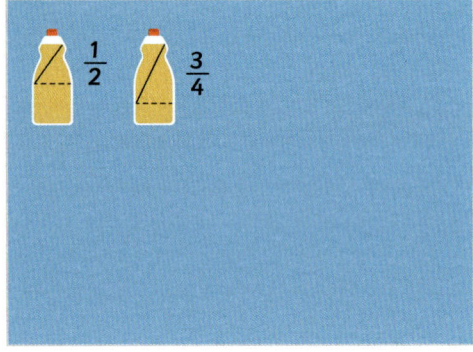

15. The total mass of 12/25 kilograms (kg) and 8/5 kg is _____ kg.

a
b
c
Enter the missing improper fraction. Don't include the units in your answer.

■ **52/25 kilograms** ■ **52/25 kg** ■ **52/25** ■ **2.08**
■ **2.08 kg** ■ **2.08 kilograms**

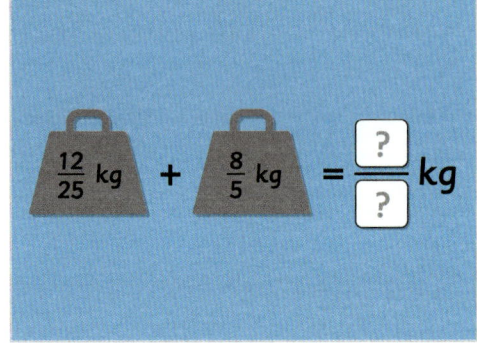

16. Jamie has a piece of wood that is 7/50 metres (m) long. Mia's piece is 3/25 m long and Alfie has a piece of wood measuring 97/100 m. If the children put their pieces of wood end-to-end, the total length will be _____ metres.

Enter the missing improper fraction. Don't include the units in your answer.

- ■ 123/100 metres ■ 123/100 ■ 1.23 m
- ■ 123/100 m ■ 1.23 metres ■ 1.23

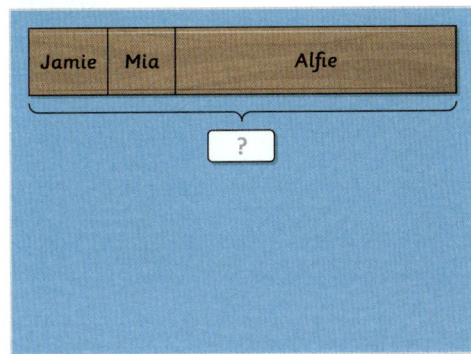

17. At an animal rescue centre, the staff use 1 1/3 bags of hay for the donkey, 1/6 of a bag of hay for the hamsters and 1 5/12 bags of hay for the rabbits. How much hay do they use altogether?

Give your answer as a mixed number fraction.

- ■ 2 11/12 ■ 35/12

18. Emma pours 1 2/10 litres (l) of water into a jug already containing 1 3/50 litres of water. How many litres of water are there in the jug in total?

Give your answer as a mixed number fraction. Don't include the units in your answer.

- ■ 2 13/50 l ■ 2 13/50 ■ 2 13/50 litres ■ 2.26 l
- ■ 2.26 ■ 2.26 litres

19. What is 9/8 + 11/4 + 5/2?
Give your answer as an improper fraction. For example, 3/2.

- ■ 51/8 ■ 51/24

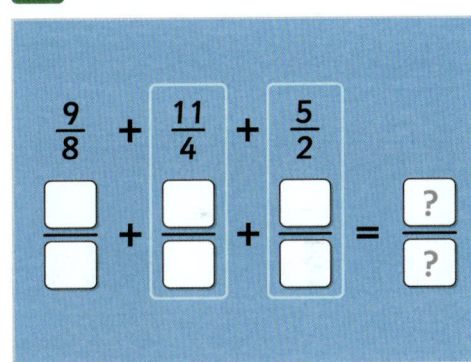

20. Calculate 2/5 + 12/15.
Give your answer as improper fraction. For example 3/2.

- ■ 6/5 ■ 18/15 ■ 90/75

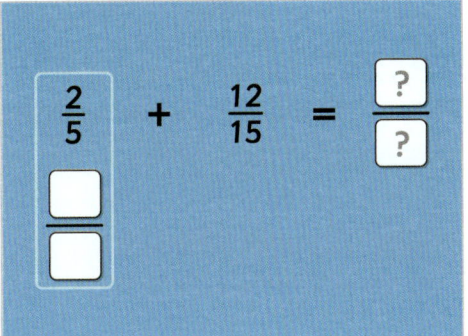

Level 3: Reasoning: Reason about adding fractions.

✱ **Required:** 5/5 ✱ **Pupil Navigation:** on

✱ **Randomised:** off

21. The large square in the image is divided into
a
b
c
2 quarters and 4 eighths. One of the quarters and one of the eighths are shaded.
Explain how you would work out the fraction of the large square that **is not shaded**.

- Open question, no set answer

22. Becky calculates 11/6 + 4/3. She draws a
a
b
c
diagram to show her calculation.
Explain what Becky has done wrong.

- Open question, no set answer

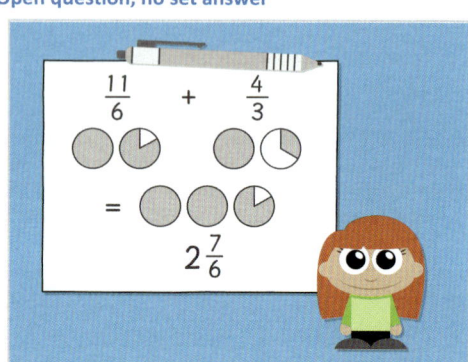

23. Which **two** fractions when added together
give a total **greater than** 40/8?

▪ 7/4 ▪ 5/2 ▪ 23/8 ▪ 11/8

2/4

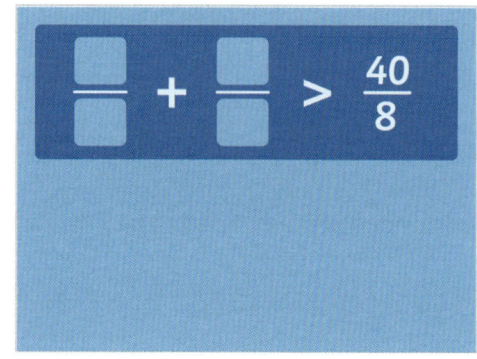

24. Two children are explaining how to calculate
a
b
c
1 4/5 + 1 7/10.
Annie says that you need to partition 1 4/5 and 1 7/10 before adding. Ethan says that you need to convert the mixed number fractions to improper fractions before adding.
Are both of these methods correct? Prove your answer.

- Open question, no set answer

25. What is the same and what is different about
a
b
c
these two calculations?

- Open question, no set answer

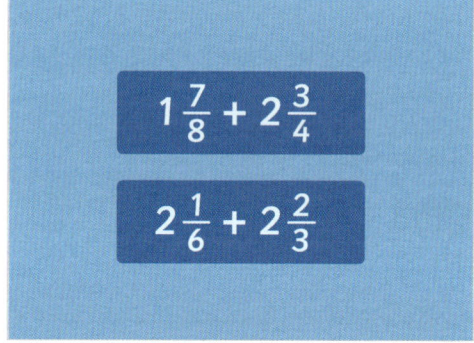

Level 4: Problem solving with greater depth: Solve multi-step problems involving adding fractions.

✳ **Required:** 5/5 ✳ **Pupil Navigation:** on
✳ **Randomised:** off

26. Bobby uses his grandma's recipe for chocolate chip biscuits. The total weight of ingredients that Bobby uses is _____ lbs.
Enter the missing fraction. Don't include the units in your answer.

■ 15/16 ■ 15/16 lbs ■ 15/16 pounds

27. Add two side-by-side fractions to find the answer to the box above.
What fraction goes in box A?
Give your answer as an improper fraction.
For example, 3/2.

■ 137/28 ■ 37/14 ■ 63/28

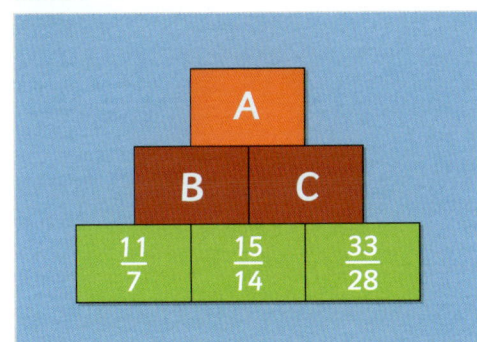

28. Dan and his friends take part in a relay race.

April runs 1 2/5 kilometres (km).
Charlie runs 1 7/20 km.
Dan runs 1 7/10 km further than April and Charlie run in total.

All of the children run a total of _____ km.
Enter the missing mixed number fraction.
Don't include the units in your answer.

■ 7 1/5 ■ 7 2/10 km ■ 7 1/5 kilometres ■ 7 1/5 km
■ 7 4/20 ■ 7 2/10 kilometres ■ 7 2/10 ■ 7 4/20 km
■ 7 4/20 kilometres

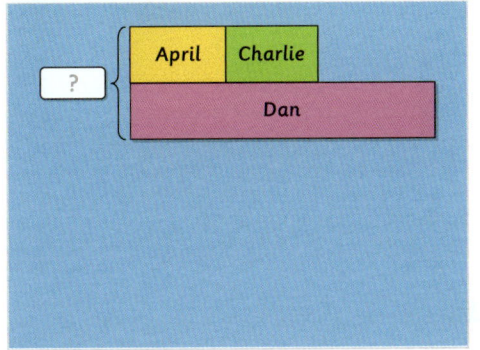

29. Each week, Rory needs to spend 1 ½ hours doing chores to earn his pocket money. So far this week, he has spent 35 minutes tidying his room and 5/6 of an hour walking the dog. In order to get his pocket money, Rory still needs to spend _____ of an hour doing chores.
Enter the missing fraction. Don't include the units in your answer.

■ 1/12 ■ 5/60

Level 4 *continued*

30. Jane pours 12/5 litres of paint into a 5-litre
bucket. If she pours another 41/15 litres into
the bucket, the paint will overflow by _____
of a litre.

*Enter the missing fraction. Don't include the
units in your answer.*

■ 2/15 l ■ 2/15 ■ 2/15 litres

Add Improper Fractions with Related Denominators

Objective: I can add improper fractions with the same denominator and denominators that are multiples of the same number.

Quick Search Ref: 11476

Level 1: Understanding: Add improper fractions with image support.

✻ **Required:** 7/10   ✻ **Pupil Navigation:** on   ✻ **Randomised:** off

1. To calculate 7/5 + 2/15 you need to make the denominators the same by finding the lowest number that is a multiple of both denominators.
What is the lowest number that is a multiple of both 5 and 15?

■ 10   ■ **15**   ■ 30

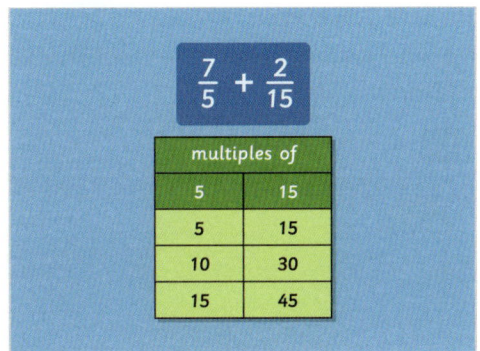

2. What is 7/5 + 2/15?
Give your answer as an improper fraction. For example, 3/2.

■ 23/30   ■ **23/15**   ■ 9/20

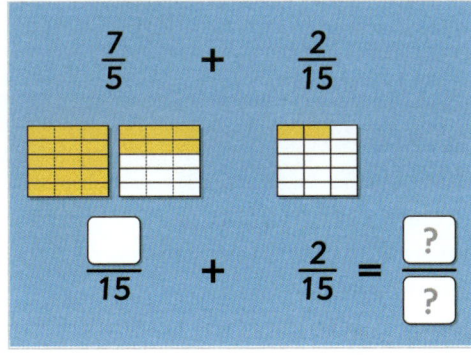

3. Find the total of 7/8 and 5/4.
Give your answer as an improper fraction. For example, 3/2.

■ **17/8**   ■ 12/12   ■ 17/16

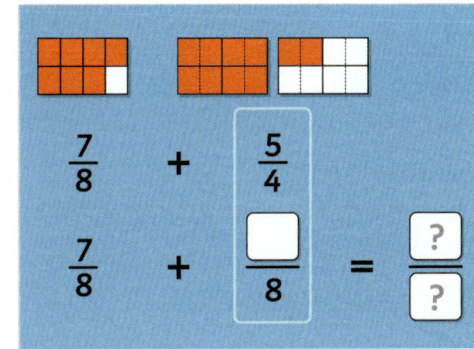

4. Calculate 7/5 + 13/10.
Give your answer as an improper fraction. For example, 3/2.

■ **27/10**   ■ 20/15   ■ 27/20

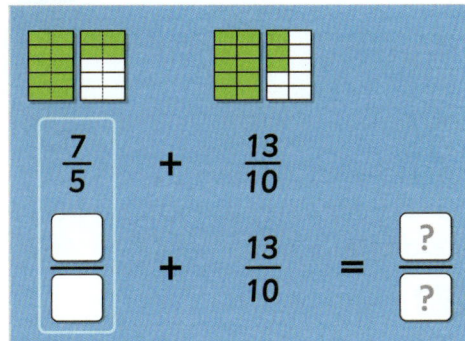

5. What is 11/8 + 9/4?
Give your answer as an improper fraction. For example, 3/2.

■ **29/8**   ■ 29/16   ■ 20/12

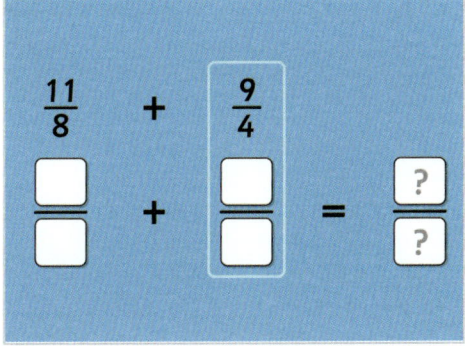

Level 1 *continued*

6. What is the total of 7/3 + 11/9?
   a
   b Give your answer as an improper fraction.
   c For example, 3/2.

   ■ **32/9** ■ 18/12 ■ 32/18

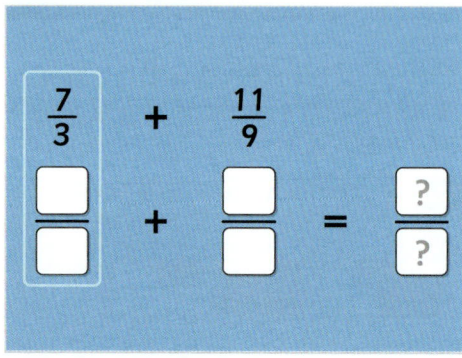

7. What is 14/2 + 7/4?
   a
   b Give your answer as an improper fraction.
   c For example, 3/2.

   ■ **35/4** ■ 21/6 ■ 35/8

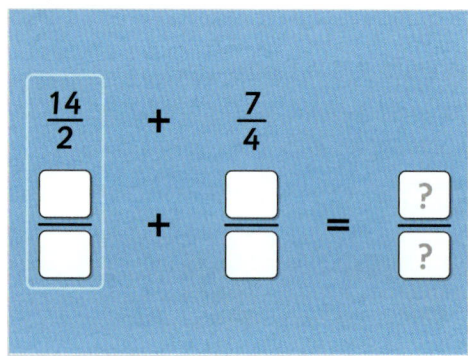

8. What is the total of 19/8 + 7/2?
   a
   b Give your answer as an improper fraction.
   c For example, 3/2.

   ■ **47/8** ■ 47/16 ■ 26/10

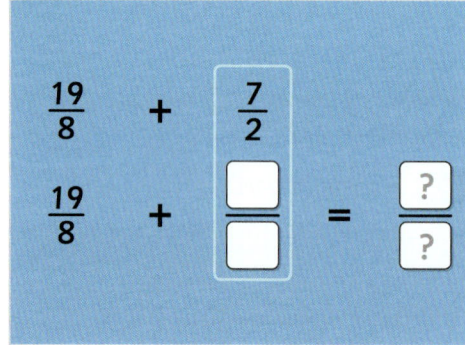

9. What is the lowest number that is a multiple
   1 of both denominators that you would use to
   2 calculate 9/4 + 19/16?
   3

   ■ **16** ■ 8 ■ 32

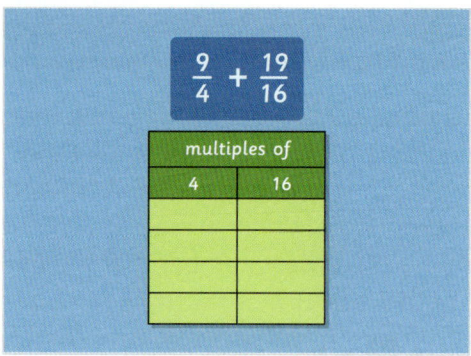

10. Calculate 7/6 + 8/3.
    a
    b Give your answer as an improper fraction.
    c For example, 3/2.

    ■ **23/6** ■ 23/12 ■ 15/9

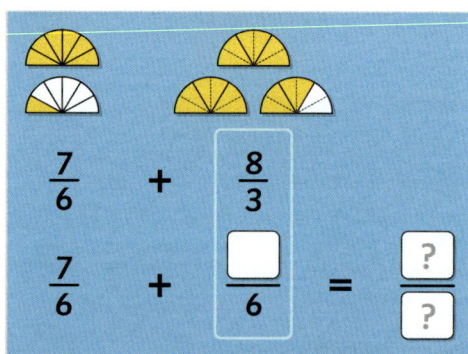

Level 2: Fluency: Add improper fractions including adding three fractions. Solve one-step word problems.

✿ **Required:** 7/10   ✿ **Pupil Navigation:** on
✿ **Randomised:** off

11. Calculate 11/7 + 20/14.
    a
    b Give your answer as an improper fraction.
    c For example, 3/2.

    ■ **42/14** ■ 3

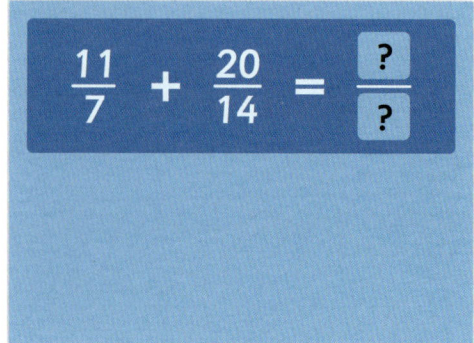

12. Find the total of 7/3 + 1/6 + 7/6.

a b c *Give your answer as an improper fraction. For example, 3/2.*

- **22/6** ■ **22/18**

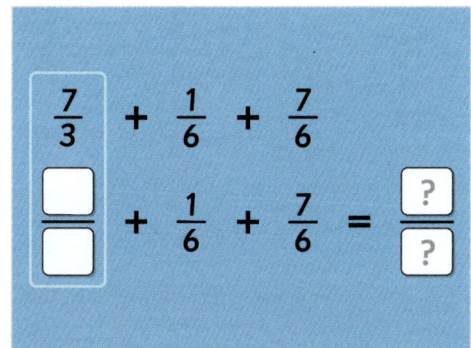

13. What is 9/8 + 11/4 + 5/2?

a b c *Give your answer as an improper fraction. For example, 3/2.*

- **51/8** ■ **51/24**

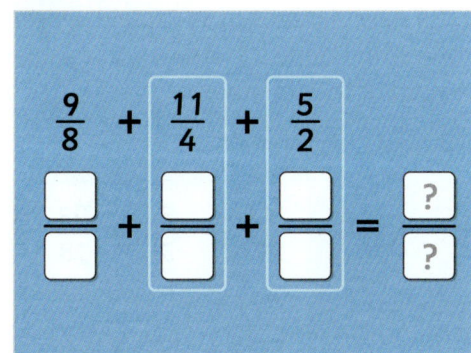

14. The total mass of 12/25 kilograms (kg) and

a b c 8/5 kg is _____ kg.

Enter the missing improper fraction. Don't include the units in your answer.

- **52/25 kilograms** ■ **52/25 kg** ■ **52/25** ■ 2.08 kg
- 2.08 ■ 2.08 kilograms

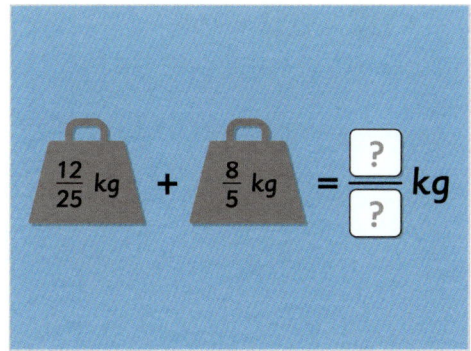

15. A jug contains 55/50 litres of water. If

a b c another 12/10 litres of water are poured into the jug, the total amount of water in the jug will be _____ litres.

Enter the missing improper fraction. Don't include the units in your answer.

- **115/50 litres** ■ **115/50 l** ■ **115/50** ■ 2.3 l ■ 2.3
- 2.3 litres

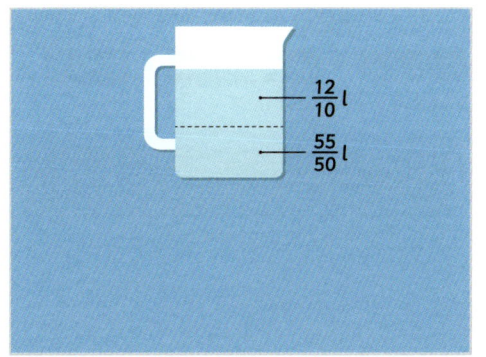

16. At an animal rescue centre, the staff use 3/2

a b c bags of hay for the donkey, 1/4 of a bag of hay for the hamsters and 9/8 bags of hay for the rabbits. How much hay do they use altogether?

Give your answer as an improper fraction. For example, 3/2.

- **23/8**

Level 2 continued

17. Jamie has a piece of wood that is 7/5 metres
a (m) long. Mia's piece is 12/10 m long and
b
c Alfie has a piece of wood measuring 97/100
m. If the children put their pieces of wood
end-to-end, the total length will be ____
metres.
*Enter the missing improper fraction. Don't
include the units in your answer.*

- ■ **357/100 m** ■ **357/100 metres** ■ **357/100** ■ 3.57
- ■ 3.57 metres ■ 3.57 m

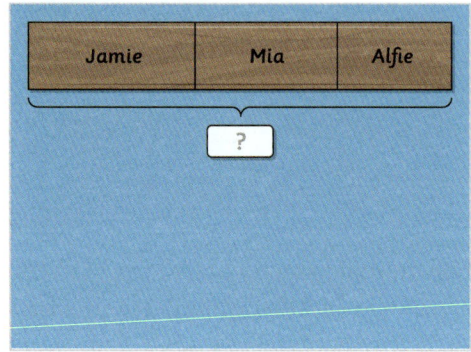

18. What is 9/4 + 5/8 + 17/16?
a *Give your answer as an improper fraction.*
b
c *For example, 3/2.*

- ■ **63/16**

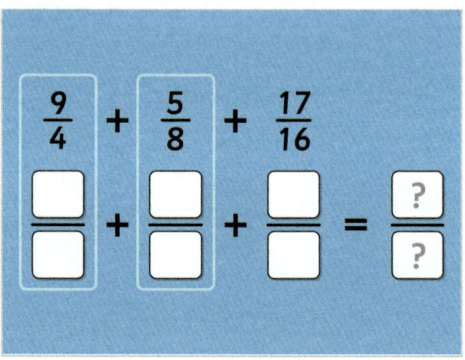

19. What is the lowest number that is a multiple
1 of all the denominators that you would use
2
3 to calculate 8/7 + 1/28 + 17/14?

- ■ **28** ■ 14 ■ 56

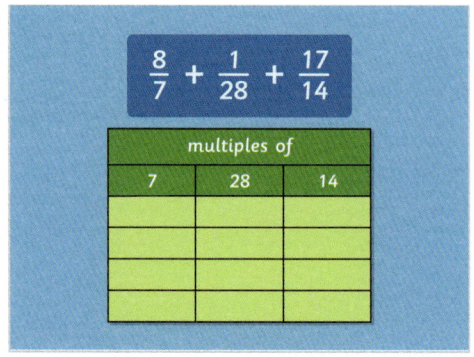

20. What is the total of 19/10 and 6/5?
a *Give your answer as an improper fraction.*
b
c ■ **31/10**

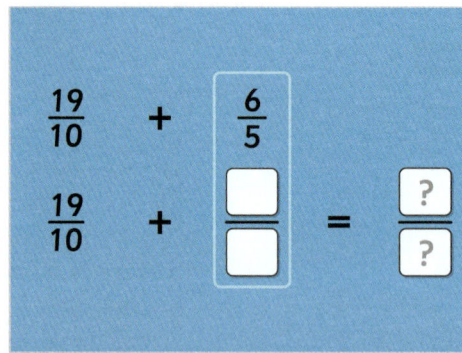

Level 3: Reasoning: Reason about adding improper
fractions with related denominators.

❋ **Required:** 5/5   ❋ **Pupil Navigation:** on
❋ **Randomised:** off

21. Poppy says that to calculate 11/9 + 8/3 +
a 23/18 she will convert all the fractions to
b
c have a denominator of 9. Explain what is
wrong with Poppy's method and calculate
the total of the fractions.

- Open question, no set answer

22. Which **two** fractions when added together
give a total **greater than** 40/8?

- ■ 7/4 ■ **5/2** ■ **23/8** ■ 11/8

2/4

Level 3 continued

23. Becky calculates 11/6 + 4/3. She draws a
diagram to show her calculation.
Explain what Becky has done wrong.

a
b
c

- Open question, no set answer

24. Abbie says, "When you add two improper
fractions, your answer will always be greater
than 2."
Is Abbie correct? Explain your answer.

a
b
c

- Open question, no set answer

25. Oscar is finding the total of 17/12 and 19/6.
He converts both fractions to have a
denominator of 72 before adding them.
Explain how Oscar can make his calculation
easier.

a
b
c

Level 4: Problem Solving with greater depth: Solve
problems involving the addition of improper
fractions with related denominators.

✷ **Required:** 5/5 ✷ **Pupil Navigation:** on
✷ **Randomised:** off

26. Hannah records how long she spends on
different activities during one day. The total
amount of time she spends eating and
playing is _____ of an hour longer than she
spends doing homework.
*Enter the missing improper fraction. Don't
include the units in your answer.*

a
b
c

■ 22/8 ■ 27/8

activity	time spent (hours)
sleeping	$11\frac{1}{2}$
eating	$1\frac{1}{4}$
lessons	$5\frac{1}{2}$
playing	$2\frac{1}{8}$
homework	$\frac{5}{8}$

Time Hanna spends on different activities each day

27. Add two side-by-side fractions to find the
answer to the box above.
What fraction goes in box A?
*Give your answer as an improper fraction.
For example, 3/2.*

a
b
c

■ 137/28 ■ 63/28 ■ 37/14

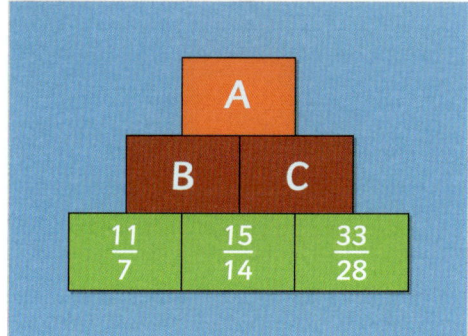

Level 4 *continued*

28. Jane pours 12/5 litres of paint into a 5-litre bucket. If she pours another 41/15 litres into the bucket, the paint will overflow by _____ of a litre.

Enter the missing fraction. Don't include the units in your answer.

- 2/15 l ▪ 2/15 ▪ 2/15 litres

29. Dan and his friends complete a 5 kilometre (km) relay race.

April runs the first 7/5 km.

Charlie runs the next 23/20 km.

How many kilometres does Dan run to complete the 5 km course?

Give your answer as an improper fraction. Don't include the units in your answer.

- 49/20 ▪ 51/20

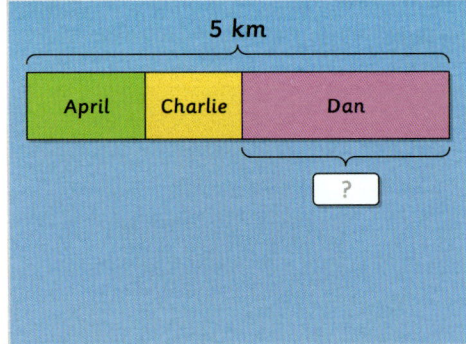

30. An electrician needs to buy some electrical wire to complete a kitchen project. He needs 9/5 metres (m) of wire for the dishwasher, 13/25 m of wire for the cooker and 223/100 m of wire for the lighting. Electrical wire is sold in whole metre lengths and costs 50 pence for each metre.

How much does the electrician pay for the wire he needs?

Include the £ sign in your answer.

- £2.50p ▪ £2.50 ▪ 2.50 ▪ 250 pence

Add Mixed Number Fractions With Related Denominators

Objective:	I can add mixed number fractions with the same denominator and denominators that are multiples of the same number.

Quick Search Ref:	11490

Level 1: Understanding: Add a proper fraction to a mixed number fraction with a related denominator. Add two mixed number fractions with the same denominator.

✿ **Required:** 7/10   ✿ **Pupil Navigation:** on   ✿ **Randomised:** off

1. What is 1 1/3 + 1/3?

a b c *Give your answer as a mixed number fraction. For example, five and one-half is 5 1/2.*

- ▪ **1 2/3** ▪ 5/3 ▪ 2/3

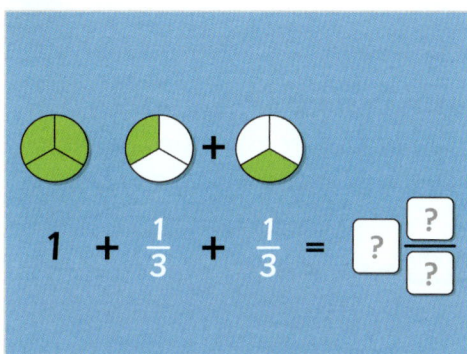

2. 1 3/8 and 5/8 = ___ wholes.

a b c
- ▪ **2** ▪ 1 8/8 ▪ 1

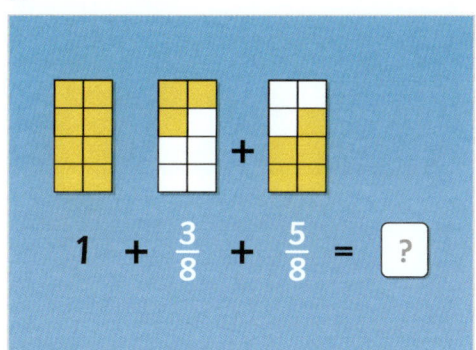

3. What is the total of 1 3/4 + 3/4?

a b c *Give your answer as a mixed number fraction. For example, five and one-half is 5 1/2.*

- ▪ 1 2/4 ▪ **2 1/2** ▪ **2 2/4** ▪ 1 6/4 ▪ 10/4

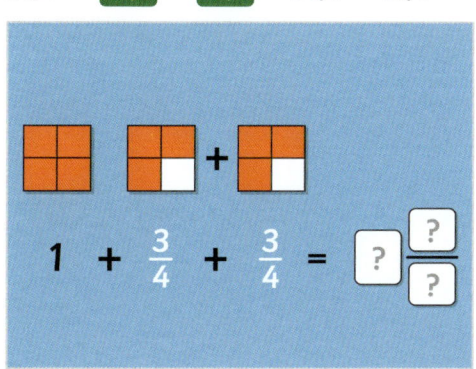

4. Calculate 1 1/3 + 1/6.

a b c *Give your answer as a mixed number fraction. For example, five and one-half is 5 1/2.*

- ▪ **1 1/2** ▪ **1 3/6** ▪ **1 9/18** ▪ 3/6 ▪ 9/6 ▪ 1/2
- ▪ 27/18

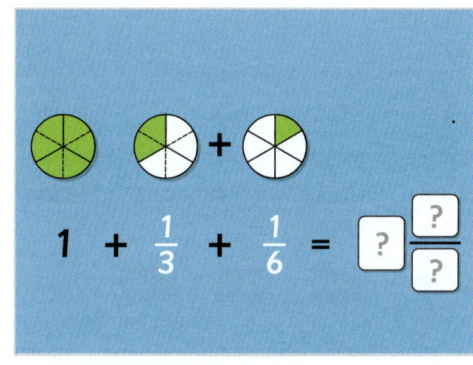

5. What is 1 5/8 + 3/4?

a b c *Give your answer as a mixed number fraction. For example, five and one-half is 5 1/2.*

- ▪ **2 12/32** ▪ **2 6/16** ▪ **2 3/8** ▪ 38/16 ▪ 19/8
- ▪ 1 3/8 ▪ 1 11/8 ▪ 76/32

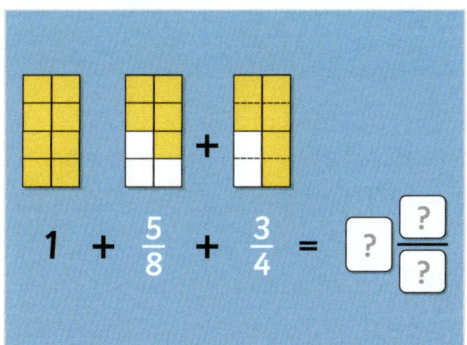

Level 1 *continued*

6. What is the total of 1 1/6 + 1 2/6?

a b c *Give your answer as a mixed number fraction. For example, five and one-half is 5 1/2.*

■ 2 1/2 ■ 2 3/6 ■ 15/6 ■ 3/6 ■ 1/2

7. What is 2 5/8 + 1 7/8?

a b c *Give your answer as a mixed number fraction. For example, five and one-half is 5 1/2.*

■ 3 12/8 ■ 4 2/4 ■ 4 1/2 ■ 4 4/8 ■ 1 1/2
■ 3 3/2 ■ 36/8 ■ 1 2/4 ■ 3 6/4 ■ 1 4/8

$$2\frac{5}{8} + 1\frac{7}{8}$$

$$\frac{5}{8} + \frac{7}{8} = \boxed{}$$

$$2 + 1 + \boxed{} = \boxed{?}\,\frac{?}{?}$$

8. What is the total of 1 3/7 + 1 5/7?

a b c *Give your answer as a mixed number fraction. For example, five and one-half is 5 1/2.*

■ 3 1/7 ■ 22/7 ■ 2 8/7 ■ 1 1/7

$$1 + \frac{3}{7} + 1 + \frac{5}{7} = \boxed{?}\,\frac{?}{?}$$

9. Calculate 1 1/2 + 3/4.

a b c *Give your answer as a mixed number fraction. For example, five and one-half is 5 1/2.*

■ 2 1/4 ■ 1 1/4 ■ 2 2/8 ■ 9/4 ■ 1 5/4 ■ 1 2/8

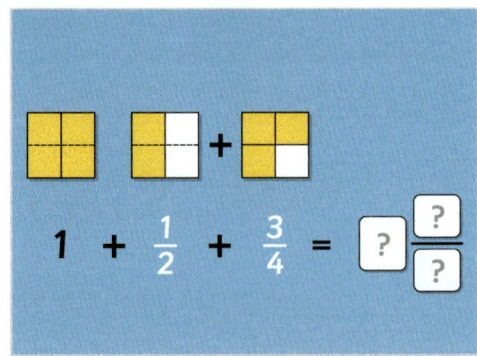

$$1 + \frac{1}{2} + \frac{3}{4} = \boxed{?}\,\frac{?}{?}$$

10. What is the total of 1 3/5 + 1/5?

a b c *Give your answer as a mixed number fraction. For example, five and one-half is 5 1/2.*

■ 1 4/5 ■ 9/5 ■ 4/5

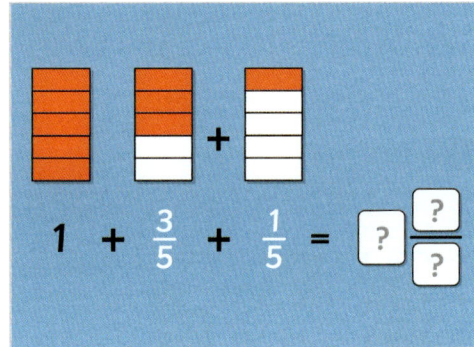

$$1 + \frac{3}{5} + \frac{1}{5} = \boxed{?}\,\frac{?}{?}$$

Level 2: Fluency: Add mixed number fractions with related denominators, including adding three fractions. Solve one-step word problems.

❋ **Required:** 7/10   ❋ **Pupil Navigation:** on
❋ **Randomised:** off

11. Find the total of 1 1/6 and 1 1/3.

a b c *Give your answer as a mixed number fraction. For example, five and one-half is 5 1/2.*

■ 2 3/6 ■ 2 9/18 ■ 2 1/2 ■ 15/6

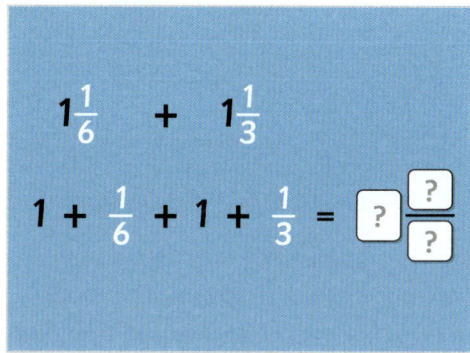

12. What is 1 3/5 + 2 7/15?

a b c *Give your answer as a mixed number fraction. For example, five and one-half is 5 1/2.*

■ 3 16/15 ■ 4 1/15 ■ 61/15

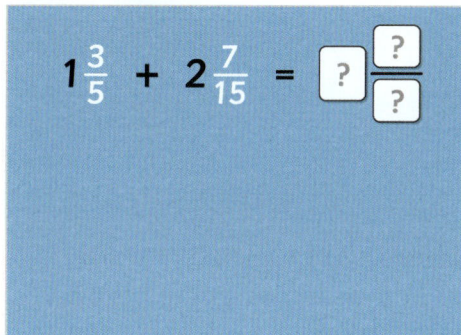

13. Find the total of 1 1/3 + 1/6 + 1 1/6.

a b c *Give your answer as a mixed number fraction. For example, five and one-half is 5 1/2.*

■ 2 2/3 ■ 2 4/6 ■ 16/6

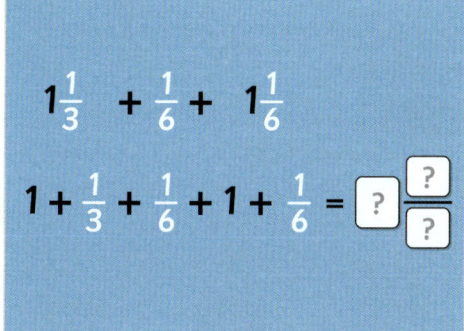

14. What is 1 1/4 + 2 3/4 + 2 1/8?

a b c *Give your answer as a mixed number fraction. For example, five and one-half is 5 1/2.*

■ 6 1/8 ■ 49/8 ■ 5 9/8

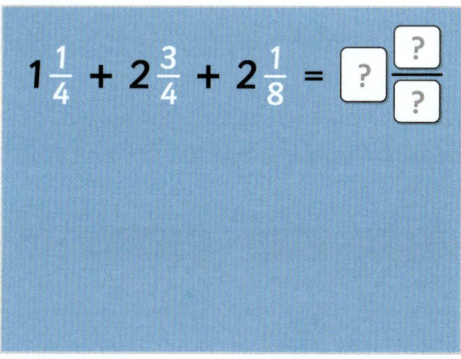

15. The total mass of 12/25 kilograms (kg) and 1 3/5 kg is _____ kg.

a b c *Enter the missing mixed number fraction. Don't include the units in your answer.*

■ 1 27/25 kilograms ■ 2 2/25 ■ 2.08 kg
■ 2 2/25 kilograms ■ 2 2/25 kg ■ 1 27/25 kg
■ 1 27/25 ■ 2.08 kilograms ■ 2.08

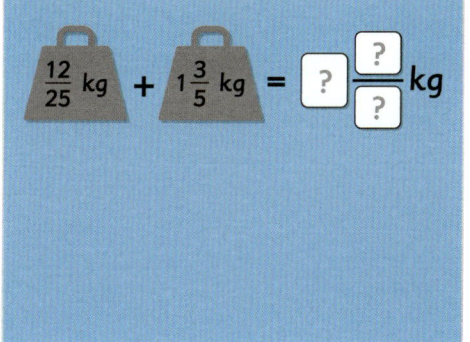

Level 2 *continued*

16. At an animal rescue centre, the staff use 1
a b c 1/3 bags of hay for the donkey, 1/6 of a bag of hay for the hamsters and 1 5/12 bags of hay for the rabbits. How much hay do they use altogether?

Give your answer as a mixed number fraction.

- ■ **2 11/12** ■ 35/12

17. Emma pours 1 2/10 litres (l) of water into a
a b c jug already containing 1 3/50 litres of water. How many litres of water are there in the jug in total?

Give your answer as a mixed number fraction. Don't include the units in your answer.

- ■ **2 13/50 l** ■ **2 13/50** ■ **2 13/50 litres** ■ 2.26
- ■ 2.26 l ■ 2.26 litres

18. The total length of 1 9/10 cm and 1 1/5 cm is
a b c _____ cm.

Enter the missing mixed number fraction fraction. Don't include the units in your answer.

- ■ **3 1/10** ■ **3 1/10 cm** ■ **3 1/10 centimetres** ■ 3.1
- ■ 31

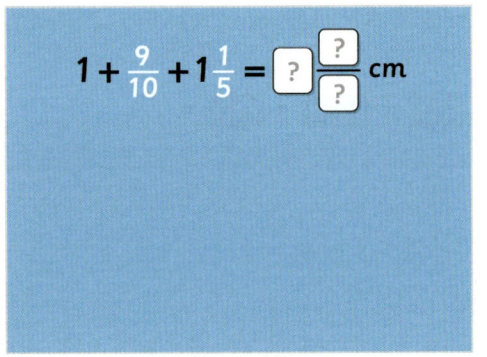

$$1 + \frac{9}{10} + 1\frac{1}{5} = \boxed{?}\,\frac{\boxed{?}}{\boxed{?}} \text{ cm}$$

19. What is 2 1/4 + 5/8 + 1 1/16?
a b c *Give your answer as a mixed number fraction. For example, five and one-half is 5 1/2.*

- ■ **3 15/16** ■ 63/16

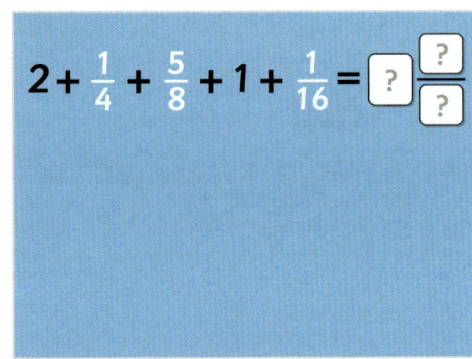

$$2 + \frac{1}{4} + \frac{5}{8} + 1 + \frac{1}{16} = \boxed{?}\,\frac{\boxed{?}}{\boxed{?}}$$

20. Calculate 1 1/8 + 1 1/4.
a b c *Give your answer as a mixed number fraction. For example, five and one-half is 5 1/2.*

- ■ **2 3/8** ■ 19/8

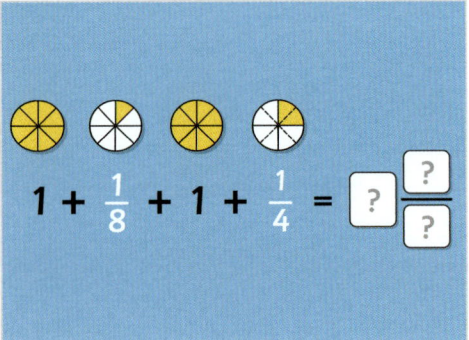

$$1 + \frac{1}{8} + 1 + \frac{1}{4} = \boxed{?}\,\frac{\boxed{?}}{\boxed{?}}$$

21. Marcus has tried to find the total of 2 8/9 and 1 2/3. Explain what Marcus has done correctly and what he needs to do to complete the calculation.

- Open question, no set answer

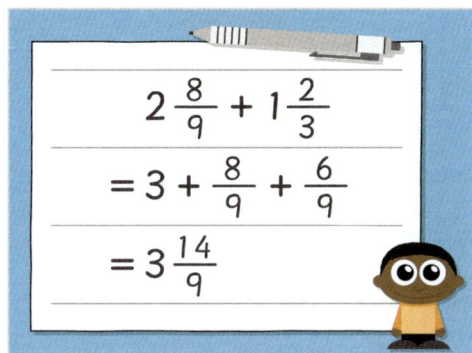

22. What is the same and what is different about these two calculations?

- Open question, no set answer

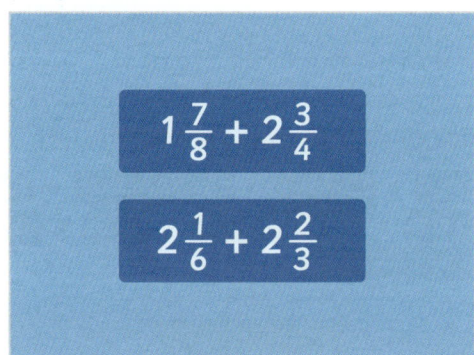

23. Select **two** mixed number fractions that add together to give a total greater than 5.

■ 2 1/16 ■ **2 1/2** ■ **2 7/8** ■ 1 3/8

2/4

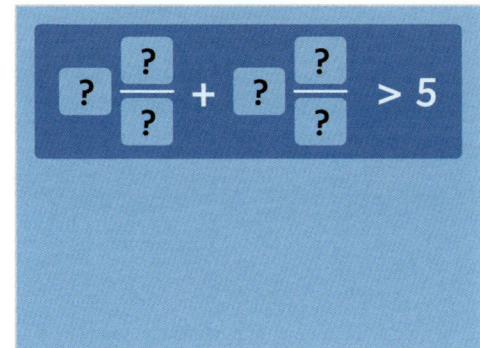

24. Abbie says, "When you add three mixed number fractions, your answer will never be a whole number." Is Abbie correct? Explain your answer and give an example.

- Open question, no set answer

25. Two children are explaining how to calculate 1 4/5 + 1 7/10.
Annie says that you need to partition 1 4/5 and 1 7/10 before adding. Ethan says that you need to convert the mixed number fractions to improper fractions before adding.
Are both of these methods correct? Prove your answer.

- Open question, no set answer

Level 4: Problem solving with greater depth: Solve problems involving adding mixed number fractions with related denominators.

� **Required:** 5/5   ✱ **Pupil Navigation:** on
✱ **Randomised:** off

26. Jane pours 2 2/5 litres of paint into a 5 litre
a
b bucket. If she pours another 2 11/15 litres
c into the bucket, the paint will overflow
by____ of a litre.
Enter the missing fraction. Don't include the units in your answer.

- ▪ 2/15 ▪ 2/15 litres ▪ 2/15 l ▪ 5 2/15 litres
- ▪ 5 2/15 l ▪ 5 2/15

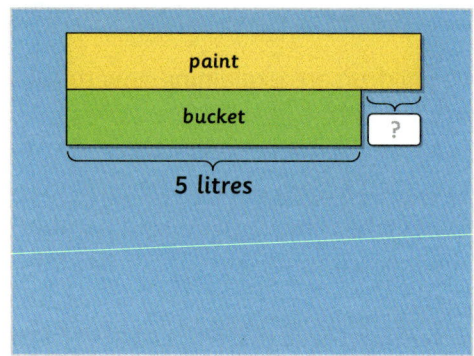

27. Hannah records how long she spends on
a
b different activities during one day. How
c many hours longer does Hannah spend
eating and playing than she spends doing
homework?
Give your answer as a whole number. Don't include the units in your answer.

- ▪ 3 ▪ 3 hours ▪ 3 hrs ▪ 3 3/8

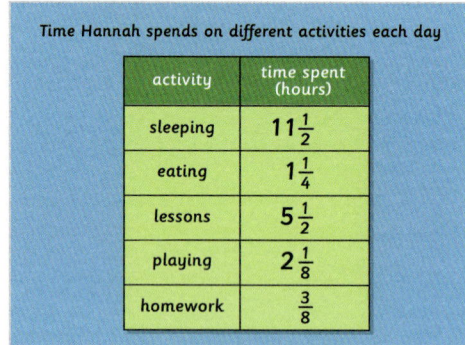

\multicolumn{2}{c}{Time Hannah spends on different activities each day}	
activity	time spent (hours)
sleeping	$11\frac{1}{2}$
eating	$1\frac{1}{4}$
lessons	$5\frac{1}{2}$
playing	$2\frac{1}{8}$
homework	$\frac{3}{8}$

28. Add two side-by-side fractions to find the
a
b answer to the box above.
c What fraction goes in box A?
Give your answer as a mixed number fraction.

- ▪ 37/14 ▪ 2 7/28 ▪ 4 25/28 ▪ 2 9/14 ▪ 137/28
- ▪ 63/28

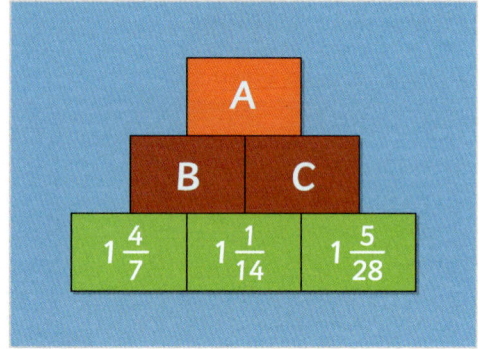

29. Dan and his friends take part in a relay race.
a
b
c April runs 1 2/5 kilometres (km).
Charlie runs 1 7/20 km.
Dan runs 1 7/10 km further than April and
Charlie run in total.

All of the children run a total of _____ km.
Enter the missing mixed number fraction. Don't include the units in your answer.

- ▪ 7 1/5 ▪ 7 2/10 km ▪ 7 1/5 kilometres ▪ 7 1/5 km
- ▪ 7 4/20 ▪ 7 2/10 kilometres ▪ 7 2/10 ▪ 7 4/20 km
- ▪ 7 4/20 kilometres

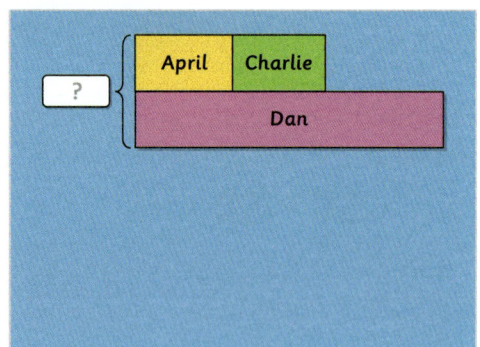

Level 4 *continued*

30. An electrician working in a school uses 1 4/5
a metres (m) of wire for the lights and he uses
b twice as much wire for the plug sockets. He
c needs 223 centimetres of wire for the
computers. If the electrician has 7 m of wire,
how many more metres of wire will he need
to complete his work?
*Give your answer as a fraction. Don't include
the units in your answer.*

- 7 63/100   - 63/100 m   - 63/100   - 63/100 metres
- 7 63/100 m   - 7 63/100 metres

Add Proper Fractions With Related Denominators

Objective: I can add fractions with the same denominator and denominators that are multiples of the same number.

Quick Search Ref: 11471

Level 1: Understanding: Add fractions with related denominators within one whole.

✱ Required: 7/10   ✱ Pupil Navigation: on   ✱ Randomised: off

1. To calculate 1/2 + 1/8 you need to make the denominators the same by finding the lowest number that is a multiple of both 2 and 8.

What is the lowest number that is a multiple of both denominators?

▪ **8** ▪ 16 ▪ 4

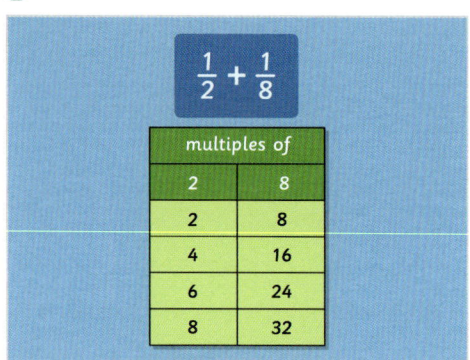

2. What is 1/2 + 1/8?

Put a forward slash (/) between the numerator and denominator. For example, one-half is 1/2.

▪ **5/8** ▪ 2/10 ▪ 5/16

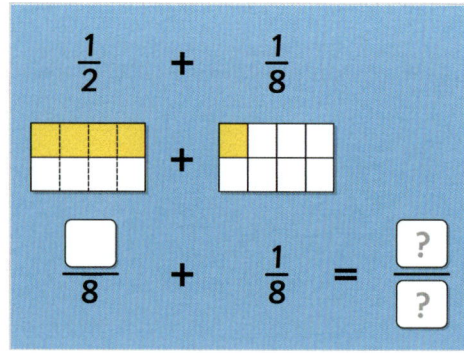

3. What is the lowest number that is a multiple of both 7 and 21 that you would use to calculate 1/7 + 1/21?

▪ **21** ▪ 14 ▪ 147 ▪ 7 ▪ 42

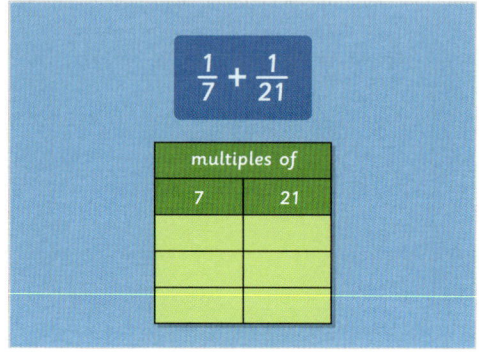

4. What is 1/12 + 5/6?

Put a forward slash (/) between the numerator and denominator. For example, one-half is 1/2.

▪ **66/72** ▪ **11/12** ▪ 11/24 ▪ 6/18

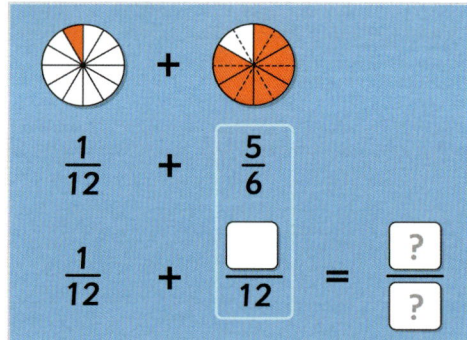

Level 1 *continued*

5. What is 2/3 + 1/6?

a
b
c
Put a forward slash (/) between the numerator and denominator. For example, one-half is 1/2.

■ 15/18 ■ 5/6 ■ 5/12 ■ 3/9

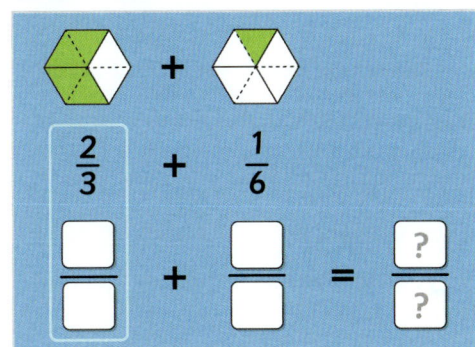

6. 1/5 + 1/20 = ___

a
b
c
Put a forward slash (/) between the numerator and denominator. For example, one-half is 1/2.

■ 1/4 ■ 25/100 ■ 2/25 ■ 5/20 ■ 5/40

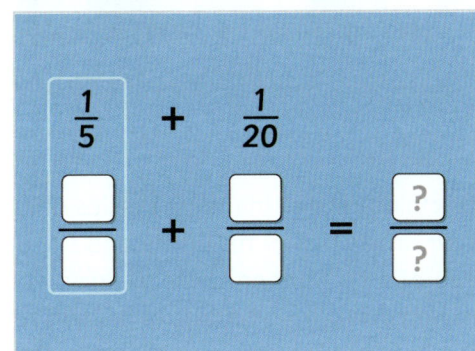

7. What is 3/16 + 1/4?

a
b
c
Put a forward slash (/) between the numerator and denominator. For example, one-half is 1/2.

■ 28/64 ■ 14/32 ■ 7/16 ■ 4/20 ■ 7/32

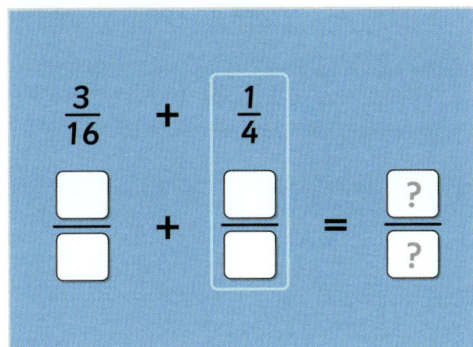

8. What is 1/10 + 1/30?

a
b
c
Put a forward slash (/) between the numerator and denominator. For example, one-half is 1/2.

■ 2/15 ■ 8/60 ■ 4/30 ■ 40/300 ■ 4/60 ■ 2/40

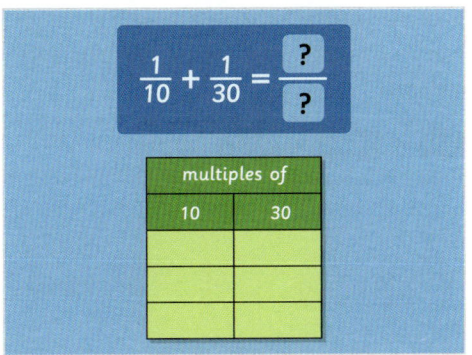

9. To calculate 1/5 + 1/25 you need to make the denominators the same by finding the lowest number that is a multiple of both 5 and 25.
What is the lowest number that is a multiple of both denominators?

1
2
3

■ 50 ■ 25 ■ 125 ■ 5

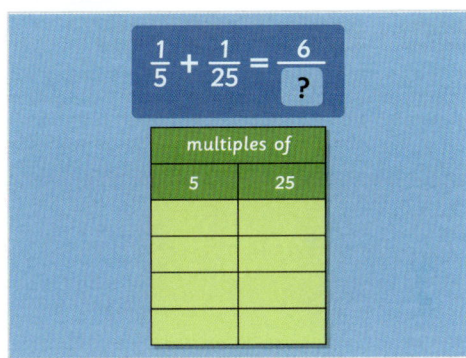

10. 3/5 + 1/10 = ___

a
b
c
Put a forward slash (/) between the numerator and denominator. For example, one-half is 1/2.

■ 35/50 ■ 4/15 ■ 7/10 ■ 7/20

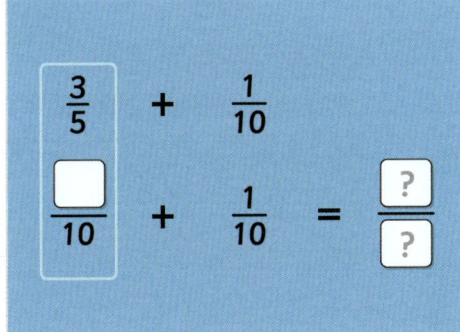

Level 2: Fluency: Add two or more proper fractions with related denominators beyond one whole.

⚙ **Required:** 7/10   ⚙ **Pupil Navigation:** on
⚙ **Randomised:** off

11. Calculate 2/5 + 12/15.
[abc] *Give your answer as improper fraction. For example 3/2.*

▪ `6/5` ▪ `18/15` ▪ `90/75`

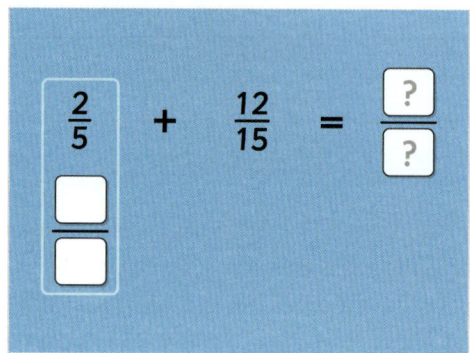

12. What is 7/24 + 5/6?
[abc] *Give your answer as improper fraction. For example 3/2.*

▪ `27/24` ▪ `162/144` ▪ `81/72` ▪ `9/8`

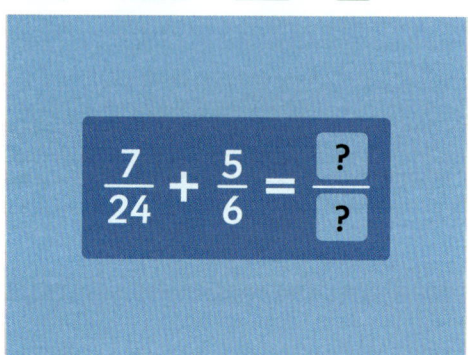

13. What is the lowest number that is a multiple of 6, 12 and 24?
[123]

▪ `24` ▪ 48 ▪ 12

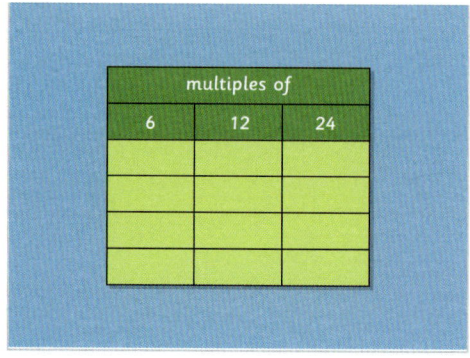

14. What is 1/6 + 1/12 + 1/24?
[abc] ▪ `7/24` ▪ 7/42

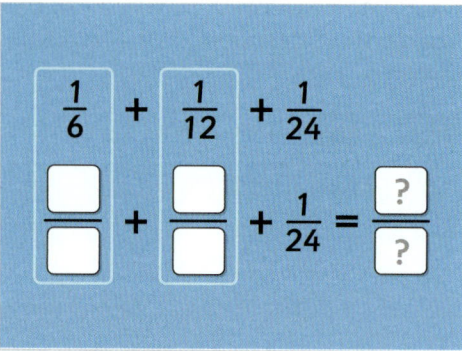

15. 2/3 + 1/6 + 7/12 = ?
[abc] *Give your answer as improper fraction. For example 3/2.*

▪ `17/12` ▪ 17/36

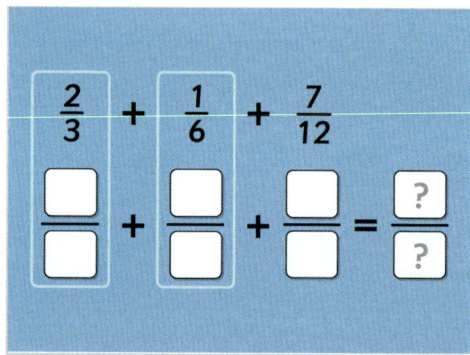

16. Two children have identical bottles of juice. Oliver drinks half of his bottle and Lily drinks three-quarters of her bottle of juice. The children drink _____ bottles of juice in total.
[abc] *Enter the missing improper fraction. Don't include the units in your answer.*

▪ `5/4`

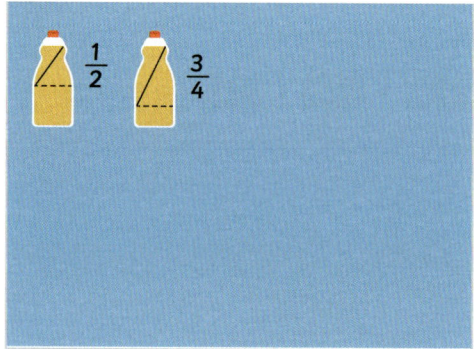

Level 2 *continued*

17. Jamie has a piece of wood that is 7/50
a metres (m) long. Mia's piece is 3/25 m long
b and Alfie has a piece of wood measuring
c 97/100 m. If the children put their pieces of
wood end-to-end, the total length will be
_____ metres.
*Enter the missing improper fraction. Don't
include the units in your answer.*

■ **123/100 metres** ■ **123/100** ■ **123/100 m** ■ **1.23**
■ **1.23 m** ■ **1.23 metres**

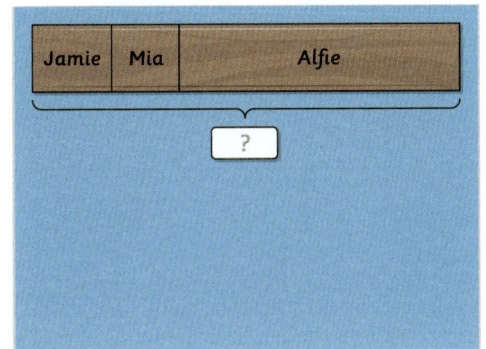

18. What is 3/4 + 1/8 + 1/16?
a *Give your answer as improper fraction. For*
b *example 3/2.*
c

■ **15/16** ■ **15/48**

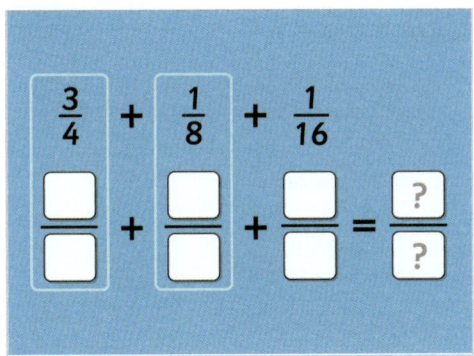

19. What is the lowest number that is a multiple
1 of all the denominators that you would use
2 to calculate 2/7 + 5/14 + 1/28?
3

■ **28** ■ **56** ■ **14**

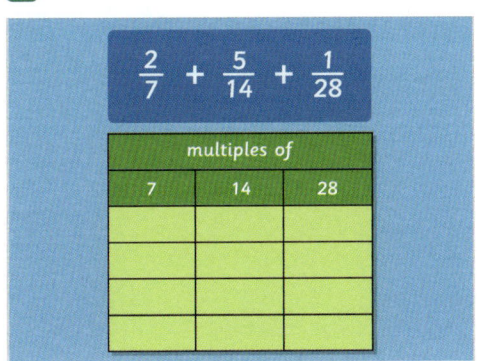

20. What is 5/7 + 9/14?
a *Give your answer as improper fraction. For*
b *example 3/2.*
c

■ **19/14**

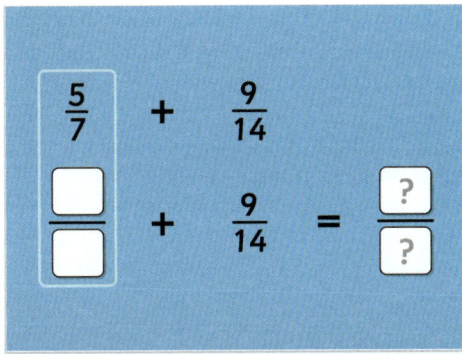

Level 3: Reasoning: Reason about adding fractions
with related denominators.

✱ **Required:** 5/5   ✱ **Pupil Navigation:** on
✱ **Randomised:** off

21. Poppy says that to calculate 4/6 + 1/2 + 1/12
a she will convert all the fractions to have a
b denominator of 6. Explain what is wrong
c with Poppy's method and calculate the total
of the fractions.

- Open question, no set answer

22. The large square in the image is divided into
a 2 quarters and 4 eighths. One of the quarters
b and one of the eighths are shaded.
c Explain how you would work out the fraction
of the large square that **is not shaded**.

- Open question, no set answer

23. Which **three** fractions give a total of 15/18 when added together?

3/5

- 1/6 ▪ **5/18** ▪ 7/9 ▪ **1/3** ▪ **2/9**

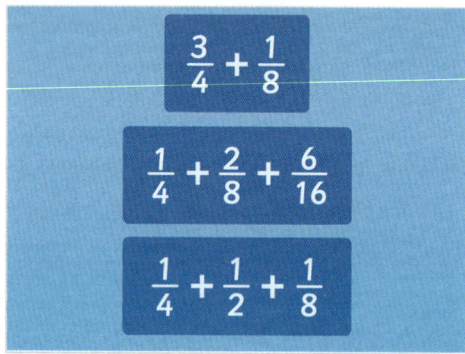

$$\frac{?}{?} + \frac{?}{?} + \frac{?}{?} = \frac{15}{18}$$

24. What is the same about these three calculations?

a b c

- Open question, no set answer

$$\frac{3}{4} + \frac{1}{8}$$

$$\frac{1}{4} + \frac{2}{8} + \frac{6}{16}$$

$$\frac{1}{4} + \frac{1}{2} + \frac{1}{8}$$

25. Oscar has added 2/7 to 5/14 to get the answer 9/28. Explain what Oscar has done wrong and calculate the correct answer.

a b c

- Open question, no set answer

$$\frac{2}{7} + \frac{5}{14}$$
$$\frac{4}{14} + \frac{5}{14} = \frac{9}{28}$$

Level 4: Problem solving with greater depth: Solve problems involving the addition of fractions with related denominators.

✹ **Required:** 5/5   ✹ **Pupil Navigation:** on
✹ **Randomised:** off

26. Bobby uses his grandma's recipe for chocolate chip biscuits. The total weight of ingredients that Bobby uses is _____ lbs. *Enter the missing fraction. Don't include the units in your answer.*

a b c

- **15/16** ▪ **15/16 lbs** ▪ **15/16 pounds**

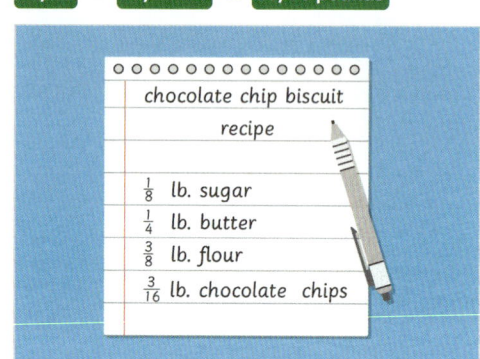

chocolate chip biscuit
recipe

$\frac{1}{8}$ lb. sugar
$\frac{1}{4}$ lb. butter
$\frac{3}{8}$ lb. flour
$\frac{3}{16}$ lb. chocolate chips

27. Miss Kay carried out a survey of weekend bed times for the children in her class. What **fraction** of the class said that they went to bed before 9.00 p.m.?

a b c

- **15/24** ▪ **5/8** ▪ 9/24 ▪ 3/8

bed time	fraction of class
before 8.00 p.m.	$\frac{3}{12}$
8.00 p.m - 8.29 p.m.	$\frac{3}{24}$
8.30 p.m - 8.59 p.m.	$\frac{1}{4}$
9.00 p.m - 9.29 p.m.	$\frac{1}{6}$
9.30 p.m - 10.00 p.m.	$\frac{1}{8}$
after 10.00 p.m.	$\frac{1}{12}$

Level 4 *continued*

28. Dom and his friends complete a 100-piece
a
b
c
jigsaw.

Aaron completes 1/5 of the jigsaw.
Beth completes 3/25 of the jigsaw.
Cara completes 25 pieces.
Dom finishes the rest of the jigsaw.

What **fraction** of the jigsaw does Dom
complete?

▪ 43/100   ▪ 57/100

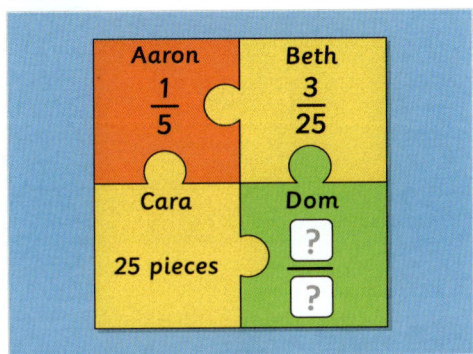

29. Add two side-by-side fractions to find the
a
b
c
answer to the box above.
What fraction goes in box A?
*Give your answer as improper fraction. For
example 3/2.*

▪ 1 3/28   ▪ 31/28   ▪ 13/28   ▪ 9/14

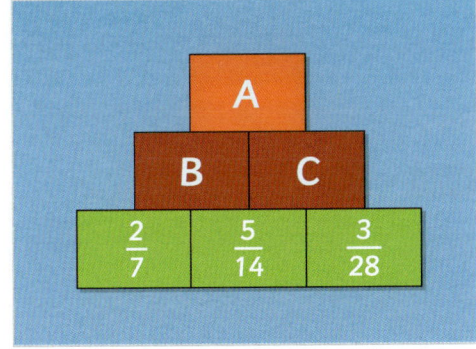

30. Each week, Rory needs to spend 1 ½ hours
a
b
c
doing chores to earn his pocket money. So
far this week, he has spent 35 minutes
tidying his room and 5/6 of an hour walking
the dog. In order to get his pocket money,
Rory still needs to spend ____ of an hour
doing chores.
*Enter the missing fraction. Don't include the
units in your answer.*

▪ 1/12   ▪ 5/60

Practise Adding Proper Fractions with Related Denominators

Objective: I can add fractions with the same denominator and denominators that are multiples of the same number.

Quick Search Ref: 10428

Level 1: I can add proper fractions with related denominators.

✹ **Required:** 17/17　　✹ **Pupil Navigation:** off　　✹ **Randomised:** on

---

1. 1/2 + 1/4

a b c ■ 3/4 ■ 6/8

2. 1/4 + 2/8

a b c ■ 1/2 ■ 4/8 ■ 16/32 ■ 8/16 ■ 2/4

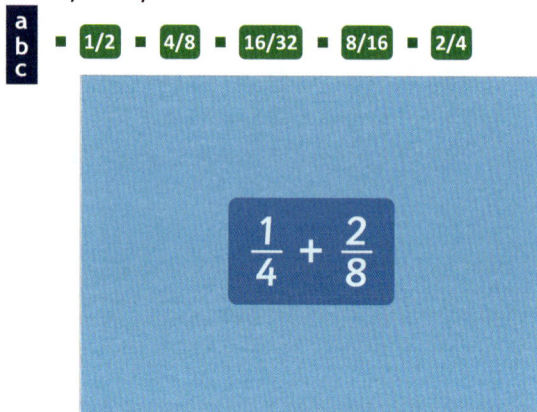

3. 3/16 + 1/8

a b c ■ 5/16 ■ 10/32 ■ 40/128 ■ 20/64

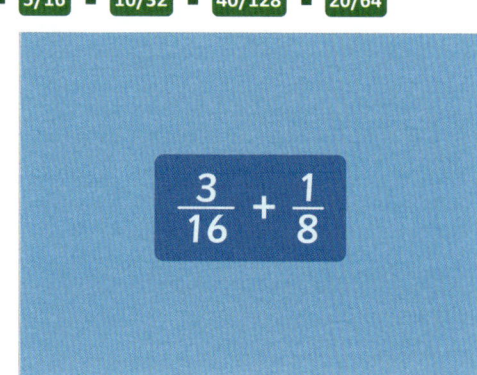

4. 1/8 + 3/4

a b c ■ 28/32 ■ 14/16 ■ 7/8

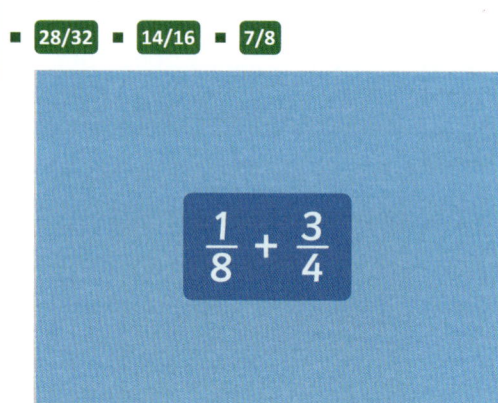

5. 1/5 + 3/15

a b c ■ 30/75 ■ 6/15 ■ 2/5

6. 3/10 + 2/5

a b c ■ 35/50 ■ 7/10

Level 1 *continued*

7. 1/9 + 2/3

a b c ▪ 7/9 ▪ 21/27

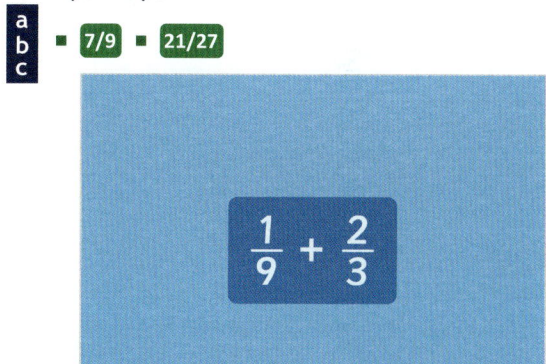

8. 6/9 + 1/36

a b c ▪ 25/36 ▪ 225/324 ▪ 75/108

9. 2/5 + 2/10

a b c ▪ 30/50 ▪ 6/10 ▪ 3/5 ▪ 15/25

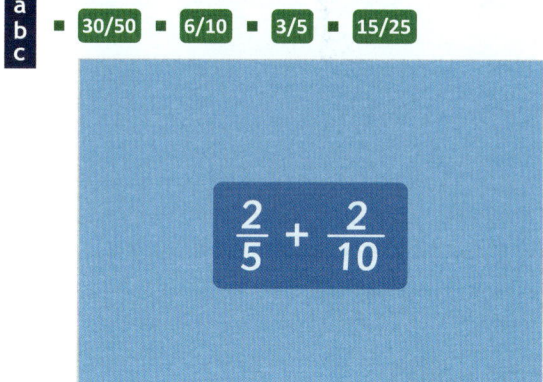

10. 1/70 + 4/7

a b c ▪ 287/490 ▪ 41/70

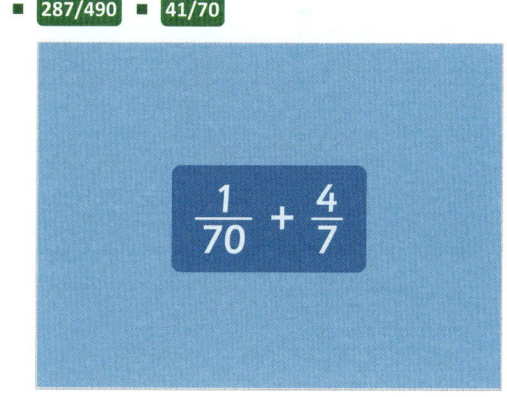

11. 13/45 + 2/9

a b c ▪ 207/405 ▪ 23/45

12. 12/15 + 1/15 + 4/30

a b c ▪ 9/9 ▪ 90/90 ▪ 450/450 ▪ 45/45 ▪ 30/30 ▪ 15/15 ▪ 225/225 ▪ 1 ▪ 1 whole

13. 5/6 + 1/18

a b c ▪ 32/36 ▪ 16/18 ▪ 8/9 ▪ 96/108

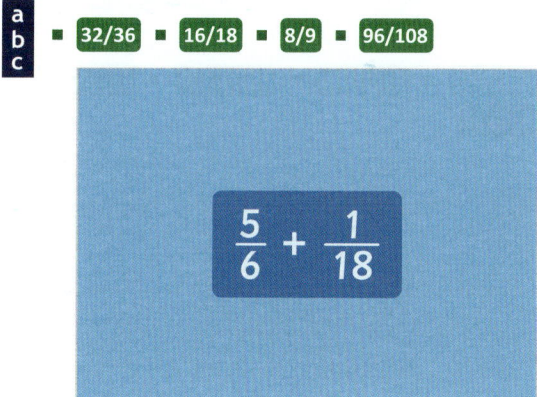

Ref:10428   Practise Adding Proper Fractions with Related D...

Level 1 *continued*

14. 5/11 + 27/99

a b c ■ 264/363 ■ 792/1,089 ■ 72/99 ■ 792/1089 ■ 88/121 ■ 8/11

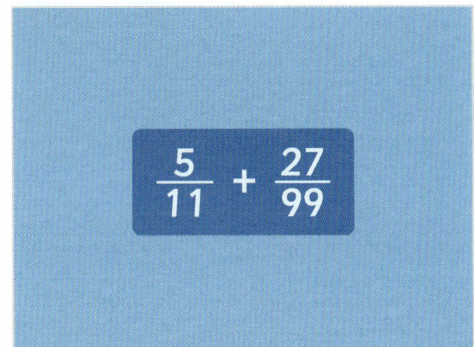
$$\frac{5}{11} + \frac{27}{99}$$

15. 5/14 + 1/7

a b c ■ 49/98 ■ 7/14 ■ 1/2

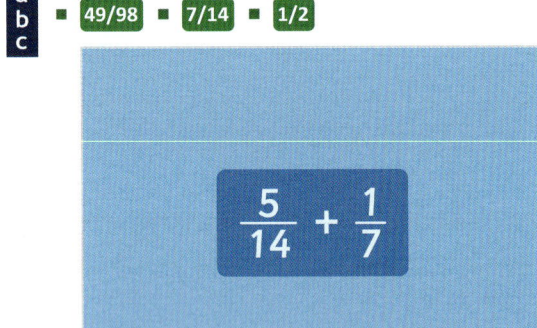
$$\frac{5}{14} + \frac{1}{7}$$

16. 45/100 + 2/10

a b c ■ 325/500 ■ 650/1000 ■ 650/1,000 ■ 13/20 ■ 65/100

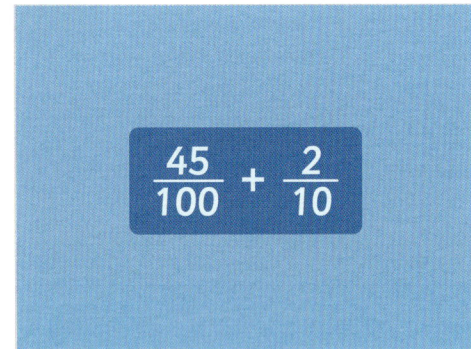
$$\frac{45}{100} + \frac{2}{10}$$

17. 2/13 + 20/26

a b c ■ 156/169 ■ 24/26 ■ 312/338 ■ 12/13

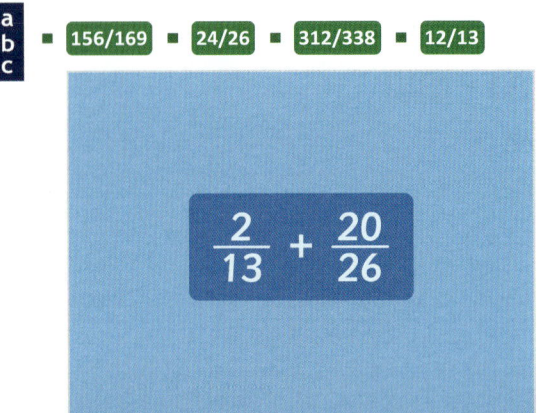
$$\frac{2}{13} + \frac{20}{26}$$

Level 2: I can add proper fractions with related denominators giving answers greater than 1.

✸ **Required:** 8/8 ✸ **Pupil Navigation:** off
✸ **Randomised:** on

18. 3/5 + 7/15
a b c *Give your answer as a mixed number fraction. For example, five and one-half is 5 1/2.*

■ 1 1/15 ■ 16/15

$$\frac{3}{5} + \frac{7}{15}$$

19. 3/4 + 5/8
a b c *Give your answer as a mixed number fraction. For example, five and one-half is 5 1/2.*

■ 11/8 ■ 1 3/8

$$\frac{3}{4} + \frac{5}{8}$$

Level 2 *continued*

20. 5/7 + 11/21

a b c *Give your answer as a mixed number fraction. For example, five and one-half is 5 1/2.*

- 26/21
- 1 5/21

21. 23/25 + 7/50

a b c *Give your answer as a mixed number fraction. For example, five and one-half is 5 1/2.*

- 53/50
- 1 3/50

22. 52/75 + 19/25

a b c *Give your answer as a mixed number fraction. For example, five and one-half is 5 1/2.*

- 1 34/75
- 109/75

23. 29/33 + 8/11

a b c *Give your answer as a mixed number fraction. For example, five and one-half is 5 1/2.*

- 53/33
- 1 20/33

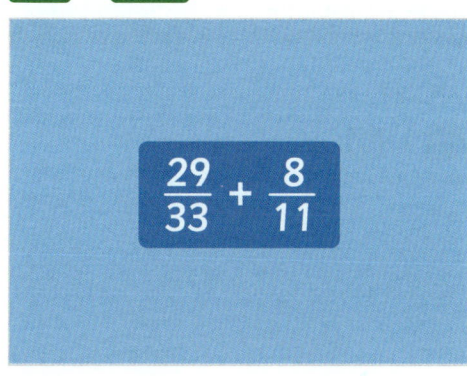

24. 29/40 + 53/80

a b c *Give your answer as a mixed number fraction. For example, five and one-half is 5 1/2.*

- 111/80
- 1 31/80

25. 23/30 + 65/90

a b c *Give your answer as a mixed number fraction. For example, five and one-half is 5 1/2.*

- 1 44/90
- 134/90

Practise Subtracting Proper Fractions with Related Denominators

Objective: I can subtract fractions with the same denominator and denominators that are multiples of the same number.

Quick Search Ref: 10431

1. $1/2 - 1/4$

a b c ▪ 2/8 ▪ 1/4

$$\frac{1}{2} - \frac{1}{4}$$

4. $7/8 - 3/4$

a b c ▪ 4/32 ▪ 1/8 ▪ 2/16

$$\frac{7}{8} - \frac{3}{4}$$

2. $1/4 - 2/8$

a b c ▪ 0/32 ▪ 0/8 ▪ 0

$$\frac{1}{4} - \frac{2}{8}$$

5. $8/15 - 1/5$

a b c ▪ 1/3 ▪ 25/75 ▪ 5/15

$$\frac{8}{15} - \frac{1}{5}$$

3. $15/16 - 1/8$

a b c ▪ 13/16 ▪ 104/128 ▪ 52/64 ▪ 26/32

$$\frac{15}{16} - \frac{1}{8}$$

6. $9/10 - 2/5$

a b c ▪ 1/2 ▪ 5/10 ▪ 25/50

$$\frac{9}{10} - \frac{2}{5}$$

Level 1 *continued*

7. 8/9 − 2/3

a b c ▪ 2/9 ▪ 6/27

$$\frac{8}{9} - \frac{2}{3}$$

8. 2/3 − 1/36

a b c ▪ 23/36 ▪ 69/108

$$\frac{2}{3} - \frac{1}{36}$$

9. 4/5 − 2/10

a b c ▪ 3/5 ▪ 6/10 ▪ 15/25 ▪ 30/50

$$\frac{4}{5} - \frac{2}{10}$$

10. 4/7 − 3/70

a b c ▪ 259/490 ▪ 37/70

$$\frac{4}{7} - \frac{3}{70}$$

11. 23/45 − 2/9

a b c ▪ 39/135 ▪ 117/405 ▪ 13/45

$$\frac{23}{45} - \frac{2}{9}$$

12. 12/15 − 1/15 − 8/30

a b c ▪ 105/225 ▪ 14/30 ▪ 30/30 ▪ 21/45 ▪ 7/15 ▪ 210/450

$$\frac{12}{15} - \frac{1}{15} - \frac{8}{30}$$

13. 5/6 − 1/18

a b c ▪ 42/54 ▪ 84/108 ▪ 21/27 ▪ 7/9 ▪ 14/18

$$\frac{5}{6} - \frac{1}{18}$$

14. 10/11 − 34/99

a b c ▪ 616/1,089 ▪ 616/1089 ▪ 56/99

$$\frac{10}{11} - \frac{34}{99}$$

Level 1 *continued*

15. 5/14 − 1/7

a b c ▪ 21/98 ▪ 3/14

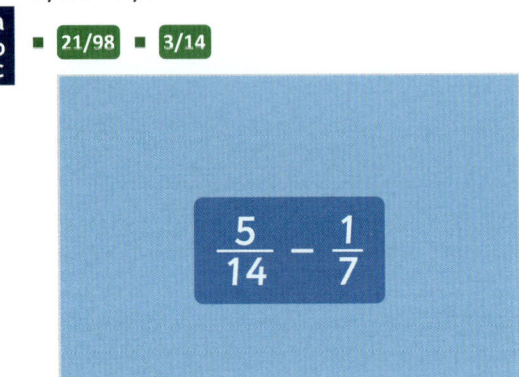

$$\frac{5}{14} - \frac{1}{7}$$

16. 45/100 − 3/10

a b c ▪ 150/1,000 ▪ 15/100 ▪ 75/500 ▪ 150/1000
▪ 3/20

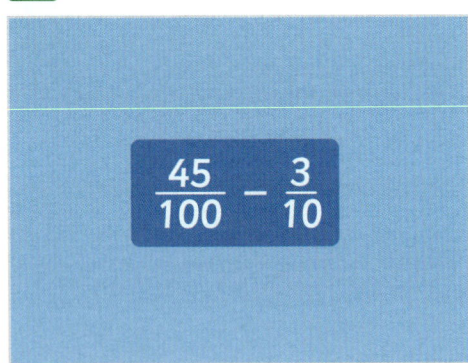

$$\frac{45}{100} - \frac{3}{10}$$

17. 20/26 − 4/13

a b c ▪ 6/13 ▪ 12/26 ▪ 78/169 ▪ 156/338

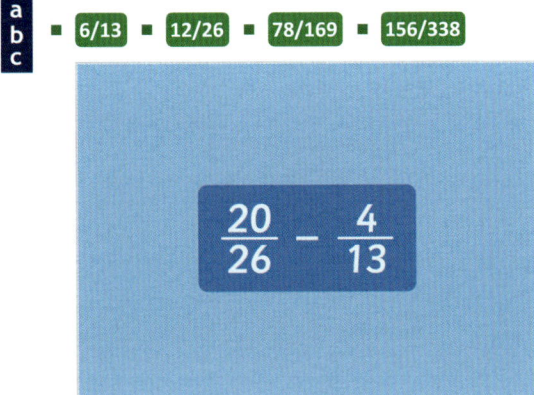

$$\frac{20}{26} - \frac{4}{13}$$

18. 71/100 − 12/50

a b c ▪ 1,175/2,500 ▪ 2,350/5,000 ▪ 1175/2500
▪ 47/100 ▪ 235/500 ▪ 2350/5000

$$\frac{71}{100} - \frac{12}{50}$$

19. 59/100 − 23/50

a b c ▪ 13/100

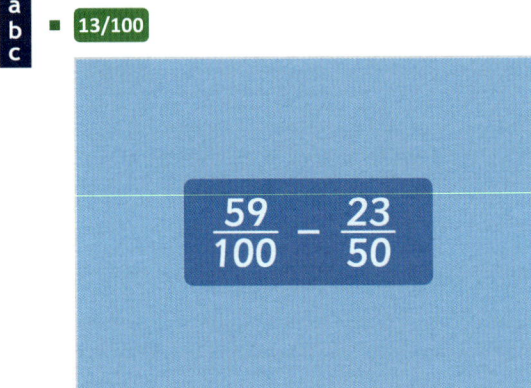

$$\frac{59}{100} - \frac{23}{50}$$

20. 56/75 − 13/25

a b c ▪ 425/1875 ▪ 85/375 ▪ 425/1,875 ▪ 17/75

$$\frac{56}{75} - \frac{13}{25}$$

Subtract Fractions with Related Denominators (Proper, Improper and Mixed Number Fractions)

Objective: I can subtract proper, improper and mixed number fractions with the same denominator and denominators that are multiples of the same number.

Quick Search Ref: 10077

Level 1: Understanding: Subtract fractions with image support.

✿ Required: 7/10　　　**✿ Pupil Navigation:** on　　　**✿ Randomised:** off

1. What is 5/6 − 1/3?

a b c　*Put a forward slash (/) between the numerator and denominator. For example, one-half is 1/2.*

■ 3/6 ■ 9/18 ■ 1/2 ■ 4/3

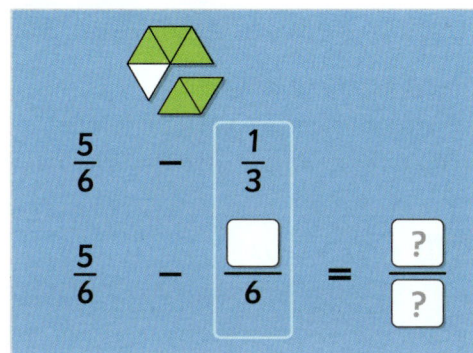

2. Calculate 7/5 − 13/10.

a b c　*Give your answer as a fraction. For example, 1/2.*

■ 1/10 ■ 27/10

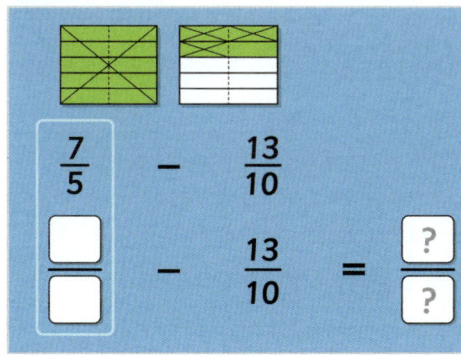

3. Calculate 2 5/6 − 1 1/6.

a b c　*Give your answer as a mixed number fraction. For example, five and one-half is 5 1/2.*

■ 1 2/3 ■ 1 4/6 ■ 10/6 ■ 5/3 ■ 2/3 ■ 4/6

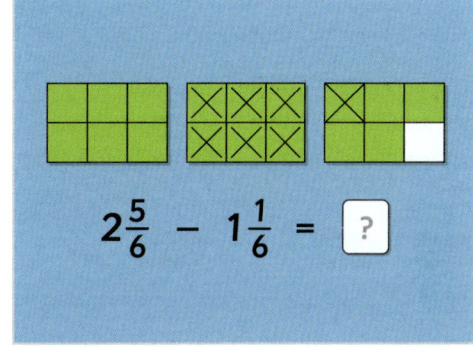

4. What is 1/4 − 1/16?

a b c　*Put a forward slash (/) between the numerator and denominator. For example, one-half is 1/2.*

■ 12/64 ■ 3/16 ■ 6/32 ■ 5/16 ■ 1/12

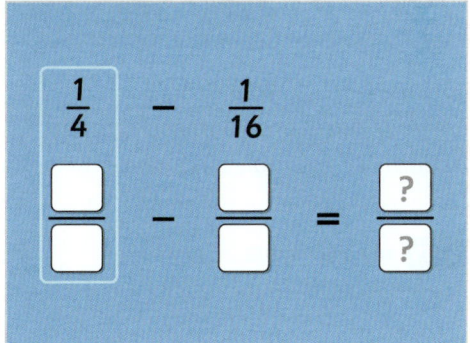

Level 1 *continued*

5. What is 19/8 − 7/4?

a b c *Give your answer as a fraction. For example, 1/2.*

- 12/4 - **5/8** - 33/8

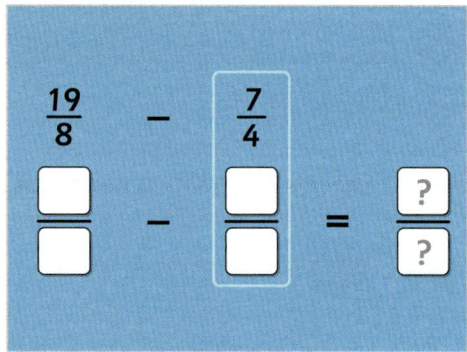

6. What is 2 1/6 − 2/3?

a b c *Give your answer as a mixed number fraction. For example, five and one-half is 5 1/2.*

- **1 3/6** - **1 1/2** - 3/2 - **1 9/18** - 9/6 - 27/18

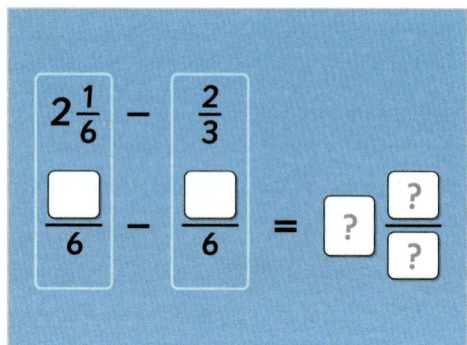

7. Subtract 1 3/5 from 3 2/5.

a b c *Give your answer as a mixed number fraction. For example, five and one-half is 5 1/2.*

- **1 4/5** - 2 1/5 - 9/5

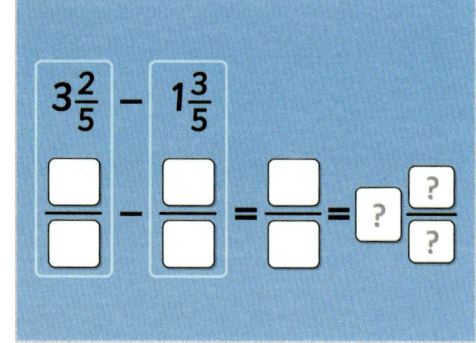

8. What is the difference between 37/8 and 7/2?

a b c *Give your answer as an improper fraction. For example, 3/2.*

- **9/8** - 30/6 - 65/8

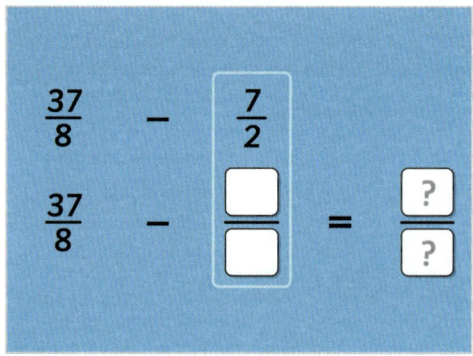

9. What is 4 1/3 − 2 2/3?

a b c *Give your answer as a mixed number fraction. For example, five and one-half is 5 1/2.*

- **1 2/3** - 5/3 - 2 1/3

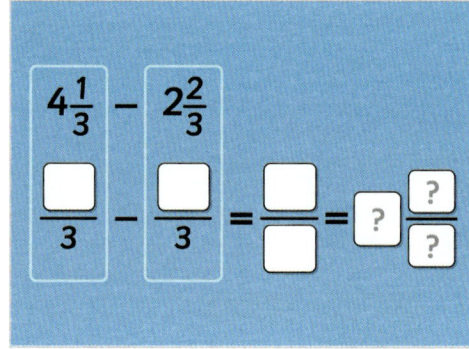

10. What is 1/4 − 1/8?

a b c *Put a forward slash (/) between the numerator and denominator. For example, one-half is 1/2.*

- **1/8** - **2/16** - **4/32** - 3/8 - 1/4

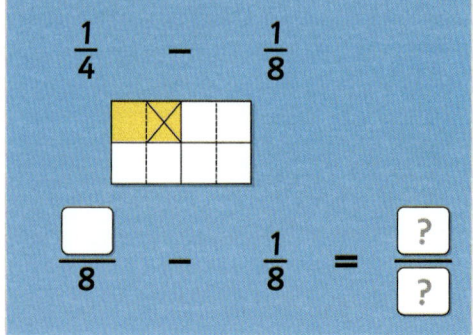

Level 2: Fluency: Subtract fractions including subtracting more than two fractions and 1-step word problems.

❊ **Required:** 7/10 ❊ **Pupil Navigation:** on
❊ **Randomised:** off

11. What is 1/3 − 1/9 − 1/27?
a
b *Put a forward slash (/) between the*
c *numerator and denominator. For example, one-half is 1/2.*

- 5/27 ■ 13/27

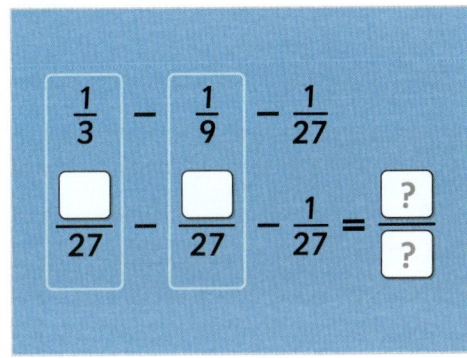

12. What is 13/4 − 3/2 − 9/8?
a
b *Give your answer as a fraction. For example,*
c *1/2.*

- 5/8

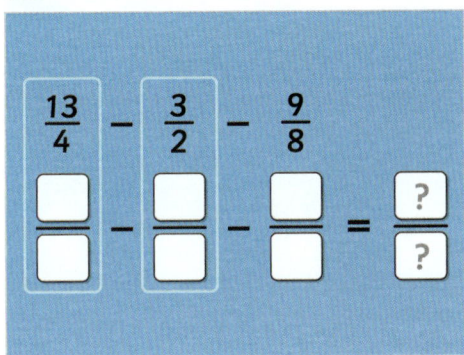

13. Find the answer to 3 1/4 − 1 1/2 − 1 1/8.
a
b  5/8
c

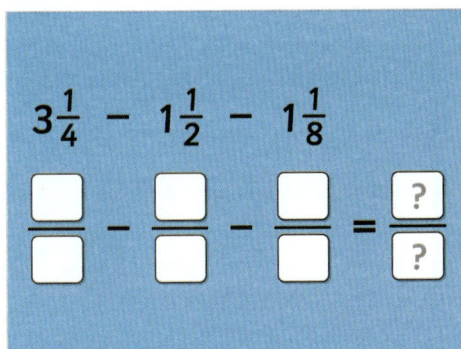

14. Calculate the answer to 3 1/6 − 2/3 − 11/12.
a
b *Give your answer as a mixed number*
c *fraction.*

- 2 3/6 ■ 1 7/12 ■ 2 1/2 ■ 19/12

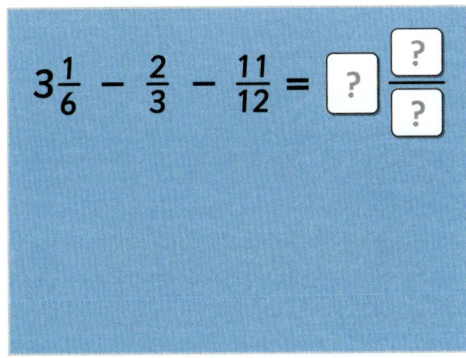

15. Adil has 5/4 kilograms (kg) of flour. He spills
a 1/8 kg on the floor. How many kilograms of
b flour does Adil have left?
c *Give your answer as an improper fraction.*
Don't include the units in your answer.

- 9/8 kilograms ■ 9/8 ■ 9/8 kg ■ 1.125
- 1.125 kilograms ■ 1.125 kg

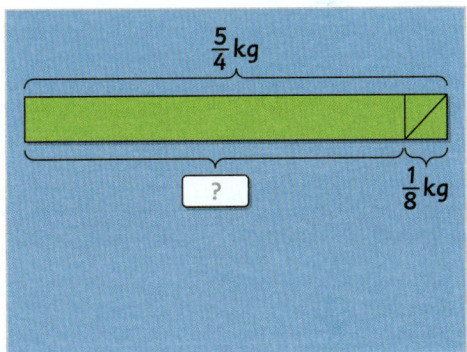

16. Josie has a full bar of chocolate. She gives
a 3/18 of the bar to Ellie and 17/36 to Izzy, and
b she eats 1/9 of the bar. What fraction of a
c bar of chocolate does Josie have left?
*Put a forward slash (/) between the
numerator and denominator. For example,
one-half is 1/2.*

- 9/36 ■ 1/4 ■ 3/12 ■ 1 19/36

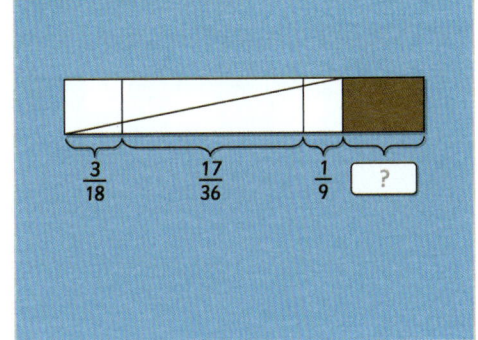

Level 2 *continued*

17. There are 2 ½ boxes of pens in the school
a cupboard. Mrs Walsh takes ¾ of a box of
b pens. How many boxes of pens are left?
c *Give your answer as a mixed number*
fraction.

- ■ **1 6/8** ■ 14/8 ■ **1 3/4** ■ 7/4 ■ 1

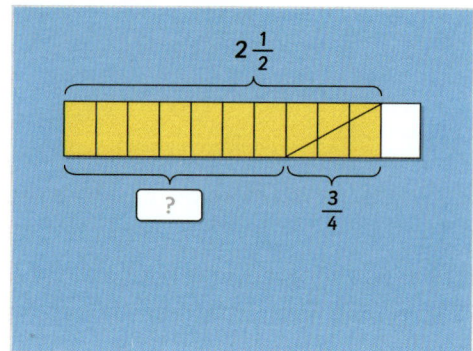

18. What is 71/16 − 9/4 − 5/8?
a *Give your answer as an improper fraction.*
b *For example, 3/2.*
c

- ■ **25/16**

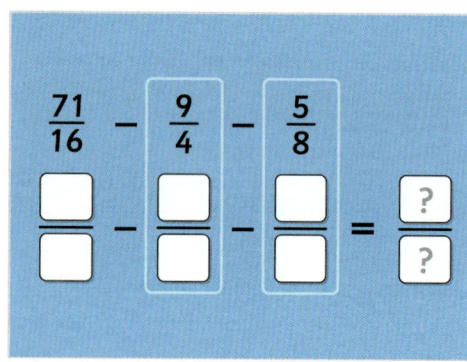

19. What is 3 1/6 − 1 1/3?
a *Give your answer as a mixed number*
b *fraction.*
c

- ■ **1 5/6** ■ 11/6

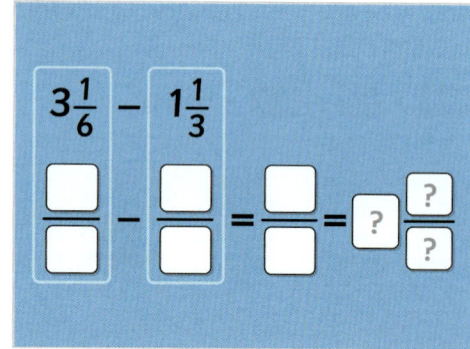

20. What is 2 4/7 − 5/14?
a *Give your answer as a mixed number*
b *fraction.*
c

- ■ **2 3/14** ■ **2 21/98** ■ 31/14 ■ 217/98

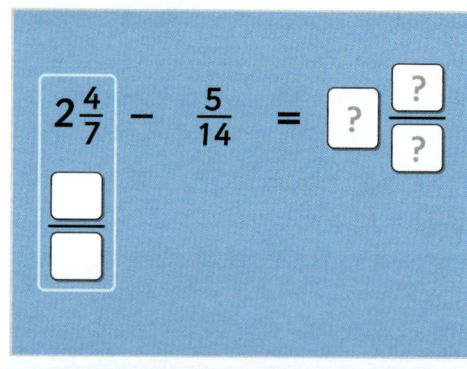

Level 3: Reasoning: Reason about subtracting
fractions with related denominators.

❋ **Required:** 5/5   ❋ **Pupil Navigation:** on
❋ **Randomised:** off

21. Use **three** of the number cards to make the
1 statement correct. What number goes in box
2 B?
3

- ■ **8** ■ 33 ■ 7

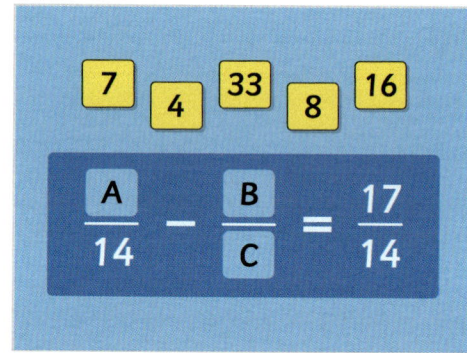

22. Harry calculates that 30/48 − 5/12 = 25/36.
a Explain what Harry has done wrong and
b explain how he can calculate the answer
c correctly.

- **Open question, no set answer**

Level 3 continued

23. Find the missing proper fraction in this
a calculation:
b
c 1 3/7 − ___ = 9/14

- 11/14

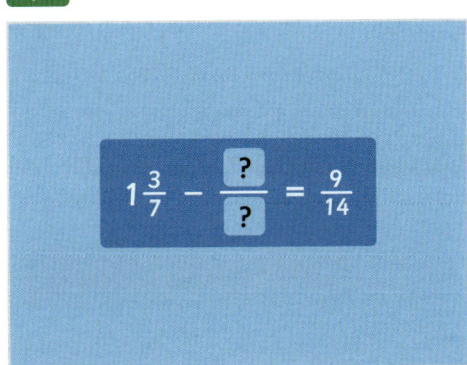

24. Select **two** mixed number fractions which
have a difference that is greater than 2 but
less than 3.

2/4 ■ 2 7/8 ■ 1 3/8 ■ 5 3/8 ■ 4 5/8

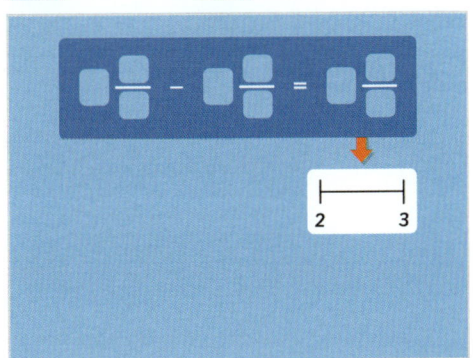

25. What is the same and what is different about
a these two calculations?
b
c - Open question, no set answer

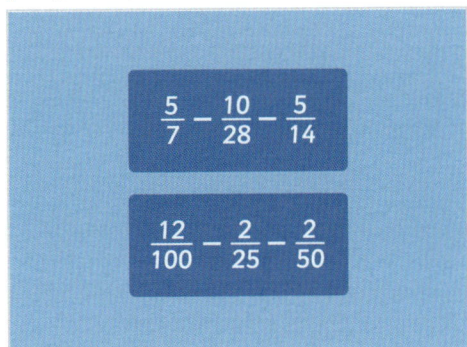

Level 4: Problem solving with greater depth: Solve multi-step problems involving subtracting fractions with related denominators.

✿ **Required:** 5/5 ✿ **Pupil Navigation:** on
✿ **Randomised:** off

26. Billy is making a model of a Viking longboat.
a He has a piece of wood that is 3/4 metre (m)
b long.
c Billy cuts off a piece for the mast which
measures 3/16 m. He cuts off another piece
for the deck which measures 1/8 m and a
final piece to make the oars which measures
3/8 m.
How many metres of wood does Billy have
left?
*Give your answer as a fraction. Don't include
the units in your answer.*

- 11/16 metres ■ 1/16 ■ 1/16 metres ■ 1/16 m
- 11/16 m ■ 11/16

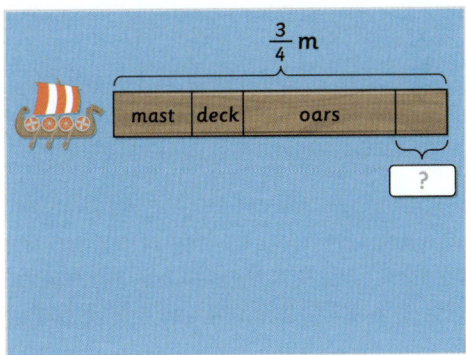

27. A triangle has a perimeter of 6 1/2 metres
a (m). One side of the triangle measures 1 3/8
b m and another side is 2 1/4 m long. The
c missing side length is ____ m long.
*Enter the missing mixed number fraction.
Don't include the units in your answer.*

- 2 7/8 ■ 2 7/8 m ■ 2 7/8 metres

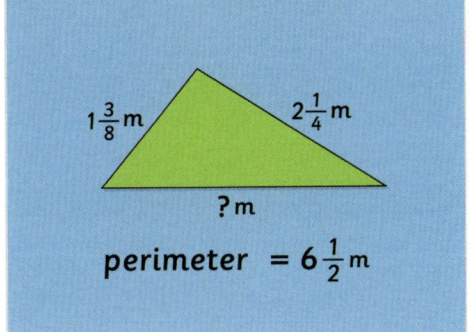

28. A group of children have 7 1/3 litres (l) of
water to use on the school garden. They
pour 7/12 litres on the carrot plants and
another 5/6 litres on the onions. The
children use less than one litre of water on
the lettuces.

If the children have exactly 5 litres of water
left, how many litres of water do they use on
the lettuces?

*Give your answer as a fraction. Don't include
the units in your answer.*

- 11/12 l ▪ 5 11/12 l ▪ 11/12 litres ▪ 11/12
- ▪ 5 11/12 litres ▪ 5 11/12

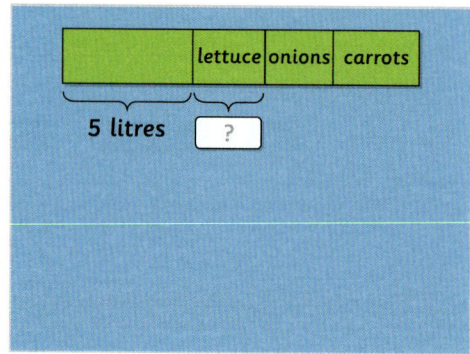

29. An animal rescue centre has 4 full bags of
hay. They use 1/4 of a bag of hay for the
hamsters and 9/8 bags of hay for the rabbits.
If there are 3/2 bags of hay left at the end of
the day, how much hay do they give to the
donkey?

*Give your answer as an improper fraction.
For example, 3/2.*

- 9/8 ▪ 21/8

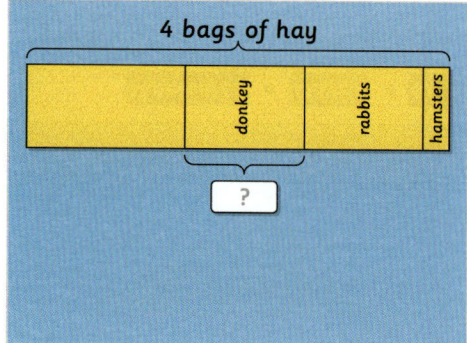

30. An electrician working in a school has 7
metres (m) of wire. He uses 1 4/5 m of wire
for the lights and 3 3/5 for the plug sockets.
If the electrician needs 223 centimetres of
wire for the computers, what **fraction** of a
metre more wire will he need to complete
his work?

*Give your answer as a fraction. Don't include
the units in your answer.*

- 63/100 m ▪ 63/100 ▪ 63/100 metres
- ▪ 7 63/100 metres ▪ 7 63/100 ▪ 7 63/100 m

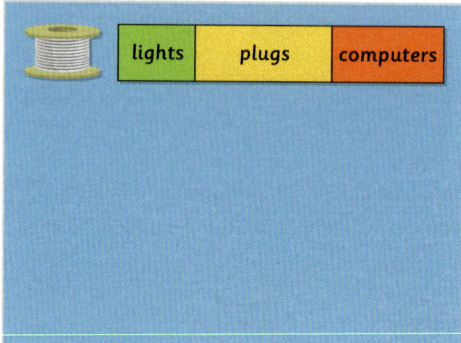

Subtract Improper Fractions with Related Denominators

Objective: I can subtract fractions with the same denominator and denominators that are multiples of the same number.

Quick Search Ref: 11482

Level 1: Understanding: Subtract improper fractions with related denominators with image support.

✿ **Required:** 7/10   ✿ **Pupil Navigation:** on   ✿ **Randomised:** off

---

1. To calculate 7/5 – 2/15 you need to make the denominators the same by finding the lowest number that is a multiple of both denominators.
What is the lowest number that is a multiple of both 5 and 15?

■ **15**   ■ 30   ■ 10

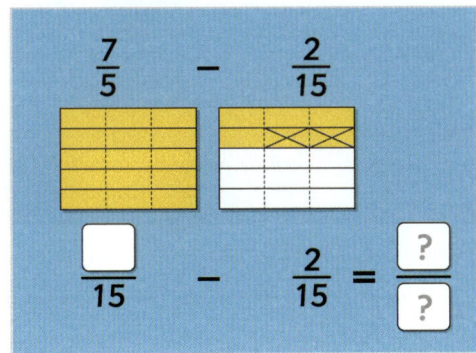

2. What is 7/5 – 2/15?
Give your answer as an improper fraction. For example, 3/2.

■ 23/15   ■ **19/15**   ■ 5/10

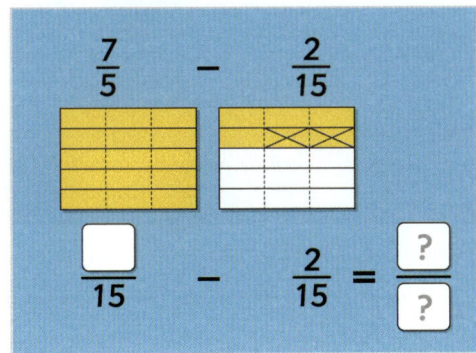

3. Find the difference between 11/8 and 1/4.
Give your answer as an improper fraction. For example, 3/2.

■ 13/8   ■ **9/8**   ■ 10/4

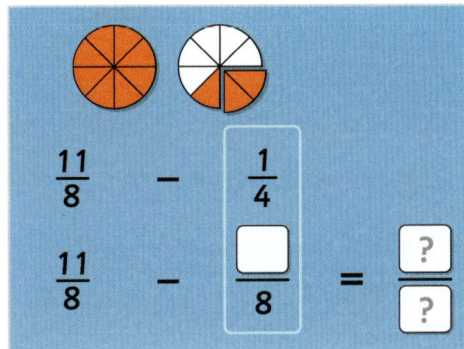

4. Calculate 7/5 – 13/10.
Give your answer as a fraction. For example, 1/2.

■ **1/10**   ■ 27/10

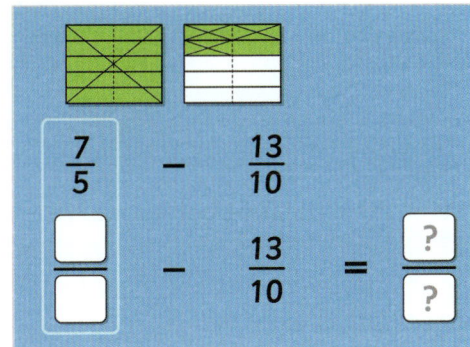

5. What is 2 – 5/4?
Give your answer as an improper fraction. For example, 3/2.

■ **3/4**   ■ 13/4

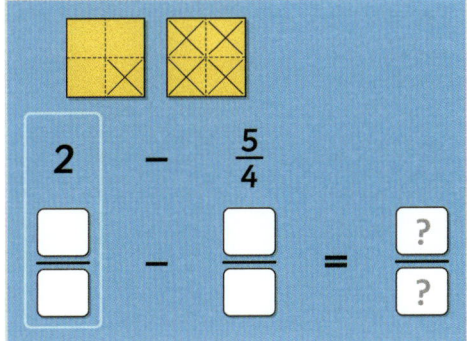

Level 1 *continued*

6. What is 19/8 – 7/4?

   a b c *Give your answer as a fraction. For example, 1/2.*

   ■ **5/8** ■ 12/4 ■ 33/8

   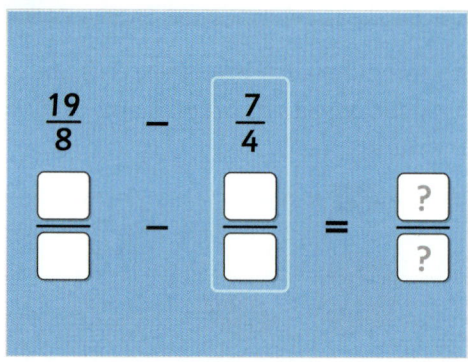

7. What is the difference between 7/3 and 11/9?

   a b c *Give your answer as an improper fraction. For example, 3/2.*

   ■ **10/9** ■ 32/9

   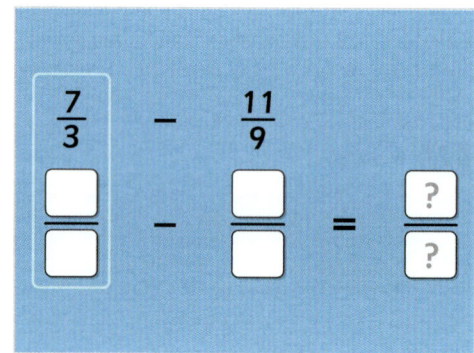

8. What is the difference between 37/8 and 7/2?

   a b c *Give your answer as an improper fraction. For example, 3/2.*

   ■ **9/8** ■ 65/8 ■ 30/6

   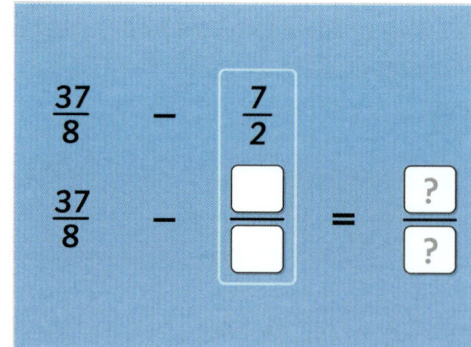

9. What is the lowest number that is a multiple of both denominators that you would use to calculate 9/4 – 19/16?

   1 2 3

   ■ **16** ■ 8 ■ 32

   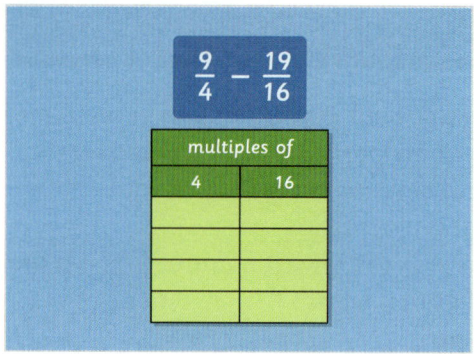

10. Calculate 15/6 – 4/3.

    a b c *Give your answer as an improper fraction. For example, 3/2.*

    ■ **7/6** ■ 23/6 ■ 11/3

    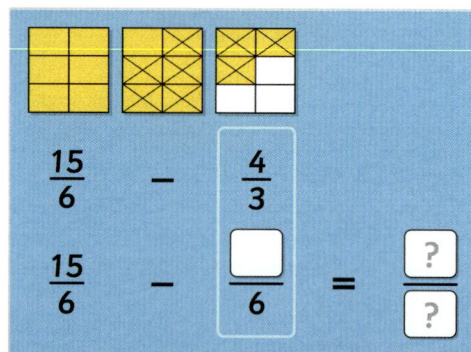

Level 2: Fluency: Subtract improper fractions with related denominators including subtracting more than one fraction. Solve one-step word problems.

✳ **Required:** 7/10   ✳ **Pupil Navigation:** on
✳ **Randomised:** off

11. Calculate 11/7 – 20/14.

    a b c *Give your answer as a fraction. For example, 1/2.*

    ■ **1/7** ■ 2/14

    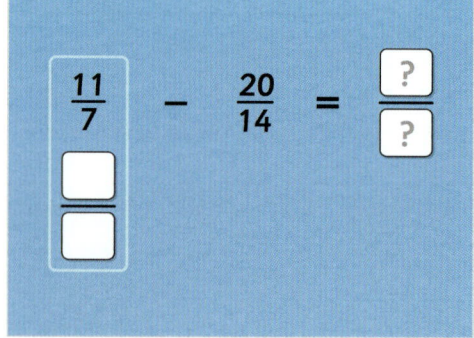

12. What is 8/3 − 7/6 − 1/3?

a b c *Give your answer as an improper fraction.*
For example, 3/2.

- 7/6

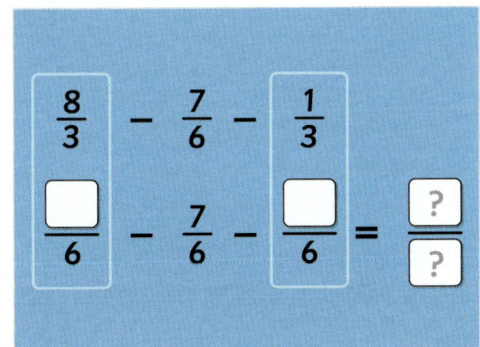

13. What is 13/4 − 3/2 − 9/8?

a b c *Give your answer as a fraction. For example,*
1/2.

- 5/8

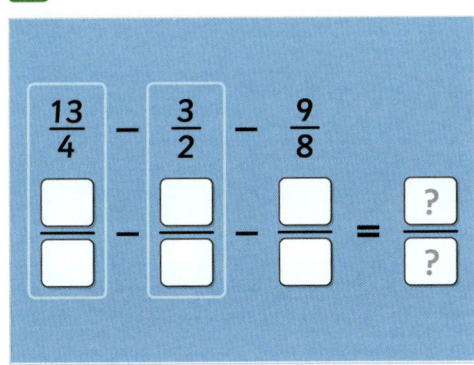

14. What is 3 − 8/5?

a b c *Give your answer as an improper fraction.*

- 7/5

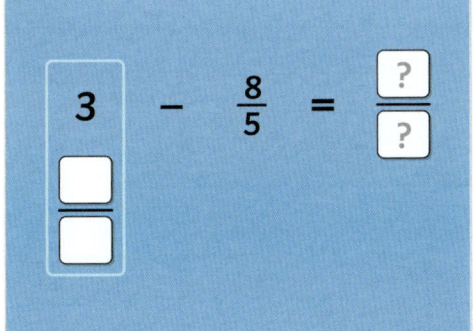

15. A jug contains 12/10 litres (l) of water. How
a b c many litres of water is left in the jug after
Emma pours out 55/50 litres?
Give your answer as a fraction. Don't include
the units in your answer.

- 5/50 l   - 5/50 litres   - 1/10   - 1/10 litres   - 0.1 l
- 1/10 l   - 5/50   - 0.1 litres   - 0.1

16. Adil has 5/4 kilograms (kg) of flour. He spills
a b c 1/8 kg on the floor. How many kilograms of
flour does Adil have left?
Give your answer as an improper fraction.
Don't include the units in your answer.

- 9/8 kilograms   - 9/8   - 9/8 kg   - 1.125 kilograms
- 1.125   - 1.125 kg

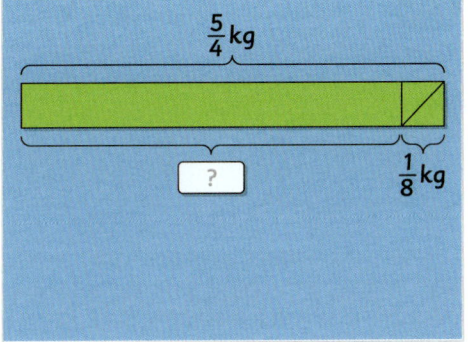

Level 2 continued

17. Three children build a tower that is 297/100
a
b metres (m) tall. Mia builds the bottom part
c of the tower, which measures 12/10 m. Alfie
adds the next section measuring 7/5 m and
Jamie builds the final part of the tower
which measures _____ m.
*Enter the missing fraction. Don't include the
units in your answer.*

- 0.37 m - **37/100 metres** - **37/100 m** - **37/100**
- 0.37 metres - 0.37

18. What is 71/16 − 9/4 − 5/8?
a
b *Give your answer as an improper fraction.*
c *For example, 3/2.*

- **25/16**

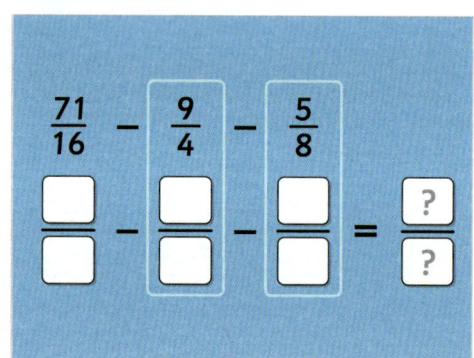

19. What is 27/6 − 2?
a
b *Give your answer as an improper fraction.*
c *For example, 3/2.*

- **15/6**

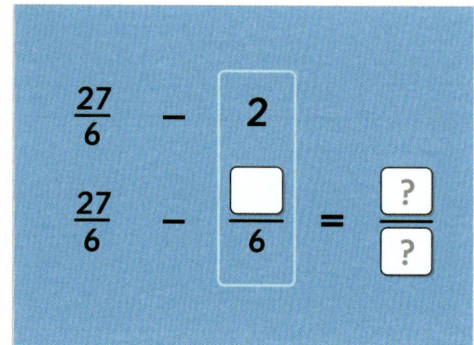

20. What is 6/5 less than 19/10?
a
b *Give your answer as a fraction.*
c
- **7/10**

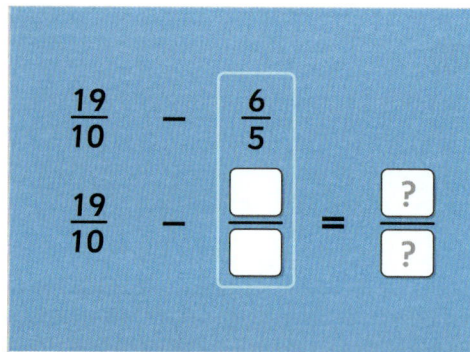

Level 3: Reasoning: Reason about subtracting
improper fractions with related
denominators.

✱ **Required:** 5/5   ✱ **Pupil Navigation:** on
✱ **Randomised:** off

21. Use **three** of the number cards to make the
1 statement correct. What number goes in box
2 B?
3

- **8** - 7 - 33

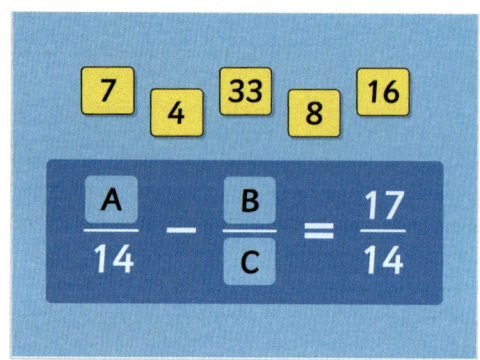

22. Becky says that 13/5 − 7/5 = 6/5, so if you
a
b subtract two improper fractions with the
c same denominator, you will always get an
answer that is an improper fraction. Is Becky
correct? Explain your answer, giving an
example.

- **Open question, no set answer**

Level 3 continued

23. Which fraction can you subtract from 11/6 to make an improper fraction?

- 4/3 ▪ 11/12 ▪ **1/3**

1/3

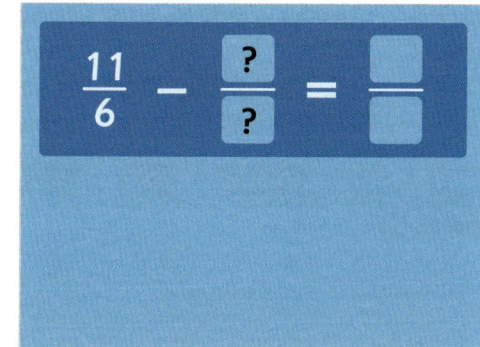

24. What is the same and what is different about these two calculations?

- Open question, no set answer

25. Oscar is subtracting 17/12 from 19/6. He converts both fractions to have a denominator of 72 before subtracting the numerators. Explain how Oscar can find the best multiple of both 12 and 6 to make his calculation easier.

- Open question, no set answer

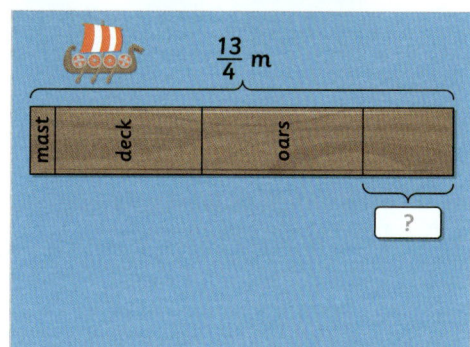

Level 4: Problem solving with greater depth: Solve problems involving the subtraction of improper fractions with related denominators.

❋ **Required:** 5/5   ❋ **Pupil Navigation:** on
❋ **Randomised:** off

26. A triangle has a perimeter of 13/2 metres (m). One side of the triangle measures 11/8 m and another side measures 9/4 m. The missing side length measures _____ metres. *Enter the missing improper fraction. Don't include the units in your answer.*

- **23/8**

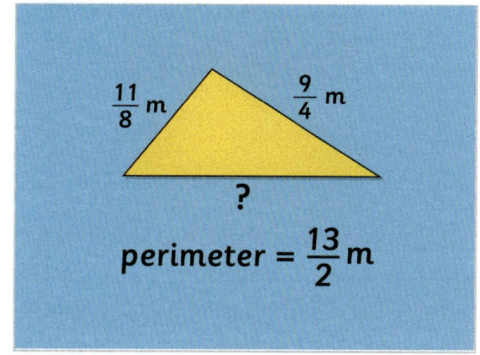

27. Billy is making a model of a Viking longboat. He has a piece of wood that is 13/4 metres (m) long. Billy cuts off a piece for the mast which measures 3/16 m. He cuts off another piece for the deck which measures 9/8 m and a final piece to make the oars which measures 5/4 m.
How many metres of wood does Billy have left?
Give your answer as a fraction. Don't include the units in your answer.

- **11/16 m** ▪ **11/16** ▪ **11/16 metres**

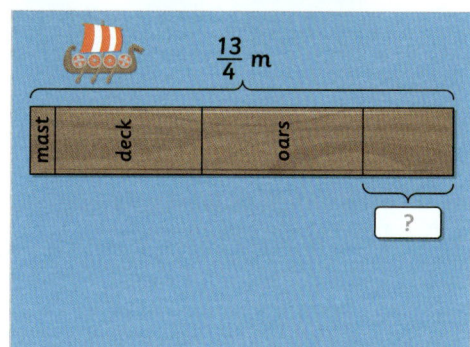

28. A rectangle has a perimeter of 31/5 metres
a (m) and a side length of 17/10 m. What does
b the shorter side length measure?
c
Give your answer as an improper fraction.
Don't include the units in your answer.

- **7/5 metres** ▪ **14/10** ▪ **14/10 metres** ▪ **7/5 m**
- **7/5** ▪ **14/10 m** ▪ 28/10 ▪ 14/5 m
- 14/5 metres ▪ 28/10 m ▪ 28/10 metres ▪ 14/5

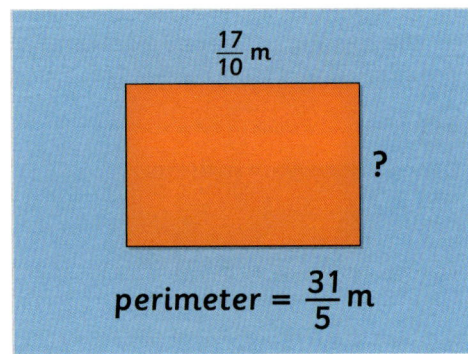

$$\frac{17}{10} \text{ m}$$

?

$$\text{perimeter} = \frac{31}{5} \text{ m}$$

29. An animal rescue centre has 4 full bags of
a hay. They use 1/4 of a bag of hay for the
b hamsters and 9/8 bags of hay for the rabbits.
c If there are 3/2 bags of hay left at the end of
the day, how much hay do they give to the
donkey?
Give your answer as an improper fraction.
For example, 3/2.

- **9/8** ▪ 21/8

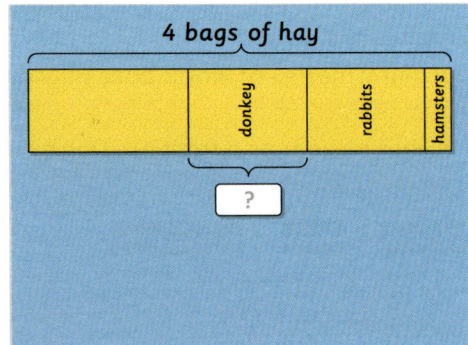

4 bags of hay

donkey | rabbits | hamsters

?

30. A group of children have 22/3 litres (l) of
a water to use on the school garden. They
b pour 17/12 litres on the carrot plants and
c another 7/6 litres on the onions. If the
children have exactly 4 litres of water left,
how many litres of water do they use on the
lettuces?
Give your answer as a fraction. Don't include
the units in your answer.

- **9/12** ▪ **3/4** ▪ **3/4 litres** ▪ **9/12 l** ▪ **3/4 l**
- **9/12 litres**

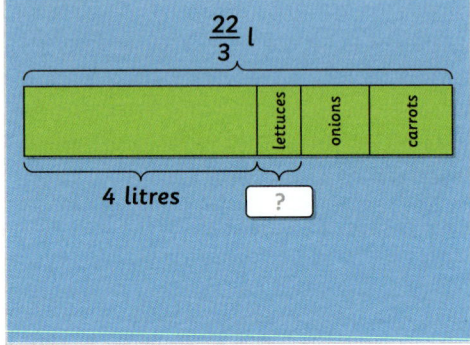

$$\frac{22}{3} \text{ l}$$

lettuces | onions | carrots

4 litres | ?

Subtract Mixed Number Fractions With Related Denominators

Objective: I can subtract mixed number fractions with the same denominator and denominators that are multiples of the same number.

Quick Search Ref: 11500

Level 1: Understanding: Subtract a proper fraction from a mixed number fraction with a related denominator. Subtract two mixed number fractions with the same denominator.

✸ **Required:** 7/10   ✸ **Pupil Navigation:** on   ✸ **Randomised:** off

---

1. What is 1 2/3 − 1/3?

a b c *Give your answer as a mixed number fraction. For example, five and one-half is 5 1/2.*

▪ **1 1/3** ▪ 1/3 ▪ 4/3 ▪ 2

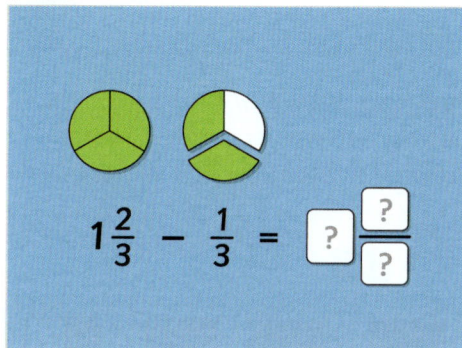

2. 2 5/8 − 1 5/8 = ___

1 2 3 ▪ **1** ▪ 2

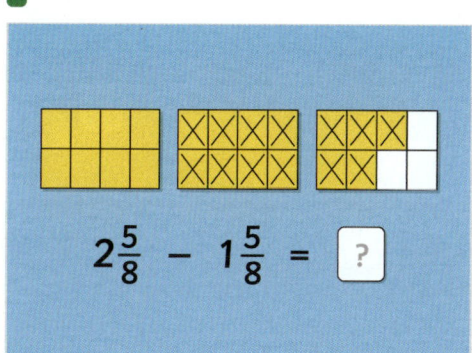

3. Calculate 2 5/6 − 1 1/6.

a b c *Give your answer as a mixed number fraction. For example, five and one-half is 5 1/2.*

▪ **1 2/3** ▪ **1 4/6** ▪ 5/3 ▪ 4/6 ▪ 10/6 ▪ 2/3

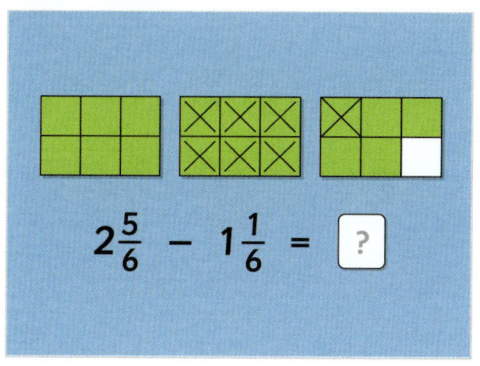

4. Calculate 1 3/8 − 1/4.

a b c *Give your answer as a mixed number fraction. For example, five and one-half is 5 1/2.*

▪ **1 1/8** ▪ 1/8 ▪ 9/8

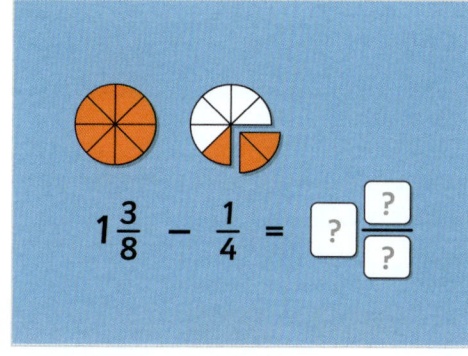

5. What is 2 1/4 − 1 3/4?

a b c *Give your answer as a fraction. For example, one-half is 1/2.*

▪ **2/4** ▪ **1/2** ▪ 1 1/2 ▪ 1 2/4

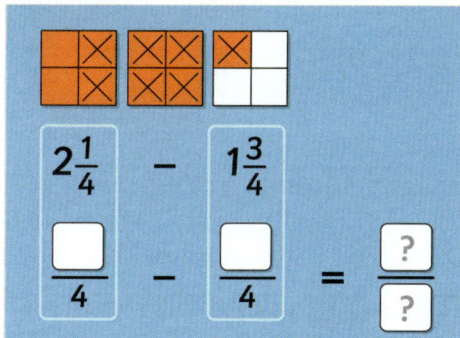

Level 1 *continued*

6. Subtract 1 3/5 from 3 2/5.

a
b
c
Give your answer as a mixed number fraction. For example, five and one-half is 5 1/2.

■ 1 4/5 ■ 9/5 ■ 2 1/5

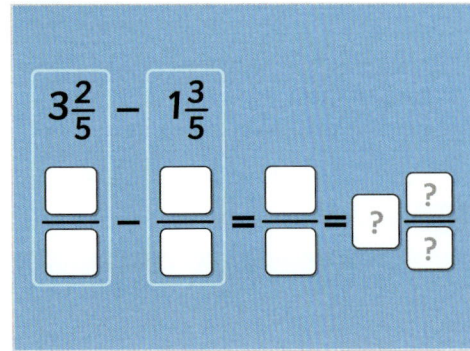

7. Calculate 1 3/5 − 7/10.

a
b
c
Give your answer as a fraction. For example, one-half is 1/2.

■ 9/10 ■ 1 1/10

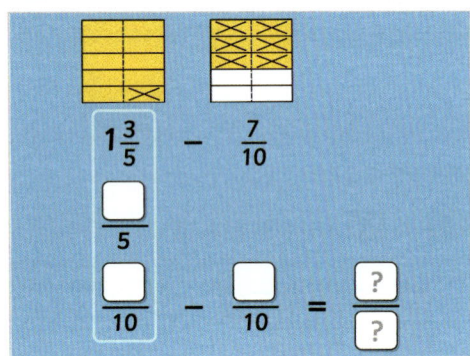

8. What is 5 1/8 − 1 3/8?

a
b
c
Give your answer as a mixed number fraction. For example, five and one-half is 5 1/2.

■ 3 3/4 ■ 3 6/8 ■ 4 2/8 ■ 30/8 ■ 15/4 ■ 4 1/4

$$5\frac{1}{8} - 1\frac{3}{8} = \frac{\square}{\square} = \boxed{?}\ \frac{?}{?}$$

9. What is 4 1/3 − 2 2/3?

a
b
c
Give your answer as a mixed number fraction. For example, five and one-half is 5 1/2.

■ 1 2/3 ■ 2 1/3 ■ 5/3

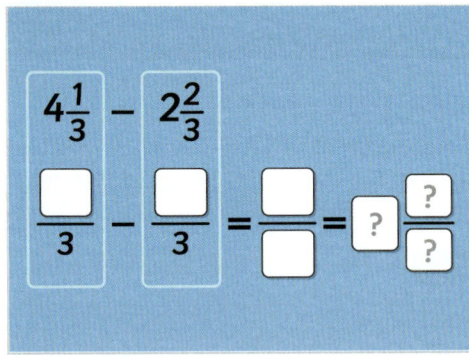

10. Calculate 3 2/4 − 1 3/4.

a
b
c
Give your answer as a mixed number fraction. For example, five and one-half is 5 1/2.

■ 7/4 ■ 1 3/4 ■ 2 1/4

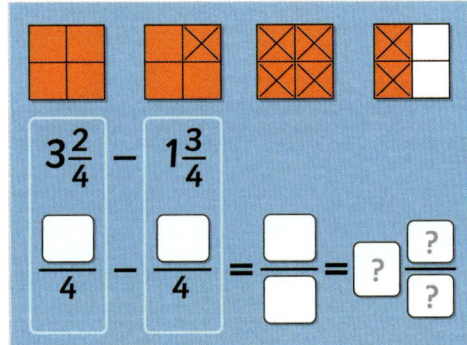

Level 2: Fluency: Subtract mixed number fractions with related denominators, including subtracting more than one fraction. Solve one-step word problems.

✱ **Required:** 7/10 ✱ **Pupil Navigation:** on
✱ **Randomised:** off

11. Subtract 1 1/6 from 1 1/2.

a
b
c
■ 1/3 ■ 2/6

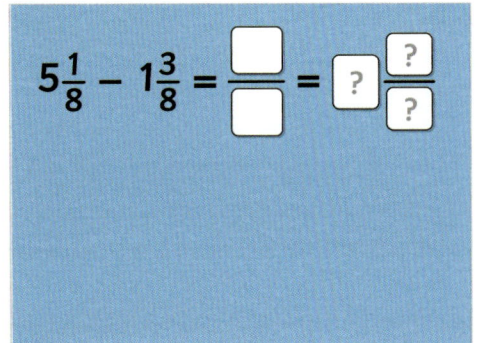

Level 2 *continued*

12. Calculate 3 3/7 − 1 9/14.

a *Give your answer as a mixed number*
b *fraction. For example, five and one-half is 5*
c *1/2.*

- **1 11/14** ▪ 25/14

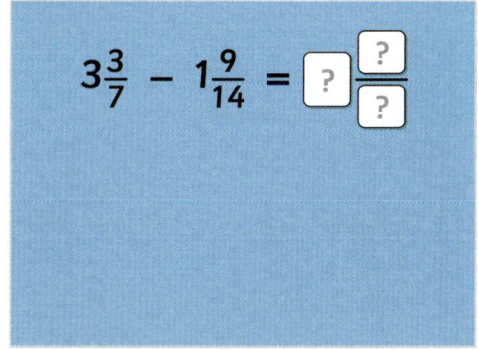

13. What is 3 1/3 − 1 1/6 − 1/3?

a *Give your answer as a mixed number*
b *fraction.*
c

- **1 5/6** ▪ 11/6

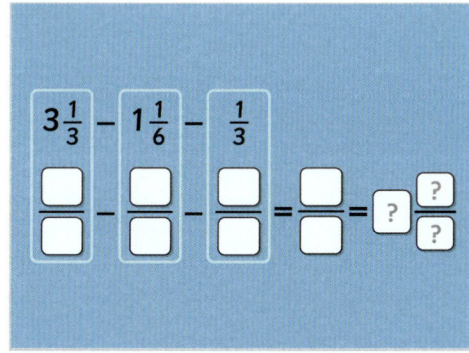

14. Find the answer to 3 1/4 − 1 1/2 − 1 1/8.

a
b ▪ **5/8**
c

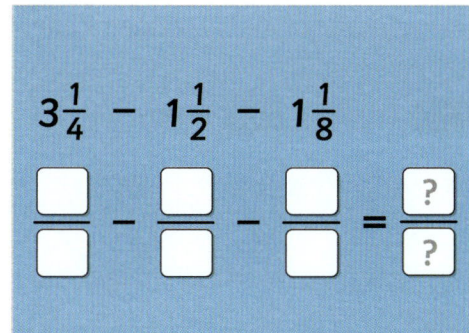

15. There are three full boxes of pens in the
school cupboard. Mr Timms takes 1 ¾ boxes
of pens. How many **full boxes** of pens are left
in the cupboard?

- **1**

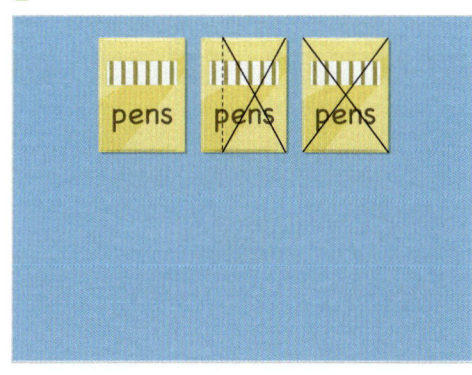

16. A jug contains 2 1/5 litres of juice. What
a **fraction** of a litre of juice is left in the jug
b after Aiden pours out 1 3/10 litres?
c *Don't include the units in your answer.*

- **9/10 litres** ▪ **9/10** ▪ **9/10 l** ▪ 0.9 l ▪ 0.9
- 0.9 litres

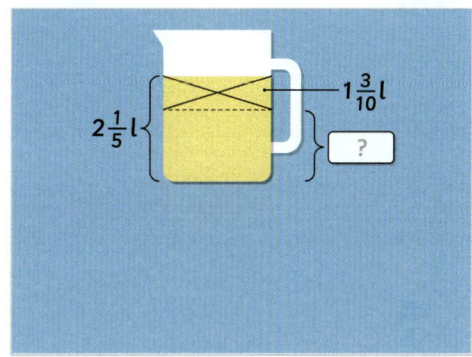

17. The shorter flower in the diagram is _____ cm
a tall.
b *Enter the missing mixed number fraction.*
c *Don't include the units in your answer.*

- **4 3/4 cm** ▪ **4 3/4 centimetres** ▪ **4 3/4**
- 4.75 centimetres ▪ 4.75 ▪ 4.75 cm

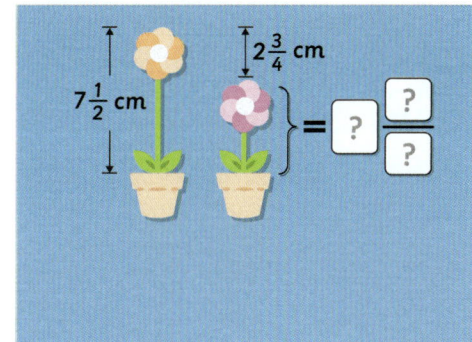

Level 2 continued

18.
a
b
c
Gemma has 2 1/4 kilograms (kg) of sugar. She spills 1 3/8 kg on the floor. What **fraction** of a kilogram of sugar does Gemma have left?

Don't include the units in your answer.

- ▪ 7/8 ▪ 7/8 kilogram ▪ 7/8 kilograms ▪ 7/8 kg

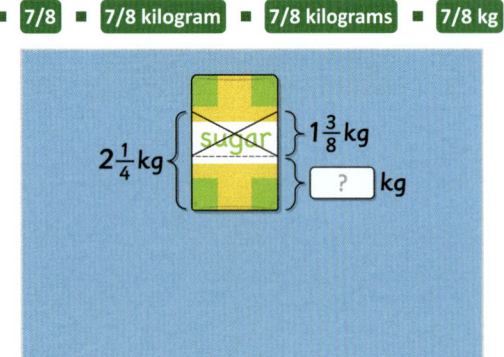

19.
a
b
c
What is 3 1/6 − 1 1/3?
Give your answer as a mixed number fraction.

- ▪ 1 5/6 ▪ 11/6

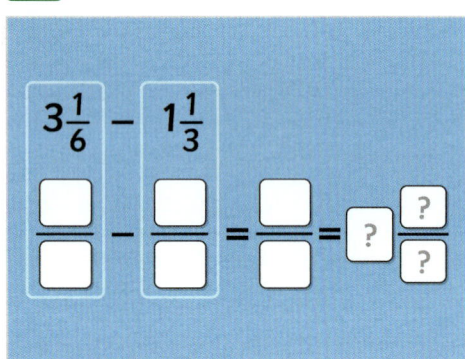

20.
a
b
c
Calculate 1 3/14 − 1 1/7.

- ▪ 1/14

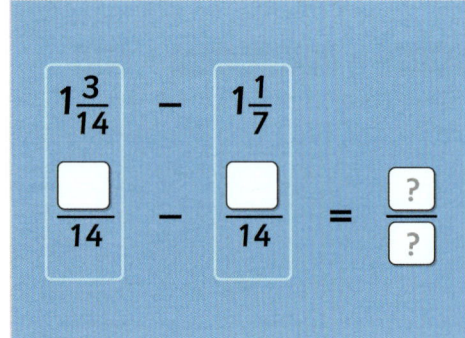

Level 3: Reasoning: Reason about subtracting mixed number fractions with related denominators.

✿ **Required:** 5/5   ✿ **Pupil Navigation:** on
✿ **Randomised:** off

21.
a
b
c
Henry has tried to subtract 1 2/3 from 3 2/9. Explain what Henry has done wrong and explain how he can calculate the answer correctly.

- Open question, no set answer

22.
a
b
c
What is the same and what is different about these calculations?

- Open question, no set answer

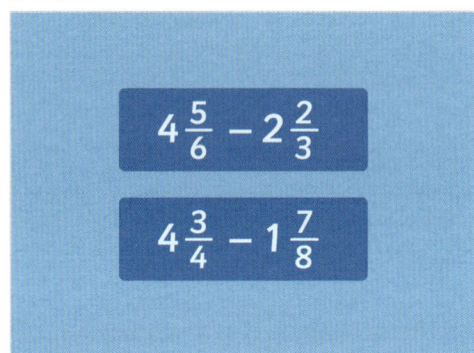

23.
Select **two** mixed number fractions which have a difference that is greater than 2 but less than 3.

2/4
- ▪ 2 7/8 ▪ 1 3/8 ▪ 5 3/8 ▪ 4 5/8

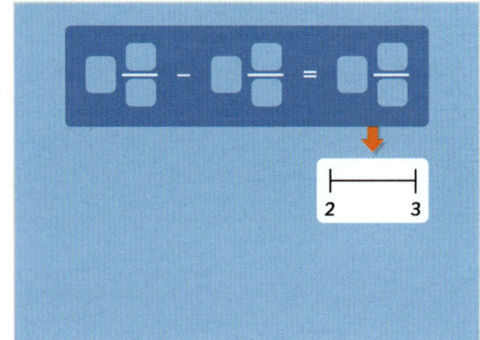

Level 3 *continued*

24. Abbie says, "When you subtract two smaller mixed number fractions from a larger mixed number fraction, your answer will **never** be a whole number." Is Abbie correct? Explain your answer and give an example.

- Open question, no set answer

$$5\tfrac{1}{4} - 1\tfrac{1}{2} - 1\tfrac{1}{8} = 2\tfrac{5}{8}$$

25. Two children are calculating 7 3/8 – 1 5/8. Grace converts both of the mixed number fractions to improper fractions before subtracting. Freddie converts 7 3/8 to 6 11/8 and then subtracts the whole numbers and fractions separately.
Calculate the answer and explain which method you prefer.

- Open question, no set answer

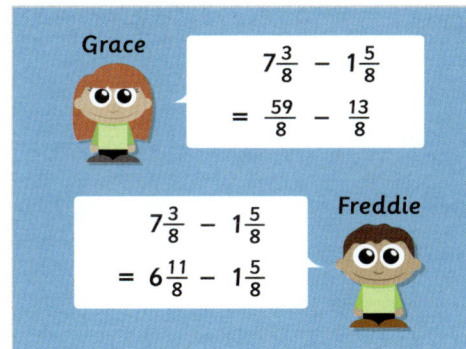

Grace

$$7\tfrac{3}{8} - 1\tfrac{5}{8}$$
$$= \tfrac{59}{8} - \tfrac{13}{8}$$

Freddie

$$7\tfrac{3}{8} - 1\tfrac{5}{8}$$
$$= 6\tfrac{11}{8} - 1\tfrac{5}{8}$$

Level 4: Problem solving with greater depth: Solve problems involving subtracting mixed number fractions with related denominators.

✳ **Required:** 5/5   ✳ **Pupil Navigation:** on
✳ **Randomised:** off

26. Jane pours 1 ¾ litres (l) of sand into a 3 ½ litre bucket. If she pours another ½ litre of sand into the bucket, how many more litres of sand can fit in the bucket?
Give your answer as a mixed number fraction. Don't include the units in your answer.

- ▪ 1 1/4 ▪ 1 1/4 l ▪ 1 3/12 ▪ 1 3/12 litres
- ▪ 1 3/12 l ▪ 1 1/4 litres ▪ 2 1/4 l ▪ 2 1/4
- ▪ 2 1/4 litres

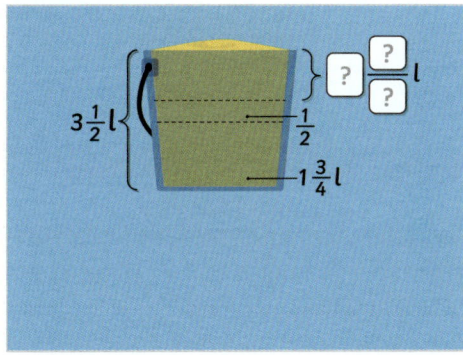

27. A triangle has a perimeter of 6 1/2 metres (m). One side of the triangle measures 1 3/8 m and another side is 2 1/4 m long. The missing side length is _____ m long.
Enter the missing mixed number fraction. Don't include the units in your answer.

- ▪ 2 7/8 ▪ 2 7/8 m ▪ 2 7/8 metres

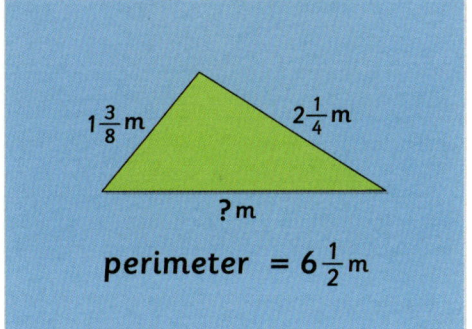

$$\text{perimeter} = 6\tfrac{1}{2}\text{ m}$$

Level 4 continued

28. On Thursday, Beth records how long she
a spends on different activities during the day.
b The next day Beth spends ½ an hour less
c playing and she goes to bed ¾ of an hour
later. On **Friday**, Beth spends _____ more
hours sleeping than playing.
Enter the missing mixed number fraction.
Don't include the units in your answer.

- ■ **9 1/8** ■ **9 1/8 hours** ■ **9 1/8 hrs** ■ 1 5/8 ■ 9 3/8
- ■ 10 3/4

Time Beth spends on different activities on Thursday

activity	time spent (hours)
sleeping	$11\frac{1}{2}$
eating	$1\frac{1}{4}$
lessons	$5\frac{1}{2}$
playing	$2\frac{1}{8}$
homework	$\frac{3}{8}$

29. A rectangle has a perimeter of 6 1/5 metres
a (m) and a side length of 1 7/10 m. The
b missing length is _____ m.
c *Enter the missing mixed number fraction.*
Don't include the units in your answer.

- ■ **1 2/5 m** ■ **1 4/10 metres** ■ **1 4/10 m**
- ■ 2 8/10 metres ■ 2 4/5 ■ **1 2/5** ■ **1 2/5 metres**
- ■ **1 4/10** ■ 2 4/5 metres ■ 2 8/10 m ■ 2 4/5 m
- ■ 2 8/10

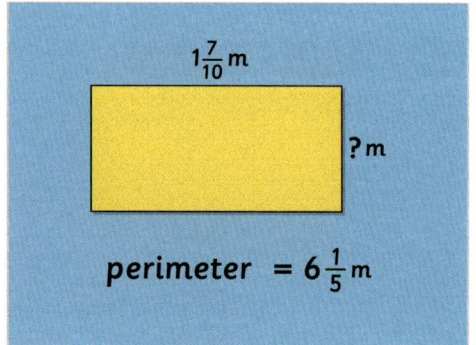

30. An electrician working in a school has 7
a metres (m) of wire. He uses 1 4/5 m of wire
b for the lights and 3 3/5 for the plug sockets.
c If the electrician needs 223 centimetres of
wire for the computers, what **fraction** of a
metre more wire will he need to complete
his work?
*Give your answer as a fraction. Don't include
the units in your answer.*

- ■ **63/100 m** ■ **63/100** ■ **63/100 metres** ■ 7 63/100
- ■ 7 63/100 metres ■ 7 63/100 m

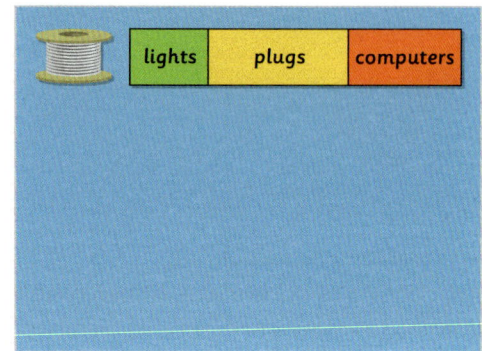

Subtract Proper Fractions from Mixed Number Fractions with Related Denominators

Objective: I can subtract proper fractions from mixed number fractions with the same denominator or with denominators that are multiples of the same number.

Quick Search Ref: 11481

Level 1: Understanding: Subtract proper fractions from mixed numbers with image support.

✱ **Required:** 7/10   ✱ **Pupil Navigation:** on   ✱ **Randomised:** off

---

1. What is 1 3/4 − 1/4?

a b c *Give your answer as a mixed number fraction. For example, five and one-half is 5 1/2.*

- 1 2/4 ▪ 24/16 ▪ 1 8/16 ▪ 1 1/2 ▪ 1/2 ▪ 3/2
- 2/4 ▪ 6/4

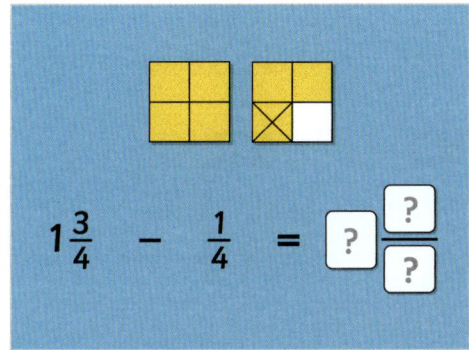

2. Calculate 2 5/6 − 1/2.

a b c *Give your answer as a mixed number fraction. For example, five and one-half is 5 1/2.*

- 2 4/12 ▪ 1/3 ▪ 2 1/3 ▪ 2 2/6 ▪ 2/6 ▪ 28/12
- 7/3

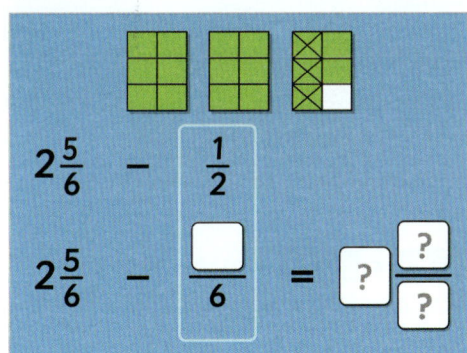

3. What is the answer to 5 3/10 − 7/10?

a b c *Give your answer as a mixed number fraction. For example, five and one-half is 5 1/2.*

- 4 6/10 ▪ 5 6/10 ▪ 46/10 ▪ 4 3/5 ▪ 5 3/5
- 23/5

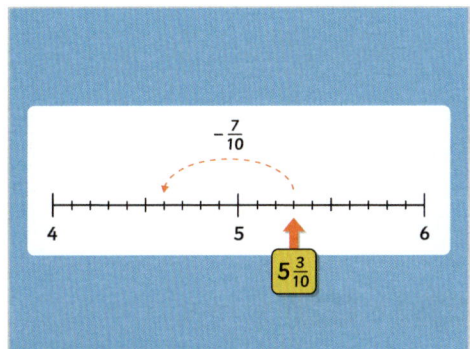

4. Subtract 1/4 from 1 1/8.

a b c
- 14/16 ▪ 1 1/8 ▪ 7/8 ▪ 28/32 ▪ 1/8

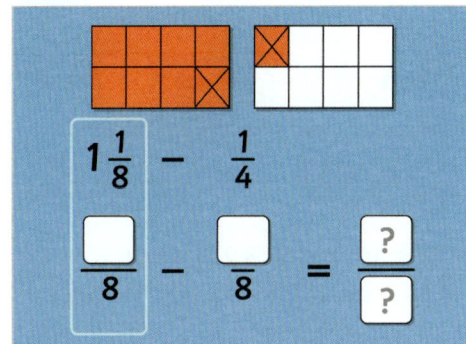

5. What is 3/10 subtracted from 3 1/5?

a b c *Give your answer as a mixed number fraction. For example, five and one-half is 5 1/2.*

- 29/10 ▪ 2 45/50 ▪ 2 9/10 ▪ 145/50

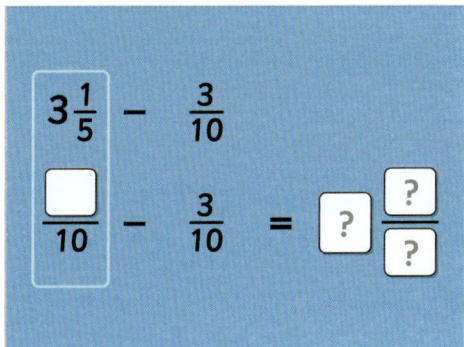

Level 1 continued

6. What is 2 1/6 − 2/3?

a b c *Give your answer as a mixed number fraction. For example, five and one-half is 5 1/2.*

■ 1 3/6 ■ 1 1/2 ■ 1 9/18 ■ 27/18 ■ 3/2 ■ 9/6

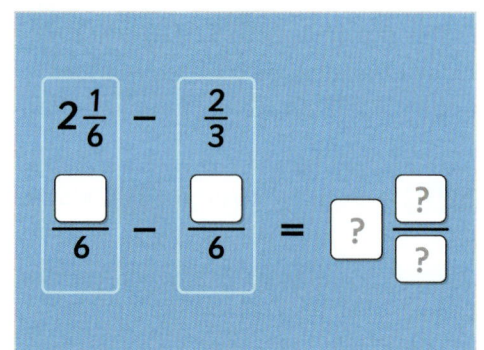

7. Calculate 4 3/5 − 7/10.

a b c *Give your answer as a mixed number fraction. For example, five and one-half is 5 1/2.*

■ 3 9/10 ■ 3 45/50 ■ 195/50 ■ 39/10

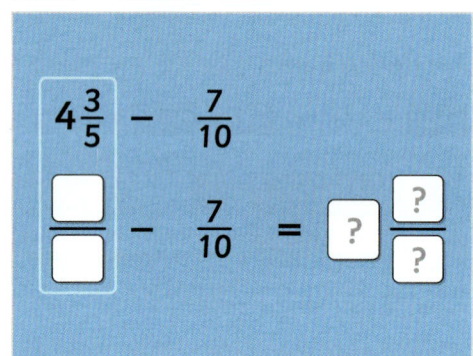

8. Find the answer to 3 2/3 − 8/9.

a b c *Give your answer as a mixed number fraction. For example, five and one-half is 5 1/2.*

■ 75/27 ■ 2 21/27 ■ 2 7/9 ■ 25/9

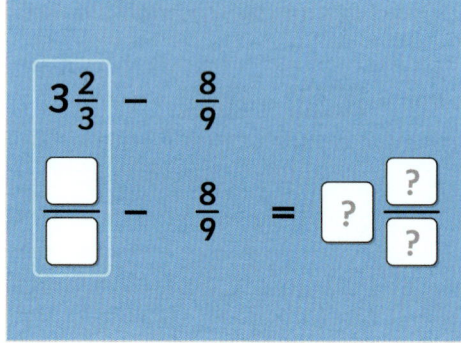

9. Find 1 1/3 − 5/12.

a b c ■ 11/12 ■ 33/36 ■ 1 1/12

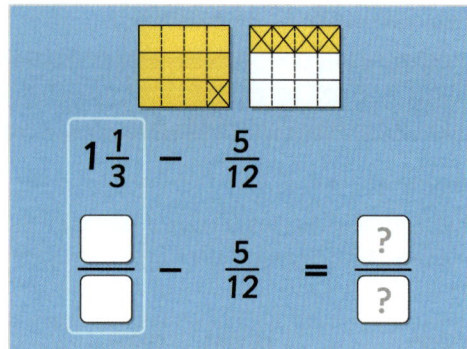

10. Calculate 1 5/8 − 1/2.

a b c *Give your answer as a mixed number fraction. For example, five and one-half is 5 1/2.*

■ 1 1/8 ■ 18/16 ■ 1 2/16 ■ 1/8 ■ 9/8

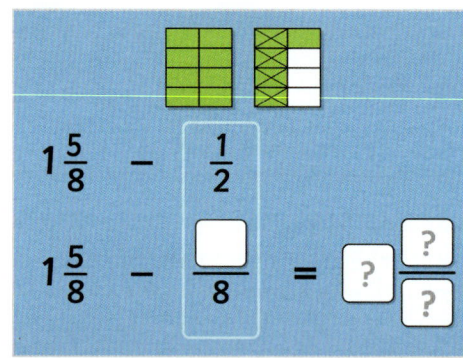

Level 2: Fluency: Subtract proper fractions from mixed numbers including one-step word problems.

✹ **Required:** 7/10   ✹ **Pupil Navigation:** on
✹ **Randomised:** off

11. What is 2 4/7 − 5/14?

a b c *Give your answer as a mixed number fraction.*

■ 2 3/14 ■ 2 21/98 ■ 217/98 ■ 31/14

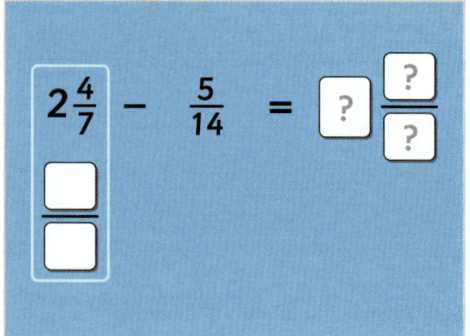

Level 2 *continued*

12. Calculate the answer to 1 3/8 – 1/2 – 5/8.

a b c ▪ 6/8 ▪ 2/8 ▪ 4/16 ▪ 1/4 ▪ 7/8 ▪ 3/4

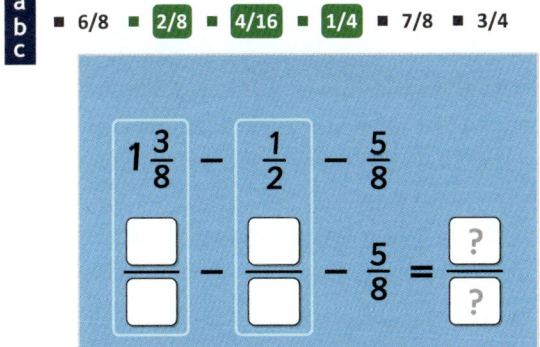

13. Find the difference between 7/9 and 3 1/3.
a b c *Give your answer as a mixed number fraction.*

▪ 69/27 ▪ 2 5/9 ▪ 2 15/27 ▪ 23/9

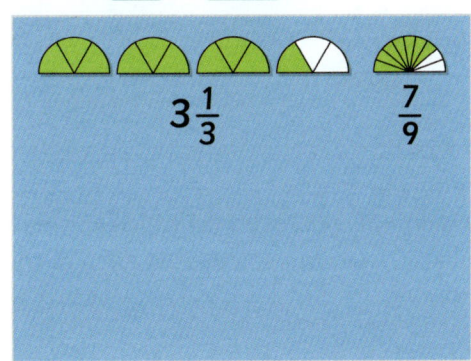

14. Calculate the answer to 3 1/6 – 2/3 – 11/12.
a b c *Give your answer as a mixed number fraction.*

▪ 1 7/12 ▪ 19/12 ▪ 2 3/6 ▪ 2 1/2

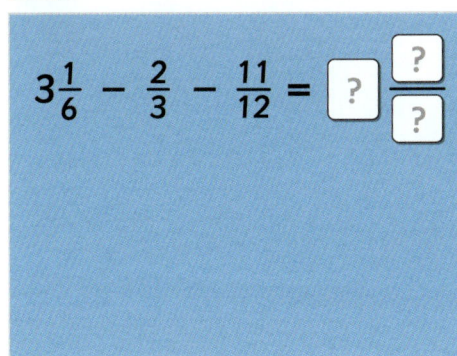

15. William has 2 ¾ bars of chocolate. He gives
a b c Sophie ½ a bar. How many bars of chocolate does William have left?
Give your answer as a mixed number fraction.

▪ 2 1/4 ▪ 2 2/8 ▪ 2 ▪ 9/4 ▪ 1/4

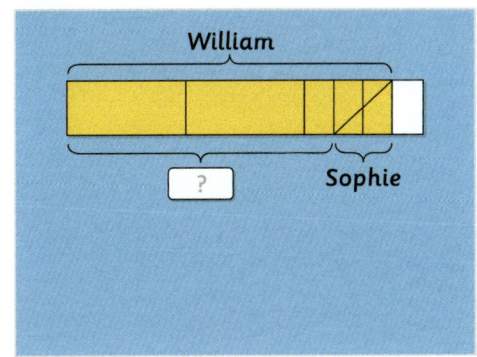

16. There are 2 ½ boxes of pens in the school
a b c cupboard. Mrs Walsh takes ¾ of a box of pens. How many boxes of pens are left?
Give your answer as a mixed number fraction.

▪ 1 6/8 ▪ 1 3/4 ▪ 1 ▪ 14/8 ▪ 7/4

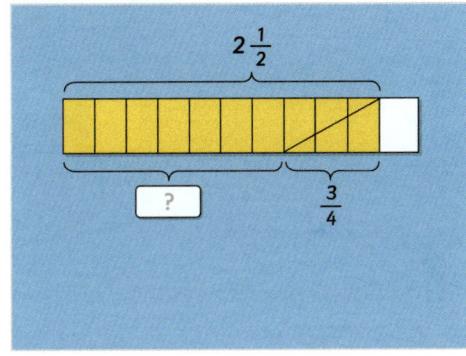

17. Mr Brown has 3 1/10 bags of corn. If Mr
1 2 3 Brown's chickens eat 3/5 of a bag of corn, how many **full bags** of corn does Mr Brown have left?

▪ 2

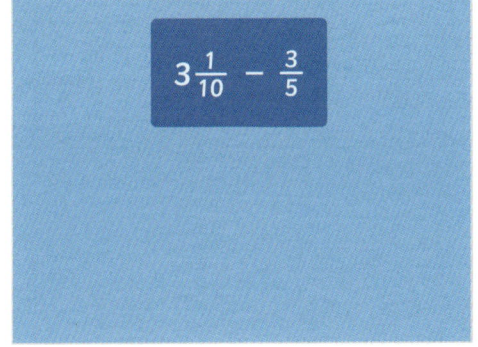

Level 2 *continued*

18. Joe has 4 1/3 packs of apples. He eats 5/6 of
a pack. How many **full packs** of apples does
Joe have left?

[1][2][3]

- **3**

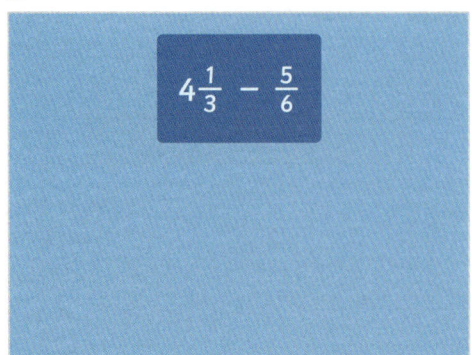

$$4\frac{1}{3} - \frac{5}{6}$$

19. Calculate the answer to 2 3/12 − 1/3 − 5/6.
[a][b][c] *Give your answer as a mixed number*
fraction.

- **1 1/12** ▪ 1 11/12 ▪ 1 5/12

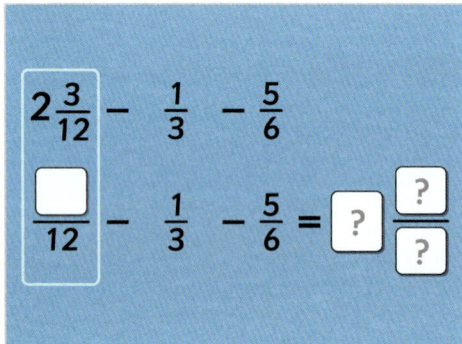

$$2\frac{3}{12} - \frac{1}{3} - \frac{5}{6}$$

$$\frac{\square}{12} - \frac{1}{3} - \frac{5}{6} = \boxed{?}\frac{\boxed{?}}{\boxed{?}}$$

20. Find the difference between 1 7/18 and 2/9.
[a][b][c] *Give your answer as a mixed number*
fraction.

- **1 3/18** ▪ 21/18

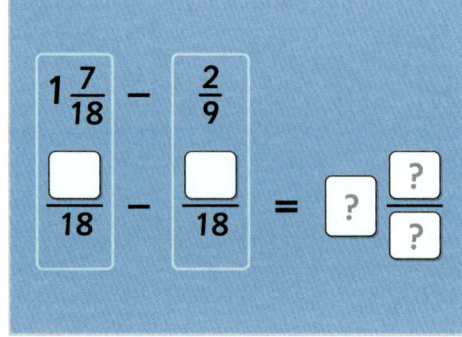

$$1\frac{7}{18} - \frac{2}{9}$$

$$\frac{\square}{18} - \frac{\square}{18} = \boxed{?}\frac{\boxed{?}}{\boxed{?}}$$

Level 3: Reasoning: Reason about subtracting proper
fractions from mixed number fractions with
related denominators.

✿ **Required:** 5/5   ✿ **Pupil Navigation:** on
✿ **Randomised:** off

21. Jenny says that if she subtracts any fraction
[a][b][c] greater than 1/6 from 3 1/6, her answer will
be less than 3.
Explain why Jenny is correct. Give an
example to show her statement is correct.

- Open question, no set answer

$$3\frac{1}{6} - \frac{1}{6}$$

$$3\frac{1}{6} -$$

22. One of these calculations has an answer that
[a][b][c] is equivalent to 3.
Explain how you can find which calculation
equals 3, without calculating each of the
answers.

- Open question, no set answer

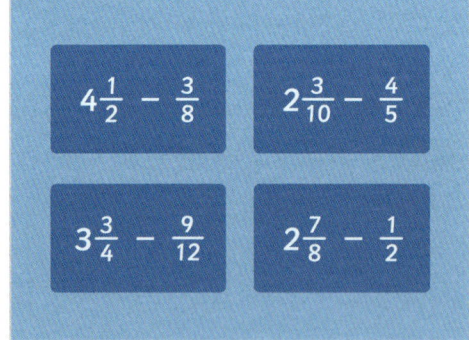

$$4\frac{1}{2} - \frac{3}{8} \qquad 2\frac{3}{10} - \frac{4}{5}$$

$$3\frac{3}{4} - \frac{9}{12} \qquad 2\frac{7}{8} - \frac{1}{2}$$

Level 3 *continued*

23. Find the missing proper fraction in this
a calculation:
b
c 1 3/7 − ___ = 9/14

- 11/14

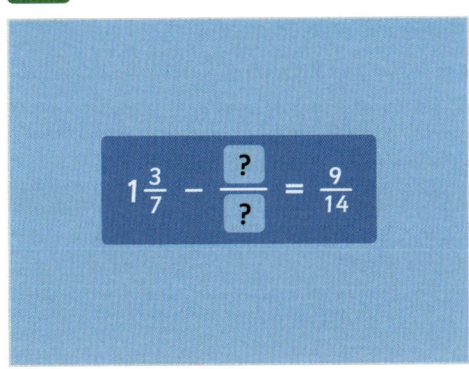

24. There are enough children in Year 5 to make
a 3 2/3 netball teams. If 5/6 of a team are ill,
b will there still be enough children to make
c three full teams? Explain your answer and
use a calculation to prove how many full
teams can be made.

- Open question, no set answer

25. Sammy has tried to calculate 3 4/5 − 7/10.
a Explain what he has done wrong and
b calculate the correct answer.
c

- Open question, no set answer

Level 4: Problem Solving with Greater Depth: Solve
problems involving the subtraction of proper
fractions from mixed number fractions with
related denominators.

✱ **Required:** 5/5   ✱ **Pupil Navigation:** on
✱ **Randomised:** off

26. The perimeter of the triangle shown is 2 1/4
a metres (m).
b The missing side length measures ____
c metres.
Enter the missing mixed number fraction.
Don't include the units in your answer.

- 1 3/8 - 1 3/8 metres - 11/8 metres - 1 3/8 m
- 11/8 m - 11/8

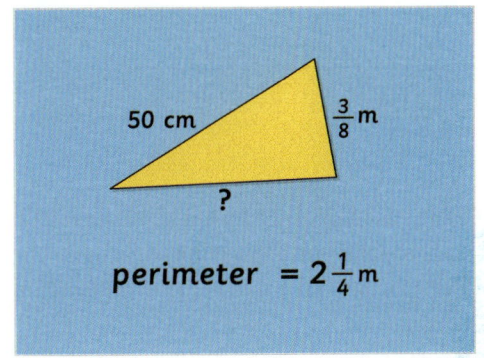

27. A group of children have 7 1/3 litres (l) of
a water to use on the school garden. They
b pour 7/12 litres on the carrot plants and
c another 5/6 litres on the onions. The
children use less than one litre of water on
the lettuces.
If the children have exactly 5 litres of water
left, how many litres of water do they use on
the lettuces?
Give your answer as a fraction. Don't include
the units in your answer.

- 11/12 l - 11/12 litres - 11/12 - 5 11/12
- 5 11/12 l - 5 11/12 litres

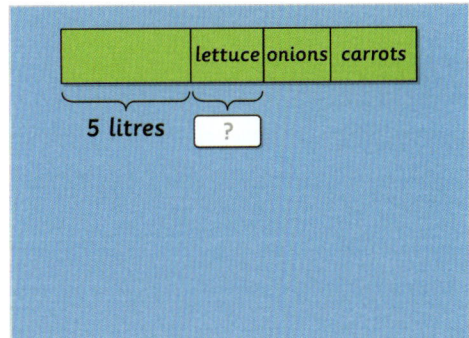

Level 4 continued

28. The table shows how many lengths three children can swim using different swimming strokes.
What is the furthest distance that a child swims doing the butterfly stroke?
Give your answer as a fraction. Don't include the units in your answer.

■ 15/16 ■ 7/8

Swimming Lengths

name	front crawl	back crawl	butterfly	total
Amy	$\frac{7}{8}$	$\frac{3}{4}$		$2\frac{1}{2}$
Sam	$\frac{11}{16}$	$\frac{5}{8}$		$2\frac{1}{4}$
Zoe	$\frac{7}{8}$	$\frac{13}{16}$	$\frac{1}{2}$	$2\frac{3}{16}$

29. Max is exactly 10 2/3 years old. Ben is 7 months younger than Max.
What fraction of a year is it until Ben's next birthday?
Give your answer as a fraction. Don't include the units in your answer.

■ 11/12 ■ 10 1/12 ■ 11

Max
Ben
$10\frac{2}{3}$ years old
7 months younger than Max

30. Chloe has a piece of ribbon 1 7/10 metres (m) long. She cuts off a piece 1/5 m long.
Ella has a piece of ribbon 1 17/20 m long. She cuts off one piece that is 3/5 m and another piece measuring 7/10 m.
How many more metres of ribbon does Chloe have left than Ella?
Give your answer as a fraction. Don't include the units in your answer.

■ 1 5/10 metres ■ 19/20 ■ 11/20 metres
■ 19/20 metres ■ 11/20 m ■ 1 1/2 ■ 19/20 m
■ 11/20 ■ 1 1/2 m ■ 1 5/10 ■ 1 5/10 m
■ 1 1/2 metres

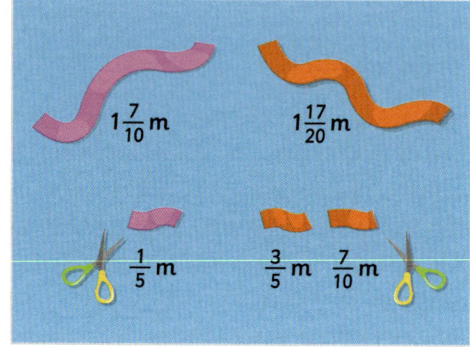

$1\frac{7}{10}$ m $1\frac{17}{20}$ m

$\frac{1}{5}$ m $\frac{3}{5}$ m $\frac{7}{10}$ m

Subtract Proper Fractions with Related Denominators

Objective: I can subtract fractions with the same denominator or with denominators that are multiples of the same number.

Quick Search Ref: 11472

Level 1: Understanding: Subtract proper fractions with related denominators with image support.

✺ Required: 7/10   **✺ Pupil Navigation:** on   **✺ Randomised:** off

---

1. To calculate $1/4 - 1/8$ you need to make the denominators the same by finding the lowest number that is a multiple of both 4 and 8.

What is the lowest number that is a multiple of both denominators?

■ 16   ■ **8**   ■ 32   ■ 2

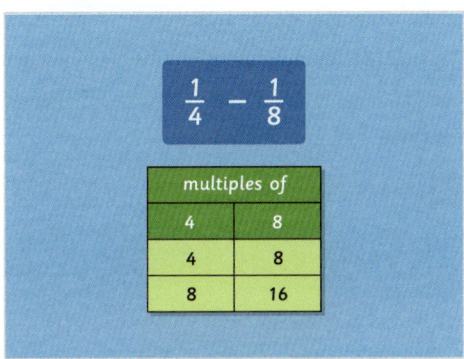

2. What is $1/4 - 1/8$?

Put a forward slash (/) between the numerator and denominator. For example, one-half is 1/2.

■ **1/8**   ■ **2/16**   ■ **4/32**   ■ 1/4   ■ 3/8

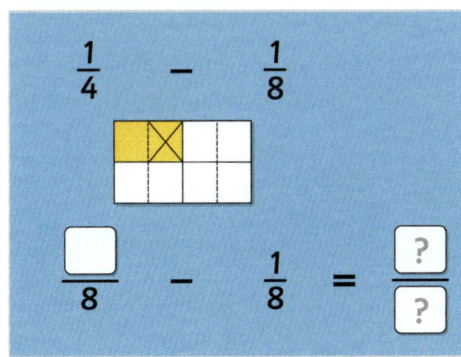

3. What is the lowest number that is a multiple of both 9 and 27 that you would use to calculate $1/9 - 1/27$?

■ 18   ■ **27**   ■ 243   ■ 54

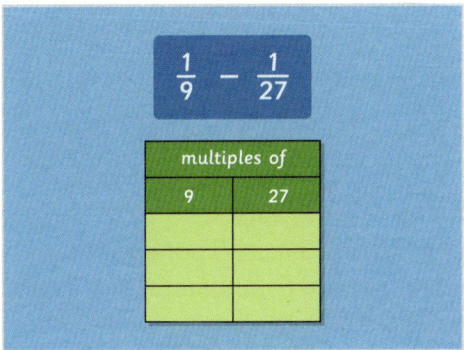

4. What is $5/6 - 1/3$?

Put a forward slash (/) between the numerator and denominator. For example, one-half is 1/2.

■ **3/6**   ■ **9/18**   ■ **1/2**   ■ 4/3

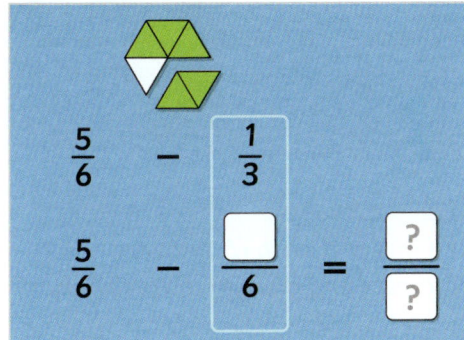

Level 1 continued

5. 1/3 − 1/9 = ___
    a b c *Put a forward slash (/) between the numerator and denominator. For example, one-half is 1/2.*

    ▪ 4/18 ▪ 6/27 ▪ 2/9 ▪ 4/9 ▪ 1/6

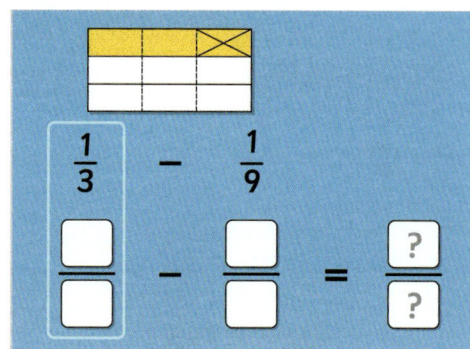

6. What is 1/4 − 1/16?
    a b c *Put a forward slash (/) between the numerator and denominator. For example, one-half is 1/2.*

    ▪ 12/64 ▪ 3/16 ▪ 6/32 ▪ 1/12 ▪ 5/16

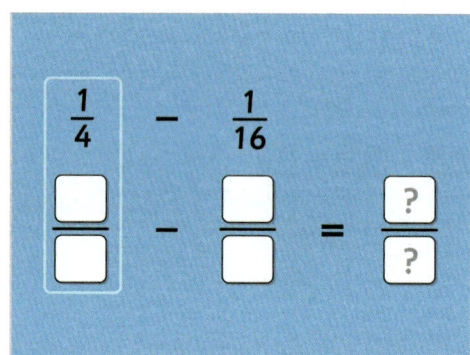

7. 19/27 − 1/3 = ___
    a b c *Put a forward slash (/) between the numerator and denominator. For example, one-half is 1/2.*

    ▪ 1 1/27 ▪ 10/27 ▪ 30/81 ▪ 28/27 ▪ 18/24

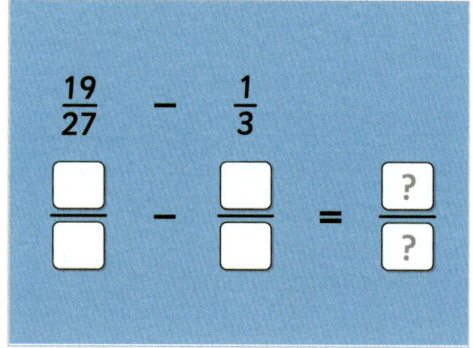

8. Calculate the answer to 1/5 − 1/25.
    a b c *Put a forward slash (/) between the numerator and denominator. For example, one-half is 1/2.*

    ▪ 4/25 ▪ 20/125 ▪ 1/20 ▪ 6/25

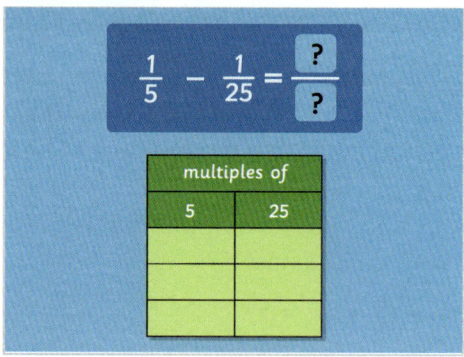

9. What is the lowest number that is a multiple of both 11 and 55 that you would use to calculate 1/11 − 1/55?
    1 2 3

    ▪ 55 ▪ 110 ▪ 605 ▪ 11

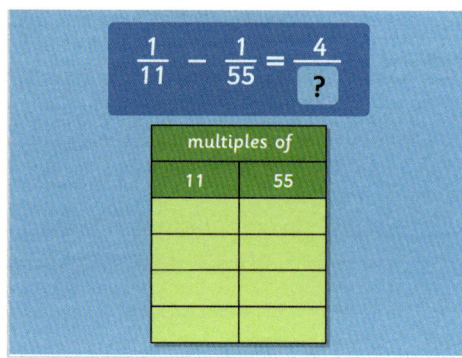

10. Calculate 7/10 − 3/5.
    a b c *Put a forward slash (/) between the numerator and denominator. For example, one-half is 1/2.*

    ▪ 1/10 ▪ 5/50 ▪ 5/10 ▪ 4/5

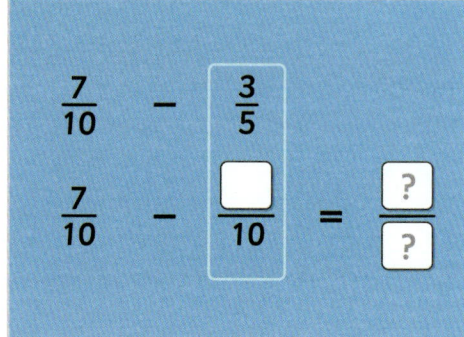

Level 2: Fluency: Subtract two or more proper fractions with related denominators.

✳ **Required:** 7/10   ✳ **Pupil Navigation:** on
✳ **Randomised:** off

11. Find the answer to 4/7 − 5/14.

a
b
c
Put a forward slash (/) between the numerator and denominator. For example, one-half is 1/2.

▪ 3/14 ▪ 21/98 ▪ 7/14 ▪ 1/14

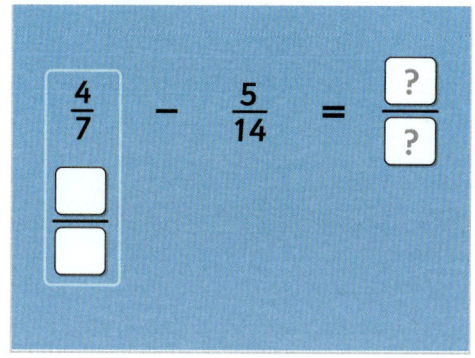

12. What is the lowest number that is a multiple of 3, 9 and 27?

1
2
3

▪ 729 ▪ 27 ▪ 54 ▪ 18

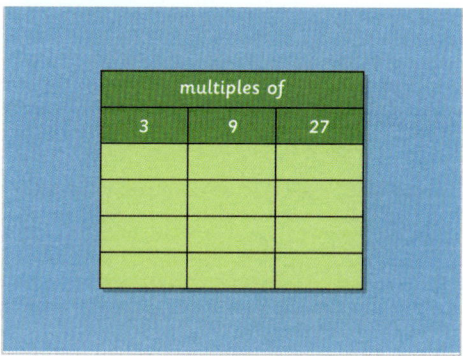

13. What is 1/3 − 1/9 − 1/27?

a
b
c
Put a forward slash (/) between the numerator and denominator. For example, one-half is 1/2.

▪ 5/27 ▪ 13/27

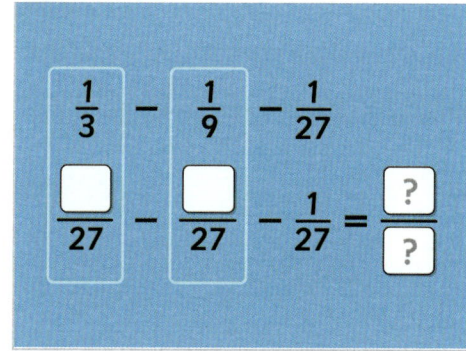

14. Calculate 11/12 − 1/24 − 5/6.

a
b
c
Put a forward slash (/) between the numerator and denominator. For example, one-half is 1/2.

▪ 1/24

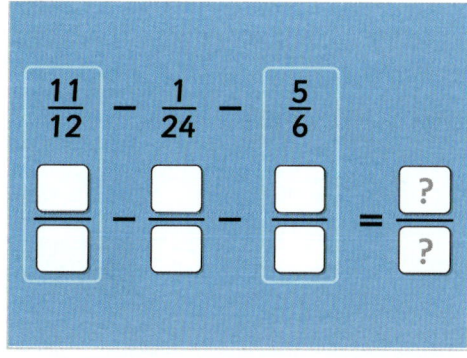

15. 14/25 of a class have school dinners. If 2/5 of the class change to packed lunches, what fraction of the class still has school dinners?

a
b
c
Put a forward slash (/) between the numerator and denominator. For example, one-half is 1/2.

▪ 4/25 ▪ 20/125 ▪ 24/25

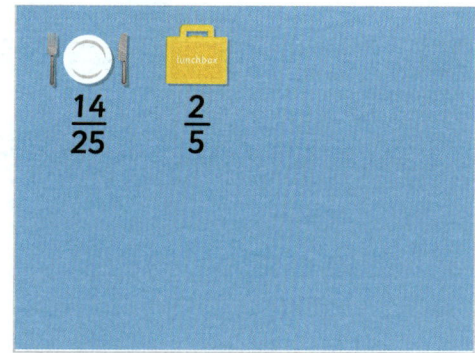

16. Adil has 4/5 of a kilogram (kg) of flour. He spills 3/10 of a kilogram on the floor. How many kilograms of flour does Adil have left?

a
b
c
Give your answer as a fraction. Don't include the units in your answer.

▪ 5/10 kg ▪ 5/10 ▪ 1/2 ▪ 1/2 kilogram ▪ 1/2 kg
▪ 5/10 kilogram

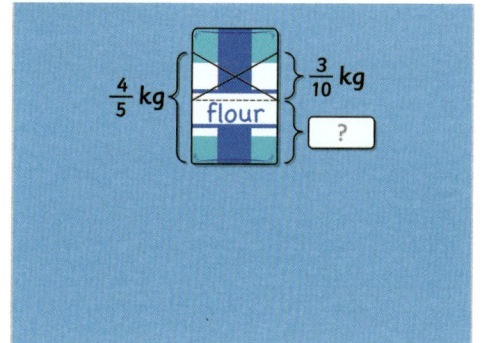

Level 2 continued

17. Josie has a full bar of chocolate. She gives
a b c 3/18 of the bar to Ellie and 17/36 to Izzy, and
she eats 1/9 of the bar. What fraction of a
bar of chocolate does Josie have left?
*Put a forward slash (/) between the
numerator and denominator. For example,
one-half is 1/2.*

■ 9/36 ■ 1/4 ■ 3/12 ■ 1 19/36

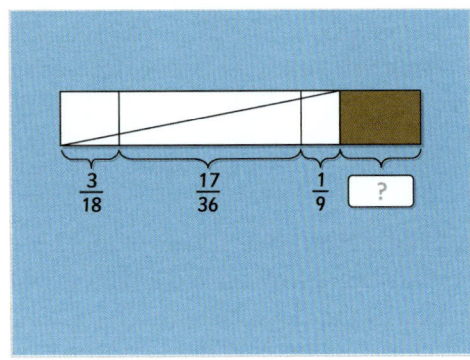

18. What is 11/14 − 5/28 − 2/7?
a b c *Put a forward slash (/) between the
numerator and denominator. For example,
one-half is 1/2.*

■ 9/28

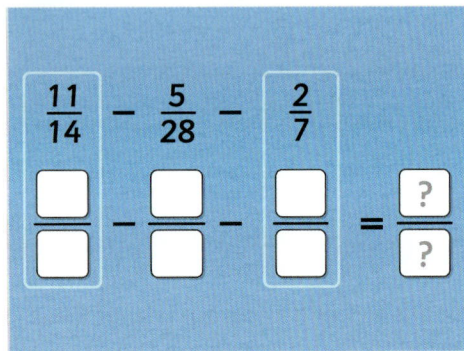

19. What is the lowest number that is a multiple
1 2 3 of all the denominators that you would use
to calculate 7/8 − 3/24 − 1/4?
*Put a forward slash (/) between the
numerator and denominator. For example,
one-half is 1/2.*

■ 24 ■ 8 ■ 768

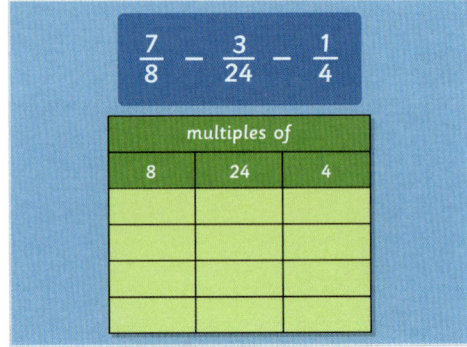

20. 11/12 − 5/24 = ___?
a b c *Put a forward slash (/) between the
numerator and denominator. For example,
one-half is 1/2.*

■ 17/24

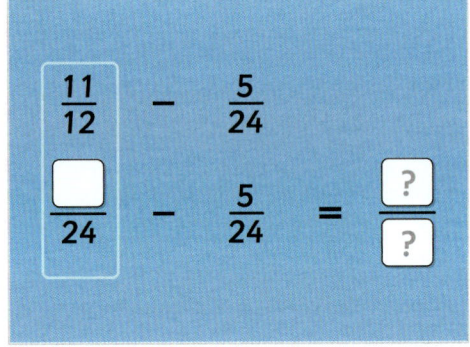

Level 3: Reasoning: Reason about subtracting fractions with related denominators.

✴ **Required:** 5/5 ✴ **Pupil Navigation:** on
✴ **Randomised:** off

21. Use three of the number cards to make the statement correct. What number goes in box C?

- ■ **6** ■ 1 ■ 15

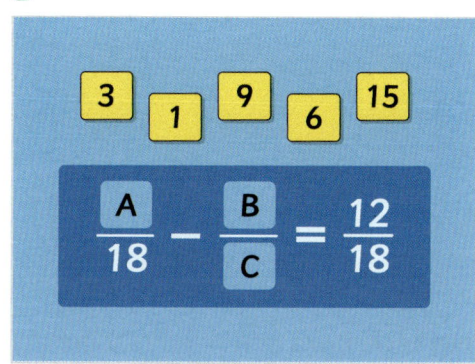

22. Dylan says he knows that 4/6 − 1/2 is less than 2/3 − 5/12 without having to complete the calculations. Explain how he might know this.

- Open question, no set answer

23. Harry calculates that 30/48 − 5/12 = 25/36. Explain what Harry has done wrong and explain how he can calculate the answer correctly.

- Open question, no set answer

24. Bella says that ½ − ¼ doesn't equal zero, so if you subtract fractions with different denominators, you will never get the answer zero.
Is Bella Correct? Explain your answer, giving an example.

- Open question, no set answer

25. What is the same and what is different about these two calculations?

- Open question, no set answer

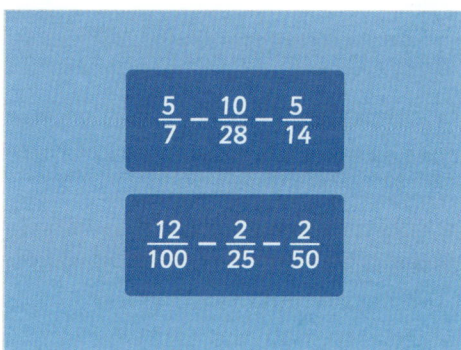

Level 4: Problem solving with greater depth: Solve problems involving the subtraction of fractions with related denominators.

❋ **Required:** 5/5   ❋ **Pupil Navigation:** on
❋ **Randomised:** off

26. Lexi's piano practice lasts for 5/6 of an hour.
a b c If she has been practising for 3/12 of an hour, what **fraction of an hour** does she have left?
Give your answer as a fraction. Don't include the units in your answer.

▪ 7/12 ▪ 35/60 ▪ 35

27. Billy is making a model of a Viking longboat.
a b c He has a piece of wood that is 3/4 metre (m) long.
Billy cuts off a piece for the mast which measures 3/16 m. He cuts off another piece for the deck which measures 1/8 m and a final piece to make the oars which measures 3/8 m.
How many metres of wood does Billy have left?
Give your answer as a fraction. Don't include the units in your answer.

▪ 1/16 ▪ 1/16 metres ▪ 1/16 m ▪ 11/16
▪ 11/16 metres ▪ 11/16 m

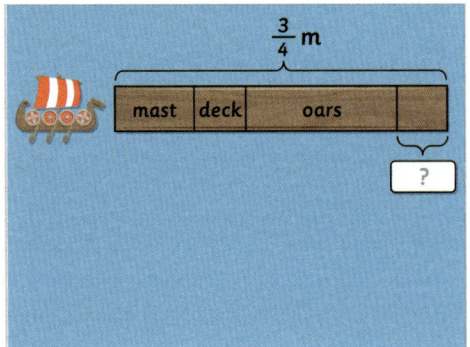

28. Ellie and Rio each have a box of chocolates.
a b c Ellie eats 4/9 of her chocolates and Rio eats 23/36 of his box.
How much more of her box of chocolates does Ellie have left than Rio?
Give your answer as a fraction.

▪ 7/36 ▪ 63/324

29. A rectangle has a perimeter of 1 metre (m)
a b c and a side length of 5/12 of a metre.
How many metres does the shorter side length measure?
Give your answer as a fraction. Don't include the units in your answer.

▪ 1/12 ▪ 1/12 m ▪ 1/12 metres ▪ 2/12 metres
▪ 2/12 ▪ 2/12 m

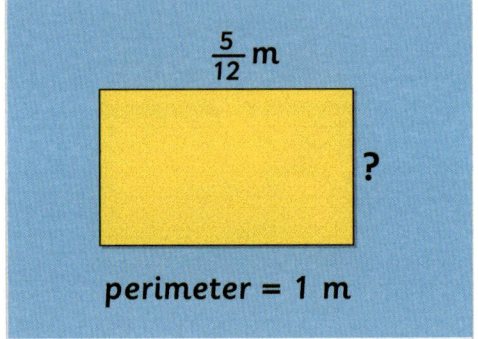

Level 4 *continued*

30. To cross the river, find the answer to each of the calculations and arrange the stepping stones starting with the smallest answer first.

- 1/3 – 5/48 ■ 45/48 – 5/6 ■ 1/2 – 5/12
- 5/6 – 11/24

Multiply Mixed Number Fractions by Whole Numbers

Objective: I can multiply mixed number fractions by whole numbers.

Quick Search Ref: 10198

Level 1: Understanding: Multiply mixed numbers by whole numbers with image support.

❋ Required: 7/10 **❋ Pupil Navigation:** on **❋ Randomised:** off

---

1. $1\frac{1}{7} \times 2 =$ ___ wholes and 2/7.

■ **2** ■ 1

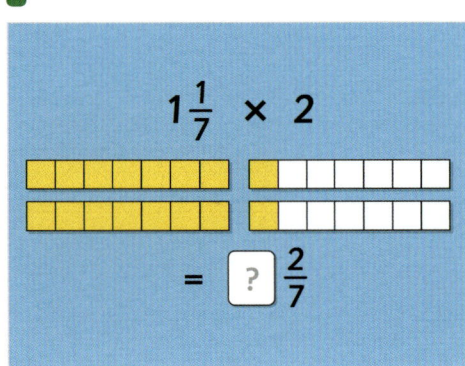

2. $2\frac{1}{4} \times 3 = 6$ wholes and ___ quarters.

■ **3**

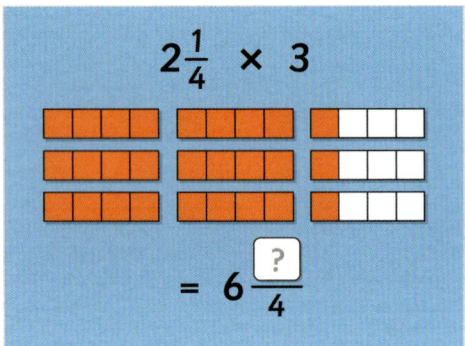

3. What is the missing mixed number fraction?
Give your answer as a mixed number fraction. For example, five and one-half is 5 1/2.

■ **4 4/9**

$$2 \times 2\frac{2}{9}$$

$$2 \times \frac{20}{9} = \frac{40}{9}$$

$$= \boxed{?}\,\frac{\boxed{?}}{\boxed{?}}$$

4. What is $1\frac{2}{5} \times 2$?
Give your answer as a mixed number fraction. For example, five and one-half is 5 1/2.

■ 7/5 ■ **2 4/5** ■ 2 2/5 ■ 2 4/10 ■ 14/5

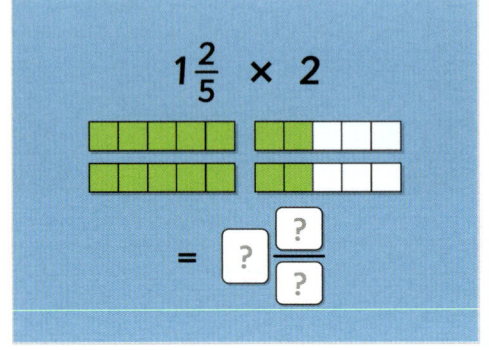

5. What is 1 1/9 multiplied by 4?
Give your answer as a mixed number fraction. For example, five and one-half is 5 1/2.

■ 10/9 ■ 40/9 ■ **4 4/9** ■ 4 1/9 ■ 4 4/36

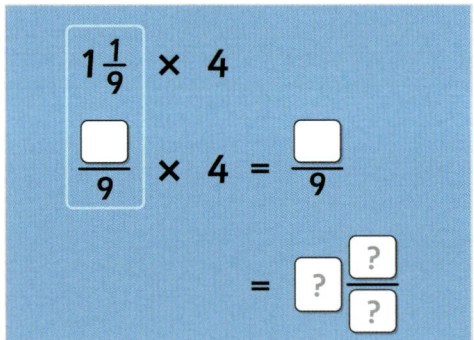

Level 1 *continued*

6. What is 3 multiplied by 2 1/4?
a
b *Give your answer as a mixed number*
c *fraction. For example, five and one-half is 5 1/2.*

▪ 9/4 ▪ **6 3/4** ▪ 6 1/4 ▪ 6 3/12

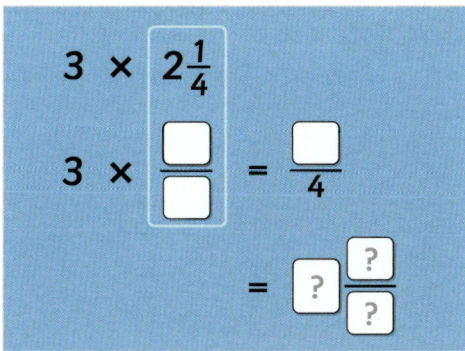

7. Calculate 5 × 1 1/6.
a
b *Give your answer as a mixed number*
c *fraction. For example, five and one-half is 5 1/2.*

▪ **5 5/6** ▪ 35/6 ▪ 5 1/6 ▪ 7/6 ▪ 5 5/30

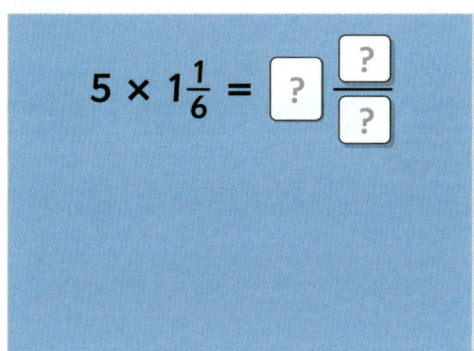

8. Find the answer to 4 2/5 × 2.
a
b *Give your answer as a mixed number*
c *fraction. For example, five and one-half is 5 1/2.*

▪ 8 4/10 ▪ **8 4/5** ▪ 44/5 ▪ 8 2/5

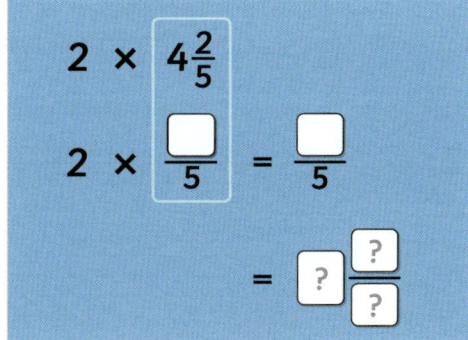

9. What is the missing mixed number fraction?
a *Give your answer as a mixed number*
b
c *fraction. For example, five and one-half is 5 1/2.*

▪ **6 4/7**

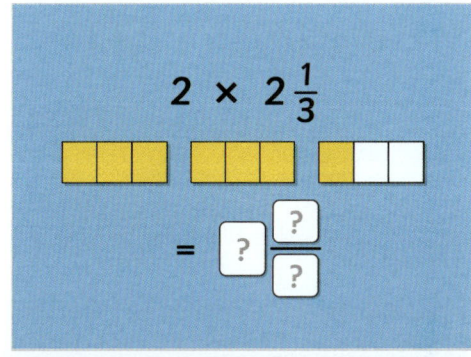

10. 2 × 2 1/3 = ____
a
b *Give your answer as a mixed number*
c *fraction. For example, five and one-half is 5 1/2.*

▪ 7/3 ▪ **4 2/3** ▪ 14/3 ▪ 4 2/6 ▪ 4 1/3

$$2 \times 2\frac{1}{3}$$

= ? ?/?

Level 2:   Fluency: Multiply mixed number fractions by whole numbers and one-step word problems.

✿ **Required:** 7/10   ✿ **Pupil Navigation:** on
✿ **Randomised:** off

11. Calculate 1 2/3 × 2.
a
b *Give your answer as a mixed number*
c *fraction. For example, five and one-half is 5 1/2.*

▪ 5/3 ▪ **3 1/3** ▪ 2 4/3 ▪ 10/3 ▪ 2 1/3 ▪ 2 2/3

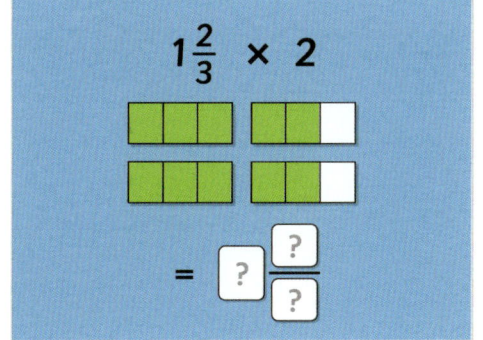

Level 2 *continued*

12. $5 \times 1\,5/6 =$ ___

Give your answer as a mixed number fraction.

- 9 1/6 ■ 55/6

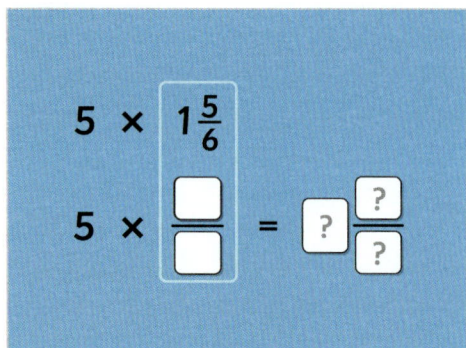

13. What is $1\,4/9$ multiplied by 3?

Give your answer as a mixed number fraction.

- 4 1/3 ■ 4 3/9 ■ 39/9

$$3 \times 1\frac{4}{9} = \boxed{?}\;\frac{\boxed{?}}{\boxed{?}}$$

14. Calculate $4 \times 2\,6/7$.

Give your answer as a mixed number fraction.

- 11 3/7 ■ 8 24/7 ■ 80/7

$$4 \times 2\frac{6}{7} = \boxed{?}\;\frac{\boxed{?}}{\boxed{?}}$$

15. Ian buys four pieces of wood which each measure 1 3/7 metres (m). How many metres of wood does Ian have in total?
Give your answer as a mixed number fraction. Don't include the units in your answer.

- 5 5/7 ■ 5 5/7metres ■ 5 5/7m ■ 5 5/7 m
- 5 5/7 metres ■ 40/7

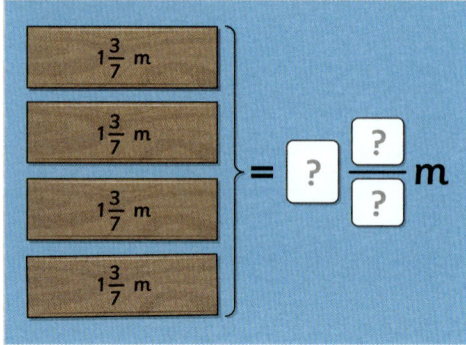

16. Maggie exercises for 1 3/4 hours three times each week. How many **hours** does she exercise for in total in one week?
Give your answer as a mixed number fraction. Don't include the units in your answer.

- 5 1/4 ■ 5 1/4hours ■ 5 1/4 hrs ■ 5 1/4 hours
- 5 1/4hrs ■ 21/4

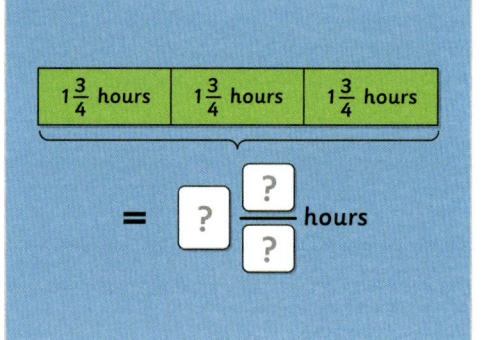

Level 2 *continued*

17. The width of the rectangle is 4 4/5 metres
a b c (m). The length of the rectangle is twice as long as its width. What is the length of the rectangle?

Give your answer as a mixed number fraction. Include the units m (metres) in your answer.

- 9 3/5m - 9 3/5metres - 9 3/5 metres - 9 3/5
- 9 3/5 m - 48/5

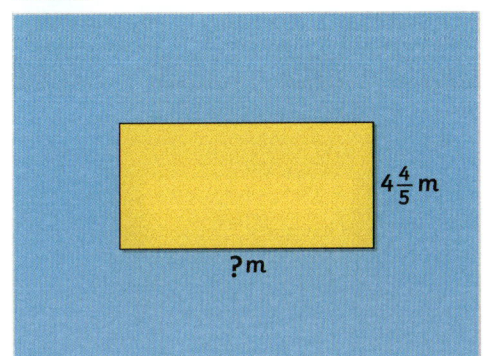

18. Each oil drum contains 2 2/3 litres (l) of oil.
a b c Four oil drums contain _____ litres of oil.
Enter the missing mixed number fraction. Don't include the units in your answer.

- 10 2/3 - 10 2/3l - 10 2/3 l - 10 2/3 litres
- 8 2/3 - 10 2/3litres - 32/3 - 2 2/3

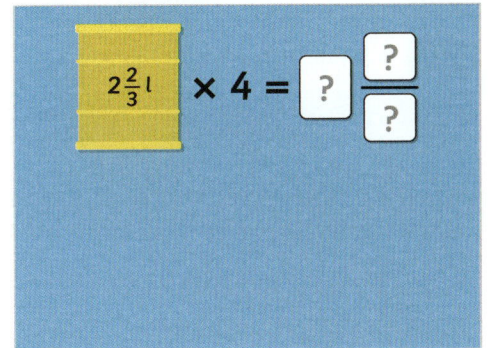

19. Calculate 4 × 1 3/5.
a b c *Give your answer as a mixed number fraction.*

- 32/5 - 6 2/5 - 2 2/5 - 4 3/5

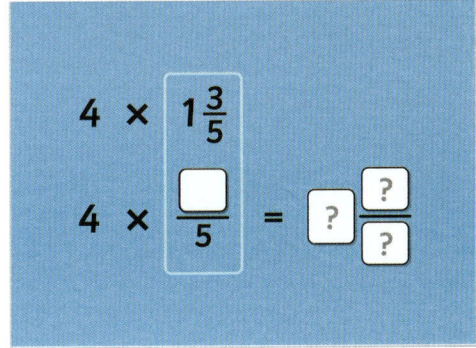

20. 4 × 2 1/3 = ____
Give your answer as a mixed number fraction.

1/5 - 8 1/3 - 9 1/3 - 28/3 - 1 1/3 - 8 4/3

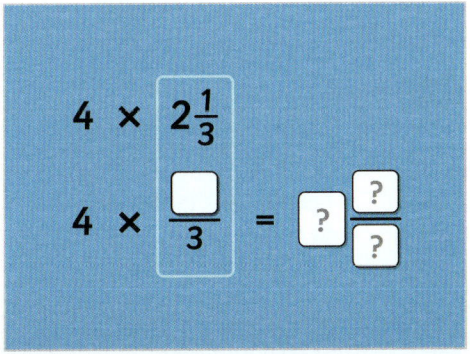

Level 3: Reasoning: Compare and reason about mixed number fraction multiplication.

✿ **Required:** 5/5   ✿ **Pupil Navigation:** on
✿ **Randomised:** off

21. Sally says that 5 × 1 3/4 = 3 3/4. Explain why
a b c Sally is incorrect and give the correct answer.

- Open question, no set answer

22. Matthew knows that 2 × 1 5/9 = 3 1/9.
a b c Explain how Matthew can use this to find 20 × 1 5/9 and calculate the answer.

- Open question, no set answer

Level 3 *continued*

23. What is the missing mixed number fraction
a that completes the following statement?
b
c ____ ÷ 5 = 1 2/9

Give your answer as a mixed number fraction.

- ■ 6 1/9 ■ 55/9

$$\boxed{?}\,\dfrac{\boxed{?}}{\boxed{?}} \div 5 = 1\dfrac{2}{9}$$

24. Two of these calculations have equivalent
a answers. Explain how to find the calculations
b with equivalent answers.
c

- Open question, no set answer

$$4 \times 2\dfrac{1}{4}$$

$$2 \times 4\dfrac{1}{4}$$

$$4 \times 2\dfrac{1}{2}$$

$$2 \times 4\dfrac{1}{2}$$

25. Amy says, "5 × 4 1/5 is greater than 4 × 5 1/4
a because the number the fraction is
b multiplied by is greater." Prove that Amy is
c incorrect.

- Open question, no set answer

$$5 \times 4\dfrac{1}{5} > 4 \times 5\dfrac{1}{4}$$

Level 4: Problem solving: Problem solving with mixed number fraction multiplication.

✸ **Required:** 5/5   ✸ **Pupil Navigation:** on
✸ **Randomised:** off

26. Julie wants to make 12 portions of coconut
a macaroons. What is the total weight in
b ounces (oz.) of the ingredients she needs?
c
Give your answer as a mixed number fraction. Don't include the units in your answer.

- ■ 25 7/8ounces ■ 25 7/8 oz. ■ 25 7/8 ounces
- ■ 25 7/8oz ■ 25 7/8 ■ 25 7/8 oz ■ 25 7/8oz.
- ■ 207/8 ■ 8 5/8

27. Complete these calculations. What is the
a smallest answer?
b
c *Give your answer as a mixed number fraction.*

- ■ 4 5/16 ■ 69/16 ■ 6 1/4 ■ 5 4/8 ■ 5 1/2

Level 4 *continued*

28. Complete these calculations. What is the
a difference between the largest answer and
b the smallest answer?
c
*Give your answer as a mixed number
fraction.*

- 4 8/64 ▪ 4 1/8 ▪ 5 4/8 ▪ 4 2/16 ▪ 5 1/2
- 2 2/8 ▪ 6 3/4 ▪ 5 2/4 ▪ 2 10/16

$$4 \times 1\frac{3}{8}$$

$$3 \times 2\frac{1}{4}$$

$$3 \times 1\frac{5}{16}$$

29. Toy cars are packed in boxes of 8. Each toy
a car weighs 2 3/5 kilograms (kg). What is the
b total weight in kilograms of seven boxes of
c toy cars?
*Give your answer as a mixed number
fraction. Don't include the units in your
answer.*

- 145 3/5 ▪ 18 1/5 ▪ 145 3/5kg
- 145 3/5kilograms ▪ 145 3/5 kilograms ▪ 145 3/5 kg
- 20 4/5 ▪ 728/5

$$2\frac{3}{5}\,kg$$

30. Use every digit card once to complete the
a fraction multiplication. What is the answer
b to the calculation?
c
*Give your answer as a mixed number
fraction.*

- 42 1/7 ▪ 295/7

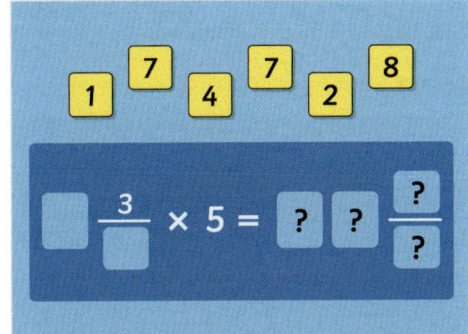

7   7
1   4   2   8

$$\square\frac{3}{\square} \times 5 = ?\ ?\ \frac{?}{?}$$

Ref:10198   Multiply Mixed Number Fractions by Whole Num...

Multiply Proper Fractions and Mixed Number Fractions by Whole Numbers

Objective: I can multiply proper fractions and mixed number fractions by whole numbers.

Quick Search Ref: 10005

Level 1: Understanding: Multiply proper fractions and mixed number fractions by whole numbers with image support.

✿ Required: 7/10   **✿ Pupil Navigation:** on   **✿ Randomised:** off

1. When you multiply 1/5 by 4, what is the numerator of your answer?

▪ 4 ▪ 5

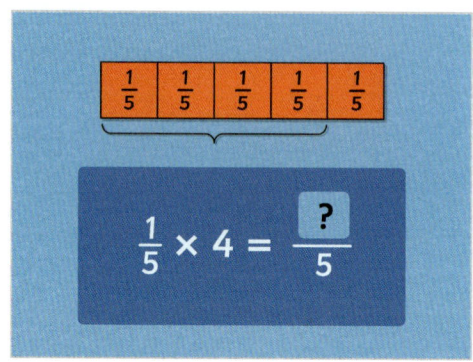

2. Calculate 9 × 1/11.

Put a forward slash (/) between the numerator and denominator. For example, one-half is 1/2.

▪ 9/11 ▪ 1/99 ▪ 9/99

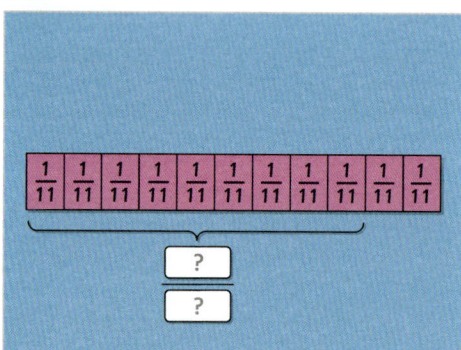

3. 1 1/7 × 2 = ___ wholes and 2/7.

▪ 2 ▪ 1

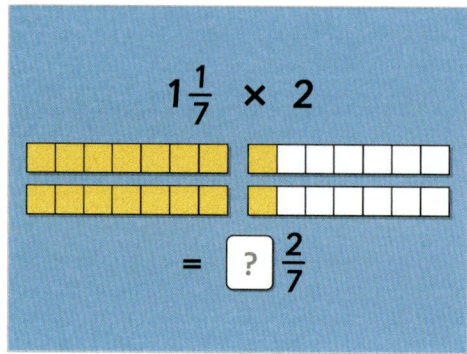

4. 2 × 2 1/3 = ___

Give your answer as a mixed number fraction. For example, five and one-half is 5 1/2.

▪ 4 2/3 ▪ 4 2/6 ▪ 7/3 ▪ 14/3 ▪ 4 1/3

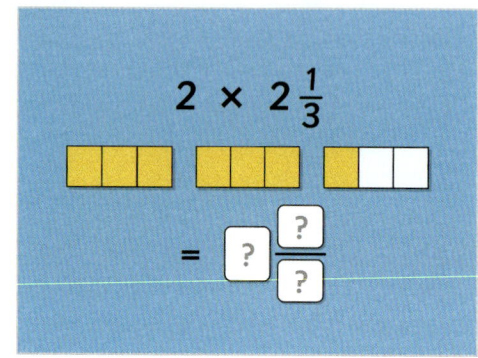

5. What is 1/5 × 12?

Put a forward slash (/) between the numerator and denominator. For example, one-half is 1/2.

▪ 12/60 ▪ 12/5 ▪ 2 2/5 ▪ 1/60

Level 1 *continued*

6. Find the answer to 4 2/5 × 2.

a
b
c
Give your answer as a mixed number fraction. For example, five and one-half is 5 1/2.

- ▪ 8 4/5 ▪ 8 2/5 ▪ 8 4/10 ▪ 44/5

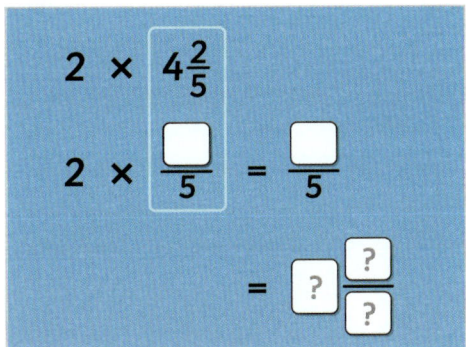

7. What is 3 multiplied by 2 1/4?

a
b
c
Give your answer as a mixed number fraction. For example, five and one-half is 5 1/2.

- ▪ 6 3/4 ▪ 6 3/12 ▪ 9/4 ▪ 6 1/4

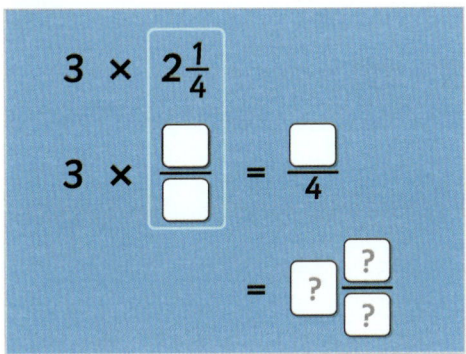

8. What is 1 1/9 multiplied by 4?

a
b
c
Give your answer as a mixed number fraction. For example, five and one-half is 5 1/2.

- ▪ 40/9 ▪ 4 4/9 ▪ 4 4/36 ▪ 10/9 ▪ 4 1/9

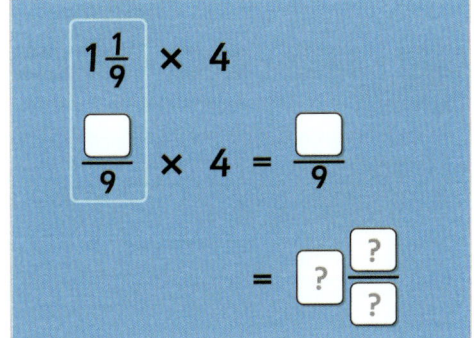

9. Calculate 5 × 1/8.

a
b
c
Put a forward slash (/) between the numerator and denominator. For example, one-half is 1/2.

- ▪ 5/8 ▪ 1/40 ▪ 5/40

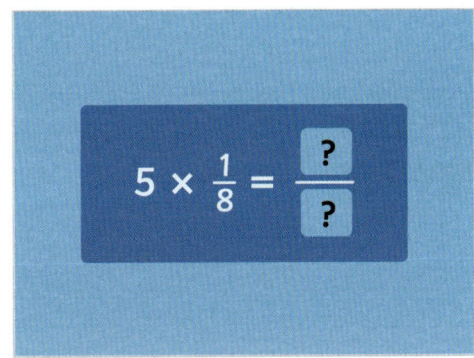

10. What is 1 2/5 × 2?

a
b
c
Give your answer as a mixed number fraction. For example, five and one-half is 5 1/2.

- ▪ 2 4/5 ▪ 2 4/10 ▪ 7/5 ▪ 2 2/5 ▪ 14/5

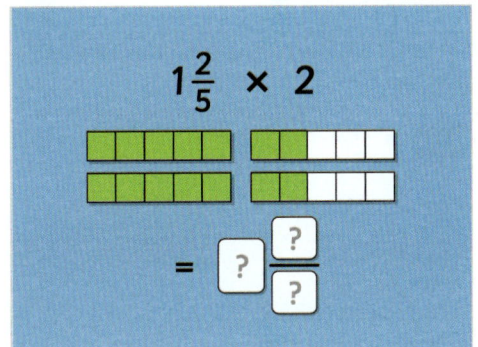

Level 2: Fluency: Multiply proper fractions and mixed number fractions including one-step word problems.

✱ **Required:** 7/10 ✱ **Pupil Navigation:** on
✱ **Randomised:** off

11. What is 4 × 4/5?

a b c *Put a forward slash (/) between the numerator and denominator. For example, one-half is 1/2.*

▪ 3 1/5 ▪ 16/5 ▪ 16/20 ▪ 4/20

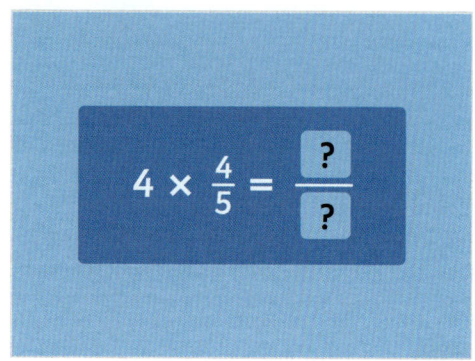

$$4 \times \frac{4}{5} = \frac{?}{?}$$

12. 5 × 1 5/6 = ____

a b c *Give your answer as a mixed number fraction.*

▪ 9 1/6 ▪ 55/6

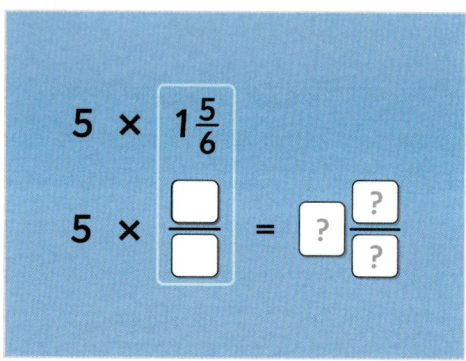

$$5 \times 1\frac{5}{6}$$
$$5 \times \frac{\square}{\square} = ?\frac{?}{?}$$

13. What is 1 4/9 multiplied by 3?

a b c *Give your answer as a mixed number fraction.*

▪ 4 1/3 ▪ 4 3/9 ▪ 39/9

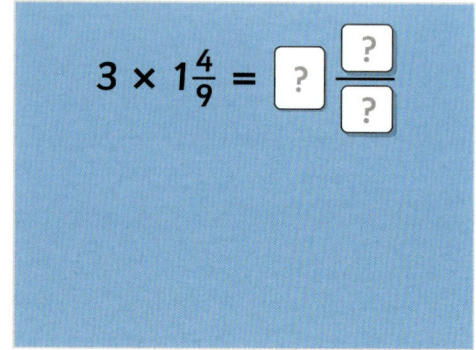

$$3 \times 1\frac{4}{9} = ?\frac{?}{?}$$

14. Izzy and George each swim 2/5 of a length of the swimming pool under water. What fraction of a length do they swim in total?

a b c *Put a forward slash (/) between the numerator and denominator. For example, one-half is 1/2.*

▪ 4/5 ▪ 4/10 ▪ 2/10

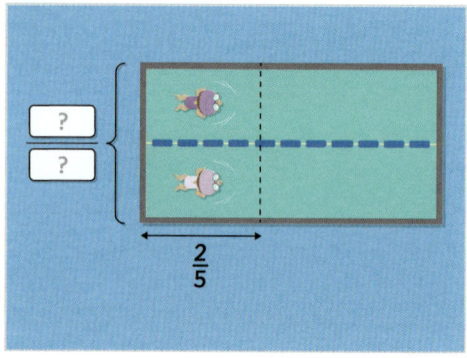

$$\frac{2}{5}$$

15. Sally has five rabbits and each rabbit eats 3/5 of a bag of hay. How many bags of hay do Sally's rabbits eat altogether?

a b c *Put a forward slash (/) between the numerator and denominator. For example, one-half is 1/2.*

▪ 3 ▪ 15/5 ▪ 15/25 ▪ 3/25

$$\frac{3}{5}$$
hay

16. Each oil drum contains 2 2/3 litres (l) of oil. Four oil drums contain ____ litres of oil.

a b c *Enter the missing mixed number fraction. Don't include the units in your answer.*

▪ 10 2/3 ▪ 10 2/3l ▪ 10 2/3 l ▪ 10 2/3 litres
▪ 10 2/3litres ▪ 2 2/3 ▪ 8 2/3 ▪ 32/3

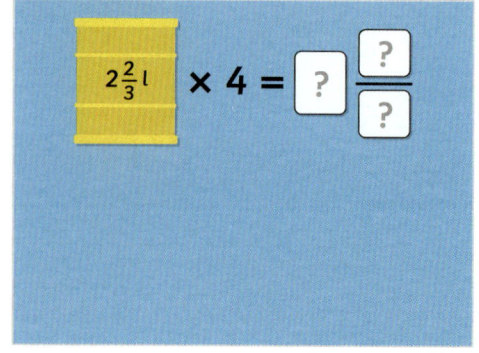

$$2\frac{2}{3}l \times 4 = ?\frac{?}{?}$$

Level 2 *continued*

17. The width of the rectangle is 4 4/5 metres
a (m). The length of the rectangle is twice as
b long as its width. What is the length of the
c rectangle?

*Give your answer as a mixed number
fraction. Include the units m (metres) in your
answer.*

- ■ 9 3/5m ■ 9 3/5metres ■ 9 3/5 metres ■ 9 3/5
- ■ 9 3/5 m ■ 48/5

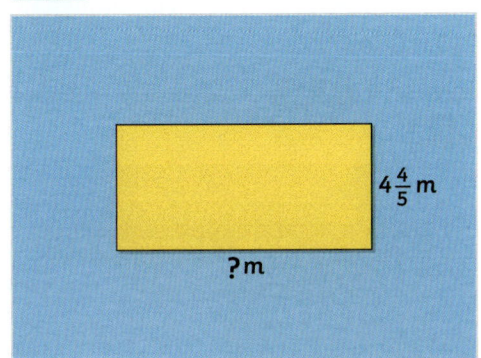

18. Dan, Luke and Ellie each eat 2/7 of a cake.
a How much of the cake do they eat in total?
b *Put a forward slash (/) between the
c numerator and denominator. For example,
one-half is 1/2.*

- ■ 6/7 ■ 6/21 ■ 2/21

19. 4 × 2 1/3 = _____
☐ *Give your answer as a mixed number
☒ fraction.*
☐

1/5
- ■ 8 1/3 ■ 9 1/3 ■ 28/3 ■ 1 1/3 ■ 8 4/3

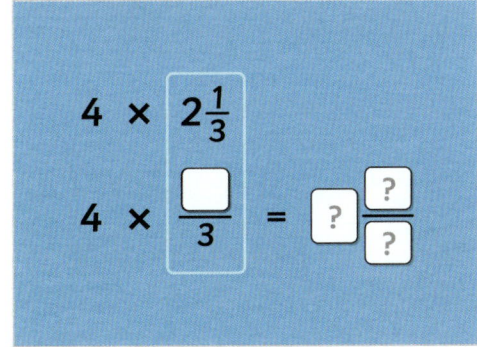

20. Calculate 7/10 × 5.
a *Put a forward slash (/) between the
b numerator and denominator. For example,
c one-half is 1/2.*

- ■ 7/2 ■ 35/10 ■ 7/50 ■ 35/50

Level 3: Reasoning: Reason about multiplying proper
and mixed number fractions by whole
numbers.

✱ **Required:** 5/5   ✱ **Pupil Navigation:** on
✱ **Randomised:** off

21. Mick and Finn each ate 2/5 of a packet of
☐ biscuits. Which two numbers do you need to
☒ multiply to work out what fraction of the
☐ packet of biscuits they ate altogether?
2/6

- ■ 2/5 ■ 5 ■ 1/2 ■ 2 ■ 5/2 ■ 1/5

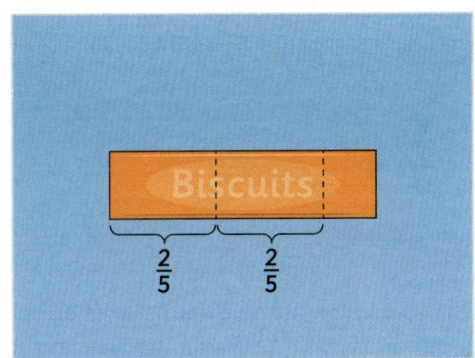

22. Sally says that 5 × 1 3/4 = 3 3/4. Explain why
a Sally is incorrect and give the correct answer.
b
c - Open question, no set answer

$$5 \times 1\frac{3}{4} = 3\frac{3}{4}$$

Level 3 *continued*

23. What is the missing mixed number fraction that completes the following statement?

a b c

____ ÷ 5 = 1 2/9

Give your answer as a mixed number fraction.

- ■ 6 1/9 ■ 55/9

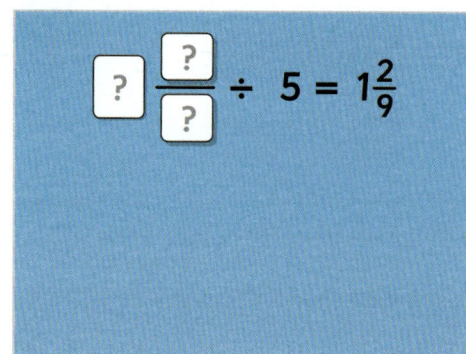

24. Sophie says that if you multiply a unit fraction by a whole number that is the same as its denominator, the answer is always 1. Is Sophie correct? Explain your answer, giving an example.

a b c

- Open question, no set answer

25. Matthew knows that 2 × 1 5/9 = 3 1/9. Explain how Matthew can use this to find 20 × 1 5/9 and calculate the answer.

a b c

- Open question, no set answer

© Learning by Questions Ltd   218   Ref:10005   Multiply Proper Fractions and Mixed Number Fr...

Level 4: Problem solving with greater depth: Solve multi-step problems involving multiplying proper and mixed number fractions by whole numbers.

✸ **Required:** 5/5   ✸ **Pupil Navigation:** on
✸ **Randomised:** off

26. Jenny cooks pizzas for children's parties. The table shows how many children she cooks for each day during the week.

2/5   If each child at the party is given 3/4 of a pizza, on which two days will Jenny have pizza left over?

- ■ Monday ■ Tuesday ■ Wednesday ■ Thursday
- ■ Friday

day	people	amount of pizza
Monday	8	
Tuesday	6	
Wednesday	12	
Thursday	4	
Friday	10	

27. Julie wants to make 12 portions of coconut macaroons. What is the total weight in ounces (oz.) of the ingredients she needs? *Give your answer as a mixed number fraction. Don't include the units in your answer.*

a b c

- ■ 25 7/8ounces ■ 25 7/8 oz. ■ 25 7/8 ounces
- ■ 25 7/8oz ■ 25 7/8 ■ 25 7/8 oz ■ 25 7/8oz.
- ■ 8 5/8 ■ 207/8

recipe

(makes 4 macaroons)

$2\frac{1}{2}$ oz. coconut
$4\frac{1}{2}$ oz. sugar
$1\frac{5}{8}$ oz. butter

Level 4 *continued*

28. A baker uses 4/5 of a bag of flour to make one cake. In the bakery, each cake is cut into 8 slices. How many **whole bags** of flour must the baker buy for an order of 56 slices of cake?

■ 6   ■ 5

29. Complete these calculations. What is the difference between the largest answer and the smallest answer?
Give your answer as a mixed number fraction.

■ 4 8/64   ■ 4 1/8   ■ 4 2/16   ■ 2 2/8   ■ 5 2/4
■ 5 4/8   ■ 5 1/2   ■ 6 3/4   ■ 2 10/16

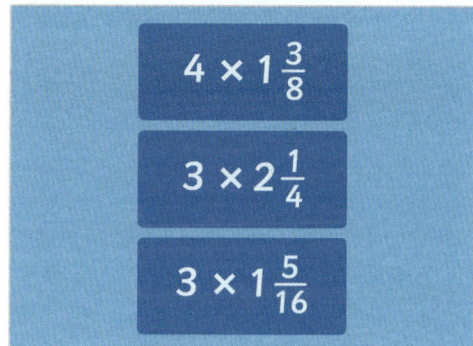

$$4 \times 1\frac{3}{8}$$

$$3 \times 2\frac{1}{4}$$

$$3 \times 1\frac{5}{16}$$

30. Toy cars are packed in boxes of 8. Each toy car weighs 2 3/5 kilograms (kg). What is the total weight in kilograms of seven boxes of toy cars?
Give your answer as a mixed number fraction. Don't include the units in your answer.

■ 145 3/5   ■ 145 3/5kg   ■ 145 3/5kilograms
■ 145 3/5 kilograms   ■ 145 3/5 kg   ■ 728/5   ■ 18 1/5
■ 20 4/5

$$2\frac{3}{5} kg$$

Multiply Proper Fractions by Whole Numbers

Objective: I can multiply proper fractions by whole numbers.

Quick Search Ref: 10045

Level 1: Understanding: Multiply unit fractions by whole numbers.

✱ **Required:** 7/10   ✱ **Pupil Navigation:** on   ✱ **Randomised:** off

---

1. Which of these fractions is a proper fraction?

   ▪ 1 1/4   ▪ **1/4**   ▪ 6/4

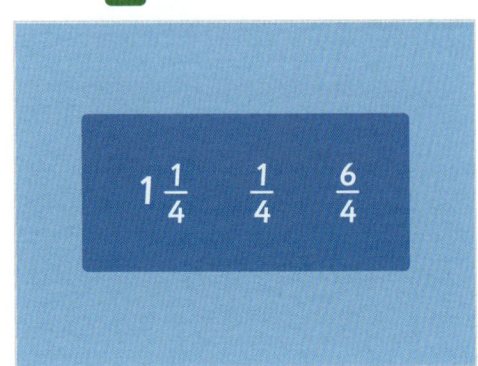

2. When you multiply 1/7 by 3, what is the denominator of your answer?

   ▪ **7**   ▪ 3   ▪ 21

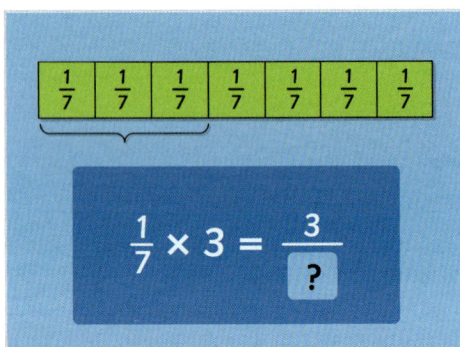

3. When you multiply 1/5 by 4, what is the numerator of your answer?

   ▪ **4**   ▪ 5

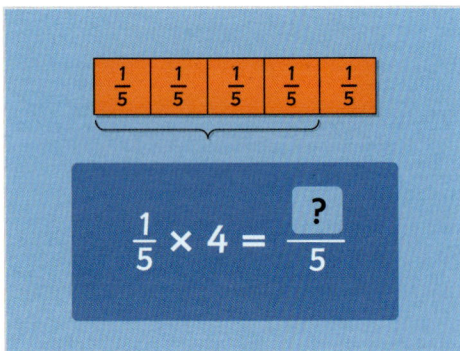

4. How many thirds are the same as $1/3 \times 2$?

   ▪ **2**

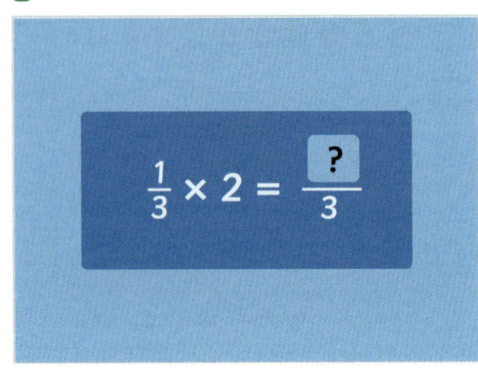

5. What is $1/10 \times 7$?

   ▪ 1/70   ▪ **7/10**   ▪ 7/70

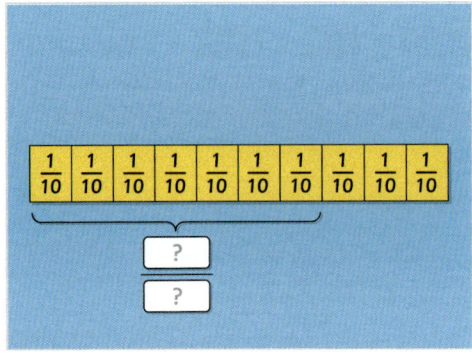

6. Calculate $5 \times 1/8$.

Put a forward slash (/) between the numerator and denominator. For example, one-half is 1/2.

   ▪ **5/8**   ▪ 5/40   ▪ 1/40

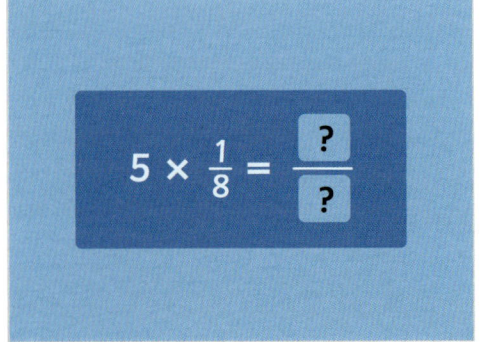

Level 1 *continued*

7. What is 1/6 × 7?

Put a forward slash (/) between the numerator and denominator. For example, one-half is 1/2.

- **7/6** · **1 1/6** · 1/42 · 7/42

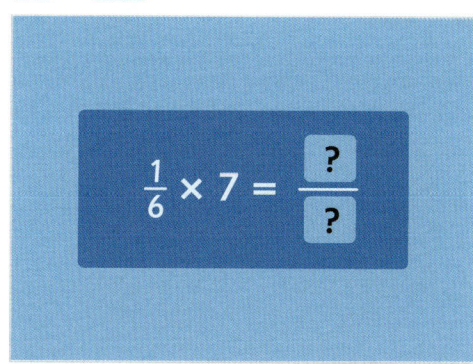

8. When you multiply 1/7 by 5, what is the numerator of your answer?

- **5** · 7

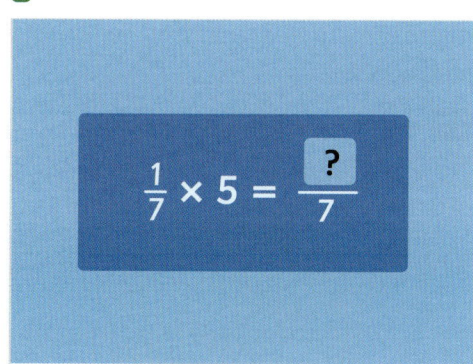

9. Calculate 9 × 1/11.

Put a forward slash (/) between the numerator and denominator. For example, one-half is 1/2.

- **9/11** · 9/99 · 1/99

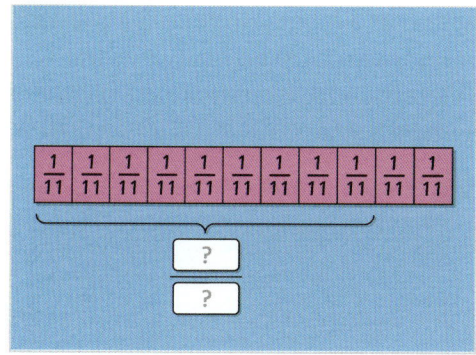

10. What is 1/5 × 12?

Put a forward slash (/) between the numerator and denominator. For example, one-half is 1/2.

- **12/5** · **2 2/5** · 12/60 · 1/60

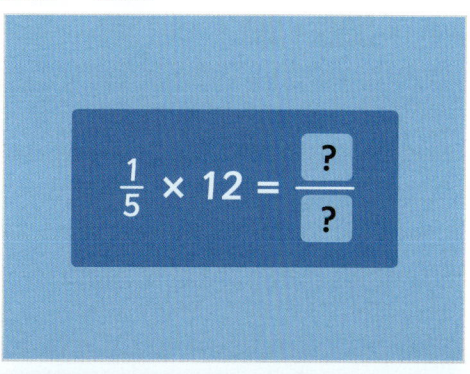

Level 2: Fluency: Multiply non-unit proper fractions by whole numbers including in context.

✱ **Required:** 7/10   ✱ **Pupil Navigation:** on
✱ **Randomised:** off

11. Calculate 2/5 × 3.

1/3
- 2/15 · **6/5** · 6/15

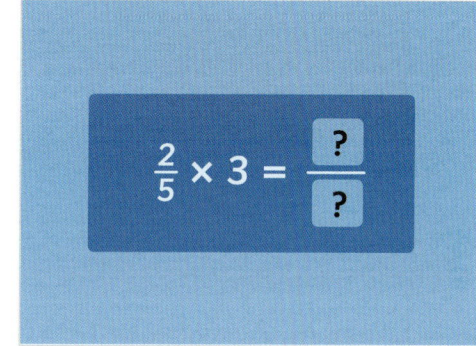

12. What is 2/10 × 4?

Put a forward slash (/) between the numerator and denominator. For example, one-half is 1/2.

- **4/5** · **8/10** · 2/40 · 8/40

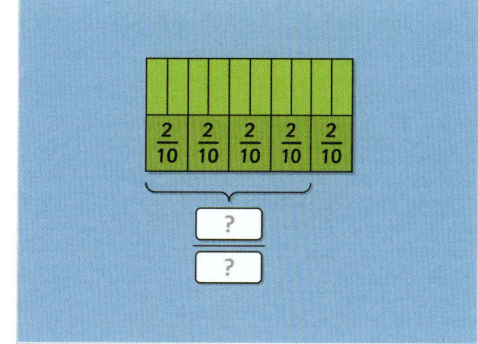

Level 2 continued

13. What is 4 × 2/3?

   a b c *Put a forward slash (/) between the numerator and denominator. For example, one-half is 1/2.*

   ■ 2/12   ■ **2 2/3**   ■ **8/3**   ■ 8/12

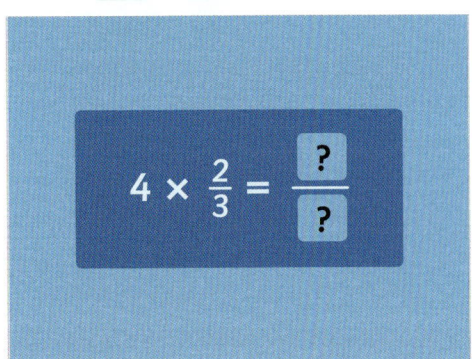

14. The length of the rectangle is twice as long as its width. What fraction of a metre is the length of the rectangle?

   ■ 1/6   ■ **2/3**   ■ 2/6

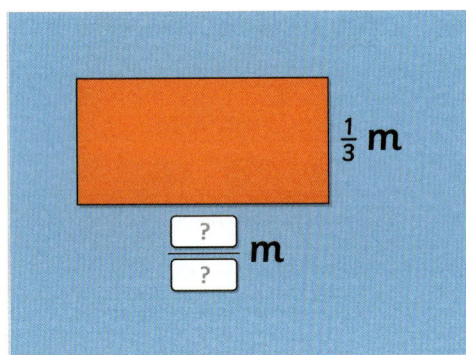

15. Three children each complete 2/9 of a puzzle. What fraction of the puzzle do they complete in total?

   a b c *Put a forward slash (/) between the numerator and denominator. For example, one-half is 1/2.*

   ■ **2/3**   ■ **6/9**   ■ 2/27   ■ 6/27

16. Sally has five rabbits and each rabbit eats 3/5 of a bag of hay. How many bags of hay do Sally's rabbits eat altogether?

   a b c *Put a forward slash (/) between the numerator and denominator. For example, one-half is 1/2.*

   ■ **3**   ■ **15/5**   ■ 3/25   ■ 15/25

17. Dan, Luke and Ellie each eat 2/7 of a cake. How much of the cake do they eat in total?

   a b c *Put a forward slash (/) between the numerator and denominator. For example, one-half is 1/2.*

   ■ **6/7**   ■ 2/21   ■ 6/21

18. Izzy and George each swim 2/5 of a length of the swimming pool under water. What fraction of a length do they swim in total?

   a b c *Put a forward slash (/) between the numerator and denominator. For example, one-half is 1/2.*

   ■ **4/5**   ■ 2/10   ■ 4/10

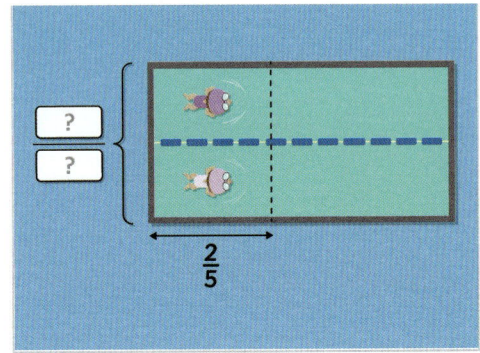

Level 2 *continued*

19. What is 4 × 4/5?

a
b
c

Put a forward slash (/) between the numerator and denominator. For example, one-half is 1/2.

- **3 1/5** ▪ **16/5** ▪ 4/20 ▪ 16/20

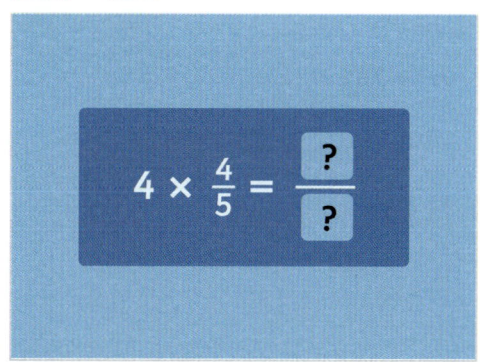

20. Calculate 7/10 × 5.

a
b
c

Put a forward slash (/) between the numerator and denominator. For example, one-half is 1/2.

- **7/2** ▪ **35/10** ▪ 35/50 ▪ 7/50

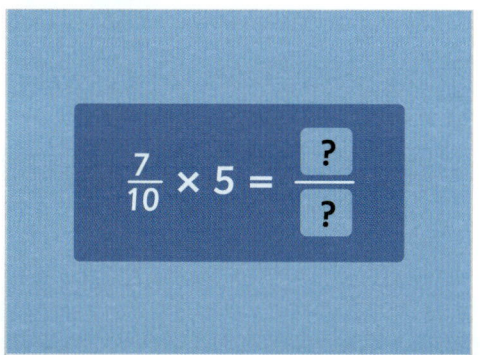

Level 3:   Reasoning: Compare and reason about fraction multiplication.

✱ **Required:** 5/5   ✱ **Pupil Navigation:** on
✱ **Randomised:** off

21. Toby has calculated 1/6 × 5 = 5/30. Calculate

a
b
c

the correct answer and explain what Toby has done wrong.

- Open question, no set answer

22. Mick and Finn each ate 2/5 of a packet of biscuits. Which two numbers do you need to multiply to work out what fraction of the packet of biscuits they ate altogether?

2/6

- **2/5** ▪ 5 ▪ 1/2 ▪ **2** ▪ 5/2 ▪ 1/5

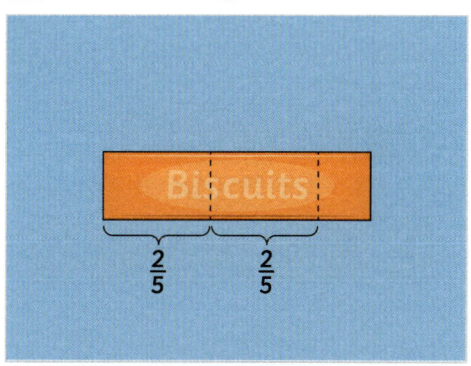

23. Sophie says that if you multiply a unit

a
b
c

fraction by a whole number that is the same as its denominator, the answer is always 1. Is Sophie correct? Explain your answer, giving an example.

- Open question, no set answer

24. Which two calculations give the same answer?

2/5

- **3/8 × 6** ▪ 3/6 × 8 ▪ 4/8 × 3 ▪ **6/8 × 3** ▪ 6/3 × 8

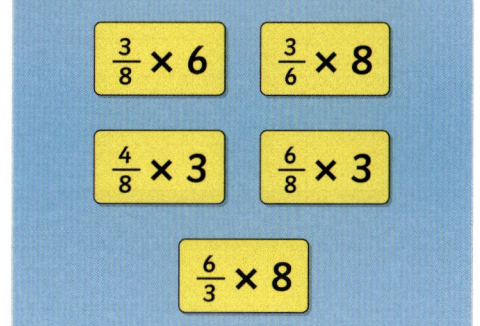

25. Use **four different** digit cards to find possible calculations and answers.

a
b
c

What is the only possible **improper fraction** answer?

Put a forward slash (/) between the numerator and denominator. For example, one-half is 1/2.

- 7 ■ 8/7 ■ 6/7 ■ 8

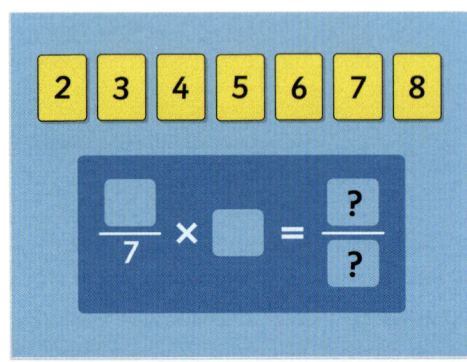

Level 4: Problem Solving: Solve problems with fraction multiplications.

❋ **Required:** 5/5   ❋ **Pupil Navigation:** on

❋ **Randomised:** off

26. Jenny cooks pizzas for children's parties. The table shows how many children she cooks for each day during the week.

2/5 If each child at the party is given 3/4 of a pizza, on which two days will Jenny have pizza left over?

- Monday ■ Tuesday ■ Wednesday ■ Thursday
- Friday

day	people	amount of pizza
Monday	8	
Tuesday	6	
Wednesday	12	
Thursday	4	
Friday	10	

27. At Greenbank School, there are six classes.

a
b
c

Each class eats 2/7 of a bag of potatoes for lunch. If two whole bags of potatoes are cooked, what fraction of a bag will be left over?

Put a forward slash (/) between the numerator and denominator. For example, one-half is 1/2.

- 12/7 ■ 2/7 ■ 14/7

28. A baker uses 4/5 of a bag of flour to make one cake. In the bakery, each cake is cut into 8 slices. How many **whole bags** of flour must the baker buy for an order of 56 slices of cake?

1
2
3

- 6 ■ 5

29. Sort the calculations by the size of their answer (smallest first).

↑
↓

- 4 × 1/20 ■ 7 × 2/10 ■ 3 × 3/5 ■ 6 × 2/5
- 3 × 9/10

$$3 \times \frac{9}{10}$$
$$3 \times \frac{3}{5}$$
$$4 \times \frac{1}{20}$$
$$6 \times \frac{2}{5}$$
$$7 \times \frac{2}{10}$$

Level 4 *continued*

30. The ingredients for one portion of a meal
a are:
b
c 1/12 kg onions
2/6 kg chicken
2/12 kg mushrooms

To make a meal for five people, what is the total weight of all of the ingredients in kilograms?
Put a forward slash (/) between the numerator and denominator. For example, one-half is 1/2. Include the units kg (kilograms) in your answer.

■ **7/12 kilograms** ■ 2 11/12 kg ■ 35/12 kg
■ 2 11/12 kilograms ■ 35/12 kilograms ■ **7/12 kg**
■ **35/12** ■ **7/12**

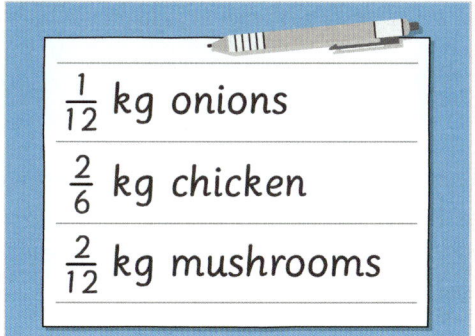

$\frac{1}{12}$ kg onions

$\frac{2}{6}$ kg chicken

$\frac{2}{12}$ kg mushrooms

Solve Problems Involving Fractions of Amounts

Objective: I can solve problems involving finding fractions of amounts, multiplying fractions and scaling by simple fractions.

Quick Search Ref: 10165

Level 1: Problem solving: Finding fractions of amounts, multiplying fractions and scaling by simple fractions.

✱ **Required:** 10/10   ✱ **Pupil Navigation:** on   ✱ **Randomised:** off

1. There are 36 children in a class. 4/6 of the children have school dinners. How many children have school dinners?

▪ **24** ▪ **12**

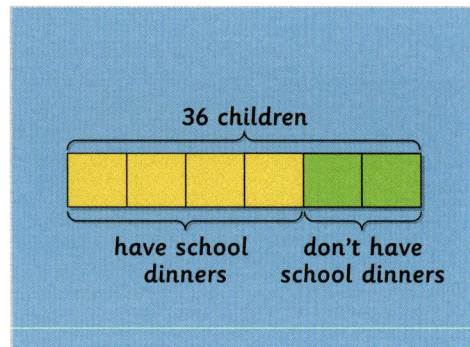

2. There are 164 sheep in field. Three-quarters of the sheep are female. How many sheep are **male**?

▪ **41** ▪ **123**

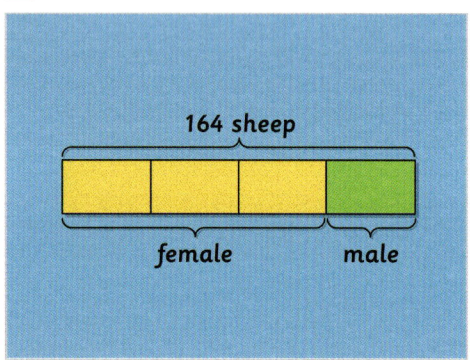

3. There are 45 children in a school play and 3/5 of them are girls. How many **boys** are in the play?

▪ **18** ▪ **27**

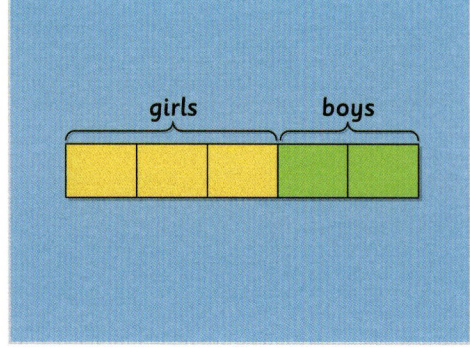

4. A box contains 120 crayons. 25 crayons are pink and 1/5 of the crayons are red. How many crayons are **green**?

▪ **49** ▪ **71** ▪ **24**

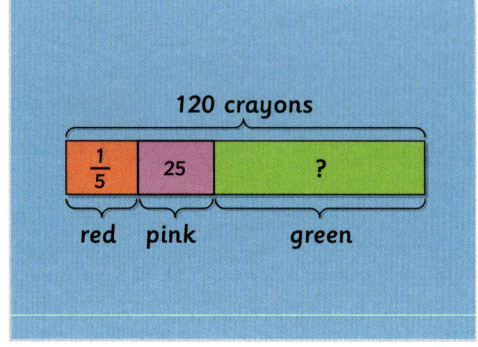

5. In square centimetres (cm²), what area of the large rectangle is **white**?
Don't include the units in your answer.

▪ **12** ▪ **6** ▪ **36**

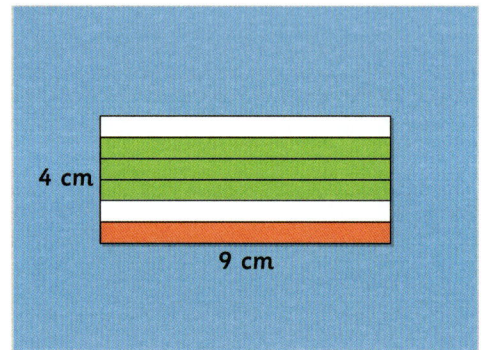

6. In a survey, 80 people are asked to choose their favourite food. How many people choose lasagne?

▪ **30**

Survey of 80 people's favourite food

food	fraction of people surveyed
fish and chips	$\frac{1}{8}$
cheeseburger	$\frac{1}{4}$
lasagne	?
salad	$\frac{1}{4}$

Level 1 *continued*

7. Julie makes nine portions of pizza dough. The total weight of the ingredients that Julie uses is _____ kilograms (kg).
Enter the missing mixed number fraction.
Don't include the units in your answer.

a
b
c

- 1 44/100
- 1 11/25kilograms
- 1 22/50
- 1 11/25
- 1 11/25 kilograms
- 1 11/25kg
- 1 11/25 kg
- 144/100
- 1.44

recipe for pizza dough

(makes 3 portions)

$\frac{9}{20}$ kg flour

$\frac{1}{100}$ kg yeast

$\frac{1}{50}$ kg butter

8. George buys a scale model of a tower. The model is 1/20 of the size of the actual tower. The model is 30 centimetres (cm) tall. How tall is the actual tower in **metres** (m)?
Include the units metres (m) in your answer.

a
b
c

- 60 m
- 6000 cm
- 60
- 60 metres
- 6,000 centimetres
- 6000 centimetres
- 6,000 cm

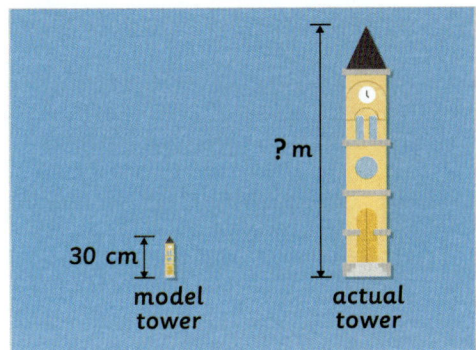

? m

30 cm

model tower

actual tower

9. In a sale, every item is reduced by 1/10 of the original price. Kian buys a jumper and a pair of trainers. He pays with two £20 notes. How much change does he get?
Include the £ sign in your answer.

a
b
c

- £10
- 13
- £13
- £13.00p
- £13.00
- £10.00
- £27.00
- 13.00
- £27

£10

£15

Sale!
$\frac{1}{10}$ off
every item

£15

£20

£50

10. During the spring holiday, Frank completes 3/7 of a computer game. In the summer holiday, he completes the remaining 96 levels. How many levels are there in Frank's computer game altogether?

1
2
3

- 168
- 24

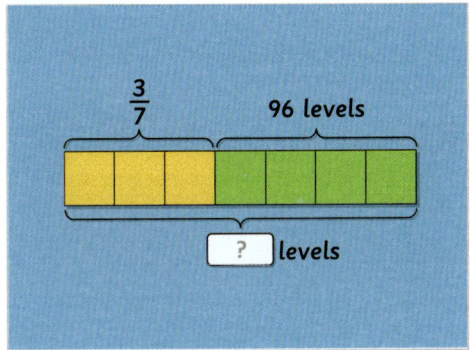

$\frac{3}{7}$

96 levels

? levels

Compare & Order Fractions with Related Denominators (Proper, Improper & Mixed Number Fractions)

Objective: I can compare and order proper fractions, improper fractions and mixed number fractions with related denominators.

Quick Search Ref: 10070

Level 1: Understanding: Compare and order fractions with related denominators with image support.

❋ **Required:** 7/10   ❋ **Pupil Navigation:** on   ❋ **Randomised:** off

---

1. To compare 11/5, 12/10 and 23/20, you need to make the denominators the same. What is the lowest number that is a multiple of the denominators?

▪ 100   ▪ **20**   ▪ 10

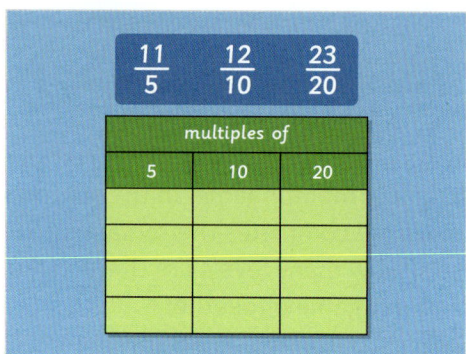

2. Which is the larger fraction, 7/10 or 3/5?
Put a forward slash (/) between the numerator and denominator.

▪ **7/10**   ▪ 3/5

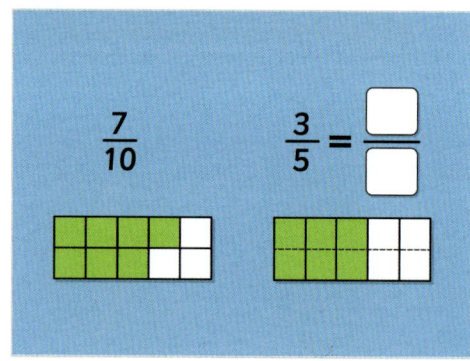

3. Which of the following is the smallest improper fraction?
9/4, 10/8, 9/8
Put a forward slash (/) between the numerator and denominator.

▪ 9/4   ▪ **9/8**   ▪ 10/8

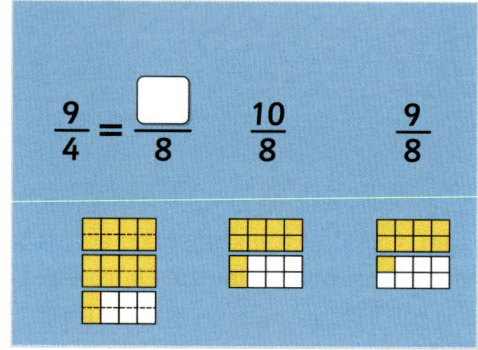

4. Which is the largest mixed number fraction?

1/3

▪ 1 2/5   ▪ 1 1/20   ▪ **1 9/10**

5. Which is the smallest mixed number fraction?

1/3

▪ 1 1/2   ▪ **1 1/8**   ▪ 1 3/4

$$1\frac{1}{2} \qquad 1\frac{1}{8} \qquad 1\frac{3}{4}$$

$$1\frac{\square}{8} \qquad 1\frac{\square}{8} \qquad 1\frac{\square}{8}$$

Level 1 *continued*

6. Sort the following fractions into ascending order (smallest first).

↑↓

- 1/4
- 1/2
- 5/8
- 3/4

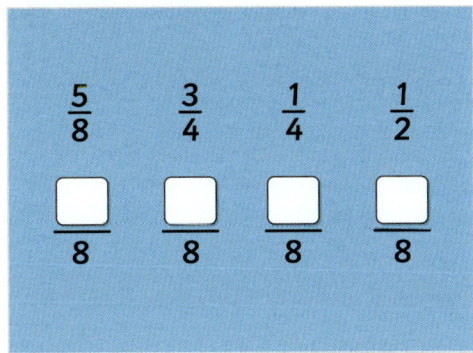

7. Sort the mixed number fractions from the largest to the smallest.

↑↓

- 2 2/3
- 2 2/6
- 1 3/12
- 1 1/12

8. Which of the following is the smallest fraction?

a
b
c

5/24, 4/6, 3/12

Put a forward slash (/) between the numerator and denominator.

- 4/6
- 5/24
- 3/12

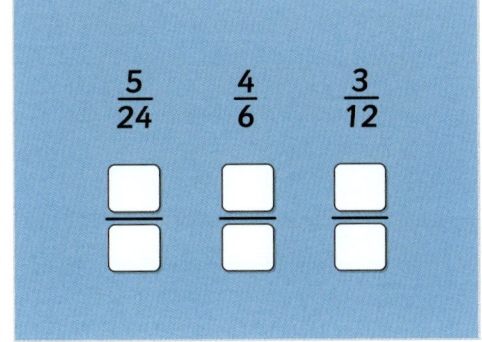

9. What is the lowest number that is a multiple of the denominators of these fractions?

1
2
3

1/5, 2/10, 2/20

- 20
- 10

10. Which of the following is the largest fraction?

a
b
c

2/3, 1/6, 3/6

Put a forward slash (/) between the numerator and denominator.

- 2/3
- 1/6
- 3/6

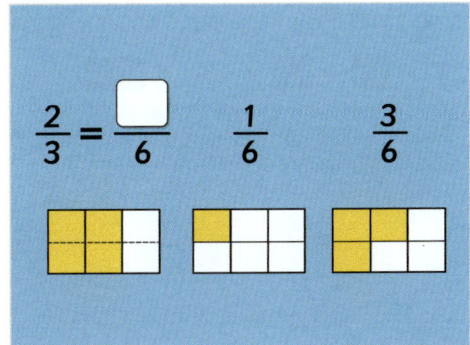

11. Which of the following is the largest fraction?

1/2, 9/20, 9/10

Put a forward slash (/) between the numerator and denominator.

■ 9/10

12. Which **three** fractions have the same value?

■ 2/4 ■ 3/4 ■ 6/8 ■ 6/16 ■ 12/16

3/5

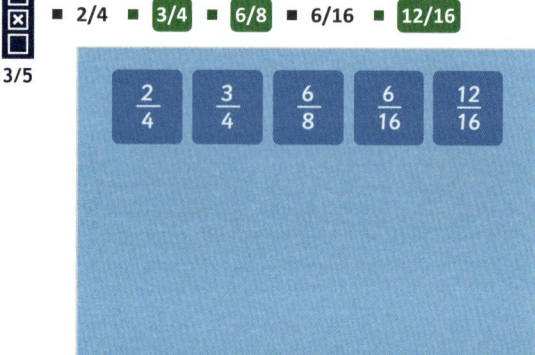

13. Sort the following fractions into descending order (largest first).

↑↓

■ 3/3 ■ 11/12 ■ 3/6 ■ 1/3

14. Sort the following fractions into ascending order (smallest first).

↑↓

■ 4 5/8 ■ 5 1/8 ■ 5 1/2 ■ 5 3/4

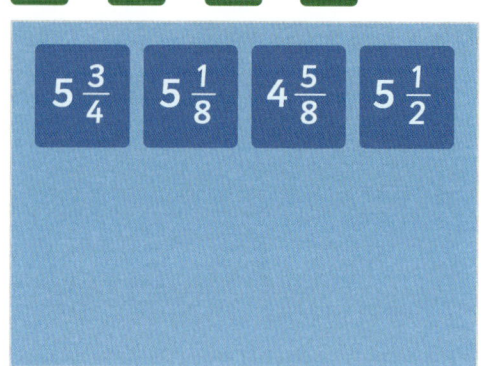

15. The durations of four television shows are recorded as fractions of one hour. Sort the shows in order, starting with the television show that lasts the longest.

↑↓

■ News at Nine ■ Drama Street ■ Cats on the Run
■ The Flee

Duration of television shows	
show	duration (hours)
Cats on the Run	$\frac{3}{15}$
News at Nine	$\frac{19}{60}$
The Flee	$\frac{5}{30}$
Drama Street	$\frac{16}{60}$

16. A farmer records the fraction of a bag of hay he uses to feed each of his animals. Which **animal** eats the **least** amount of hay?

■ cow ■ sheep ■ pig

Hay used by animals on a farm	
animal	bags of hay
pig	$\frac{11}{3}$
cow	$\frac{13}{6}$
sheep	$\frac{29}{12}$

17. Class 5K have a competition to see who can
a swim the furthest distance in an hour. What
b fraction of a mile does the child who swims
c the furthest complete?
*Put a forward slash (/) between the
numerator and denominator.*

■ 1 4/10 ■ 1 2/5 ■ 7/5 ■ 14/10 ■ Charlie

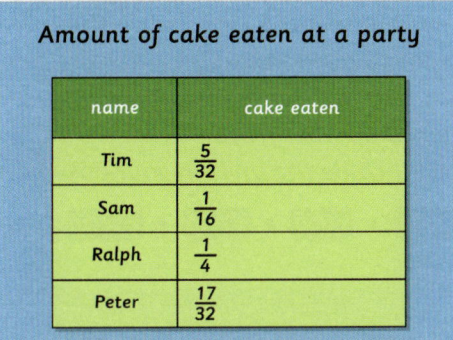

Class 5K swimming competition

name	miles completed
Jess	$1\frac{1}{10}$
Emma	$1\frac{1}{5}$
Charlie	$\frac{14}{10}$

18. Four friends record what fraction of a
a birthday cake they each eat. Which **child**
b eats the least amount of cake?
c

■ 1/16 ■ Sam ■ Peter ■ Ralph ■ Tim

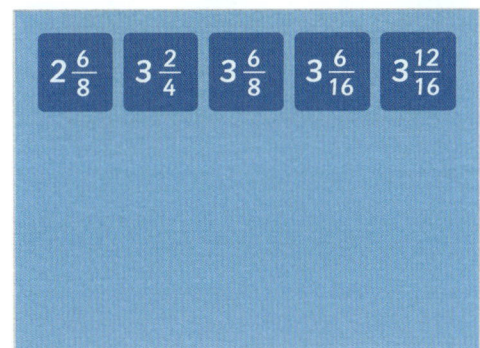

Amount of cake eaten at a party

name	cake eaten
Tim	$\frac{5}{32}$
Sam	$\frac{1}{16}$
Ralph	$\frac{1}{4}$
Peter	$\frac{17}{32}$

19. Which **two** mixed number fractions have the
same value?

2/5

■ 2 6/8 ■ 3 2/4 ■ 3 6/8 ■ 3 6/16 ■ 3 12/16

$2\frac{6}{8}$ $3\frac{2}{4}$ $3\frac{6}{8}$ $3\frac{6}{16}$ $3\frac{12}{16}$

20. Which of the following is the smallest
a improper fraction?
b
c 9/2, 29/20, 19/10
*Put a forward slash (/) between the
numerator and denominator.*

■ 29/20 ■ 19/10 ■ 9/2

$\frac{9}{2}$ $\frac{29}{20}$ $\frac{19}{10}$

Level 3: Reasoning: Reason about comparing and
ordering fractions.

✸ **Required:** 5/5 ✸ **Pupil Navigation:** on
✸ **Randomised:** off

21. Lorna has tried to arrange the fractions in
a order from largest to smallest, but she has
b made a mistake.
c Which fraction is in the wrong position?
*Put a forward slash (/) between the
numerator and denominator.*

■ 4/5 ■ 16/20

$\frac{9}{10}, \frac{9}{20}, \frac{3}{10}, \frac{4}{5}, \frac{1}{5}$

Level 3 *continued*

22. The fractions shown are in order from
[a][b][c] smallest to largest. If you add 2 to each
numerator, will the fractions still be in the
correct order?
Explain and prove your answer.

- Open question, no set answer

23. Using two of the number cards, what is the
[a][b][c] **smallest** fraction you can make?
*Put a forward slash (/) between the
numerator and denominator.*

- 2/100 - 1/50

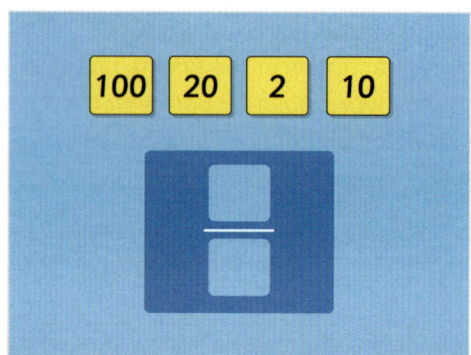

24. Three of these calculations have the same
[a][b][c] answer. Which calculation has a **different**
answer? Explain your answer.

- Open question, no set answer

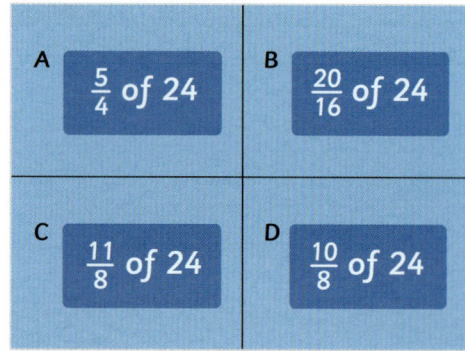

25. Jackson is comparing 11/3 with 1 1/3. He
[a][b][c] says, "An improper fraction is always worth
more than a mixed number fraction."
Is Jackson's statement correct? Explain your
answer and give at least two examples.

- Open question, no set answer

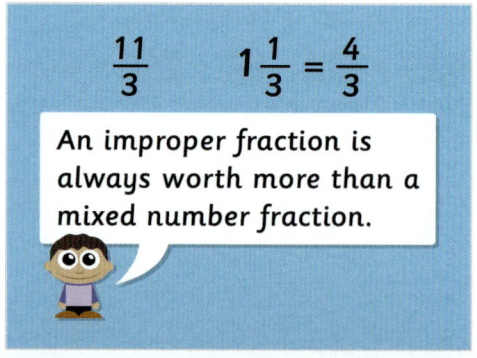

Level 4: Problem solving with greater depth: Solve
problems involving comparing and ordering
fractions with related denominators.

❋ **Required:** 5/5   ❋ **Pupil Navigation:** on
❋ **Randomised:** off

26. Five children record what fraction of a mile
[a][b][c] they each swim. Find the two children that
swim the furthest. How far do these two
children swim in total?
*Give your answer as an improper fraction or
a mixed number fraction.*
Don't include the units in your answer.

- 34/20 - 17/10 - 1 7/10 - 1 14/20 - 9/10
- 18/20

Distance children swim	
child	distance (miles)
Dale	$\frac{2}{10}$
Archie	$\frac{13}{20}$
Asif	$\frac{4}{5}$
Zaina	$\frac{11}{20}$
Georgia	$\frac{9}{10}$

Level 4 continued

27. Three children record how many packets of cereal they ate in a month. Packets of cereal weigh 250 grams (g) each. Find the child who ate the least amount of cereal. How many grams of cereal did that child eat?
Include the units g (grams) in your answer.

- 1 1/10 ■ **275 g** ■ **275 grams** ■ 11/10 ■ 275

Packets of cereal eaten in a month

child	packets
Azra	$\frac{3}{2}$
Teddy	$1\frac{1}{10}$
Evie	$1\frac{4}{5}$

28. Three children eat different fractions of a fruit cake weighing 500 grams (g). Find the child who eats the least amount of cake. How many **grams** of cake does that child eat?
Include the units g (grams) in your answer.

- **150 g** ■ **150 grams** ■ 150

Fruit cake eaten by children

child	fraction of cake eaten
Ava	$\frac{17}{50}$
Habib	$\frac{9}{25}$
Josh	$\frac{30}{100}$

29. Macy's grandma sorts all of her buttons into colours and records them as fractions. Find the fraction of red buttons she has, and then sort the colours in order, starting with the colour she has the most of.

- purple ■ red ■ yellow ■ green ■ orange

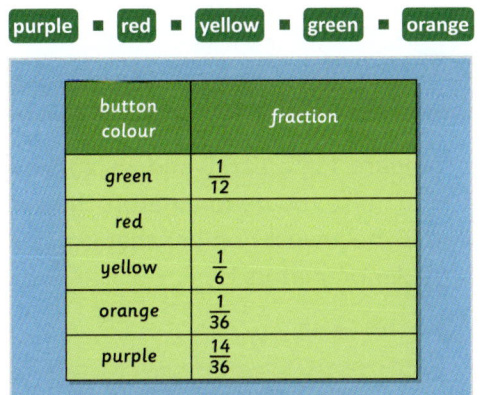

button colour	fraction
green	$\frac{1}{12}$
red	
yellow	$\frac{1}{6}$
orange	$\frac{1}{36}$
purple	$\frac{14}{36}$

30. Captain John is painting the sides of his ship with different colours.
He uses the following amounts of paint:
- 3/2 tins of **yellow** paint
- 3/4 of a tin of **blue** paint
- 1 3/8 tins of **orange** paint
- 1 tin of **green** paint
The rest of the paint is **black**.
If he uses six tins of paint in total, which **colour** paint does he use the **most** tins of?

- 3/2 ■ yellow ■ 1 3/8 ■ 11/8 ■ 1 1/2 ■ 1 4/8

Compare and Order Improper and Mixed Number Fractions with Related Denominators

Objective: I can compare and order improper fractions and mixed number fractions with related denominators.

Quick Search Ref: 11493

Level 1: Understanding: Compare and order improper fractions and mixed number fractions with related denominators using image support.

❋ **Required:** 7/10 　　❋ **Pupil Navigation:** on 　　❋ **Randomised:** off

---

1. To compare 11/5, 12/10 and 23/20, you need to make the denominators the same. What is the lowest number that is a multiple of the denominators?

- ■ 20　■ 100　■ 10

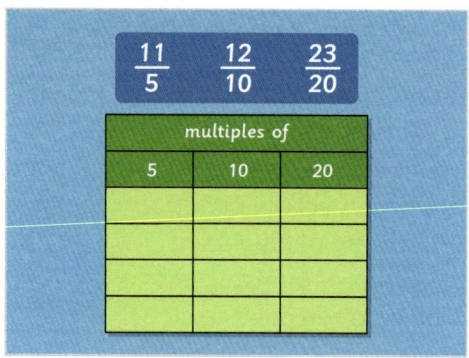

2. Which of the following is the smallest fraction?
11/12, 2/3, 2/6
Put a forward slash (/) between the numerator and denominator.

- ■ 2/6　■ 11/12　■ 2/3

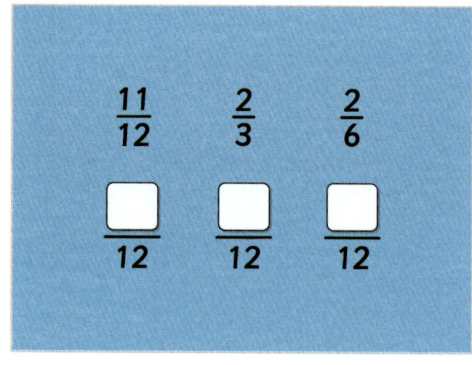

3. Which is the largest fraction, 6/5 or 11/10?
Put a forward slash (/) between the numerator and denominator.

- ■ 6/5　■ 11/10

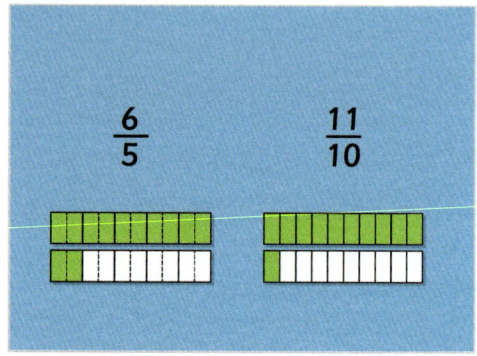

4. Which of the following fractions is smaller than 7/4?
9/4, 9/8, 5/2
Put a forward slash (/) between the numerator and denominator.

- ■ 9/8　■ 9/4　■ 5/2

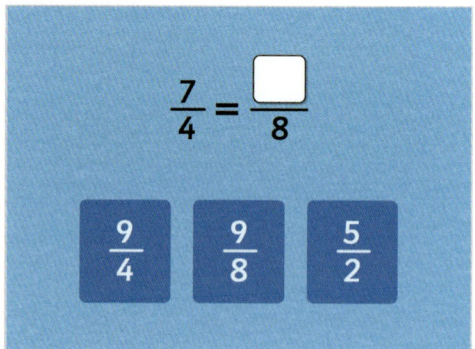

Level 1 *continued*

5. Which is the smallest mixed number fraction?

1/3

▪ 1 1/2   ▪ **1 1/8**   ▪ 1 3/4

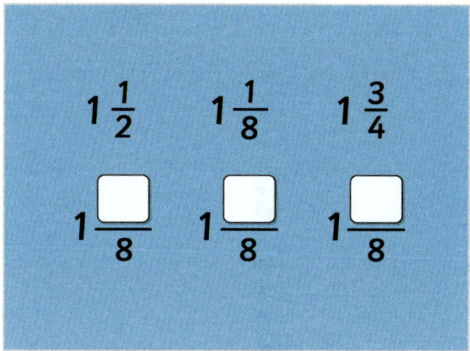

6. Sort the mixed number fractions from largest to the smallest.

↑
↓

▪ **2 3/20**   ▪ **1 3/10**   ▪ **1 1/5**

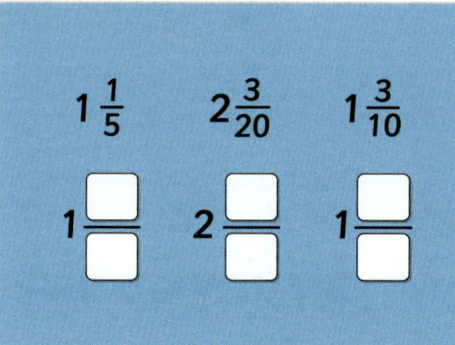

7. Sort the following improper fractions into ascending order (smallest first).

↑
↓

▪ **3/2**   ▪ **15/8**   ▪ **11/4**   ▪ **20/4**

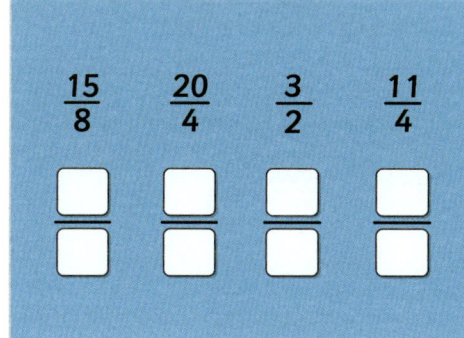

8. Sort the mixed number fractions from the largest to the smallest.

↑
↓

▪ **2 2/3**   ▪ **2 2/6**   ▪ **1 3/12**   ▪ **1 1/12**

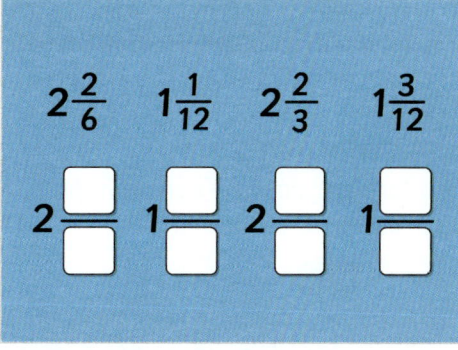

9. Which is the largest mixed number fraction?

1/3

▪ 1 2/5   ▪ 1 1/20   ▪ **1 9/10**

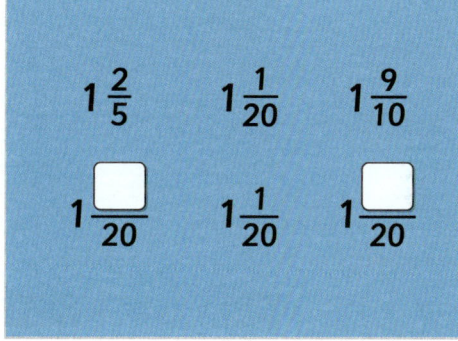

10. Which of the following is the smallest improper fraction?
9/4, 10/8, 9/8
Put a forward slash (/) between the numerator and denominator.

a
b
c

▪ **9/8**   ▪ 9/4   ▪ 10/8

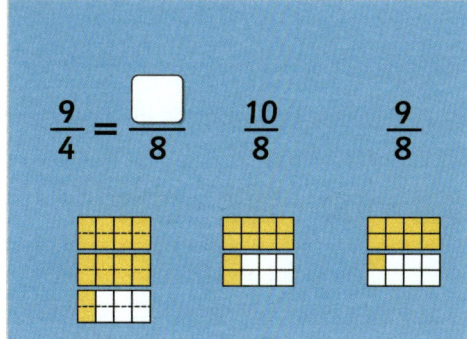

Level 2: Fluency: Compare and order improper fractions and mixed number fractions with related denominators.

✱ **Required: 7/10**   ✱ **Pupil Navigation: on**
✱ **Randomised: off**

11. Which of the following is the smallest fraction?
a b c
14/10, 26/20, 29/20, 5/2
Put a forward slash (/) between the numerator and denominator.

■ 14/10 ■ **26/20** ■ 5/2 ■ 29/20

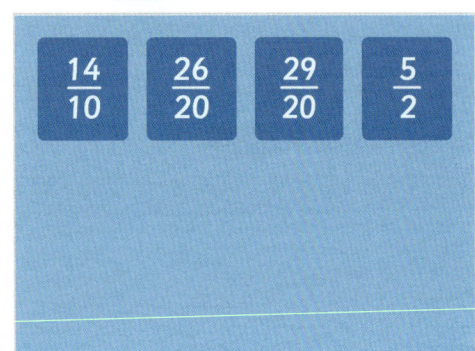

12. Which **two** mixed number fractions have the same value?
2/5

■ 2 6/8 ■ 3 2/4 ■ **3 6/8** ■ 3 6/16 ■ **3 12/16**

13. Sort the following fractions into descending order (largest first).
↑↓

■ **12/7** ■ **10/7** ■ **19/14** ■ **31/28**

14. Sort the following fractions into ascending order (smallest first).
↑↓

■ **2 5/10** ■ **3 1/10** ■ **3 3/20** ■ **3 2/5**

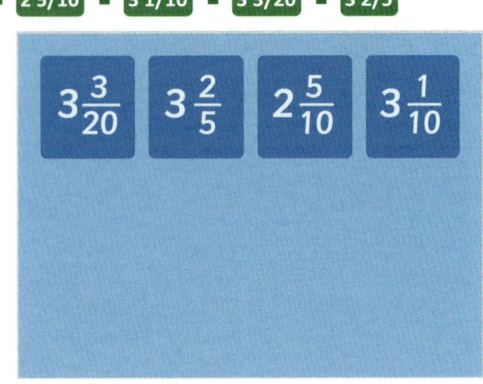

15. Which of the following statements is correct?
1/3

■ A: 6/4 < 18/12 ■ **B: 1 3/5 > 1 4/10** ■ C: 17/10 < 1 1/2

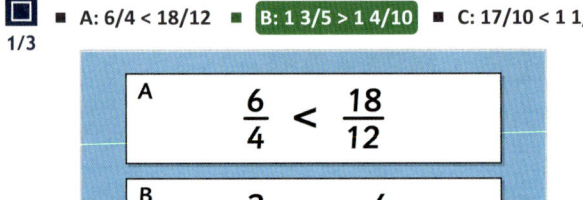

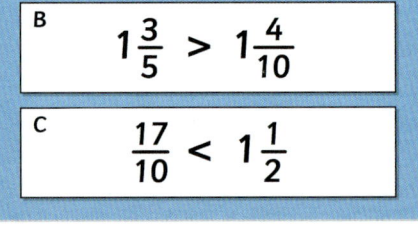

16. Which of the following is the largest fraction?
a b c
1 1/3, 11/6, 7/3, 1 4/6
Put a forward slash (/) between the numerator and denominator.

■ 1 1/3 ■ **7/3** ■ 1 4/6 ■ 11/6

Level 2 *continued*

17. Class 5K have a competition to see who can swim the furthest distance in an hour. What fraction of a mile does the child who swims the furthest complete?
Put a forward slash (/) between the numerator and denominator.

- 1 4/10 - 1 2/5 - 7/5 - 14/10 - Charlie

Class 5K swimming competition

name	miles completed
Jess	$1\frac{1}{10}$
Emma	$1\frac{1}{5}$
Charlie	$\frac{14}{10}$

18. A farmer records the fraction of a bag of hay he uses to feed each of his animals. Which **animal** eats the **least** amount of hay?

- cow - pig - sheep

Hay used by animals on a farm

animal	bags of hay
pig	$\frac{11}{3}$
cow	$\frac{13}{6}$
sheep	$\frac{29}{12}$

19. Sort the following fractions into ascending order (smallest first).

- 4 5/8 - 5 1/8 - 5 1/2 - 5 3/4

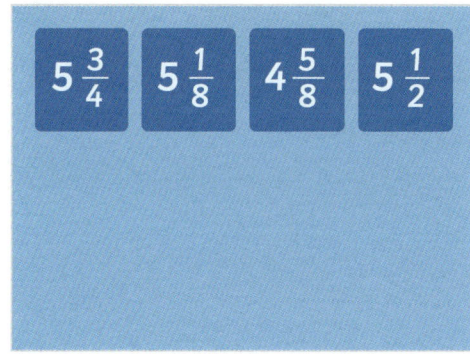

$5\frac{3}{4}$ $5\frac{1}{8}$ $4\frac{5}{8}$ $5\frac{1}{2}$

20. Which of the following is the smallest improper fraction?
9/2, 29/20, 19/10
Put a forward slash (/) between the numerator and denominator.

- 29/20 - 9/2 - 19/10

$\frac{9}{2}$ $\frac{29}{20}$ $\frac{19}{10}$

Level 3: Reasoning: Reason about converting, ordering and comparing fractions.

✴ **Required:** 5/5 ✴ **Pupil Navigation:** on
✴ **Randomised:** off

21. Luke has tried to arrange the fractions in order from smallest to largest, but he has made a mistake.
Which fraction is in the incorrect position?
Put a forward slash (/) between the numerator and denominator.

- 21/20

$\frac{6}{5}$, $\frac{15}{10}$, $\frac{21}{20}$, $\frac{8}{5}$, $\frac{17}{10}$

Level 3 *continued*

22. Jackson is comparing 11/3 with 1 1/3. He
a b c says, "An improper fraction is always worth
more than a mixed number fraction."
Is Jackson's statement correct? Explain your
answer and give at least two examples.

- Open question, no set answer

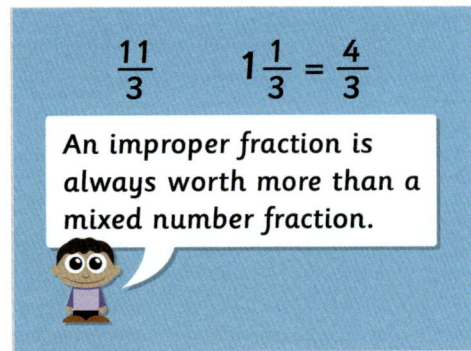

$\dfrac{11}{3}$ $1\dfrac{1}{3} = \dfrac{4}{3}$

An improper fraction is always worth more than a mixed number fraction.

23. Three of these calculations have the same
a b c answer. Which calculation has a **different**
answer? Explain your answer.

- Open question, no set answer

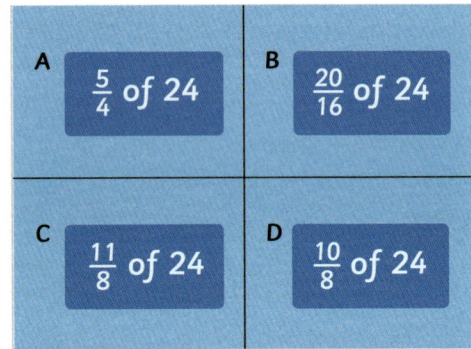

| A $\dfrac{5}{4}$ of 24 | B $\dfrac{20}{16}$ of 24 |
| C $\dfrac{11}{8}$ of 24 | D $\dfrac{10}{8}$ of 24 |

24. The mixed number fractions shown are in
a b c order from smallest to largest. If you add 2
to the numerator of each fraction, will the
mixed number fractions still be in the correct
order?
Explain and prove your answer.

- Open question, no set answer

$2\dfrac{1}{5}$ $2\dfrac{3}{10}$ $2\dfrac{2}{5}$ $2\dfrac{5}{10}$

25. What is the largest improper fraction you
a b c can make using two of the number cards?
*Put a forward slash (/) between the
numerator and denominator.*

- 100/2 - 10/20

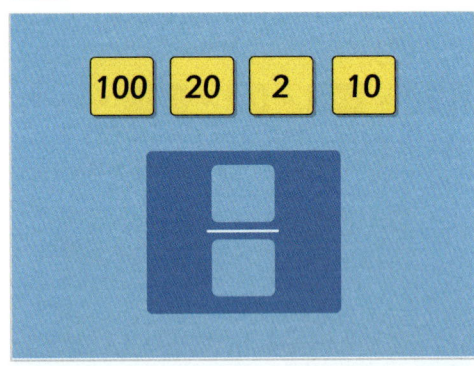

100   20   2   10

Level 4: Problem solving with greater depth: Solve
problems involving comparing and ordering
proper fractions.

✸ **Required:** 5/5   ✸ **Pupil Navigation:** on
✸ **Randomised:** off

26. What is the difference between the smallest
a b c fraction and largest fraction?
*Put a forward slash (/) between the
numerator and denominator.*

- 5/14 - 10/28

$\dfrac{31}{28}$ $\dfrac{37}{28}$ $\dfrac{10}{7}$ $\dfrac{15}{14}$

Level 4 continued

27. Five children record the distances that they
a each swim as fractions of one mile. Find the
b two children that swim the furthest. How far
c do these two children swim in total?
*Give your answer as an improper fraction or
a mixed number fraction.*

- **4 2/20** - **4 1/10** - **41/10** - 2 3/10 - **82/20**
- 2 6/20 - 46/20 - 23/10

child	distance (miles)
Distance five children swim	
Logan	$1\frac{1}{10}$
Sandy	$\frac{9}{5}$
Tom	$1\frac{3}{20}$
Briony	$\frac{7}{10}$
Paula	$2\frac{3}{10}$

28. Morgan records how many packets of raisins
↑   he eats each day. Sort the days in order,
↓   starting with the day when Morgan eats the
    most packets of raisins.

- **Friday** - **Tuesday** - **Wednesday** - **Monday**
- **Thursday**

day	packets
Packets of raisins eaten	
Mon	$\frac{21}{15}$
Tues	$1\frac{27}{30}$
Wed	$\frac{95}{60}$
Thurs	$1\frac{2}{60}$
Fri	$\frac{62}{30}$

29. Three children record how many packets of
a cereal they ate in a month. Packets of cereal
b weigh 250 grams (g) each. Find the child who
c ate the least amount of cereal. How many
    grams of cereal did that child eat?
Include the units g (grams) in your answer.

- **275 g** - **275 grams** - 275 - 1 1/10 - 11/10

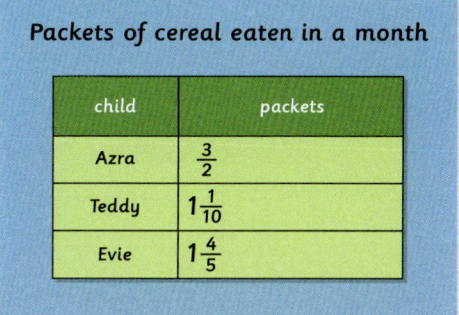

child	packets
Packets of cereal eaten in a month	
Azra	$\frac{3}{2}$
Teddy	$1\frac{1}{10}$
Evie	$1\frac{4}{5}$

30. Captain John is painting the sides of his ship
a with different colours.
b He uses the following amounts of paint:
c
- 3/2 tins of **yellow** paint
- 3/4 of a tin of **blue** paint
- 1 3/8 tins of **orange** paint
- 1 tin of **green** paint
The rest of the paint is **black**.
If he uses six tins of paint in total, which
colour paint does he use the **most** tins of?

- **yellow** - 11/8 - 1 4/8 - 3/2 - 1 3/8 - 1 1/2

Compare and Order Proper Fractions with Related Denominators

Objective: I can compare and order fractions with related denominators.

Quick Search Ref: 11480

Level 1: Understanding: Compare and order proper fractions with related denominators with image support.

❋ **Required:** 7/10   ❋ **Pupil Navigation:** on   ❋ **Randomised:** off

---

1. Select the **three** numbers that are multiples of both 4 and 8.

☐☒☐
3/6

- 4 - 8 - 12 - 16 - 20 - 24

multiples of	
4	8
4	8
8	16
12	24
16	32
20	40
24	48

2. What is the lowest number that is a multiple of 2, 4 and 8?

1 2 3

- 8 - 64

multiples of		
2	4	8
2	4	8
4	8	16
6	12	24
8	16	32

3. What is the lowest number that is a multiple of the denominators of these fractions?
1/5, 2/10, 2/20

1 2 3

- 20 - 10

4. Which is the larger fraction, 7/10 or 3/5?
a b c *Put a forward slash (/) between the numerator and denominator.*

- 7/10 - 3/5

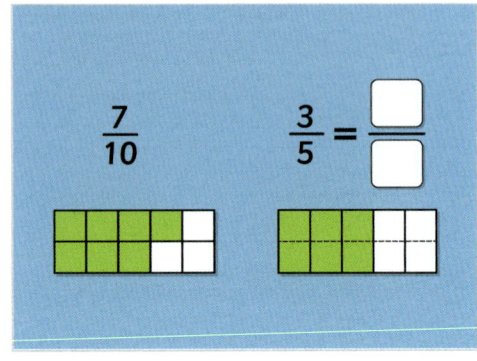

5. Which of the following fractions is smaller than ½?
a b c 4/8, 3/4, 3/8
Put a forward slash (/) between the numerator and denominator.

- 3/4 - 3/8 - 4/8

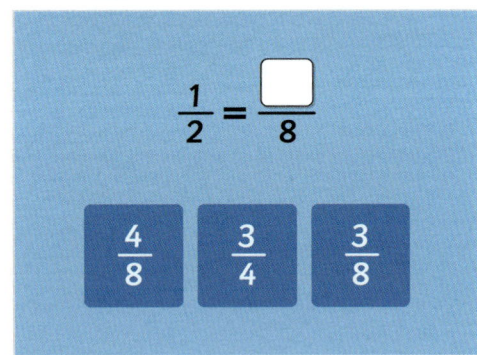

6. Sort the following fractions from the smallest to the largest.

↑ ↓

- 1/3 - 3/6 - 11/12

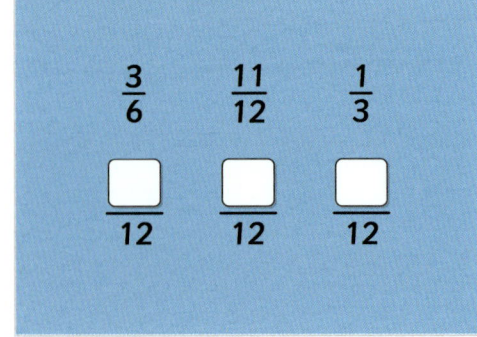

Level 1 continued

7. Sort the following fractions from the largest to the smallest.

↑↓

▪ 3/5 ▪ 7/20 ▪ 3/10 ▪ 1/5

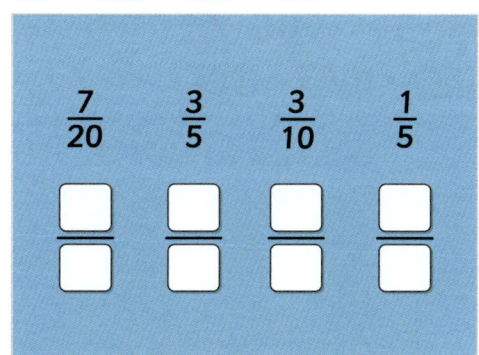

8. Which of the following is the smallest fraction?

a
b
c

5/24, 4/6, 3/12

Put a forward slash (/) between the numerator and denominator.

▪ 5/24 ▪ 4/6 ▪ 3/12

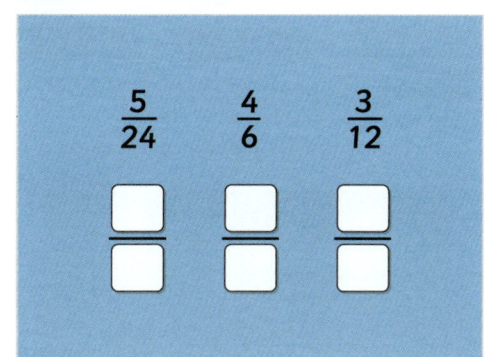

9. Sort the following fractions into ascending order (smallest first).

↑↓

▪ 1/4 ▪ 1/2 ▪ 5/8 ▪ 3/4

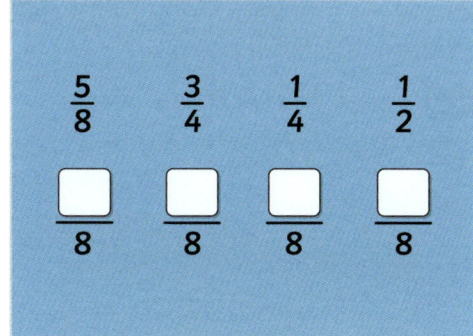

10. Which of the following is the largest fraction?

a
b
c

2/3, 1/6, 3/6

Put a forward slash (/) between the numerator and denominator.

▪ 2/3 ▪ 3/6 ▪ 1/6

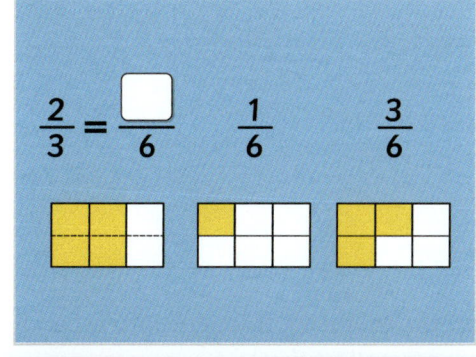

Level 2:   Fluency: Compare and order proper fractions with related denominators.

✱ **Required:** 7/10   ✱ **Pupil Navigation:** on
✱ **Randomised:** off

11. Which of the following is the smallest fraction?

a
b
c

6/20, 4/10, 19/20, 1/2

Put a forward slash (/) between the numerator and denominator.

▪ 6/20

12. Which **three** fractions have the same value?

☐
☒
☐

3/5

▪ 2/4 ▪ 3/4 ▪ 6/8 ▪ 6/16 ▪ 12/16

Level 2 continued

13. Sort the following fractions into descending order (largest first).

■ 6/7 ■ 9/14 ■ 11/28 ■ 2/7

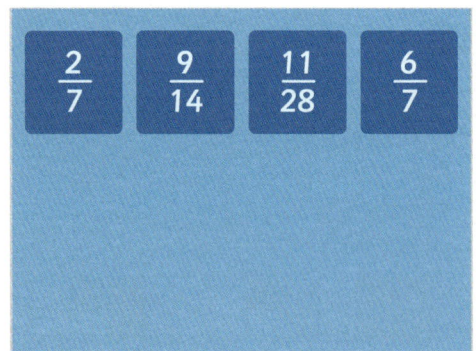

14. Sort the following fractions into ascending order (smallest first).

■ 1/10 ■ 3/20 ■ 2/5 ■ 5/10

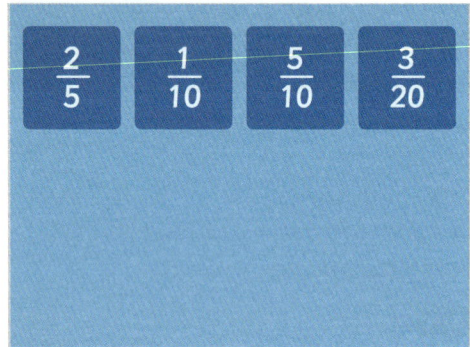

15. Which of the following statements is correct?

1/3

■ A: 3/4 < 11/16 ■ B: 3/4 = 11/16 ■ C: 3/4 > 11/16

A	$\frac{3}{4} < \frac{11}{16}$
B	$\frac{3}{4} = \frac{11}{16}$
C	$\frac{3}{4} > \frac{11}{16}$

16. Class 5 have a running competition to see who can run the furthest distance in 5 minutes. The child who runs the furthest distance completes _____ of a mile.
Enter the missing fraction. Don't include the units in your answer.

■ 42/50 ■ 84/100 ■ 21/25 ■ Abdul

Class 5 running competition

name	miles completed
Lara	$\frac{2}{5}$
Abdul	$\frac{42}{50}$
Jessica	$\frac{79}{100}$
Kane	$\frac{34}{50}$

17. The durations of four television shows are recorded as fractions of one hour. Sort the shows in order, starting with the television show that lasts the longest.

■ News at Nine ■ Drama Street ■ Cats on the Run ■ The Flee

Duration of television shows

show	duration (hours)
Cats on the Run	$\frac{3}{15}$
News at Nine	$\frac{19}{60}$
The Flee	$\frac{5}{30}$
Drama Street	$\frac{16}{60}$

18. Four friends record what fraction of a birthday cake they each eat. Which **child** eats the least amount of cake?

■ Sam ■ Ralph ■ 1/16 ■ Peter ■ Tim

Amount of cake eaten at a party

name	cake eaten
Tim	$\frac{5}{32}$
Sam	$\frac{1}{16}$
Ralph	$\frac{1}{4}$
Peter	$\frac{17}{32}$

Level 2 *continued*

19. Sort the following fractions into descending order (largest first).

■ 3/3 ■ 11/12 ■ 3/6 ■ 1/3

20. Which of the following is the largest fraction?
1/2, 9/20, 9/10
Put a forward slash (/) between the numerator and denominator.

■ 9/10

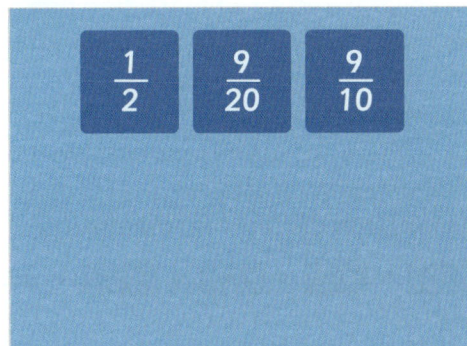

Level 3: Reasoning: Reason about converting, ordering and comparing fractions.

❋ **Required:** 5/5   ❋ **Pupil Navigation:** on
❋ **Randomised:** off

21. Lorna has tried to arrange the fractions in order from largest to smallest, but she has made a mistake.
Which fraction is in the wrong position?
Put a forward slash (/) between the numerator and denominator.

■ 4/5 ■ 16/20

22. "The larger the denominator, the larger the fraction."

Do you agree with this statement? Explain your answer and give at least two examples.

- Open question, no set answer

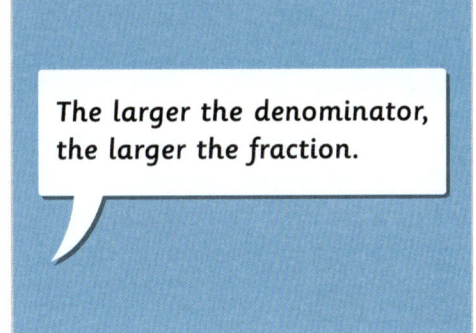

Level 3 continued

23. Three of these calculations have the same answer. Which calculation has a **different** answer?

1/4   ▪ A ▪ **B** ▪ C ▪ D

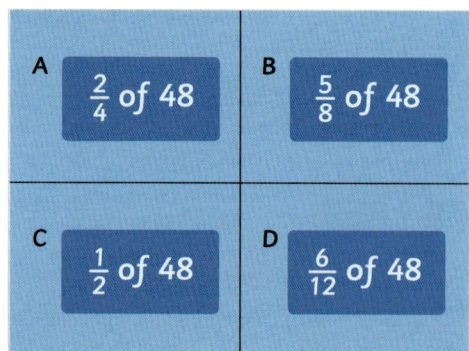

A $\frac{2}{4}$ of 48

B $\frac{5}{8}$ of 48

C $\frac{1}{2}$ of 48

D $\frac{6}{12}$ of 48

24. The fractions shown are in order from smallest to largest. If you add 2 to each numerator, will the fractions still be in the correct order?
Explain and prove your answer.

- Open question, no set answer

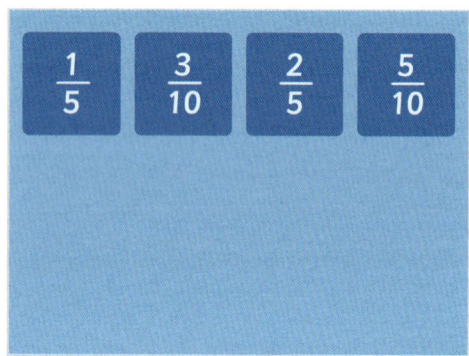

$\frac{1}{5}$ $\frac{3}{10}$ $\frac{2}{5}$ $\frac{5}{10}$

25. Using two of the number cards, what is the **smallest** fraction you can make?
Put a forward slash (/) between the numerator and denominator.

▪ **2/100** ▪ **1/50**

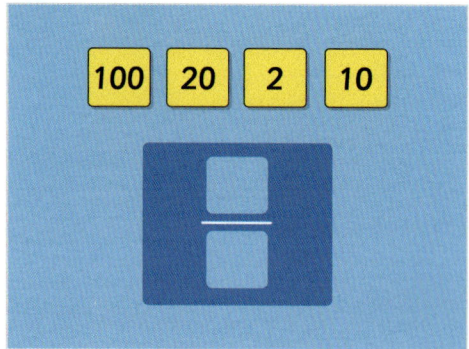

100 20 2 10

Level 4: Problem solving with greater depth: Solve problems involving comparing and ordering proper fractions.

❋ Required: 5/5   ❋ Pupil Navigation: on
❋ Randomised: off

26. What is the difference between the smallest fraction and largest fraction?
Put a forward slash (/) between the numerator and denominator.

▪ **5/8** ▪ **15/24**

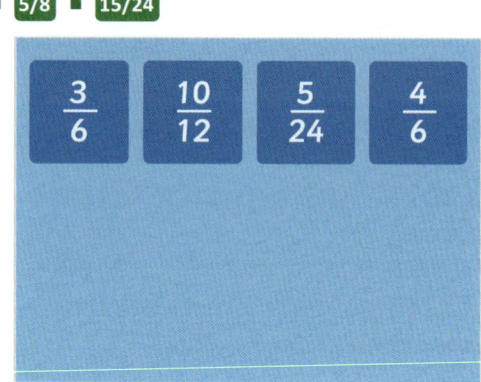

$\frac{3}{6}$ $\frac{10}{12}$ $\frac{5}{24}$ $\frac{4}{6}$

27. Five children record what fraction of a mile they each swim. Find the two children that swim the furthest. How far do these two children swim in total?
Give your answer as an improper fraction or a mixed number fraction.
Don't include the units in your answer.

▪ **34/20** ▪ **17/10** ▪ **1 7/10** ▪ **1 14/20** ▪ **18/20**
▪ **9/10**

Distance children swim	
child	distance (miles)
Dale	$\frac{2}{10}$
Archie	$\frac{13}{20}$
Asif	$\frac{4}{5}$
Zaina	$\frac{11}{20}$
Georgia	$\frac{9}{10}$

16. If Ruby has 2 5/11 packets of sweets and
a b c Jane gives her another 7/11 of a packet, how many packets of sweets will Ruby have in total?
Give your answer as a mixed number fraction. For example, five and one-half is 5 1/2.

- ■ 3 1/11 ■ 34/11 ■ 2 12/11

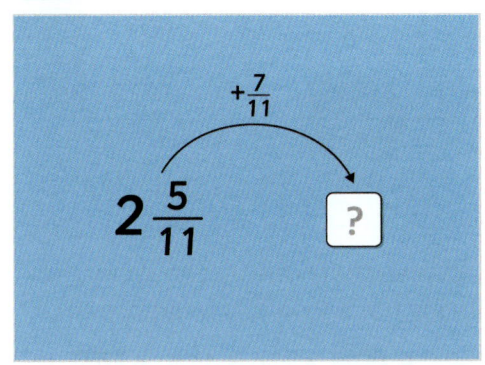

17. Grace runs 1 ¼ kilometres (km). If Freddie
a b c runs ¾ km less than Grace, what fraction of a kilometre does he run?
Give your answer as a fraction. Don't include the units in your answer.

- ■ 1/2 km ■ 2/4 kilometres ■ 1/2 kilometres ■ 1/2
- ■ 2/4 km ■ 2/4 ■ 2

18. What mixed number fraction will you reach if
a b c you count on four fifths from 2 2/5?
Give your answer as a mixed number fraction. For example, five and one-half is 5 1/2.

- ■ 3 1/5 ■ 16/5 ■ 2 6/5

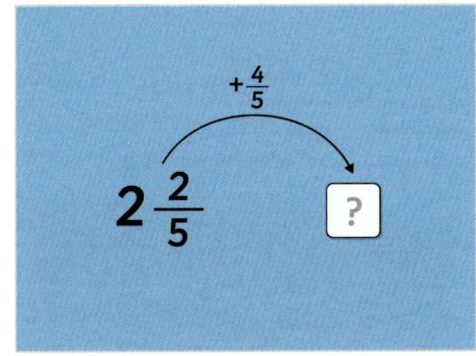

19. Select the **three** missing values in this sequence.

3/5
- ■ 3 1/9 ■ 3 7/9 ■ 4 ■ 4 1/9 ■ 5 1/9

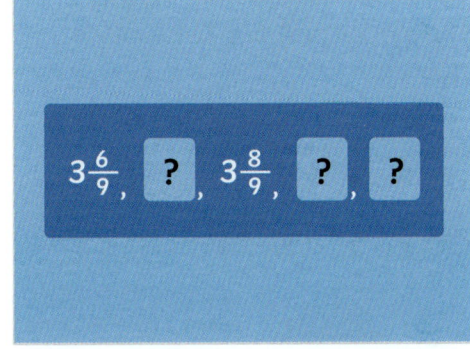

20. What is the missing mixed number fraction
a b c on the number line?
Give your answer as a mixed number fraction. For example, five and on-half is 5 1/2.

- ■ 2 2/4 ■ 2 1/2 ■ 10/4 ■ 5/2

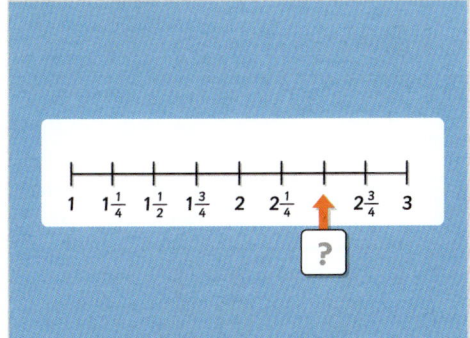

Level 3: Reasoning: Reason about fractions in sequences and in relation to equivalent fractions with related denominators.

✱ **Required:** 5/5   ✱ **Pupil Navigation:** on
✱ **Randomised:** off

21. This sequence is going down in sevenths. Explain where the sequence is incorrect and give the correct sequence.

a b c

- Open question, no set answer

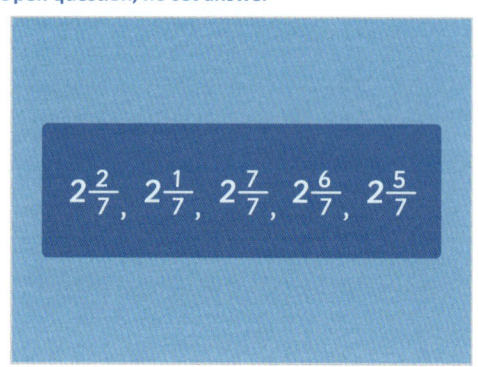

$2\frac{2}{7}$, $2\frac{1}{7}$, $2\frac{7}{7}$, $2\frac{6}{7}$, $2\frac{5}{7}$

22. Toby is counting in thirds. Explain what value Toby has missed out from his sequence and complete the next two parts of the sequence.

a b c

- Open question, no set answer

$\frac{1}{3}$, $\frac{2}{3}$, $1\frac{1}{3}$, $1\frac{2}{3}$, ☐, ☐

23. Which letter shows the position of 2 ½ on the number line?

1/3   ■ A   ■ **B**   ■ C

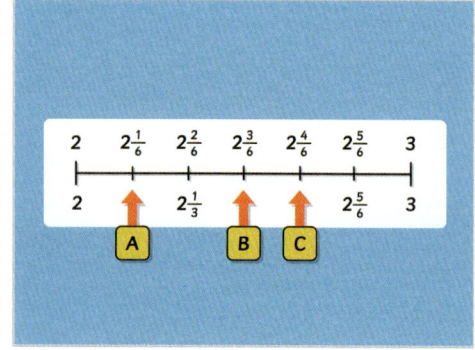

24. Lola has written a number sequence. She says her sequence is going up in tenths. Billy says that she has made a mistake because her denominators aren't all 10. Is Billy correct? Explain your answer.

a b c

- Open question, no set answer

$\frac{7}{10}$, $\frac{4}{5}$, $\frac{9}{10}$, 1, $1\frac{1}{10}$,

25. Select **two** fractions that could be the missing fraction in this sequence.

2/5   ■ 14/8   ■ **1 1/2**   ■ 4/8   ■ 1 1/4   ■ **12/8**

$\frac{14}{8}$, $1\frac{5}{8}$, **?**, $1\frac{3}{8}$, $\frac{10}{8}$

Level 4: Problem solving: Solve multi-step problems involving counting in fractions.

✱ **Required:** 5/5   ✱ **Pupil Navigation:** on
✱ **Randomised:** off

26. Freddie has eaten 4/5 of one pizza. If he eats two whole pizzas in total, how many more fifths will he eat?

1 2 3

■ **6**   ■ 1

Level 4 *continued*

27. What fraction do you reach when you count
a on twenty-seven ninths from 1 5/9?
b
c *Give your answer as a mixed number
fraction. For example, five and one-half is 5
1/2.*

■ **4 5/9** ■ **41/9**

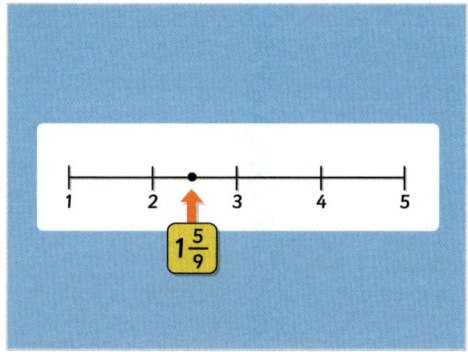

28. The arrow is halfway along the number line.
a What fraction is the arrow pointing to?
b
c

■ **1 3/4** ■ **7/4** ■ **7**

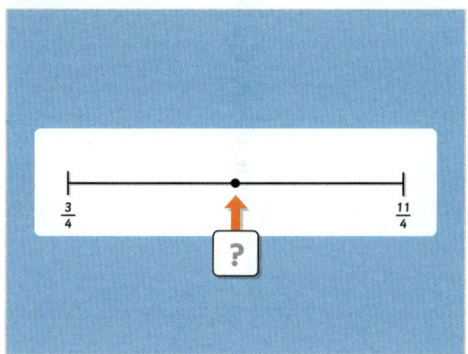

29. Jo counts up the chart in thirds from 2/3
a until she gets to **D**, then counts back in thirds
b
c to **A**. What is the value of **A**?

■ **6/3** ■ **3/3** ■ **1** ■ **9/3** ■ **2** ■ **3**

30. Emma makes some cakes. She gives 1/12 of
1 the cakes to Sam. If Sam has two cakes, how
2
3 many cakes does Emma make in total?

■ **24**

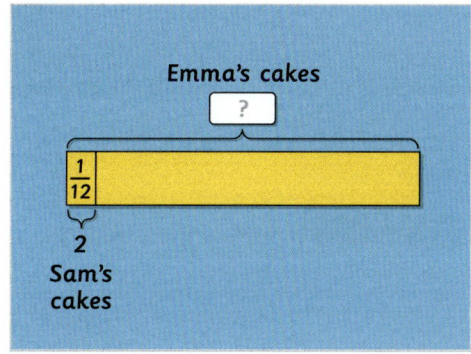

Count Up and Down in Thousandths (Fractions)

Objective:	I can count up and down in thousandths using fractions and relate thousandths to hundredths and tenths.

Quick Search Ref: **11565**

Level 1: Understanding: Know what 1/1,000 means and be able to count on and back in thousandths within 1 whole. Relate thousandths to hundredths and tenths.

✳ **Required:** 7/10　　✳ **Pupil Navigation:** on　　✳ **Randomised:** off

---

1. How many thousandths are there in one whole?

a b c

- ■ **1,000**　■ **1000**　■ 0.001

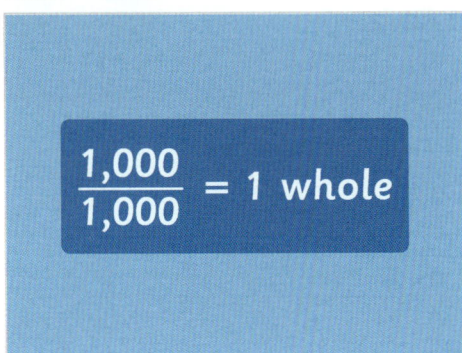

2. The large cube has been split into 1,000 equal pieces. What is one piece worth as a fraction?
Put a forward slash between the numerator and the denominator. For example, one-half is 1/2.

a b c

- ■ **1/1,000**　■ **1/1000**　■ 1

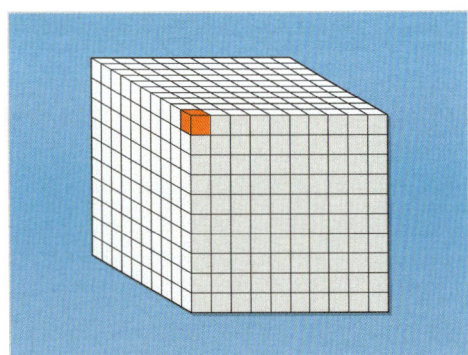

3. The large cube has been split into 1,000 equal pieces.
What fraction of the cube is shaded?
Put a forward slash between the numerator and the denominator. For example, one-half is 1/2.

a b c

- ■ **9/1,000**　■ **9/1000**　■ 1/1,000　■ 9

4. The large square has been split into 100 equal pieces. One piece of the large square has then been split into 10 equal rectangles. What is one rectangle worth as a fraction of the large square?
Put a forward slash between the numerator and the denominator. For example, one-half is 1/2.

a b c

- ■ **1/1000**　■ 1/10　■ **1/1,000**　■ 1/100

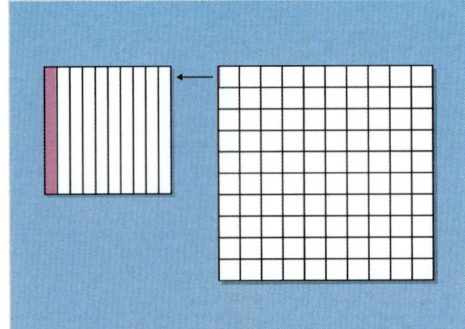

Level 1 *continued*

5. 400/1,000 is the same as how many tenths?

■ 400 ■ **4** ■ 40

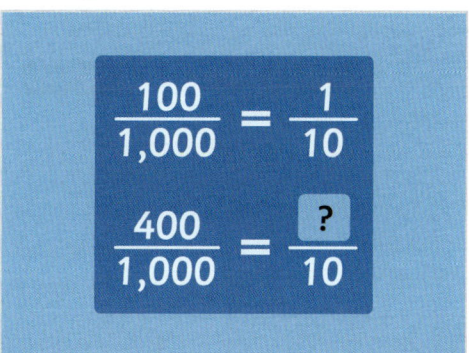

6. What is the next fraction in the sequence?
Put a forward slash between the numerator and the denominator. For example, one-half is 1/2.

■ **399/1000** ■ **399/1,000** ■ 399 ■ 401/1,000

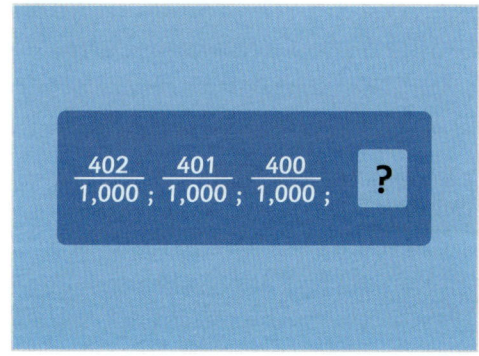

7. What is the missing fraction on the number line?
Put a forward slash between the numerator and the denominator. For example, one-half is 1/2.

■ **120/1,000** ■ **120/1000** ■ 120

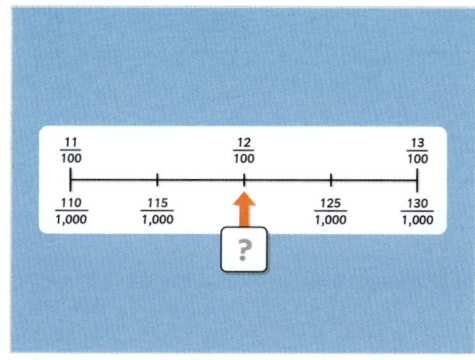

8. 400/1,000 is the same as how many hundredths?

■ **40** ■ 4 ■ 400

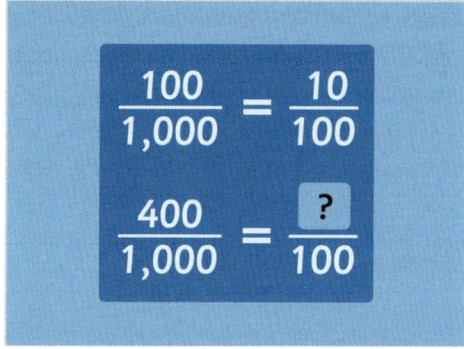

9. What is the next fraction in the sequence?
Put a forward slash between the numerator and the denominator. For example, one-half is 1/2.

■ **18/1,000** ■ **18/1000** ■ 18

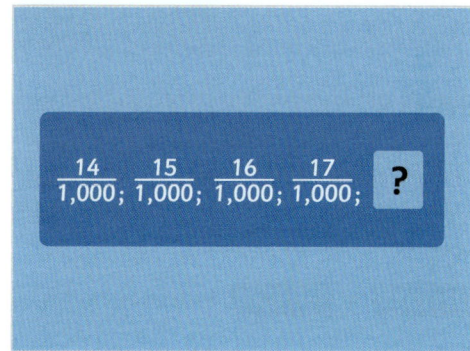

10. How many parts of the cube would you need to shade to show 15/1,000?

■ **15** ■ 1

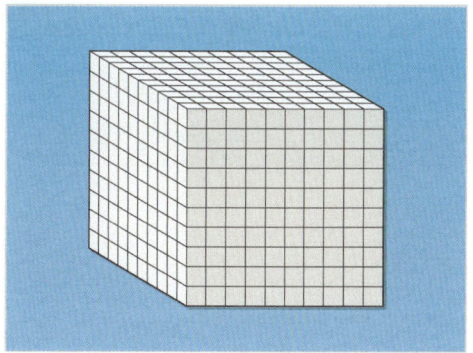

Level 2: Fluency: Count on and back in thousandths including beyond 1 whole.

✱ Required: 7/10   ✱ Pupil Navigation: on

✱ Randomised: off

11. What fraction will you reach if you count
    back seven thousandths from 54/1,000?
    *Put a forward slash between the numerator
    and the denominator. For example, one-half
    is 1/2.*

■ 47/1,000 ■ 61/1,000 ■ 47/1000 ■ 61/1000

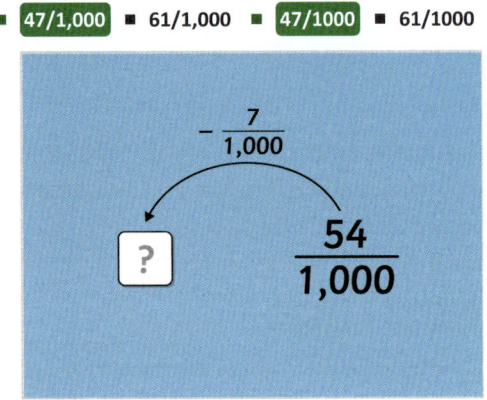

12. What is the missing fraction on the number
    line?
    *Put a forward slash between the numerator
    and the denominator. For example, one-half
    is 1/2.*

■ 725/1000 ■ 725/1,000 ■ 725/100 ■ 725

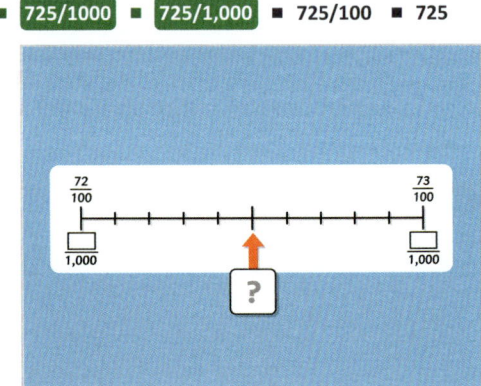

13. What is the missing fraction on the number
    line?
    *Give your answer as a mixed number
    fraction. For example, five and one-half is 5
    1/2.*

■ 4/1000 ■ 1 4/1000 ■ 1 4/10 ■ 1 4/1,000
■ 4/1,000 ■ 1 4/100

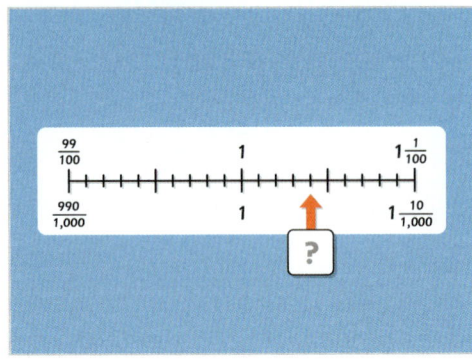

14. What mixed number fraction will you reach if
    you count on nine thousandths from 2
    415/1,000?
    *Give your answer as a mixed number
    fraction. For example, five and one-half is 5
    1/2.*

■ 2 406/1000 ■ 2 424/1,000 ■ 2 406/1,000
■ 2 424/1000 ■ 424/1000 ■ 424/1,000

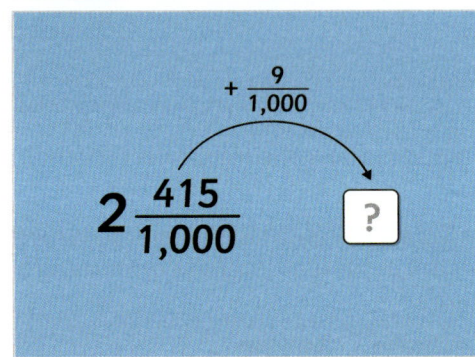

15. Select the **three** missing fractions in this
    sequence.

3/5

■ 1 16/1,000 ■ 1 18/1,000 ■ 1 21/1,000
■ 1 23/1,000 ■ 1 2/100

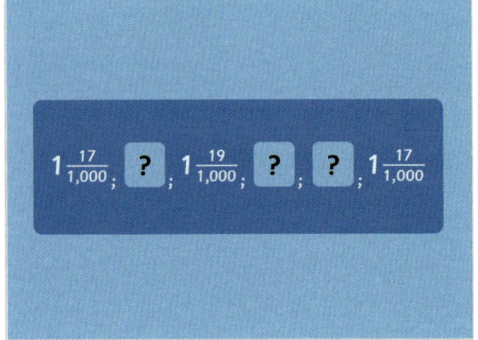

16. Jack runs 782/1,000 of a kilometre (km). If April runs 11/1,000 km less than Jack, what fraction of a kilometre does she run?
Give your answer as a fraction. Don't include the units in your answer.

- 793/1,000 ■ 771/1000 ■ 771/1,000 ■ 771
- 0.771 ■ 793/1000

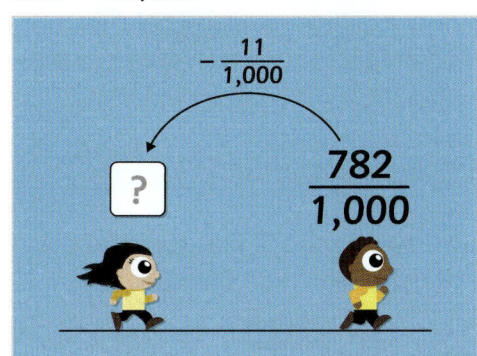

17. David is playing a computer game and he scores 595 points out of a possible 1,000 points. If Katie scores 9 more points than David, what fraction of the maximum score does she reach?
Put a forward slash between the numerator and the denominator. For example, one-half is 1/2.

- 604/1000 ■ 604/1,000 ■ 604

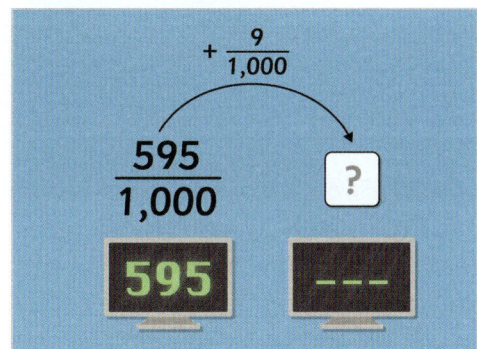

18. Count on in thousandths to find the last missing fraction in the sequence.
Put a forward slash between the numerator and the denominator. For example, one-half is 1/2.

- 603/1000 ■ 603/1,000 ■ 603

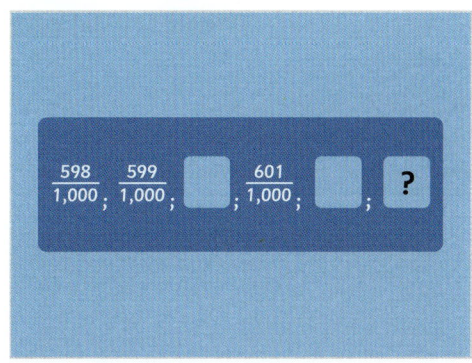

19. What fraction will you reach if you count on eight thousandths from 325/1,000?
Put a forward slash between the numerator and the denominator. For example, one-half is 1/2.

- 333/1000 ■ 333/1,000

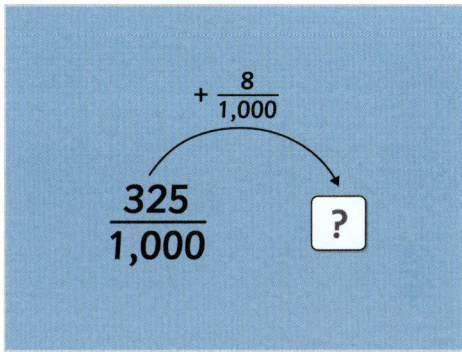

20. What is the missing fraction on the number line?
Put a forward slash between the numerator and the denominator. For example, one-half is 1/2.

- 16/10 ■ 151/1,000 ■ 151/1000 ■ 151
- 151/100

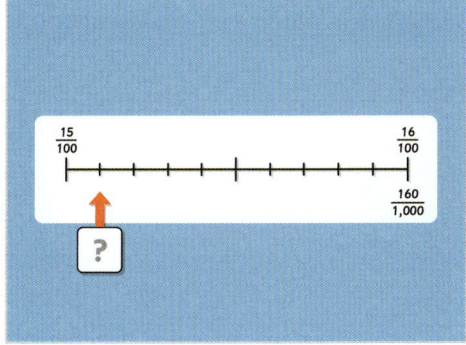

Level 3: Reasoning: Reason about thousandths in sequences and in relation to hundredths and tenths.

❋ **Required:** 5/5   ❋ **Pupil Navigation:** on

❋ **Randomised:** off

21. Rafi says, "The last fraction in the sequence
a b c is 1 whole." Is Rafi correct? Explain your answer.

- Open question, no set answer

22. Jake says, "8/1,000 has the same value as 8
a b c thousands." Is Jake correct? Explain your answer.

- Open question, no set answer

23. Which letter shows the position of
□
☒ 650/1,000 on the number line?
□
□

1/4   ■ A   ■ B   ■ C   ■ D

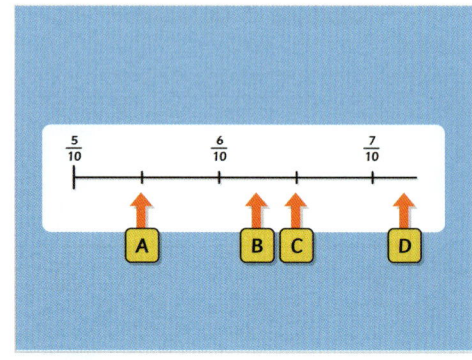

24. Erin has written a number sequence. Do you
a b c think Erin's sequence is correct? Explain your answer.

- Open question, no set answer

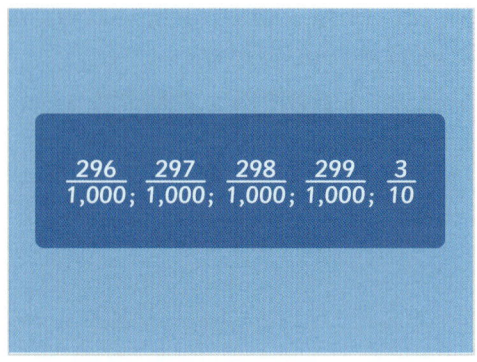

$$\frac{296}{1,000}; \frac{297}{1,000}; \frac{298}{1,000}; \frac{299}{1,000}; \frac{3}{10}$$

25. Select **two** fractions that could be the
□
☒ missing fraction in this sequence.
□

2/5   ■ 300/1,000   ■ 1 30/100   ■ 2 3/10   ■ 1 30/1,000

■ 1 3/10

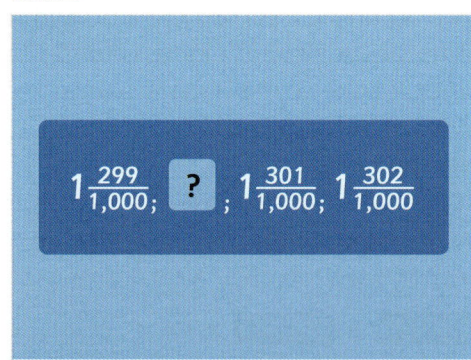

$$1\frac{299}{1,000}; \;?\;; 1\frac{301}{1,000}; 1\frac{302}{1,000}$$

Level 4: Problem Solving: Solve multi-step problems involving thousandths.

❋ **Required:** 5/5   ❋ **Pupil Navigation:** on

❋ **Randomised:** off

26. The arrow is halfway along the number line.
a b c What fraction is the arrow pointing to?
Put a forward slash between the numerator and the denominator. For example, one-half is 1/2.

■ 1 5/1000   ■ 1,005/1,000   ■ 1 5/1,000

■ 1005/1000   ■ 1,005

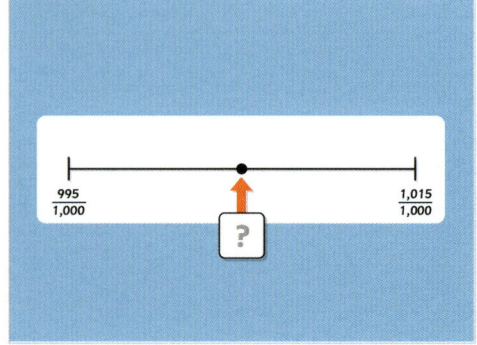

Level 4 continued

27. Jamie counts up the chart from 411/1,000 until he gets to **D**, then counts back in thousandths to **A**. What is the value of **A**?

a b c

- 418/1,000 ■ 412/1000 ■ 418/1000 ■ 412/1,000
- 415/1000 ■ 415/1,000

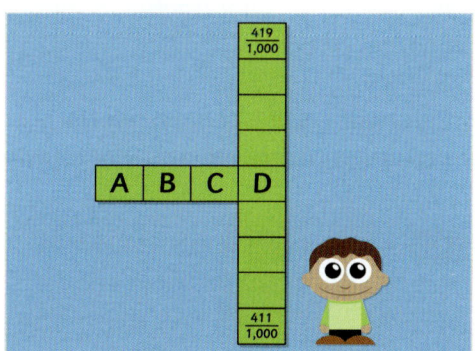

28. What fraction do you reach when you count on three-thousand thousandths from 1 743/1,000?

a b c

Give your answer as a mixed number fraction. For example, five and one-half is 5 1/2.

- 4 743/1000 ■ 4 743/1,000 ■ 4742/1,000
- 4742/1000

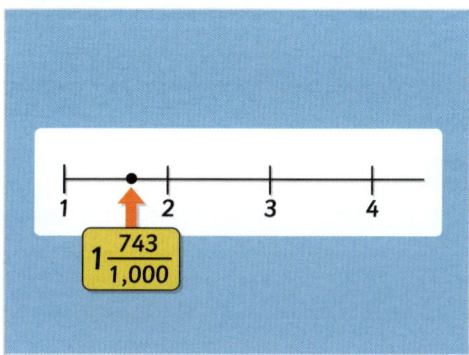

29. Nathan represents thousandths by putting ten counters on each part of a hundred square. How many counters does he need to remove if he wants to show 97/100?

1 2 3

- 30

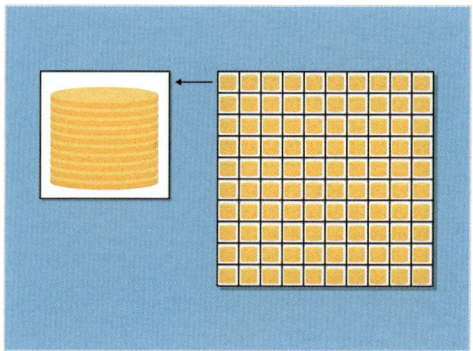

30. Millie wins £5 in a competition, which is 1/1,000 of the total prize money. How much prize money is there in total?

a b c

Include the £ (pound) sign in your answer.

- £5,000 ■ 5000.00 ■ £5000.00 ■ £5,000.00
- £5000 ■ 5,000.00 ■ 5,000 ■ 5000

Understand Linear Number Sequences With Fractions

Objective: I can describe and continue linear number sequences with fractions.

Quick Search Ref: 10297

Level 1: Understanding: Find the rule to complete and continue linear sequences with fractions.

✳ **Required:** 7/10　　　✳ **Pupil Navigation:** on　　　✳ **Randomised:** off

---

1. What is the difference between each fraction in this sequence?

1/6, 2/6, 3/6, 4/6

1/4　　▪ **1/6** ▪ 2/6 ▪ 3/6 ▪ 4/6

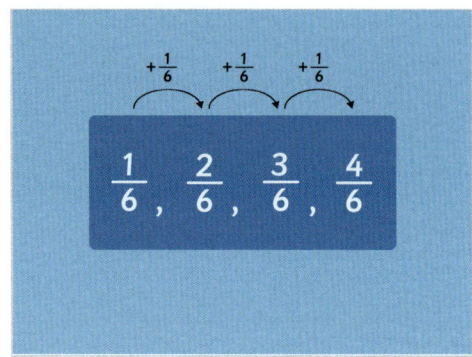

2. The fractions in this sequence decrease by _____ each time.

Give your answer as a fraction. Put a forward slash (/) between the numerator and denominator. For example, one-half is 1/2.

▪ 4/4 ▪ **1/4** ▪ 0.25

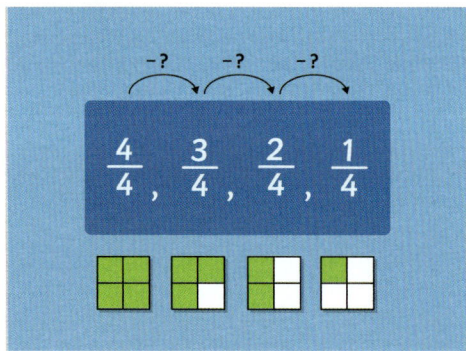

3. What is the next fraction in this sequence?

2/10, 4/10, 6/10, ____

▪ 2/10 ▪ 7/10 ▪ **8/10** ▪ 8/18

1/4

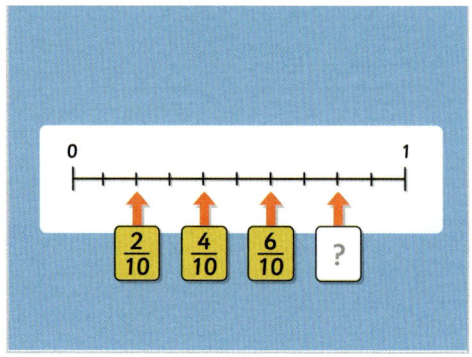

4. In this sequence, what fraction is hidden behind the cloud?

11/5, 10/5, ____, 8/5

Put a forward slash (/) between the numerator and denominator. For example, one-half is 1/2.

▪ **1 4/5** ▪ -1/5 ▪ **9/5** ▪ 1/5

5. What fraction comes next in this sequence?

1/4, 3/8, 1/2, ____

Put a forward slash (/) between the numerator and denominator. For example, one-half is 1/2.

▪ **5/8** ▪ 1/8

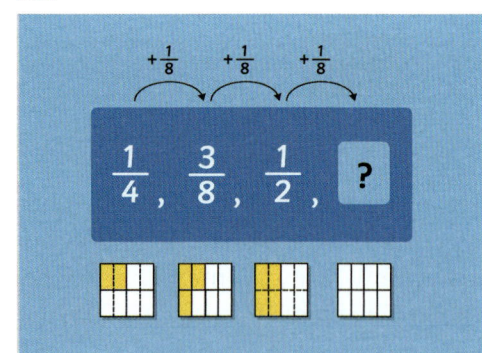

---

6. What fraction is missing from this sequence?
1/6, 1/3, ____, 8/12
Put a forward slash (/) between the numerator and denominator. For example, one-half is 1/2.

▪ **6/12** ▪ **1/2** ▪ **3/6** ▪ 1/6

7. In this sequence, what fraction is hidden behind the paint?
4/8, 1, 1 1/2, 2, ____
Put a forward slash (/) between the numerator and denominator. For example, one-half is 1/2.

▪ **5/2** ▪ **2 1/2** ▪ 1/2

8. What fraction is missing from this sequence?

▪ **5/9** ▪ 2/9

9. The fractions in this sequence increase by ____ each time.
Give your answer as a fraction. Put a forward slash (/) between the numerator and denominator. For example, one-half is 1/2.

▪ **1/8** ▪ 5/8

10. What is the next fraction in this sequence?
5/7, 4/7, 3/7, ___
Put a forward slash (/) between the numerator and denominator. For example, one-half is 1/2.

▪ **2/7** ▪ 1/7 ▪ -1/7

Level 2: Fluency: Understand and continue linear number sequences with fractions.

❋ **Required:** 7/10 ❋ **Pupil Navigation:** on
❋ **Randomised:** off

11. What is the next fraction in this sequence?
[abc] *Put a forward slash (/) between the numerator and denominator.*

▪ 10/12 ▪ 5/6 ▪ 1/6

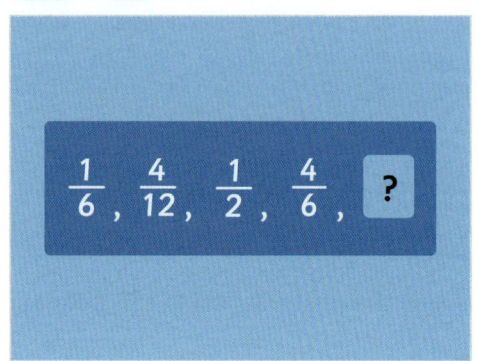

12. What fraction comes next in this sequence?
[abc] *Put a forward slash (/) between the numerator and denominator.*

▪ 9/3 ▪ 3 ▪ 2/3

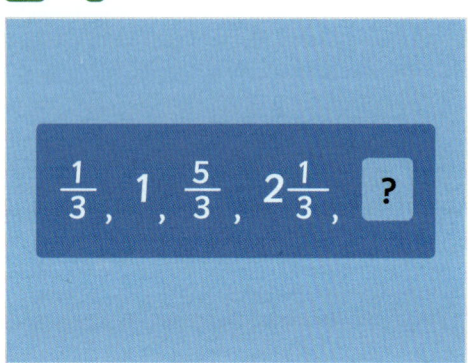

13. What fraction is hidden behind the paint?
[abc] *Put a forward slash (/) between the numerator and denominator.*

▪ 2/5 ▪ 8/20 ▪ 4/10

14. Which two fractions are missing from this sequence?
2/5

▪ 4/11 ▪ 8/15 ▪ 6/10 ▪ 7/10 ▪ 12/10

15. Which shape completes the sequence above the line?
1/4

▪ shape A ▪ shape B ▪ shape C ▪ shape D

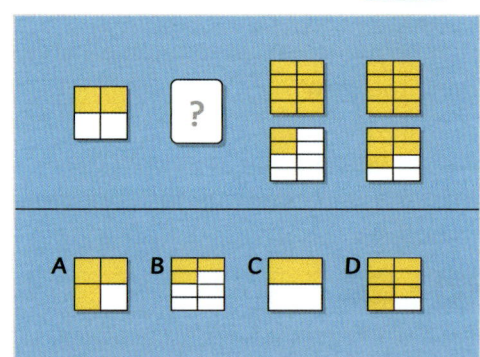

16. Each morning, Bradley eats 2/14 of a pack of cereal. If he has 1 1/7 packs of cereal on Monday evening, what **fraction** of a full pack does he have left on Friday evening?
[abc] *Put a forward slash (/) between the numerator and denominator.*

▪ 4/7 ▪ 8/14 ▪ 1/7 ▪ 2/14

Mon	Tues	Wed	Thurs	Fri
$1\frac{1}{7}$				?

Level 2 *continued*

17. On Monday, Chloe eats 1 1/5 bags of grapes.
a b c For the next four days, she eats 3/10 of a bag of grapes each day. How many bags of grapes does Chloe eat in total from Monday to Friday?
Give your answer as a mixed number fraction. For example, five and one-half is 5 1/2.

■ 2 2/5　■ 2　■ 2 4/10　■ 24/10

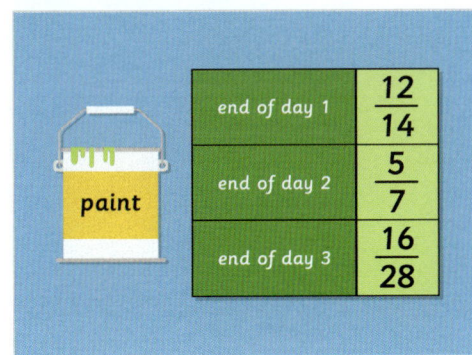

18. Jamal is a decorator and uses the same
a b c amount of paint each day. He records what fraction of his paint is left at the end of each day. What **fraction** of his paint does he use each day?
Put a forward slash (/) between the numerator and denominator.

■ 2/14　■ 4/28　■ 1/7

end of day 1	$\frac{12}{14}$
end of day 2	$\frac{5}{7}$
end of day 3	$\frac{16}{28}$

19. Select the **two** fractions that are missing
from this sequence.

2/5

■ 11/16　■ 1/2　■ 5/8　■ 12/16　■ 1

$$\frac{1}{4} , \frac{3}{8} , ? , \frac{10}{16} , ? , \frac{7}{8}$$

20. What fraction is hidden behind the paint?
a b c *Put a forward slash (/) between the numerator and denominator.*

■ 1/3　■ 4/12　■ 2/6　■ 3/12　■ 1/4

$$\text{(paint splat)} , \frac{7}{12} , \frac{5}{6} , 1\frac{1}{12} , \frac{4}{3}$$

Level 3: Reasoning: Reason about fractions in linear number sequences.

✿ **Required:** 5/5　　✿ **Pupil Navigation:** on
✿ **Randomised:** off

21. What fraction in this sequence has the
a b c **greatest value** but is **less than 50**?
Give your answer as a mixed number fraction. For example, five and one-half is 5 1/2.

■ 50　■ 49 8/10　■ 49 4/5　■ 249/5　■ 50 1/5

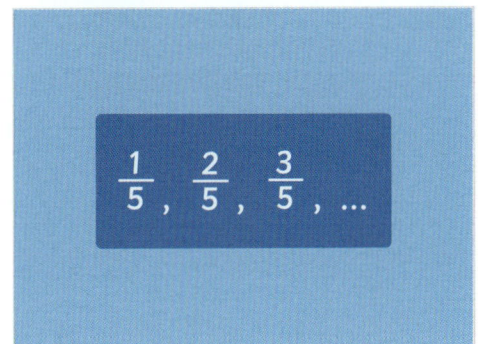

$$\frac{1}{5} , \frac{2}{5} , \frac{3}{5} , ...$$

Level 3 *continued*

22. What is the same and what is different about these sequences?

a b c

- Open question, no set answer

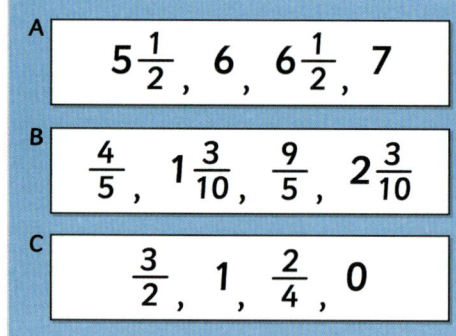

A $5\frac{1}{2}$, 6, $6\frac{1}{2}$, 7

B $\frac{4}{5}$, $1\frac{3}{10}$, $\frac{9}{5}$, $2\frac{3}{10}$

C $\frac{3}{2}$, 1, $\frac{2}{4}$, 0

23. Petra says that if she multiplies all the denominators of each fraction in this sequence by the same number, the fractions will still increase by the same amount each time.

a b c

Do you agree with Petra? Give reasons for your answer and give at least two examples.

- Open question, no set answer

$\frac{2}{8}$, $\frac{1}{2}$, $\frac{3}{4}$

24. Which two of these fractions will **not** appear in this sequence?

2/5

■ **7/13** ■ 8/13 ■ **9/13** ■ 12/13 ■ 100/13

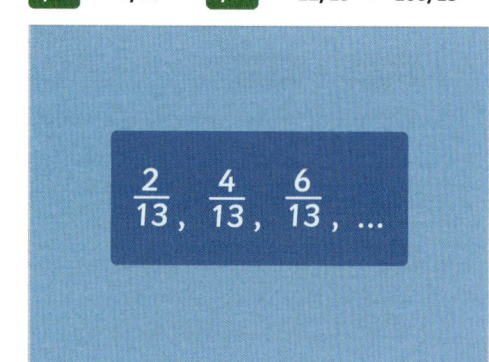

$\frac{2}{13}$, $\frac{4}{13}$, $\frac{6}{13}$, ...

25. Pete has created a fraction sequence, but he has made a mistake calculating one of the fractions. Find Pete's mistake and calculate the correct fraction.

a b c

- Open question, no set answer

$\frac{1}{4}$, $\frac{3}{8}$, $\frac{1}{2}$, $\frac{3}{4}$, $\frac{7}{8}$

Level 4: Problem solving with greater depth: Use rules to find fractions in linear number sequences in a range of contexts.

✱ **Required:** 5/5 ✱ **Pupil Navigation:** on
✱ **Randomised:** off

26. To find the next fraction in the sequence, add 18/4 each time. What is the **fourth** fraction in the sequence?

a b c

Give your answer as a mixed number fraction. For example, five and one-half is 5 1/2.

■ **14 1/8** ■ 41/8 ■ 5 1/8 ■ 9 5/8 ■ 77/8 ■ 113/8

$\frac{5}{8}$, ☐ , ☐ , **?**

27. A ship travels 3 2/5 miles each hour. Fuel costs £40 for each mile. How much does fuel cost for a journey that lasts for **five hours**?
Include the £ sign in your answer.

a
b
c

- 17 - £680 - £680.00 - 680 - £544

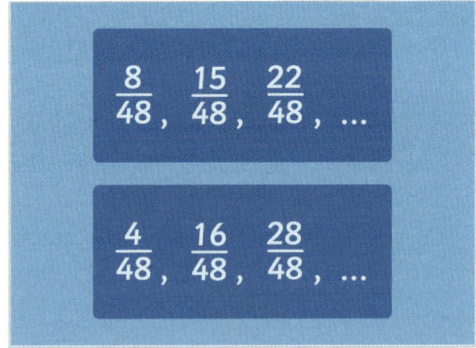

28. If you continue both of these fraction sequences, what is the first fraction that will appear in **both sequences**?
Give your answer as a mixed number fraction.

a
b
c

- 1 4/12 - 1 1/3 - 1 8/24 - 1 16/48 - 64/48
- 32/24 - 16/12

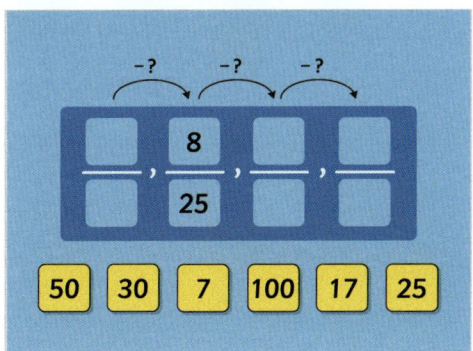

29. Use each number card once to make a sequence that **decreases** by the same amount each time. What is the **difference** between each fraction in the sequence?

a
b
c

- -2/100 - 2/100 - -1/50 - 1/50

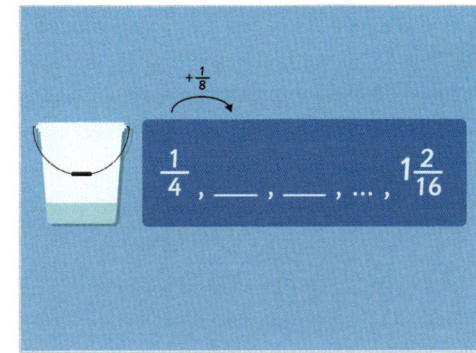

30. Tyler puts a bucket containing ¼ of a litre of water outside. Every 30 minutes, 1/8 of a litre of rainwater collects in the bucket. When she takes the bucket back inside, there are 1 2/16 litres of water in the bucket. How many minutes was the bucket outside?

1
2
3

- 210 - 240

Identify and Find Equivalent Fractions

Objective: I can identify and find equivalent fractions of a given fraction, including tenths and hundredths.

Quick Search Ref: 10153

Level 1: Understanding: Identify and find equivalent fractions with image support.

❋ **Required:** 7/10   ❋ **Pupil Navigation:** on   ❋ **Randomised:** off

1. What is the denominator in the fraction 2/5?

■ **5** ■ 2

2. Which image shows the correct way to create an equivalent fraction?

■ image A ■ **image B** ■ image C

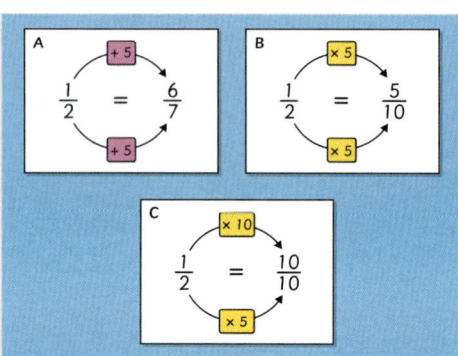

3. Which of the following is an equivalent fraction of ¾?

1/4, 3/40, 6/8

Put a forward slash (/) between the numerator and denominator.

■ **6/8** ■ 1/4 ■ 3/40

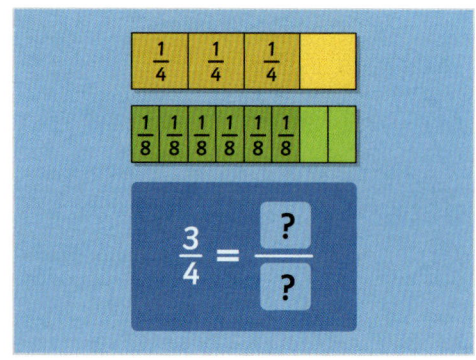

4. 1/3 of shape A is shaded. How many parts of shape B do you need to shade to make it equivalent to 1/3?

■ **2**

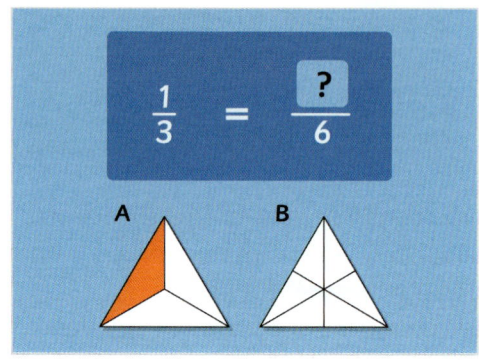

5. What is the missing denominator?

1/5 = 5/___

■ **25** ■ 10

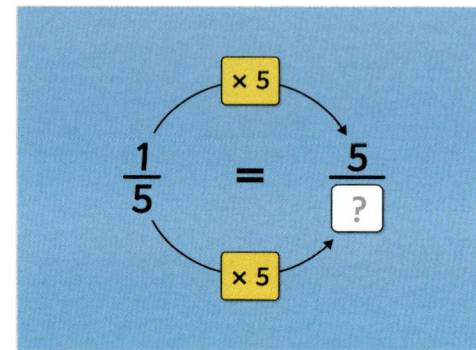

6. What is the missing numerator?

1/6 = ___/24

■ **4** ■ 5

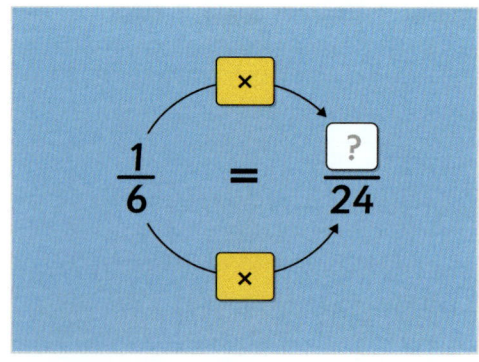

Level 1 *continued*

7. 12/72 = ____/12

1
2
3 ■ 6 ■ **2** ■ 72

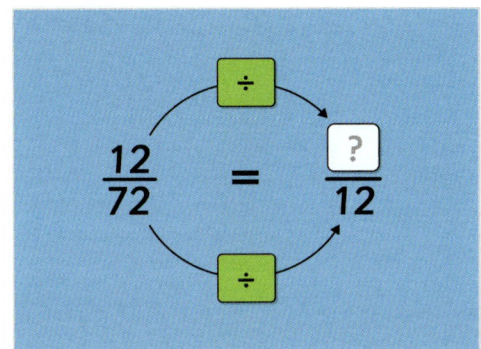

8. 6/8 = ____/24

1
2
3 ■ **18** ■ 3

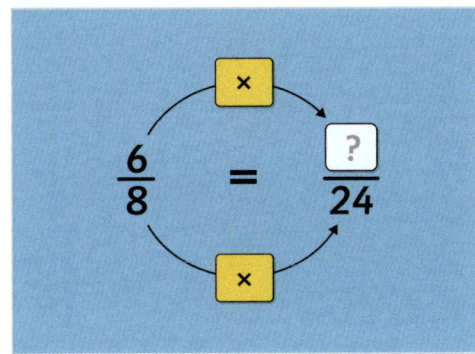

9. What is the missing denominator?

1
2
3 1/8 = 7/____

■ **56** ■ 15

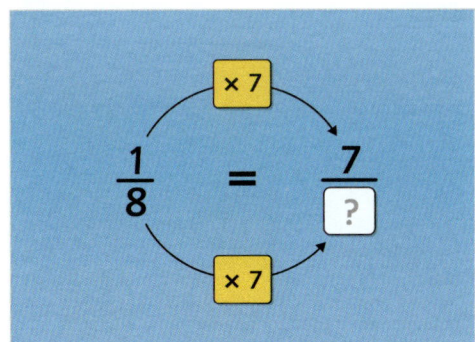

10. What is the missing numerator?

1
2
3 1/9 = ____/72

■ **8** ■ 9

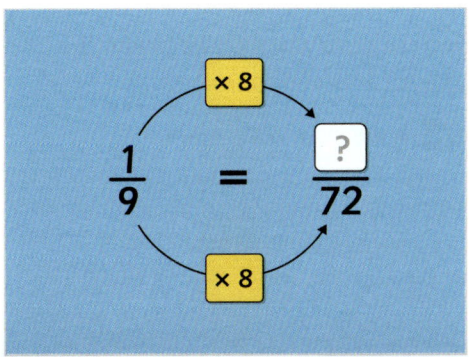

Level 2: Fluency: Find equivalent fractions and identify families of equivalent fractions.

✱ **Required:** 7/10   ✱ **Pupil Navigation:** on
✱ **Randomised:** off

11. 2/3 = 10/____

1
2
3 ■ **15** ■ 5

12. What is the missing numerator?

1
2
3 ____/5 = 40/50

■ **4** ■ 10

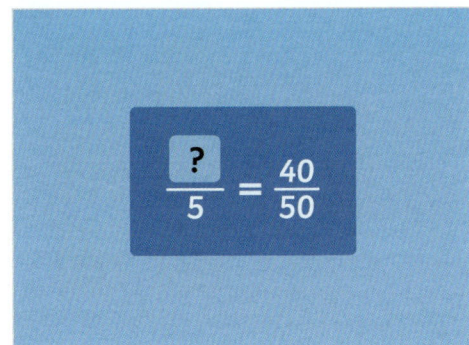

Level 2 *continued*

13. Which of the following fraction is 4/10 in its simplest form?

a b c

2/5, 40/10, 2/10, 4/5

Put a forward slash (/) between the numerator and denominator.

■ 2/5 ■ 40/10 ■ 2/10 ■ 4/5

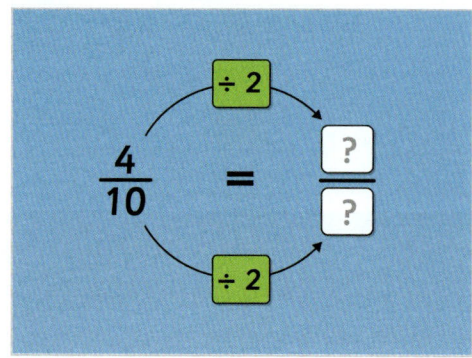

14. ___/63 = 6/7

1 2 3

■ 54

15. Which three of these fractions are equivalent to 1/10?

☐ ☒ ☐

3/6

■ 9/90 ■ 5/10 ■ 35/350 ■ 9/27 ■ 2/40 ■ 2/20

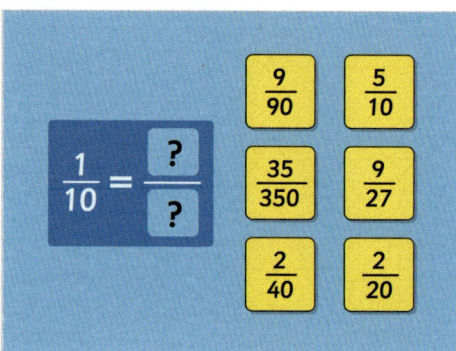

16. What is the missing **word**?

Three-quarters is equal to _____-eighths.

a b c

■ six-eighths ■ six ■ three ■ 6

17. What is 14/20 in its simplest form?

a b c

■ 7/10

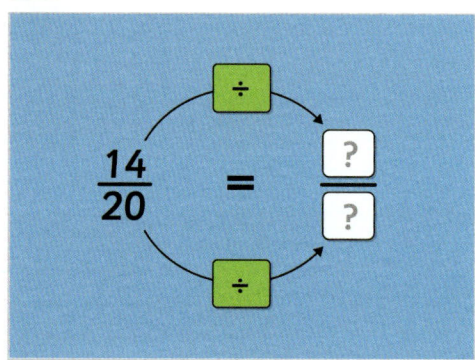

18. Which three of these fractions are equal to ¾?

☐ ☒ ☐

1/6

■ 9/12 ■ 5/50 ■ 300/400 ■ 9/27 ■ 3/12

■ 60/80

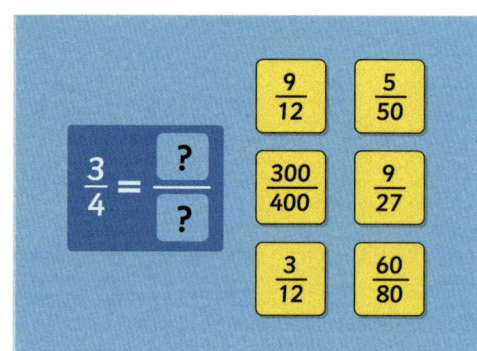

Level 2 continued

19. Which of the following is equivalent to 3/9?
a b c 1/3, 6/9, 12/15
Put a forward slash (/) between the numerator and denominator.

- ■ **1/3** ■ 12/15 ■ 6/9

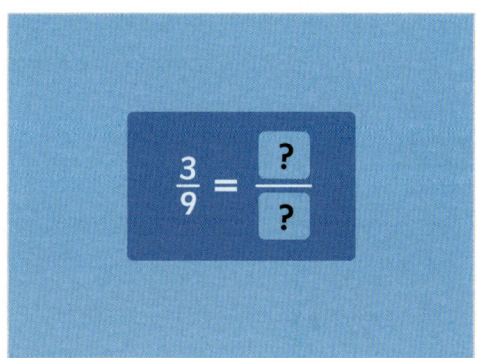

20. 7/11 = 35/___
1 2 3 ■ **55** ■ 5

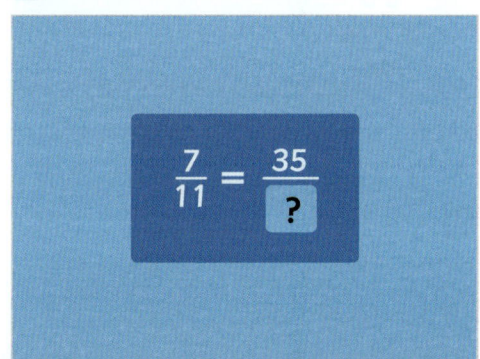

Level 3: Reasoning: Reason about equivalent fractions using fractional number lines, shapes and digit cards.

✱ **Required:** 5/5 ✱ **Pupil Navigation:** on
✱ **Randomised:** off

21. Which three fractions are equivalent to the fraction shown by the arrow?
☐☒☐
3/6
- ■ **4/16** ■ **1/4** ■ 8/16 ■ **25/100** ■ 20/100
- ■ 3/18

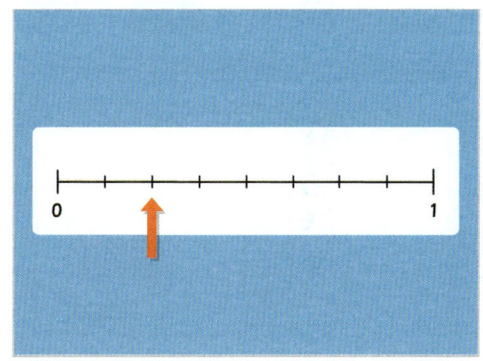

22. Jack says, "You can only find equivalent
a b c fractions for fractions that have an even numerator and denominator."
Explain why Jack is incorrect and give at least one example to prove that he's incorrect.

- **Open question, no set answer**

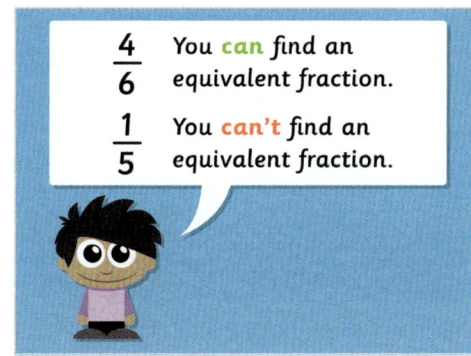

23. How many more parts of the circle do you
1 2 3 need to shade to make it equivalent to the rectangle?

- ■ **5** ■ 8

24. Which fraction is the odd one out?
☐☒☐
1/6
- ■ 4/6 ■ 16/24 ■ **9/12** ■ 20/30 ■ 12/18 ■ 50/75

Level 3 *continued*

25. At the shop, Ben spends 1/4 of his pocket
a money and Sadiq spends 25/100 of his
b pocket money. Do they both spend the same
c amount of money? Explain your answer.

- *Open question, no set answer*

Level 4: Problem Solving: Solve problems involving
equivalent fractions.

✷ **Required:** 5/5   ✷ **Pupil Navigation:** on
✷ **Randomised:** off

26. Use **every** digit card once to make an
a equivalent fraction of 75/100.
b
c ■ 63/84 ■ 6/8 ■ 36/48 ■ 3/4

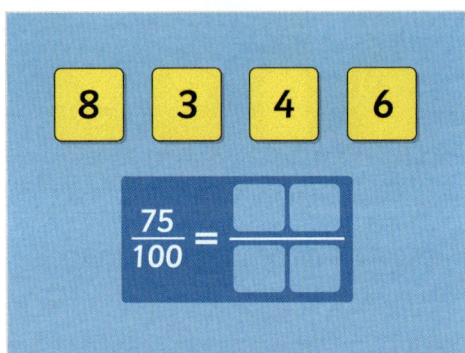

27. Sabina is thinking of a fraction that is
1 equivalent to 80/100. The fraction has a
2 denominator less than 30. How many
3 possible fractions could she be thinking of?

■ 5 ■ 6

28. Norah is thinking of an equivalent fraction to
a 3/15.
b
c The numerator is an even number between
20 and 30.
The sum of the denominator's digits is 4.

What is Norah's fraction?
*Put a forward slash (/) between the
numerator and denominator. For example,
one-half is 1/2.*

■ 26/130 ■ 21/105 ■ 22/110 ■ 30/150 ■ 26/15
■ 28/140 ■ 24/120

29. The total of x and z is 20. What is the value
1 of y?
2
3 ■ 12 ■ 16

30. Which three coloured parts of the large
rectangle are equivalent fractions?

3/7

■ green (A) ■ purple (B) ■ pink (C) ■ blue (D)
■ orange (E) ■ yellow (F) ■ red (G)

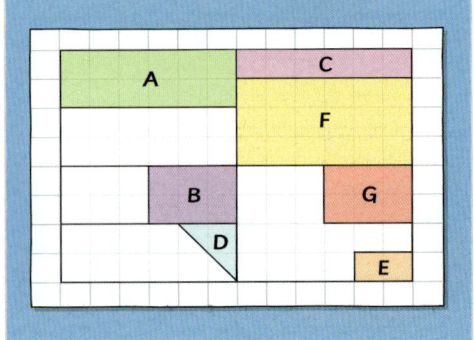

Convert Between Mixed Number Fractions and Improper Fractions

Objective: I can convert between mixed number fractions and improper fractions.

Quick Search Ref: 10278

Level 1: Understanding

✹ **Required:** 7/10   ✹ **Pupil Navigation:** on   ✹ **Randomised:** off

1. Which of these fractions is an improper fraction?

1/4

■ 13/17 ■ 3/6 ■ 9/4 ■ 1 1/2

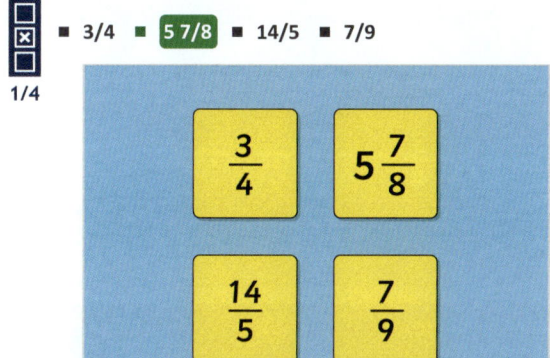

2. Which of these is a mixed number fraction?

1/4

■ 3/4 ■ 5 7/8 ■ 14/5 ■ 7/9

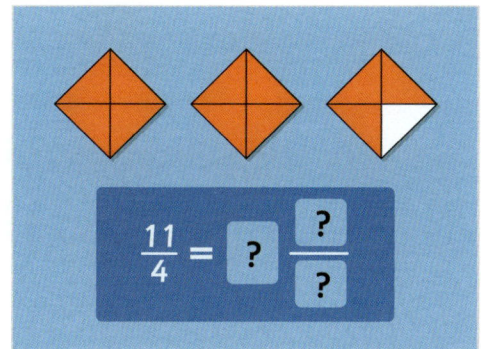

3. Convert 11/4 to a mixed number fraction.

a b c *Give your answer as a mixed number fraction. For example, five and one-half is 5 1/2.*

■ 1/4 ■ 2 3/4 ■ 2

$$\frac{11}{4} = \boxed{?} \; \frac{?}{?}$$

4. What mixed number fraction is equivalent to 12/5?

a b c *Give your answer as a mixed number fraction. For example, five and one-half is 5 1/2.*

■ 2 2/5 ■ 2 ■ 2/5

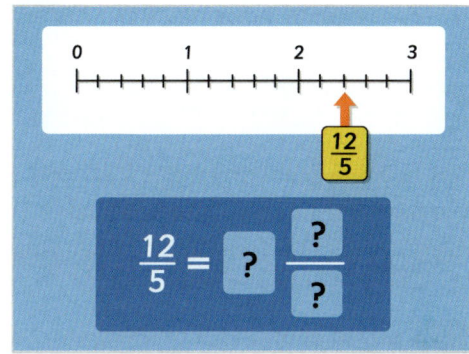

$$\frac{12}{5} = \boxed{?} \; \frac{?}{?}$$

5. What is 25/6 as a mixed number fraction?

a b c *Give your answer as a mixed number fraction. For example, five and one-half is 5 1/2.*

■ 4 1/6 ■ 1/6 ■ 4

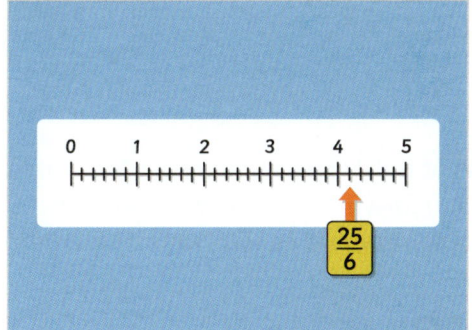

Level 1 *continued*

6. Convert 2 2/3 to an improper fraction.
   a
   b *Put a forward slash (/) between the*
   c *numerator and denominator. For example,*
      3/2.

   ■ **8/3** ■ 6/3 ■ 8

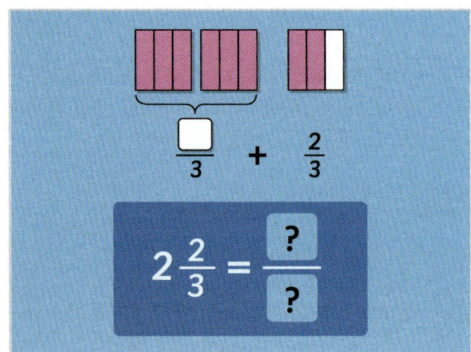

7. What is 4 3/7 as an improper fraction?
   a
   b *Put a forward slash (/) between the*
   c *numerator and denominator. For example,*
      3/2.

   ■ **31/7** ■ 28/7 ■ 31

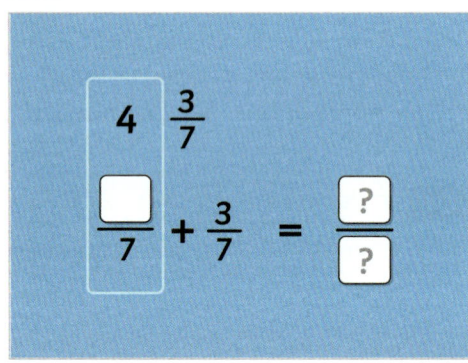

8. What is 4 1/6 as an improper fraction?
   a
   b *Put a forward slash (/) between the*
   c *numerator and denominator. For example,*
      3/2.

   ■ **25/6** ■ 24/6 ■ 25

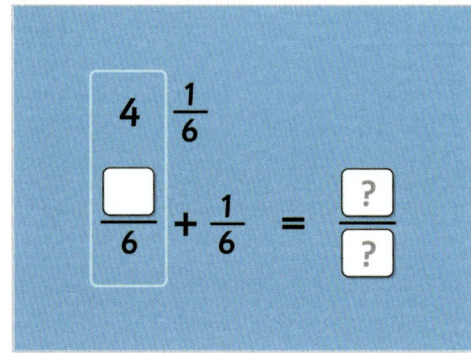

9. Convert 8/3 to a mixed number fraction.
   a
   b *Give your answer as a mixed number*
   c *fraction. For example, five and one-half is 5*
      1/2.

   ■ **2 2/3** ■ 2/3 ■ 2

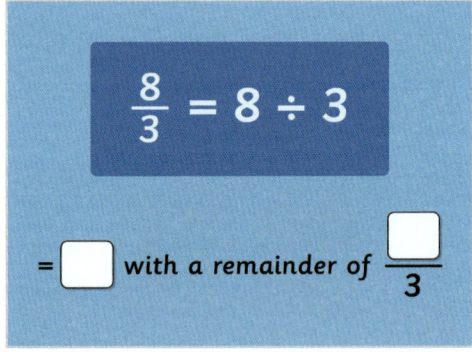

10. There are 4 quarters in one whole. How
   1
   2   many quarters are there in 2 3/4?
   3

   ■ **11** ■ 4 ■ 8

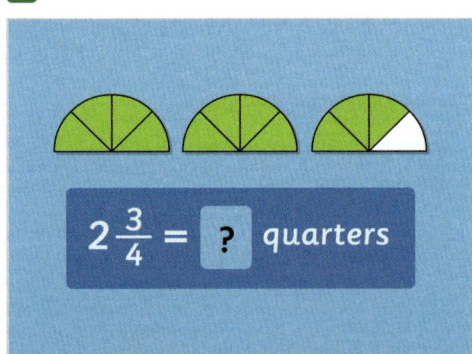

Level 2: Fluency

✱ **Required:** 7/10   ✱ **Pupil Navigation:** on
✱ **Randomised:** off

11. What is 19/6 as a mixed number fraction?
   a
   b *Give your answer as a mixed number*
   c *fraction. For example, five and one-half is 5*
      1/2.

   ■ **3 1/6**

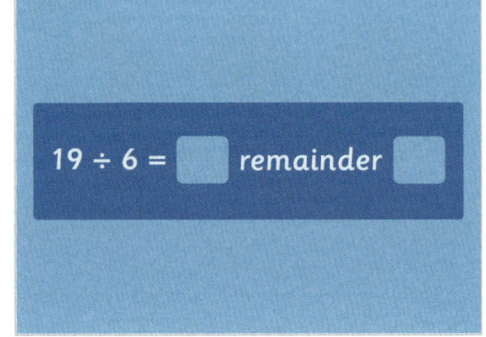

Level 2 continued

12. What improper fraction is equivalent to 2 3/7?

a
b
c

Put a forward slash (/) between the numerator and denominator. For example, 3/2.

- 17/7

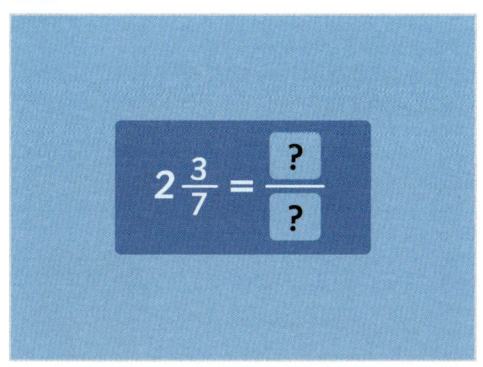

13. Convert 14/5 to a mixed number fraction.

a
b
c

Give your answer as a mixed number fraction. For example, five and one-half is 5 1/2.

- 2 4/5

14. Convert 2 7/8 to an improper fraction.

a
b
c

Put a forward slash (/) between the numerator and denominator.

- 23/8

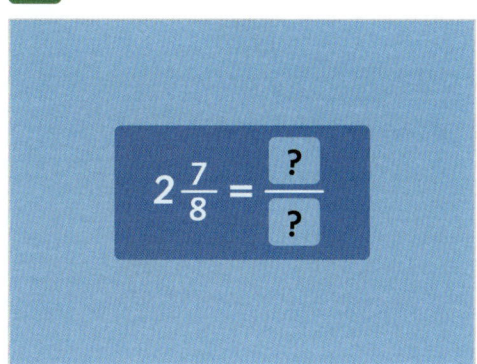

15. There are 9 ½ pairs of socks in the lost property cupboard. How many socks are there in total?

1
2
3

- 19   - 2

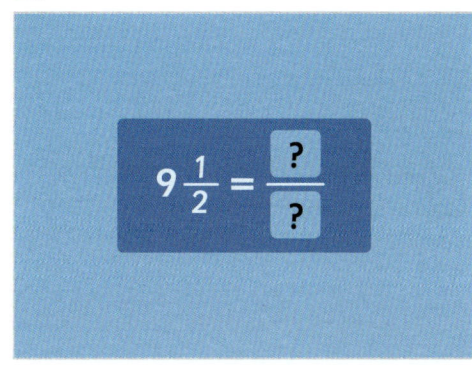

16. A school race is marked with flags every 1/10 kilometre (km) after the starting line. Eloise runs 1 3/10 km. How many flags does Eloise run past?

1
2
3

- 13   - 10

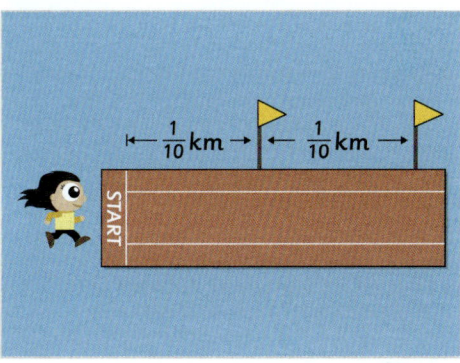

17. Two pizzas are each cut into 3 equal pieces. If Hannah eats 5 pieces, what fraction of a pizza does she eat?

a
b
c

Give your answer as a mixed number fraction. For example, five and one-half is 5 1/2.

- 1 2/3   - 5/3

Level 2 *continued*

18. One month is 1/12 of a year. How many months are there in 3 5/12 years?

1
2
3

- **41** - 12

$3\frac{5}{12}$ years = ☐ months

19. There are 19 identical socks in the lost property cupboard. How many complete pairs of socks are there?

a
b
c

- **9** - 9 1/2 - 19/2 - 2/19

$\frac{19}{☐} = ?\frac{?}{?}$

20. What is 16/3 as a mixed number fraction?
Give your answer as a mixed number fraction. For example, five and one-half is 5 1/2.

a
b
c

- **5 1/3**

$\frac{16}{3} = ?\frac{?}{?}$

Level 3: Reasoning

- **Required: 5/5**   - **Pupil Navigation: on**
- **Randomised: off**

21. Alex has tried to convert 21/10 to a mixed number fraction. Explain what is wrong with her answer and what she needs to do to complete the conversion.

a
b
c

- Open question, no set answer

$\frac{21}{10} = 1\frac{11}{10}$

22. What is the next whole number that is greater than 53/10?

a
b
c

- **6** - 5

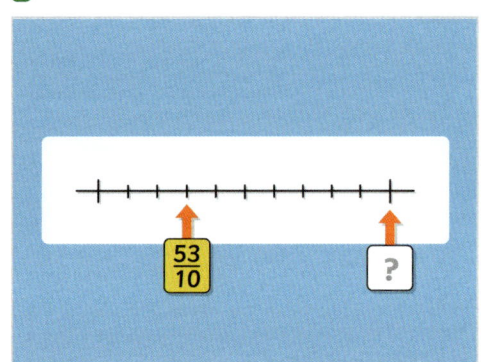

23. Four fractions have been arranged in ascending order (starting with the smallest):
27/16, ____, 34/16, 49/16
Which of these fractions could **not** be the missing fraction?

☐
☒
☐

1/3

- 1 3/4 - 2 1/16 - **1 3/8**

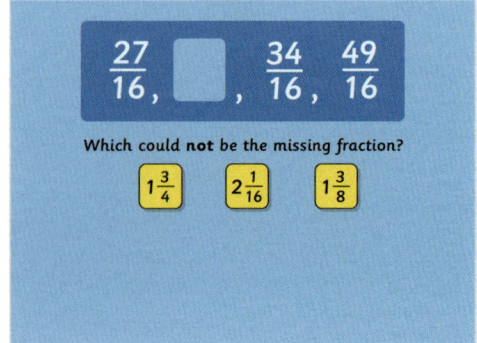

$\frac{27}{16}$, ☐, $\frac{34}{16}$, $\frac{49}{16}$

Which could **not** be the missing fraction?

$1\frac{3}{4}$ $2\frac{1}{16}$ $1\frac{3}{8}$

Level 3 continued

24. When you convert these improper fractions to mixed number fractions, which answer is the odd one out and why?

a b c

- Open question, no set answer

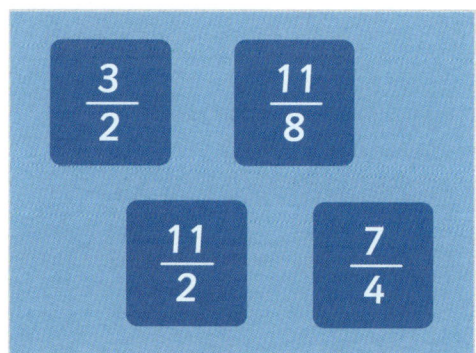

25. If Katie has 2 2/5 bunches of bananas, Katie must have 12 bananas.
Is this statement **always**, **sometimes** or **never** true? Explain your answer and give examples.

a b c

- Open question, no set answer

Level 4: Problem solving with greater depth

✸ **Required:** 5/5   ✸ **Pupil Navigation:** on
✸ **Randomised:** off

26. What is the denominator of these equivalent fractions?
8 3/__ = 35/__

1 2 3

■ **4** ■ 32

27. Use two of the number cards to make the largest possible improper fraction.
What is this fraction as a mixed number fraction?

a b c

■ **4 1/5** ■ 1 4/17 ■ 21/5

28. Mrs Brown feeds her chickens 1/10 kilogram (kg) of seed each day. How many days will a 3 4/5 kilogram (kg) bag of seed last?

1 2 3

■ **38** ■ 19

29. Mrs Gray makes some fruit juice for the class party. She pours 3/4 litre (l) of fruit juice and 15/8 l of water into a jug with a capacity of 5 litres. How much more drink can she fit into the jug?
Give your answer as a mixed number fraction. Include the units l (litres) in your answer.

a b c

■ 2 3/8 ■ **2 3/8 litres** ■ 19/8 l ■ **2 3/8 l**
■ 19/8 litres ■ 19/8

Level 4 *continued*

30. Sets of pencils come in boxes of 36. Class 5A
 has 1 11/36 boxes of pencils, Class 5B has 42
pencils and Class 5C has 2 7/36 boxes.
Mr Brown collects all the pencils together
and shares them equally between the three
classes. How many pencils does each class
get?

■ **56** ■ 79 ■ 168 ■ 47

class	boxes	pencils
5A	$1\frac{11}{36}$	
5B		42
5C	$2\frac{7}{36}$	

Convert Improper Fractions to Mixed Number Fractions

Objective: I can convert improper fractions to mixed number fractions.

Quick Search Ref: 11465

Level 1: Understanding: Convert improper fractions to mixed number fractions with image support.

✹ **Required:** 7/10   ✹ **Pupil Navigation:** on   ✹ **Randomised:** off

1. Which of these fractions is an improper fraction?

1/4

- 13/17   - 3/6   - **9/4**   - 1 1/2

2. Select the fraction that is a mixed number fraction.

1/4

- 2/3   - 18/5   - **1 5/8**   - 7/9

3. Select **two** fractions that are equivalent to whole numbers.

2/4

- 7/8   - **24/3**   - 15/12   - **7/7**

4. There are 4/4 in one circle. How many whole circles can you make with 9/4?

1
2
3

- **2**   - 9   - 8   - 4

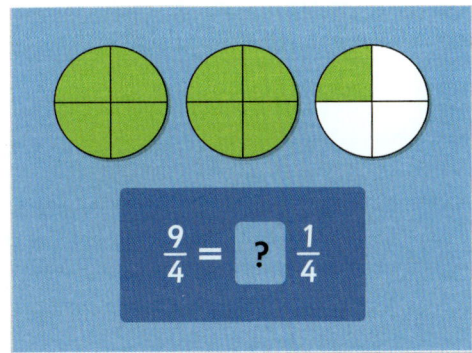

5. Each triangle is divided into 3/3. Three whole triangles can be made from 11/3 with what fraction remaining?

a
b
c

Put a forward slash (/) between the numerator and denominator. For example, one-half is 1/2.

- 3 2/3   - **2/3**   - 3   - 1/3

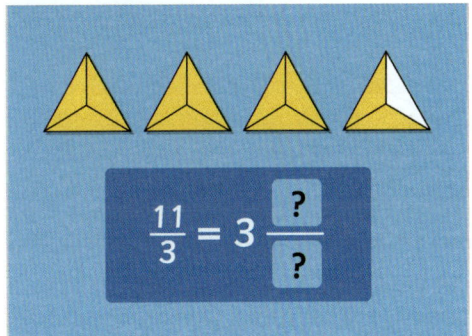

Level 1 *continued*

6. What mixed number fraction is equivalent to 12/5?
a
b
c *Give your answer as a mixed number fraction. For example, five and one-half is 5 1/2.*

- **2 2/5** ▪ 2/5 ▪ 2

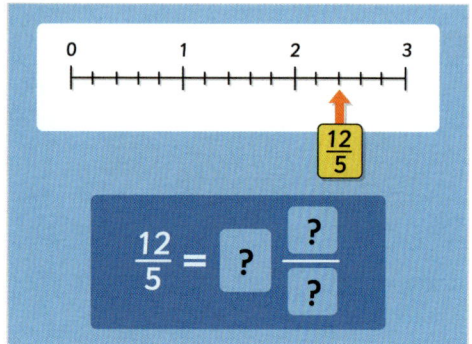

7. What is 27/8 as a mixed number fraction?
a
b
c *Give your answer as a mixed number fraction. For example, five and one-half is 5 1/2.*

- ▪ 3 ▪ **3 3/8** ▪ 3/8

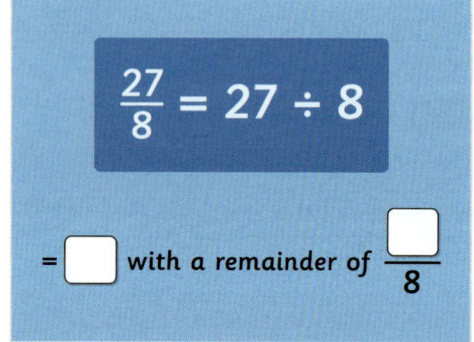

8. Convert 8/3 to a mixed number fraction.
a
b
c *Give your answer as a mixed number fraction. For example, five and one-half is 5 1/2.*

- **2 2/3** ▪ 2 ▪ 2/3

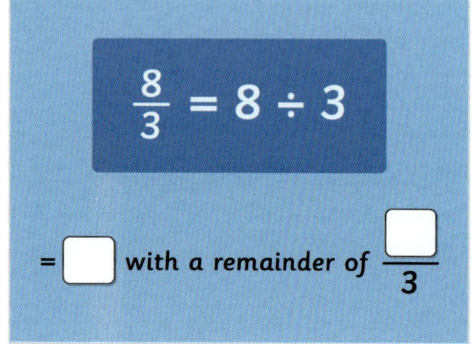

9. What is 25/6 as a mixed number fraction?
a
b
c *Give your answer as a mixed number fraction. For example, five and one-half is 5 1/2.*

- **4 1/6** ▪ 4 ▪ 1/6

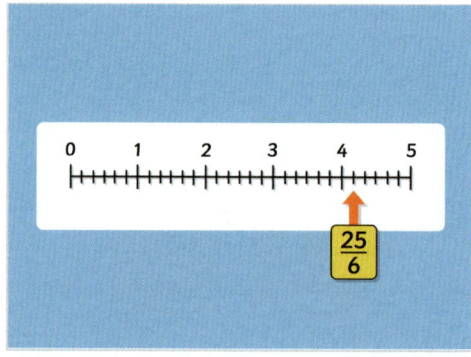

10. Convert 11/4 to a mixed number fraction.
a
b
c *Give your answer as a mixed number fraction. For example, five and one-half is 5 1/2.*

- **2 3/4** ▪ 1/4 ▪ 2

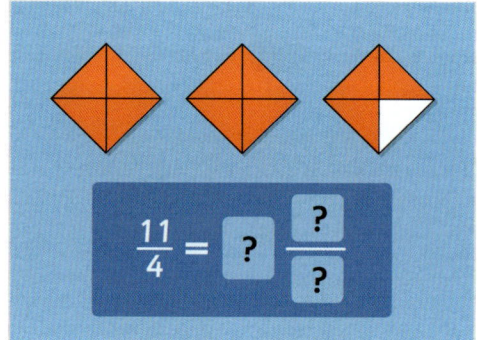

Level 2: Fluency: Convert improper fractions to mixed number fractions including one-step problems.

✹ **Required:** 7/10 ✹ **Pupil Navigation:** on
✹ **Randomised:** off

11. What mixed number fraction is equivalent to
a
b 17/7?
c
Give your answer as a mixed number fraction. For example, five and one-half is 5 1/2.

▪ 2 3/7

12. Convert 14/5 to a mixed number fraction.
a
b *Give your answer as a mixed number*
c *fraction. For example, five and one-half is 5 1/2.*

▪ 2 4/5

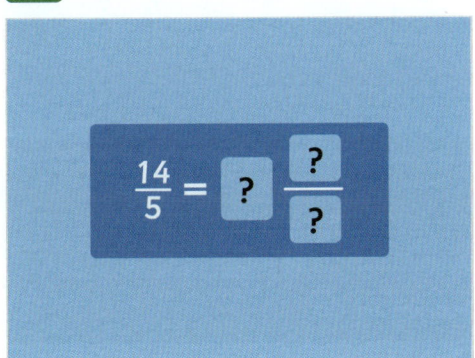

13. What is 16/3 as a mixed number fraction?
a
b *Give your answer as a mixed number*
c *fraction. For example, five and one-half is 5 1/2.*

▪ 5 1/3

14. Two pizzas are each cut into 3 equal pieces.
a If Hannah eats 5 pieces, what fraction of a
b pizza does she eat?
c
Give your answer as a mixed number fraction. For example, five and one-half is 5 1/2.

▪ 1 2/3 ▪ 5/3

15. Find the mixed number fraction that is
a equivalent to 41/13.
b
c ▪ 3 2/13

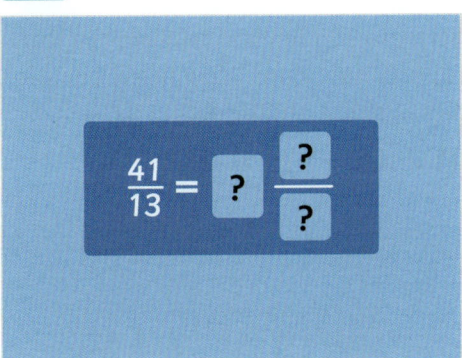

Level 2 *continued*

16.
a
b
c
A school race is marked with flags every 1/10 kilometre (km) after the starting line. Eloise runs to the thirteenth flag. How far does Eloise run?
Give your answer as a mixed number fraction. For example, five and one-half is 5 1/2. Include the units km (kilometres) in your answer.

- 1 3/10 ▪ 1 3/10 km ▪ 1 3/10 kilometres
- 1.3 kilometres ▪ 13/10 km ▪ 1.3 ▪ 13/10
- 13/10 kilometres ▪ 1.3 km

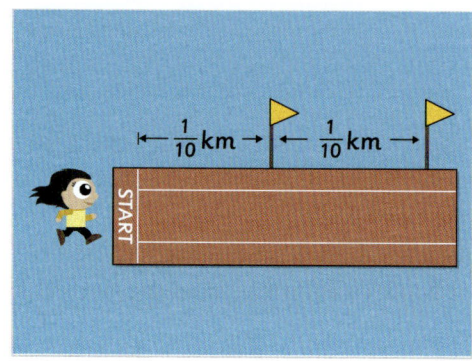

17.
a
b
c
Apples are sold in packets of 3. Harry has 7 apples. How many whole packets of apples does Harry have?

- 7/3 ▪ 2 packets ▪ 2 ▪ 2 1/3 ▪ 3/7

18.
a
b
c
There are 19 identical socks in the lost property cupboard. How many complete pairs of socks are there?

- 9 ▪ 19/2 ▪ 9 1/2 ▪ 2/19

19.
a
b
c
Convert 23/8 to a mixed number fraction.
Give your answer as a mixed number fraction. For example, five and one-half is 5 1/2.

- 2 7/8

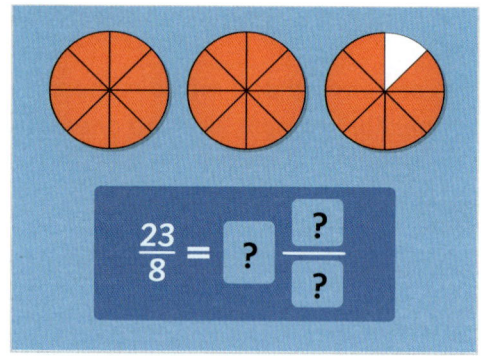

20.
a
b
c
What is 19/6 as a mixed number fraction?
Give your answer as a mixed number fraction. For example, five and one-half is 5 1/2.

- 3 1/6

Level 3: Reasoning: Reason about converting between improper fractions and mixed number fractions.

✳ **Required:** 5/5 ✳ **Pupil Navigation:** on
✳ **Randomised:** off

21.
a
b
c
Alex has tried to convert 21/10 to a mixed number fraction. Explain what is wrong with her answer and what she needs to do to complete the conversion.

- Open question, no set answer

Level 3 continued

22. When you convert these improper fractions to mixed number fractions, which answer is the odd one out and why?

abc

- Open question, no set answer

23. Tennis ball cartons each hold 4 tennis balls. Jordan has 17 tennis balls. Explain how to calculate the number of cartons that Jordan can fill completely.

abc

- Open question, no set answer

24. What is the next whole number that is greater than 53/10?

abc

 ■ **6** ■ 5

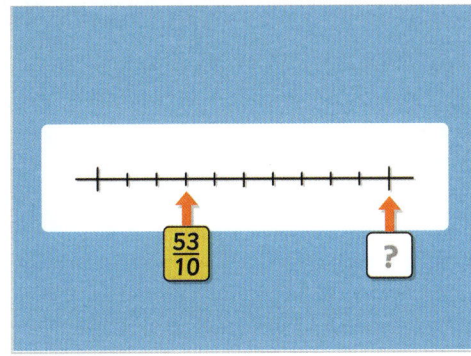

25. An improper fraction with a numerator that is a multiple of its denominator is never equal to a mixed number fraction. Explain why this statement is true.

abc

- Open question, no set answer

Level 4: Problem solving with greater depth: Solve problems that involve converting improper fractions to mixed number fractions.

✱ **Required:** 5/5   ✱ **Pupil Navigation:** on
✱ **Randomised:** off

26. What is the denominator in these equivalent fractions?

1 2 3

35/___ = 8 3/___

■ **4**

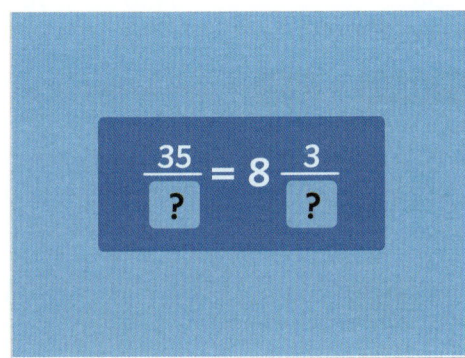

27. Use two of the number cards to make the largest possible improper fraction. What is this fraction as a mixed number fraction?

abc

■ **4 1/5** ■ 21/5 ■ 1 4/17

28. Three children are sorting marbles into sets
a
b of 36.
c Amy has 19/36 of a set of marbles, Dylan has
42 marbles and Saffy has 16/36 of a set.
How many full sets of marbles do the
children have altogether?

- **2** ■ **77/36** ■ **2 5/36**

29. Mrs Gray makes some fruit juice for the class
a
b party. She pours 3/4 litre (l) of fruit juice and
c 15/8 l of water into a jug with a capacity of 5
litres. How much more drink can she fit into
the jug?
*Give your answer as a mixed number
fraction. Include the units l (litres) in your
answer.*

- **2 3/8 litres** ■ **2 3/8 l** ■ **19/8** ■ **2 3/8** ■ **19/8 l**
- **19/8 litres**

30. The Shah family are going on holiday. The
a
b weight limit allowed for their suitcase is 20
c kilograms (kg).
Mr Shah's clothes weigh 6 3/5 kg and Mrs
Shah has 7 1/5 kg of luggage. Their children's
clothes weigh 8 4/5 kg.
What weight of luggage must they take out
of their case?
*Give your answer as a mixed number
fraction. Include the units kg (kilograms) in
your answer.*

- **22 3/5 kilograms** ■ **13/5 kilograms** ■ **2 3/5 kg**
- **2 3/5 kilograms** ■ **13/5** ■ **2 3/5** ■ **13/5 kg**
- **22 3/5** ■ **22 3/5 kg**

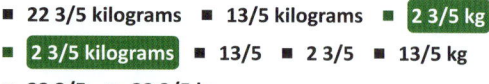

 Ref:11465 Convert Improper Fractions to Mixed Number Fr...

Convert Mixed Number Fractions to Improper Fractions

Objective: I can convert mixed number fractions to improper fractions.

Quick Search Ref: 11489

Level 1: Understanding: Convert mixed number fractions to improper fractions with image support.

✿ **Required:** 7/10　　　✿ **Pupil Navigation:** on　　　✿ **Randomised:** off

---

1. Which of these is a mixed number fraction?

■ 3/4　■ **5 7/8**　■ 14/5　■ 7/9

1/4

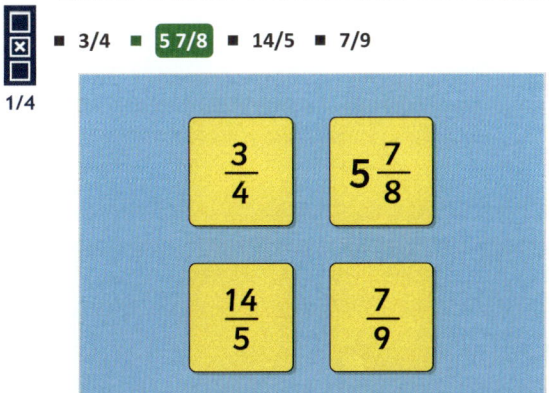

2. Which of these fractions is an improper fraction?

■ 9/17　■ 1 2/3　■ 27/100　■ **11/4**

1/4

3. Select **two** fractions that are equivalent to 3 wholes.

■ 3/5　■ **9/3**　■ 3/9　■ **18/6**

2/4

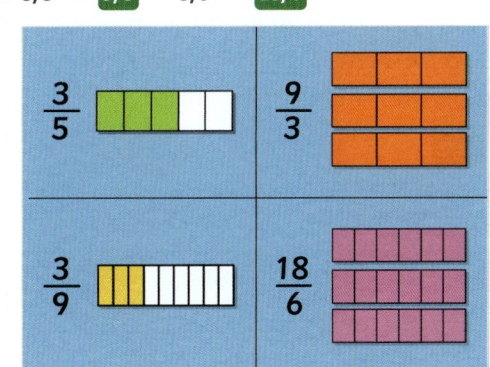

4. There are 4 quarters in one circle. How many quarters are there in 2 1/4 circles?

■ **9**　■ 4　■ 8

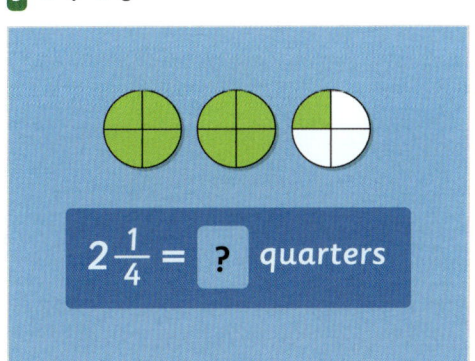

$2\frac{1}{4} = \boxed{?}$ quarters

5. The image shows 3 2/3 triangles. What is the missing denominator?

■ **3**　■ 11

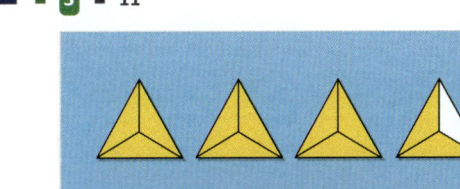

$3\frac{2}{3} = \frac{11}{\boxed{?}}$

6. What improper fraction is equivalent to 2 2/5?

Put a forward slash (/) between the numerator and denominator. For example, 3/2.

■ 12　■ **12/5**　■ 10/5

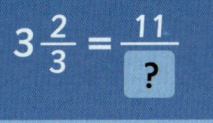

$\boxed{}{5} + \frac{2}{5}$

$2\frac{2}{5} = \frac{\boxed{?}}{\boxed{?}}$

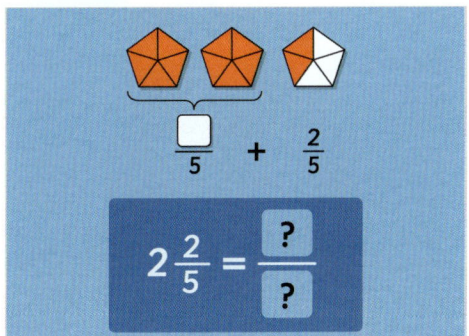

Level 1 *continued*

7. What is 4 3/7 as an improper fraction?
a
b *Put a forward slash (/) between the*
c *numerator and denominator. For example, 3/2.*

■ **31/7** ■ 31 ■ 28/7

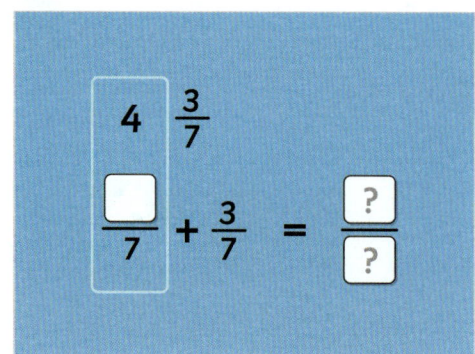

8. What is 4 1/6 as an improper fraction?
a
b *Put a forward slash (/) between the*
c *numerator and denominator. For example, 3/2.*

■ **25/6** ■ 25 ■ 24/6

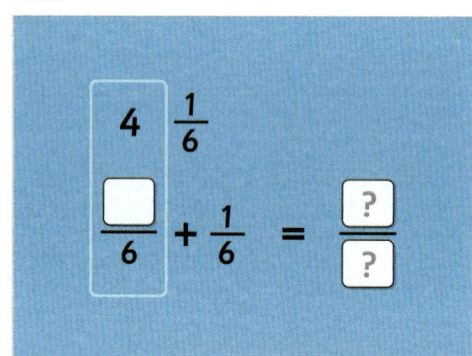

9. Convert 2 2/3 to an improper fraction.
a
b *Put a forward slash (/) between the*
c *numerator and denominator. For example, 3/2.*

■ **8/3** ■ 8 ■ 6/3

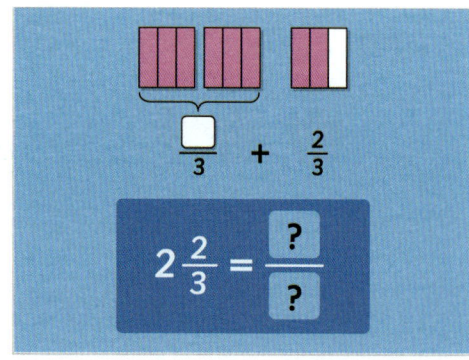

10. There are 4 quarters in one whole. How
1
2 many quarters are there in 2 3/4?
3

■ **11** ■ 8 ■ 4

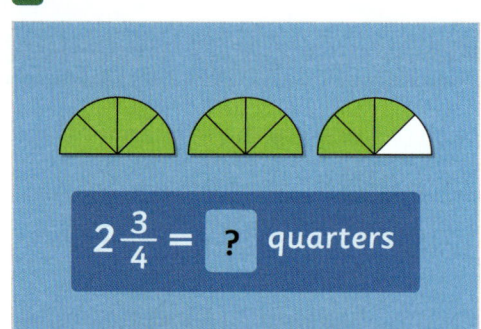

Level 2: Fluency: Convert mixed number fractions to improper fractions including in context and one-step problems.

✿ **Required:** 7/10 ✿ **Pupil Navigation:** on
✿ **Randomised:** off

11. What improper fraction is equivalent to 2
a 3/7?
b
c *Put a forward slash (/) between the*
numerator and denominator. For example, 3/2.

■ **17/7**

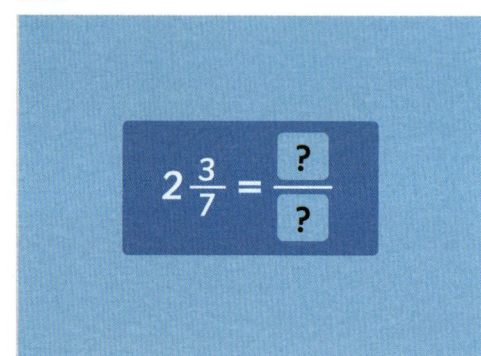

12. Convert 2 7/8 to an improper fraction.
a *Put a forward slash (/) between the*
b
c *numerator and denominator.*

■ **23/8**

Level 2 continued

13. One month is 1/12 of a year. How many months are there in 3 5/12 years?

[1 2 3]

▪ **41** ▪ 12

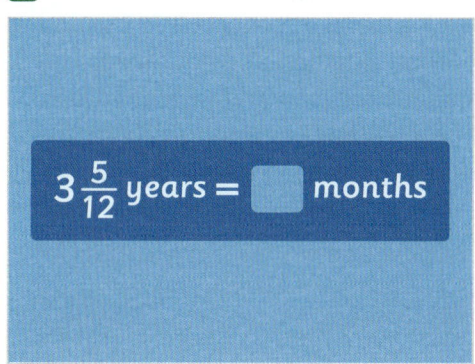

$$3\frac{5}{12} \text{ years} = \boxed{} \text{ months}$$

14. A school race is marked with flags every 1/10 kilometre (km) after the starting line. Eloise runs 1 3/10 km. How many flags does Eloise run past?

[1 2 3]

▪ **13** ▪ 10

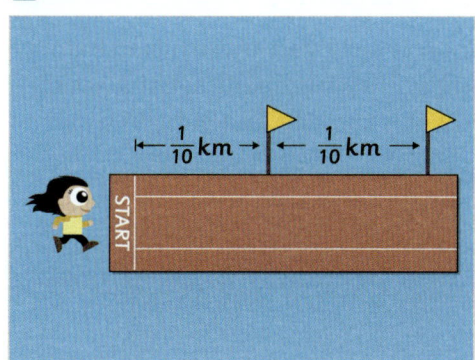

15. How many fifths of a pizza are there in 2 4/5 pizzas?

[a b c]

▪ **14** ▪ **14/5** ▪ 5

$$2\frac{4}{5} \text{ pizzas} = \boxed{?} \text{ fifths}$$

16. Apples are sold in packets of 3. Harry has two whole packets and one loose apple. What fraction of one packet of apples does Harry have?

[a b c]

Give your answer as an improper fraction. Don't include the units in your answer.

▪ **7/3** ▪ 2 1/3

17. 3 7/10 cm is _____ cm as an improper fraction.

[a b c]

Enter the missing improper fraction. Don't include the units in your answer.

▪ **3.7 cm** ▪ **37/10 cm** ▪ **37/10 centimetres** ▪ **37/10**
▪ **3.7 centimetres** ▪ **3.7**

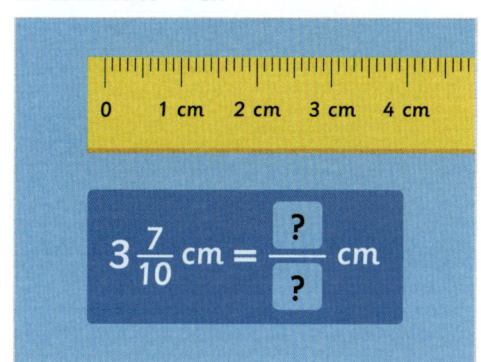

$$3\frac{7}{10} \text{ cm} = \frac{?}{?} \text{ cm}$$

18. There are 9 ½ pairs of socks in the lost property cupboard. How many socks are there in total?

[1 2 3]

▪ **19** ▪ 2

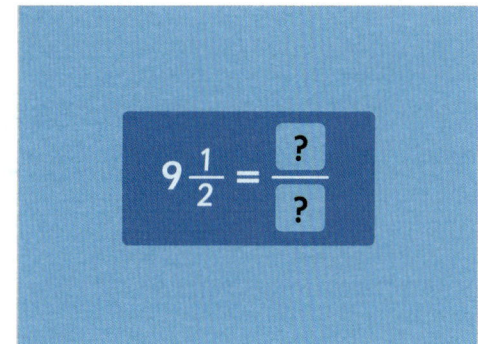

$$9\frac{1}{2} = \frac{?}{?}$$

Level 2 continued

19. Two pizzas are each cut into 3 equal pieces. If Hannah eats 1 2/3 pizzas, how many pieces of pizza does she eat?

1
2
3

- ■ **5** ■ 3

20. What is 2 3/8 as an improper fraction?

a
b
c

Put a forward slash (/) between the numerator and denominator. For example, 3/2.

- ■ **19/8**

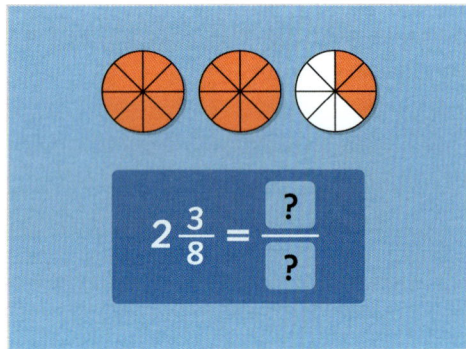

Level 3: Reasoning: Reason about converting mixed number fractions to improper fractions.

❋ **Required:** 5/5   ❋ **Pupil Navigation:** on
❋ **Randomised:** off

21. Alex has tried to convert 2 3/7 to an improper fraction and got the answer 5/7. Explain what is wrong with her answer and what she needs to do to calculate the correct answer.

a
b
c

- Open question, no set answer

22. When you convert these mixed number fractions to improper fractions, which answer is the odd one out and why?

a
b
c

- Open question, no set answer

23. Tennis ball cartons each hold 8 tennis balls. Jordan has 4 1/4 cartons of tennis balls. Explain how to use an improper fraction to calculate how many tennis balls Jordan has in total.

a
b
c

- Open question, no set answer

Level 3 continued

24. Four fractions have been arranged in ascending order (starting with the smallest):
27/16, ___, 34/16, 49/16

1/3 Which of these fractions could **not** be the missing fraction?

- 1 3/4 - 2 1/16 - **1 3/8**

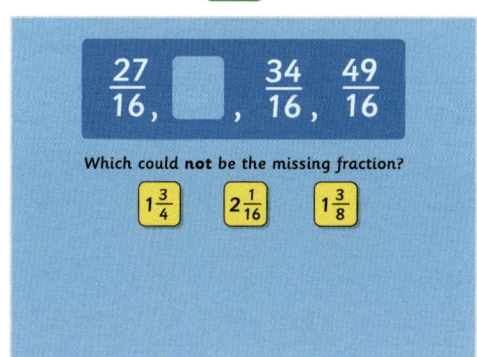

25. If Katie has 2 2/5 bunches of bananas, Katie must have 12 bananas.
Is this statement **always**, **sometimes** or **never** true? Explain your answer and give examples.

- Open question, no set answer

Level 4: Problem Solving with greater depth: Solve problems involving converting mixed number fractions to improper fractions.

❋ **Required:** 5/5 ❋ **Pupil Navigation:** on
❋ **Randomised:** off

26. What is the denominator of these equivalent fractions?

8 3/__ = 35/__

- **4** - 32

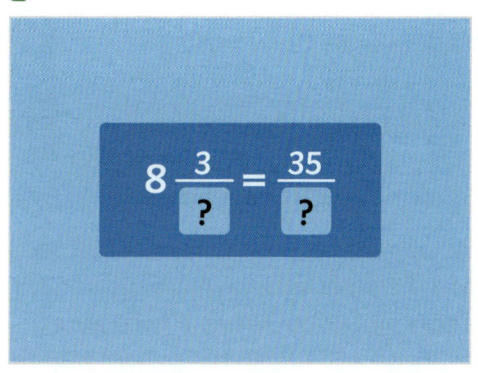

27. Use three of the digit cards to make the smallest possible mixed number fraction.
What is this fraction as an improper fraction?

- **22/9** - 49/5 - 2 4/9

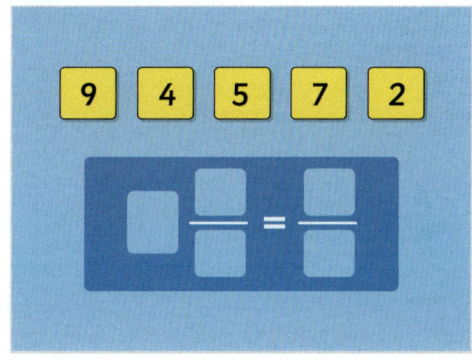

28. Mrs Gray makes some jugs of fruit juice for the class party. Each jug holds enough juice to fill 9 cups. If Mrs Gray uses 4 1/3 jugs of juice, how many cups does she fill?

- **39** - 36 - 13

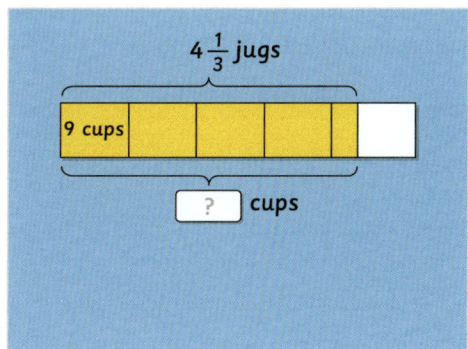

Level 4 *continued*

29. Mrs Brown feeds her chickens 1/10 kilogram
(kg) of seed each day. How many days will a
3 4/5 kilogram (kg) bag of seed last?

- **38** - 19

30. Sets of pencils come in boxes of 36. Class 5A
has 1 11/36 boxes of pencils, Class 5B has 42
pencils and Class 5C has 2 7/36 boxes.
Mr Brown collects all the pencils together
and shares them equally between the three
classes. How many pencils does each class
get?

- **56** - 168 - 79 - 47

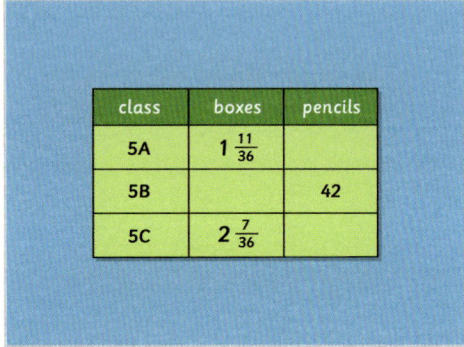

class	boxes	pencils
5A	$1\frac{11}{36}$	
5B		42
5C	$2\frac{7}{36}$	

Convert Decimal Numbers to Fractions

Objective: I can read and write decimal numbers as fractions.

Quick Search Ref: 10246

Level 1: Understanding: Convert decimals with tenths and hundredths to fractions.

❋ **Required:** 7/10   ❋ **Pupil Navigation:** on   ❋ **Randomised:** off

1. What is the decimal part of the number 0.123?

- ▪ **123**   ▪ 0.123

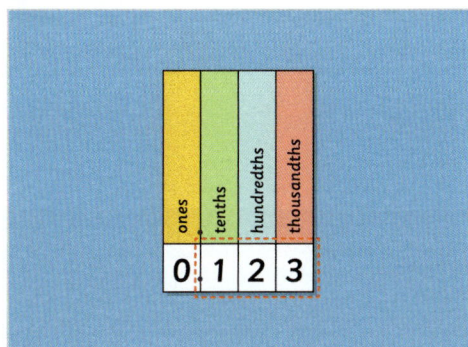

2. What is the value of the 6 in the number 0.16?

- ▪ 6 ones   ▪ 6 tenths   ▪ **6 hundredths**
- ▪ 6 thousandths

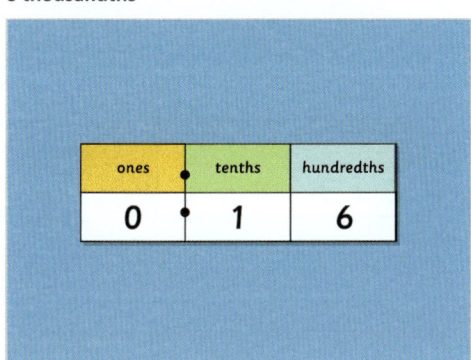

3. What is 0.3 as a fraction?

- ▪ **3/10**   ▪ 3/100   ▪ 3/1,000

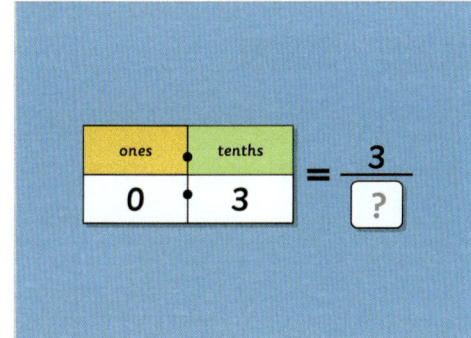

4. What is 0.7 as a fraction?
Put a forward slash (/) between the numerator and denominator. For example, one-half is 1/2.

- ▪ **7/10**

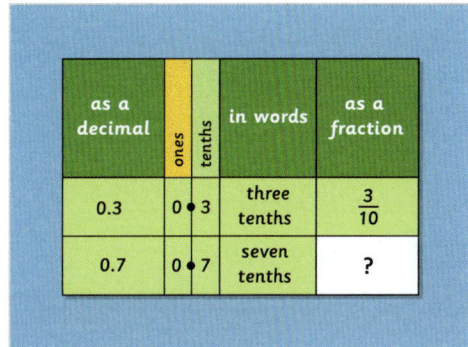

5. Convert 0.4 to a fraction.
Put a forward slash (/) between the numerator and denominator. For example, one-half is 1/2.

- ▪ **2/5**   ▪ **4/10**

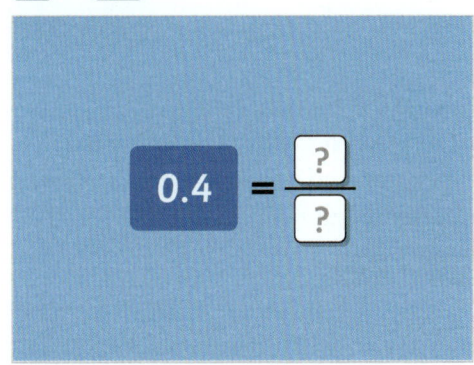

6. Which fraction has the same value as 0.21?

- ▪ 21/10   ▪ **21/100**   ▪ 21/1,000

as a decimal	ones	tenths	hundredths	in words	as a fraction
0.15	0	1	5	fifteen hundredths	15/100
0.21	0	2	1	twenty-one hundredths	?
0.35	0	3	5	thirty-five hundredths	35/100

Level 1 *continued*

7. What is 0.71 as a fraction?

a b c *Put a forward slash (/) between the numerator and denominator. For example, one-half is 1/2.*

- **71/100** - **71/10**

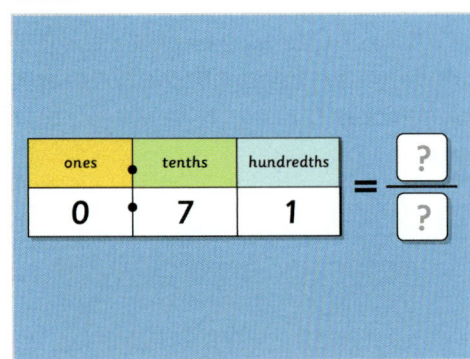

8. What is 0.1 as a fraction?

☐☒☐ **1/3**

- **1/10** - **1/100** - **1/1,000**

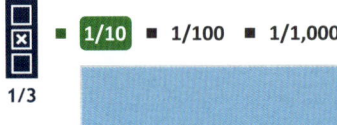

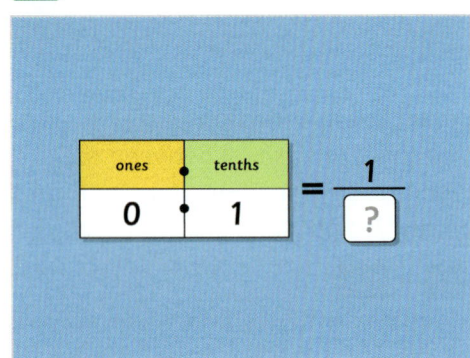

9. Convert 0.8 to a fraction.

a b c *Put a forward slash (/) between the numerator and denominator. For example, one-half is 1/2.*

- **4/5** - **8/10**

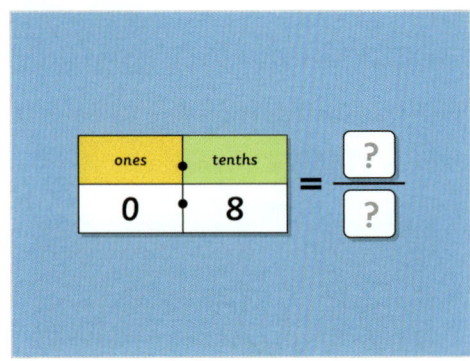

10. What is 0.37 as a fraction?

a b c *Put a forward slash (/) between the numerator and denominator. For example, one-half is 1/2.*

- **37/100**

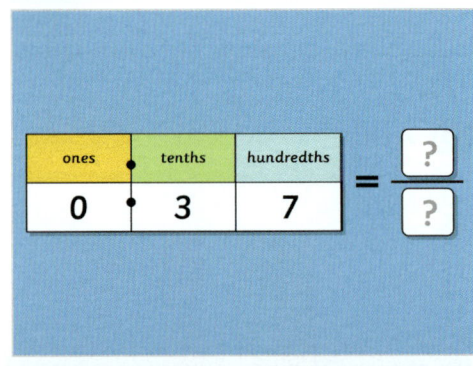

Level 2: Fluency: Convert decimal numbers to fractions including thousandths and placeholder zeros.

✿ Required: 7/10 **✿ Pupil Navigation: on**
✿ Randomised: off

11. What is 0.341 as a fraction?

☐☒☐ **1/3**

- **341/10** - **341/100** - **341/1,000**

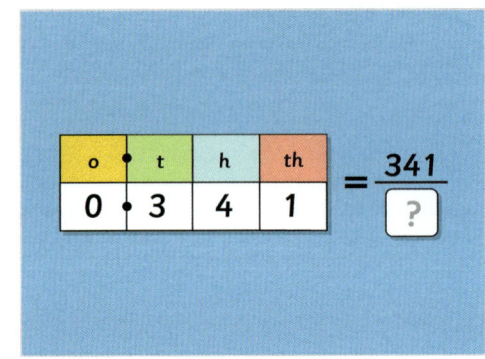

12. Convert 0.237 to a fraction.

a b c *Put a forward slash (/) between the numerator and denominator. For example, one-half is 1/2.*

- **237/1000** - **237/1,000** - **237/100** - **237/10**

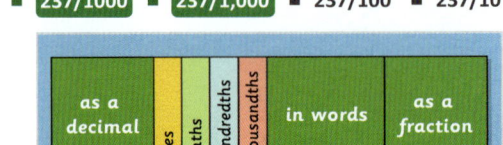

as a decimal	ones	tenths	hundredths	thousandths	in words	as a fraction
0.237	0	2	3	7	two hundred and thirty-seven thousandths	?
0.397	0	3	9	7	three hundred and ninety-seven thousandths	397 / 1,000
0.439	0	4	3	9	four hundred and thirty-nine thousandths	439 / 1,000

Level 2 continued

13. 0.563 as a fraction is _____.

a b c *Put a forward slash (/) between the numerator and denominator. For example, one-half is 1/2.*

- **563/1000** - 563/100 - **563/1,000** - 563/10

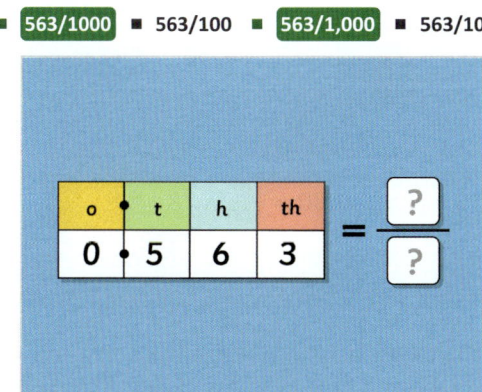

14. Which fraction is equal to 0.043?

- 43/10 - 43/100 - **43/1,000** - 043/1,000
- 430/1,000

1/5

15. What is the missing denominator?

a b c - 24/1000 - **1,000** - **1000** - 24/1,000

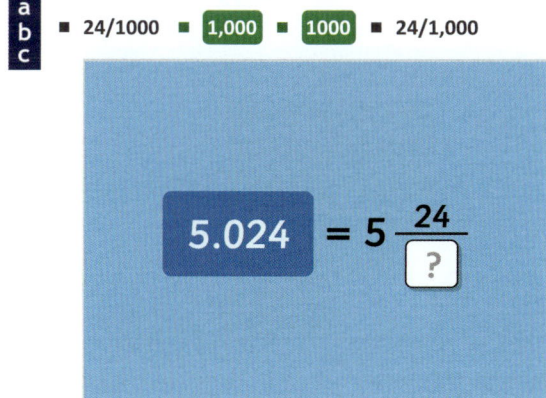

16. What is 0.08 as a fraction?

a b c *Put a forward slash (/) between the numerator and denominator. For example, one-half is 1/2.*

- 08/100 - 8/10 - **8/100** - 8/1000 - 8/1,000

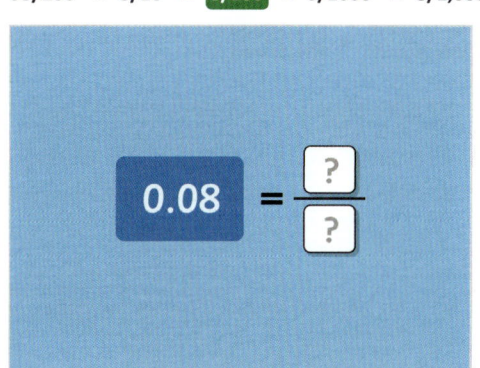

17. What is 0.001 as a fraction?

a b c *Put a forward slash (/) between the numerator and denominator. For example, one-half is 1/2.*

- **1/1,000** - **1/1000** - 001/1000 - 001/1,000

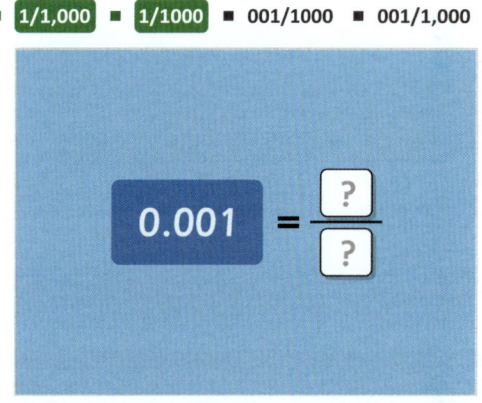

18. What is the missing denominator?

a b c - **1,000** - **1000** - 107/1000 - 107/1,000

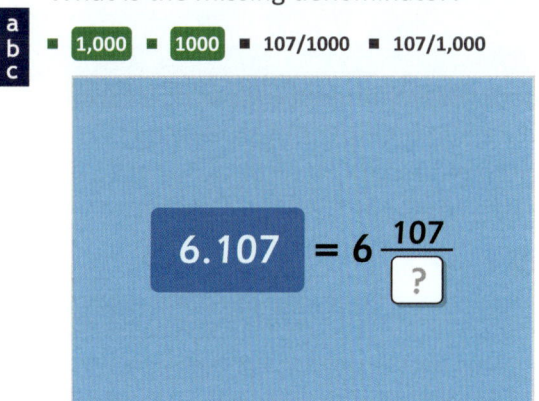

Level 2 *continued*

19. Convert 0.873 to a fraction.

a b c *Put a forward slash (/) between the numerator and denominator. For example, one-half is 1/2.*

■ 873/1,000 ■ 873/1000 ■ 873/10 ■ 873/100

as a decimal	ones	tenths	hundredths	thousandths	in words	as a fraction
0.397	0	3	9	7	three hundred and ninety-seven thousandths	$\frac{397}{1,000}$
0.439	0	4	3	9	four hundred and thirty-nine thousandths	$\frac{439}{1,000}$
0.873	0	8	7	3	eight hundred and seventy-three thousandths	?

20. What is 0.193 as a fraction?

1/3

■ 193/10 ■ 193/1,000 ■ 193/100

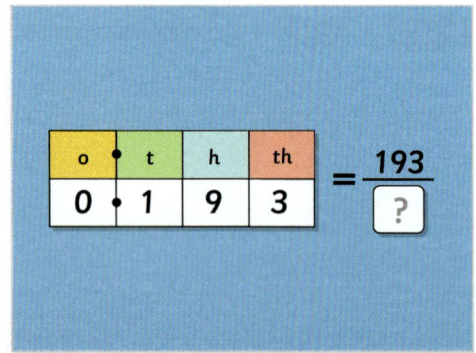

Level 3: Reasoning: Compare decimals and fractions and reason about fraction and decimal equivalents.

✸ **Required:** 5/5   ✸ **Pupil Navigation:** on
✸ **Randomised:** off

21. Which symbol makes the following statement true?

4/10 ___ 0.42

1/3   ■ < ■ = ■ >

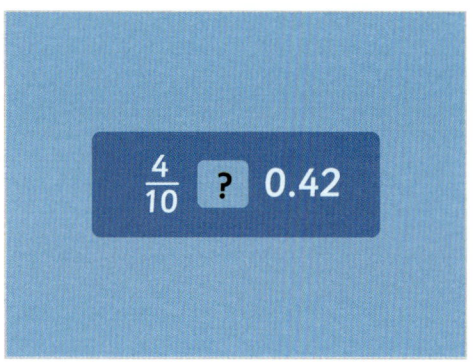

22. Charlie says that 0.30 = 3/10 and Becca says
a b c 0.30 = 30/100.

Who is correct? Explain your answer.

- Open question, no set answer

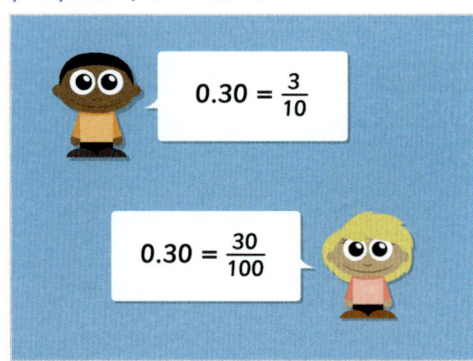

23. Which **two** fractions have the same value as the decimal number shown by the arrow?

2/4

■ 8/10 ■ 8/100 ■ 80/100 ■ 0.8/100

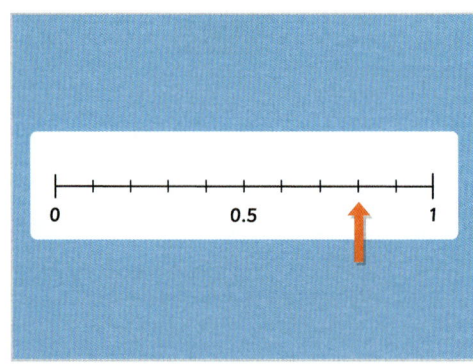

24. What is the missing decimal in the table
1 2 3 shown?

■ 0.007

decimal	fraction
0.7	$\frac{7}{10}$
0.07	$\frac{7}{100}$
?	$\frac{7}{1,000}$

Level 3 continued

25. Which symbol makes the following statement true?

1/4 ____ 0.23

1/3 · < · = · **>**

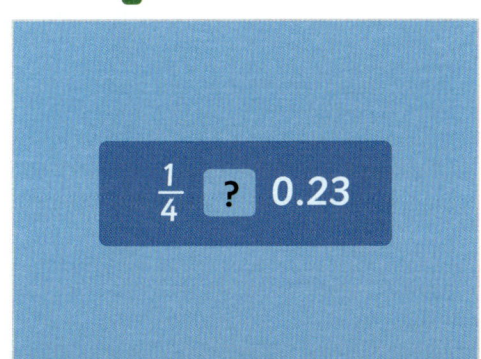

$\frac{1}{4}$ **?** 0.23

Level 4: Problem Solving: Solve problems with fraction and decimal equivalents.

❋ **Required:** 5/5   ❋ **Pupil Navigation:** on

❋ **Randomised:** off

26. Using every digit card **once**, complete the statement and enter the digit that goes in box C.

· 4 · **3** · 3/4

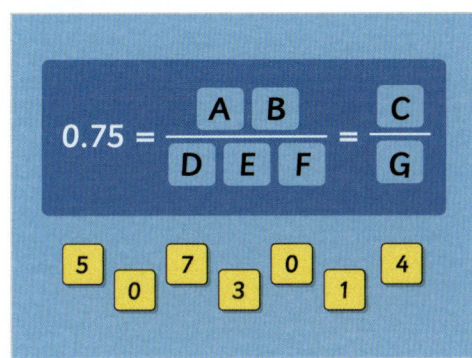

$0.75 = \frac{\boxed{A}\ \boxed{B}}{\boxed{D}\ \boxed{E}\ \boxed{F}} = \frac{\boxed{C}}{\boxed{G}}$

5 0 7 3 0 1 4

27. Sort the fractions and decimals in ascending order (smallest first).

· 0.09 · 24/100 · 0.55 · 0.7 · 9/10

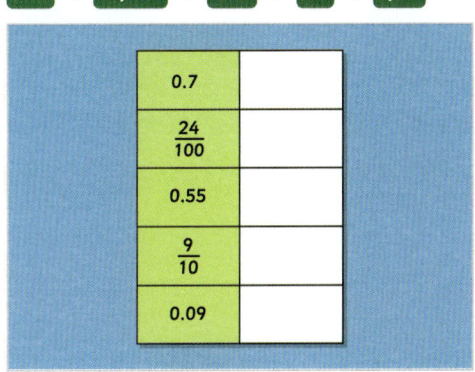

0.7	
$\frac{24}{100}$	
0.55	
$\frac{9}{10}$	
0.09	

28. Three children each have a pack of raisins. Bret eats 0.7 of his raisins, Cara eats 28/100 of her raisins and Celina eats 0.04 of her raisins. The child with the most raisins left over gives **half** of their remaining raisins to Riaz. What fraction of a pack does Riaz get?
Put a forward slash (/) between the numerator and denominator. For example, one-half is 1/2.

· **12/25** · **48/100** · **24/50** · 48/50 · 0.48

· 96/100

name	raisins eaten	raisins remaining
Brett	0.7	
Cara	$\frac{28}{100}$	
Celina	0.04	

29. Dale is thinking of a decimal number.

The ones digit is less than 1.
The tenths digit is an even number between 0 and 5, but it is **not a prime number**.
The hundredths digit is double the tenths digit.
The thousandths digit is a square number between 5 and 10.

He converts his decimal number to a fraction. What is Dale's **fraction**?
Put a forward slash (/) between the numerator and denominator. For example, one-half is 1/2.

· 249/1,000 · **489/1000** · **489/1,000** · 249/1000

· 0.489

Level 4 *continued*

30. The sum of each side of the triangle is the
a
b same. What is the total of the two missing
c fractions?

Give your answer as a fraction and put a
forward slash (/) between the numerator and
denominator. For example, one-half is 1/2.

- 20/25 - 80/100 - 8/10 - 40/50 - 4/5

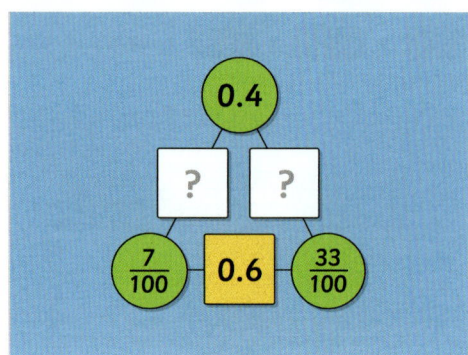

Recognise and Understand Percentages

Objective: I can recognise and understand percentages as number of parts per hundred.

Quick Search Ref: 10034

Level 1: Understanding: Recognise the per cent symbol and find percentages with image support.

✸ **Required:** 7/10 ✸ **Pupil Navigation:** on ✸ **Randomised:** off

1. What is a **percentage**?

1/3

- a whole number that divides exactly into another whole number without a remainder
- an amount expressed as a number of parts per 100
- a fraction with a denominator of 10

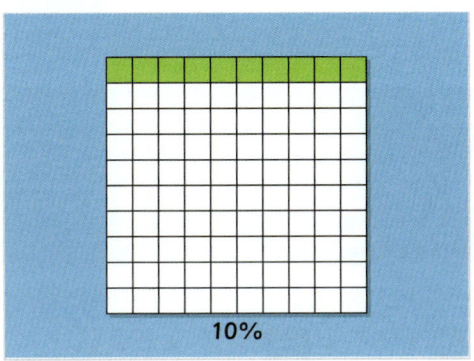

10%

2. Which symbol means **per cent**?

1/4

- A ▪ **B** ▪ C ▪ D

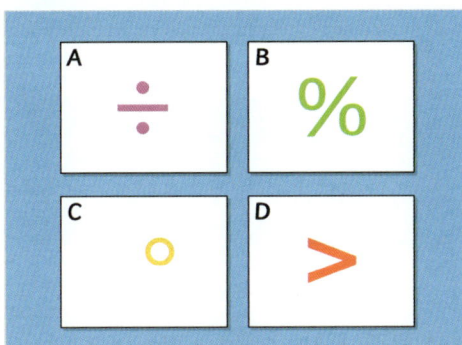

3. _____ part per hundred is shaded, which is 1% of the whole square.

1
2
3

- **1** ▪ 99 ▪ 100

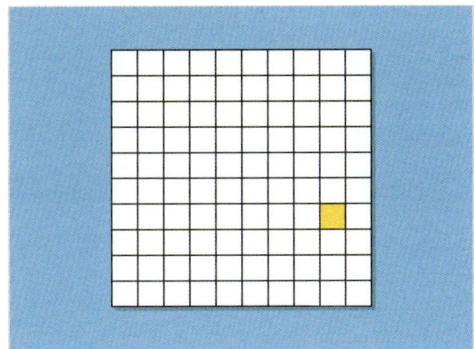

4. 8 parts per hundred are shaded, which is _____% of the whole square.

a
b
c

- **8%** ▪ 92 ▪ **8** ▪ 8/100 ▪ 92%

5. What percentage of the circle is shaded?

a
b
c

- **50** ▪ **50%** ▪ 1/2 ▪ 1% ▪ 1

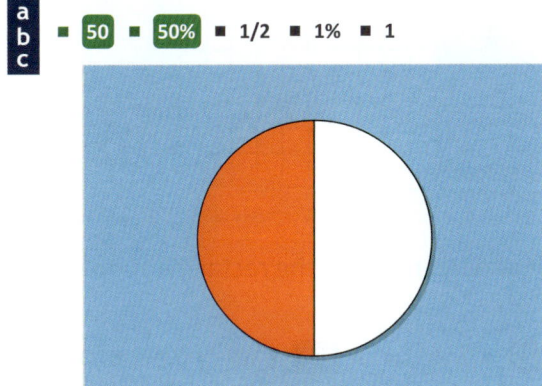

6. What percentage of the rectangle is **not** shaded?

a
b
c

- 3/4 ▪ **75** ▪ 3% ▪ **75%** ▪ 3 ▪ 25% ▪ 25

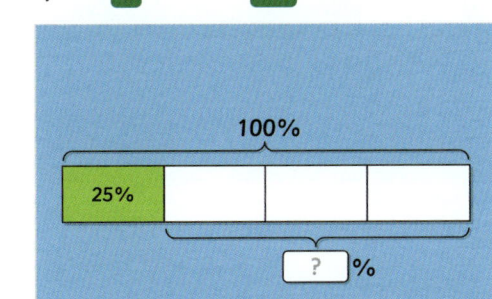

Level 1 *continued*

7. What percentage of the rectangle is shaded?

a b c ▪ 7/10 ▪ **70** ▪ 7 ▪ **70%** ▪ 30% ▪ 7% ▪ 30

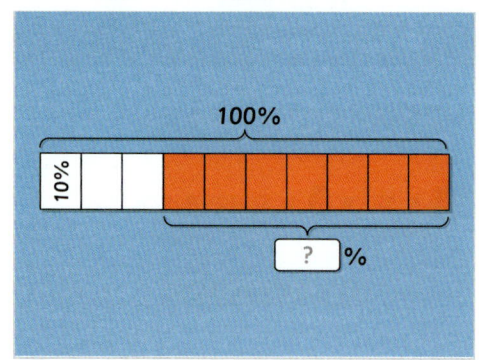

8. What percentage of the hundred square is shaded?

a b c ▪ **46%** ▪ 46/100 ▪ **46** ▪ 54% ▪ 54

9. What percentage of the rectangle is **not** shaded?

a b c ▪ **25** ▪ 1 ▪ 1/4 ▪ **25%** ▪ 1% ▪ 75% ▪ 75

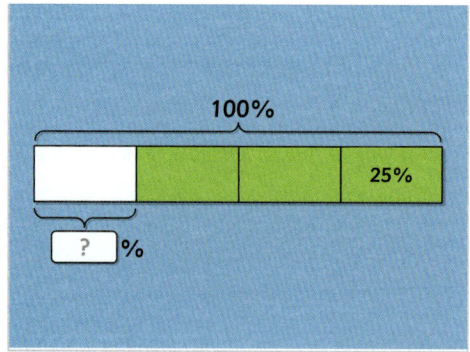

10. 72 parts per hundred are shaded, which is ____% of the whole square.

a b c ▪ **72** ▪ 72/100 ▪ 28

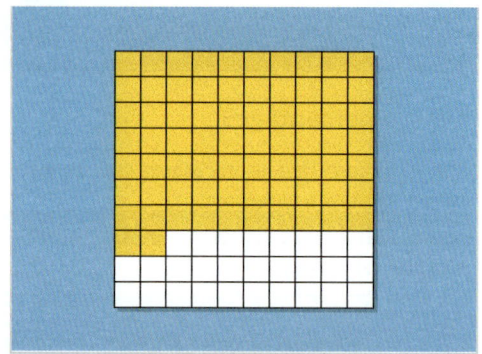

Level 2: Fluency: Find percentages of shapes and find simple percentages in context.

❋ **Required:** 7/10   ❋ **Pupil Navigation:** on
❋ **Randomised:** off

11. What percentage of the rectangle is shaded?

a b c ▪ **40** ▪ 60% ▪ **40%** ▪ 60 ▪ 4 ▪ 4%

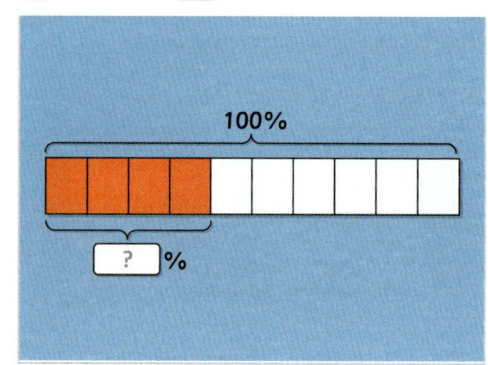

12. ____% of the circle is shaded.

a b c ▪ 3 ▪ **75%** ▪ 3% ▪ **75** ▪ 25% ▪ 25

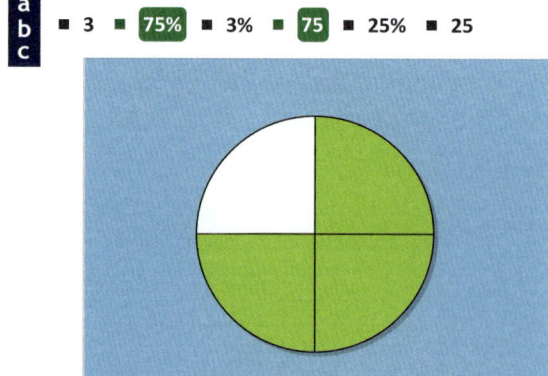

13. What percentage of the quilt is patterned?

a b c

■ 70 ■ 30 ■ 30% ■ 70%

14. There are 100 building blocks in a box. George uses 26 of them to make a tower. What percentage of the blocks does he use?

a b c

■ 26 ■ 74% ■ 26% ■ 74

15. What number makes the following statement true?

1 2 3

_____% = 2/10

■ 20 ■ 2

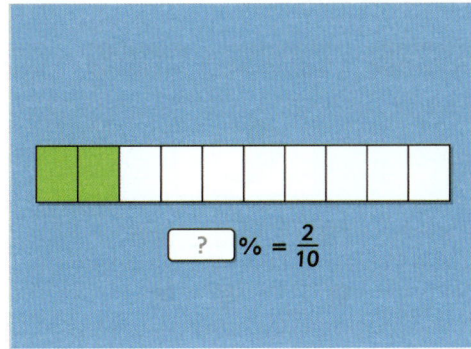

16. There are 100 guests at a party and 59 of them are children. What percentage are adults?

a b c

■ 41 ■ 41% ■ 59% ■ 59

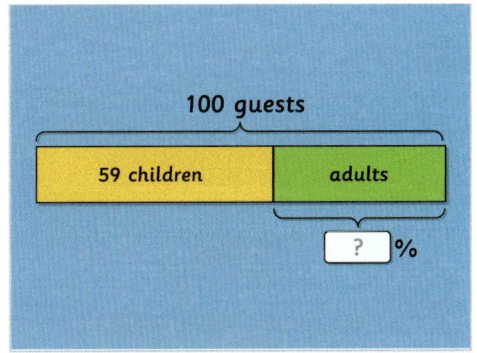

17. There are 100 grapes in a pack and Anisha eats 48 of them. What percentage of the grapes are left?

a b c

■ 52% ■ 52 ■ 48 ■ 48%

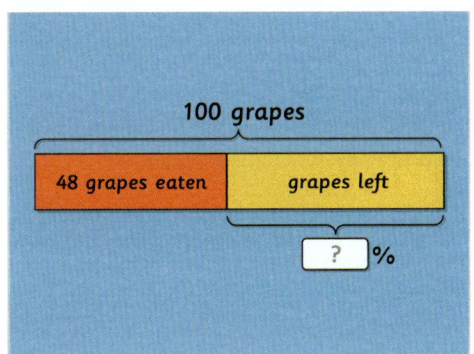

18. Three children answer 100 questions using Learning by Questions. Ellie scores three points less than Harriet. What percentage of the questions does Ellie answer correctly?

a b c

■ 87 ■ 87% ■ 90 ■ 90%

child	score	percentage
Jo	58 out of 100	58%
Ellie		?
Harriet	90 out of 100	90%

Level 2 *continued*

19. Dale uses 100 tiles in his bathroom. The plan shows the numbers of blue and white tiles that he uses. What percentage of the tiles are blue?

a
b
c

- **60%** ▪ 40% ▪ **60** ▪ 40

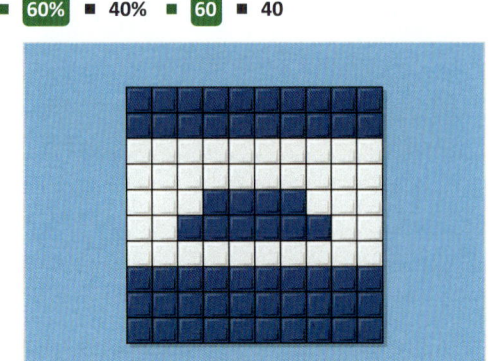

20. What percentage of the rectangle is shaded?

a
b
c

▪ 8% ▪ **80** ▪ **80%** ▪ 8 ▪ 20% ▪ 20

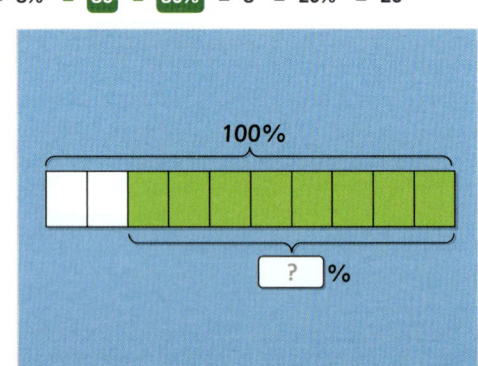

Level 3: Reasoning: Order, compare and reason about percentages.

❋ **Required:** 5/5   ❋ **Pupil Navigation:** on
❋ **Randomised:** off

21. *If 50 parts of a shape are shaded, 50% of the shape is shaded.*

a
b
c

Is this statement **always** true, **sometimes** true or **never** true? Explain your answer.

- Open question, no set answer

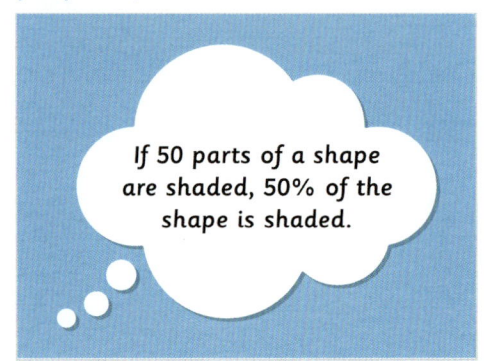

If 50 parts of a shape are shaded, 50% of the shape is shaded.

22. Which diagram is the **odd one out**?

☐
☒
1/3

▪ **diagram A** ▪ diagram B ▪ diagram C

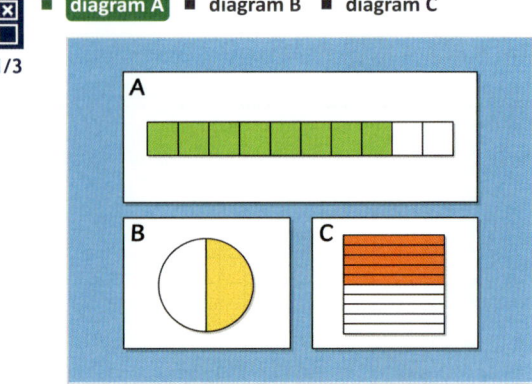

23. Josh says, *"Three parts of the shape are shaded. This means that 3% of the whole shape is shaded."*

a
b
c

Explain why Josh is incorrect and find the correct percentage of the shape that is shaded.

- Open question, no set answer

3 %

24. Sort the diagrams in descending order, starting with the diagram that has the largest percentage coloured.

↑
↓

▪ **diagram D** ▪ **diagram B** ▪ **diagram A** ▪ **diagram C**

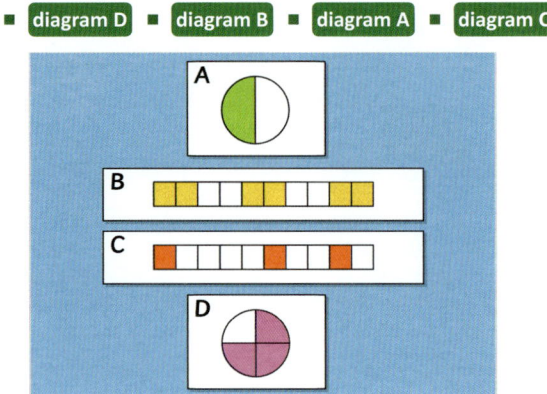

25. Part of a hundred square representing a percentage has been covered with paint.

a b c

Zack says, "Less than 50% of the hundred square is shaded."

Is Zack correct? Give reasons for your answer.

- Open question, no set answer

Level 4: Problem solving with greater depth: Solve problems involving finding percentages.

❋ **Required:** 5/5   ❋ **Pupil Navigation:** on

❋ **Randomised:** off

26. The watch has been reduced by ____% in the sale.

a b c

  ▪ 10% ▪ 10 ▪ £10.00 ▪ £10

27. Kiara has ten books, and she reads three of them. What percentage of her books has she **not** read?

a b c

  ▪ 7/10 ▪ 70 ▪ 70% ▪ 30 ▪ 30% ▪ 7

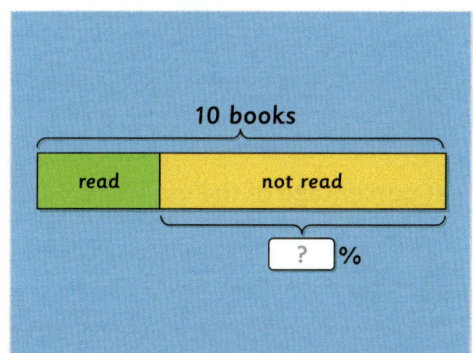

28. Will buys a pack of 100 sheets of paper. He uses half of the pack to print a school project and 2/10 of the pack to make paper aeroplanes. What **percentage of the pack** does he have left?

a b c

  ▪ 50 ▪ 3/10 ▪ 30% ▪ 30 ▪ 50% ▪ 70% ▪ 70

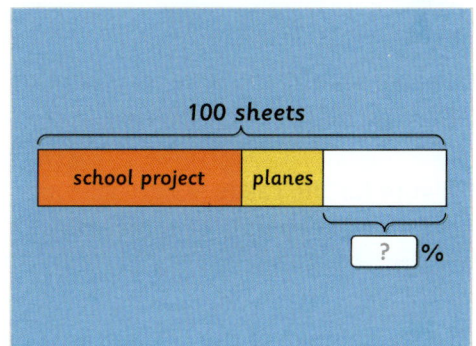

Level 4 *continued*

29. Three children answer 100 questions using Learning by Questions.

a b c

Ali scores half as many points as Ben.
Ben scores 6% less than Orla.
Orla's score is a multiple of 8 between 60 and 70.

What **percentage** of the questions does Ali answer correctly?

■ 29% ■ 29 ■ 64 ■ 58 ■ 58% ■ 64%

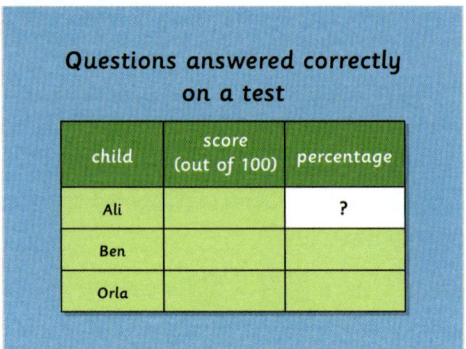

Questions answered correctly on a test

child	score (out of 100)	percentage
Ali		?
Ben		
Orla		

30. 100 children in Year 5 vote for where they want to go for their school trip.

a b c

50% vote for the castle.
The theme park gets 10% more votes than the farm.

What **percentage** of the children vote for the farm?

■ 20% ■ 50% ■ 20 ■ 40 ■ 50 ■ 30% ■ 40%
■ 30

Votes for Year 5 school trip

place	percentage
castle	50%
theme park	
farm	?

Fractions Topic Review

Objective: I can answer questions involving fractions from the Year 5 curriculum.

Quick Search Ref: 10736

Level 1: Understanding

✿ **Required:** 7/10   ✿ **Pupil Navigation:** on   ✿ **Randomised:** off

---

1. What is the missing denominator?

1/8 = 7/____

- ■ **56** ■ 15

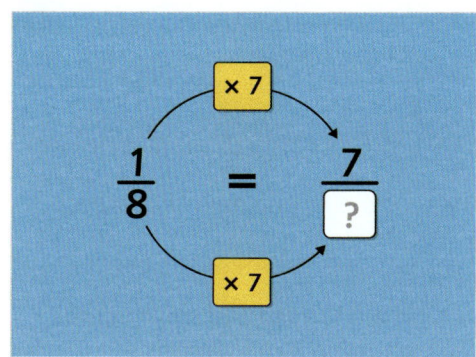

2. What is 1/2 + 1/8?

Put a forward slash (/) between the numerator and denominator. For example, one-half is 1/2.

- ■ **5/8** ■ 5/16 ■ 2/10

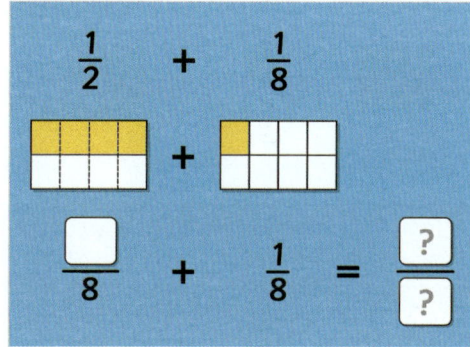

3. Which of the following is the smallest improper fraction?

9/4, 10/8, 9/8

Put a forward slash (/) between the numerator and denominator.

- ■ **9/8** ■ 10/8 ■ 9/4

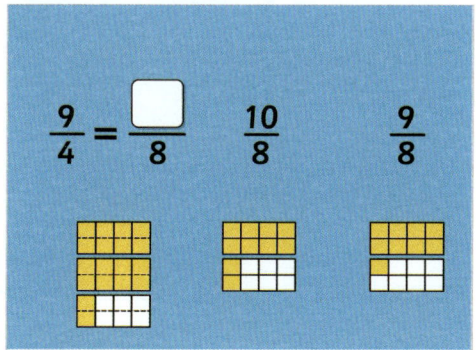

4. Convert 2 2/3 to an improper fraction.

Put a forward slash (/) between the numerator and denominator. For example, 3/2.

- ■ **8/3** ■ 6/3 ■ 8

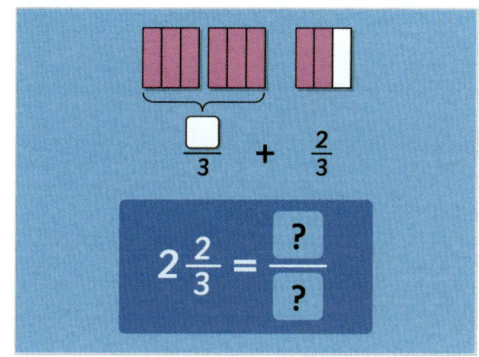

5. What is the next fraction in this sequence?

5/7, 4/7, 3/7, ____

Put a forward slash (/) between the numerator and denominator. For example, one-half is 1/2.

- ■ **2/7** ■ -1/7 ■ 1/7

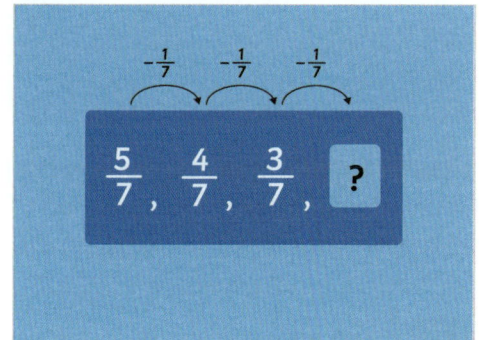

Level 1 *continued*

6. What is 1 2/5 × 2?
a
b *Give your answer as a mixed number*
c *fraction. For example, five and one-half is 5 1/2.*

- **2 4/5** ■ 7/5 ■ 14/5 ■ 2 4/10 ■ 2 2/5

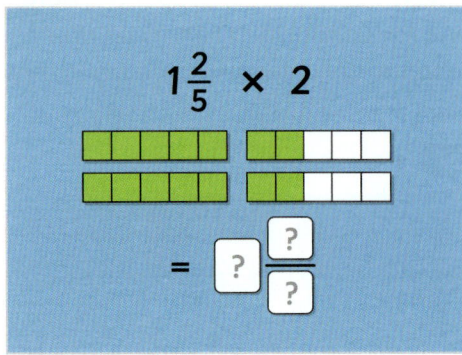

7. What is 5 1/8 − 1 3/8?
a
b *Give your answer as a mixed number*
c *fraction. For example, five and one-half is 5 1/2.*

- **3 3/4** ■ **3 6/8** ■ 30/8 ■ 4 1/4 ■ 4 2/8 ■ 15/4

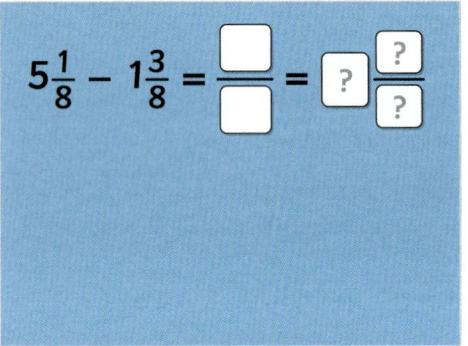

8. Convert 11/4 to a mixed number fraction.
a
b *Give your answer as a mixed number*
c *fraction. For example, five and one-half is 5 1/2.*

- **2 3/4** ■ 2 ■ 1/4

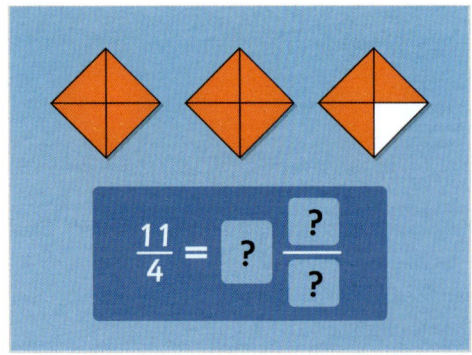

9. Calculate 15/6 − 4/3.
a
b *Give your answer as an improper fraction.*
c *For example, 3/2.*

- **7/6** ■ 11/3 ■ 23/6

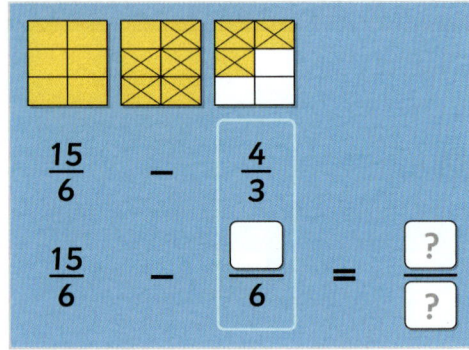

10. Which of the following is the largest
a fraction?
b
c 2/3, 1/6, 3/6
Put a forward slash (/) between the numerator and denominator.

- **2/3** ■ 1/6 ■ 3/6

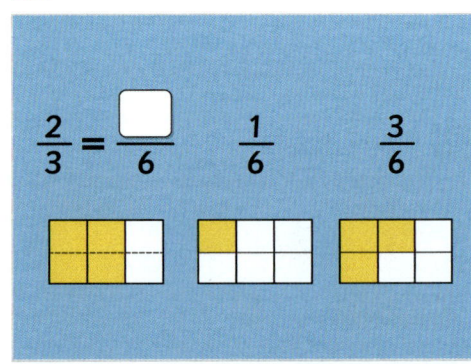

Level 2: Fluency

✱ **Required:** 7/10   ✱ **Pupil Navigation:** on
✱ **Randomised:** off

11. Calculate 7/10 × 5.
a
b *Put a forward slash (/) between the*
c *numerator and denominator. For example,*
one-half is 1/2.

- **7/2** ■ **35/10** ■ 7/50 ■ 35/50

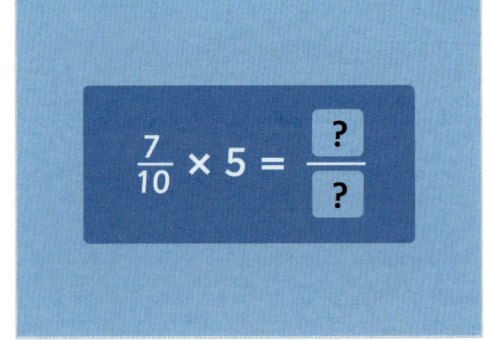

Level 2 continued

12. What is 11/14 − 5/28 − 2/7?

a
b
c
Put a forward slash (/) between the numerator and denominator. For example, one-half is 1/2.

- 9/28

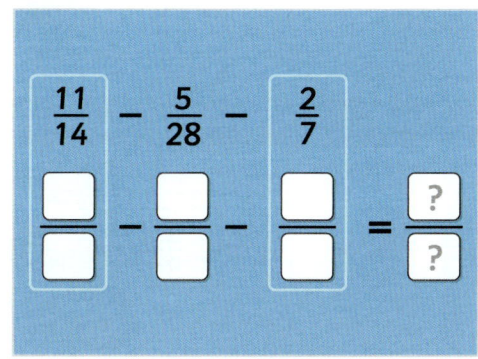

13. Convert 2 7/8 to an improper fraction.

a
b
c
Put a forward slash (/) between the numerator and denominator.

- 23/8

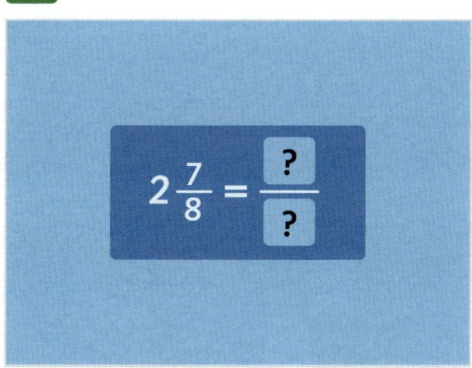

14. What fraction comes next in this sequence?

a
b
c
Put a forward slash (/) between the numerator and denominator.

- 9/3 ▪ 3 ▪ 2/3

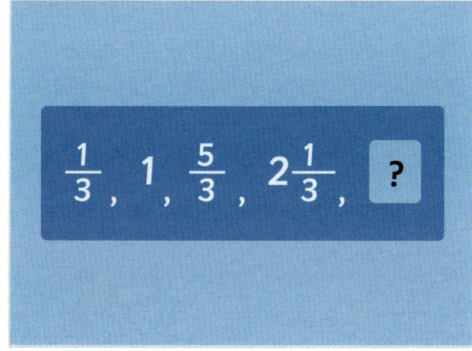

15. What is 2 1/4 + 5/8 + 1 1/16?

a
b
c
Give your answer as a mixed number fraction. For example, five and one-half is 5 1/2.

- 3 15/16 ▪ 63/16

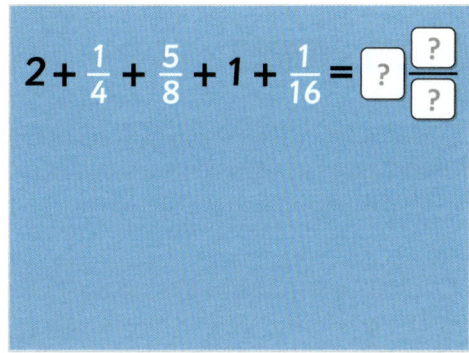

16. Four friends record what fraction of a birthday cake they each eat. Which **child** eats the least amount of cake?

a
b
c

- Sam ▪ 1/16 ▪ Tim ▪ Ralph ▪ Peter

Amount of cake eaten at a party	
name	cake eaten
Tim	$\frac{5}{32}$
Sam	$\frac{1}{16}$
Ralph	$\frac{1}{4}$
Peter	$\frac{17}{32}$

17. There are 36 children in a class. 4/6 of the children have school dinners. How many children have school dinners?

1
2
3

- 24 ▪ 12

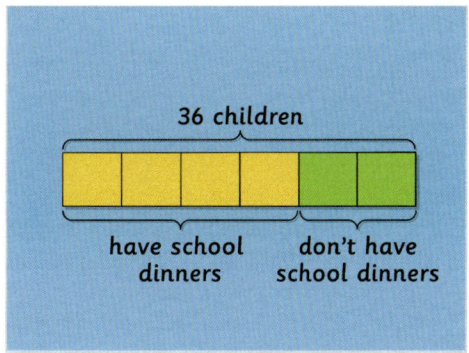

Level 2 *continued*

18. What is the missing fraction on the number line?

Put a forward slash between the numerator and the denominator. For example, one-half is 1/2.

- **151/1,000** - **151/1000** - 151/100 - 16/10
- 151

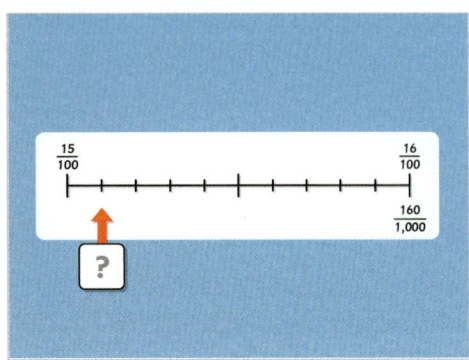

19. What is 9/4 + 5/8 + 17/16?

Give your answer as an improper fraction. For example, 3/2.

- **63/16**

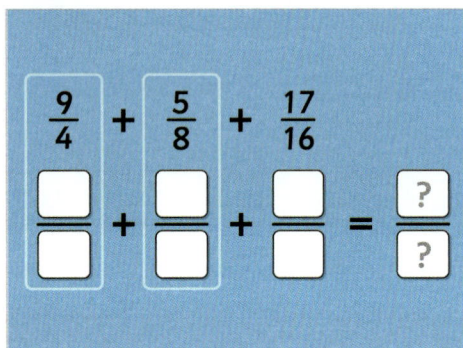

20. 7/11 = 35/___

- **55** - 5

21. Harry calculates that 30/48 − 5/12 = 25/36. Explain what Harry has done wrong and explain how he can calculate the answer correctly.

- Open question, no set answer

22. How many more parts of the circle do you need to shade to make it equivalent to the rectangle?

- **5** - 8

23. Charlie says that 0.30 = 3/10 and Becca says 0.30 = 30/100. Who is correct? Explain your answer.

- Open question, no set answer

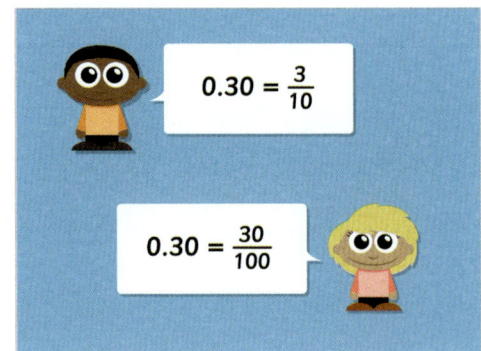

Level 3 *continued*

24. Using two of the number cards, what is the **smallest** fraction you can make?

Put a forward slash (/) between the numerator and denominator.

- **2/100** - **1/50**

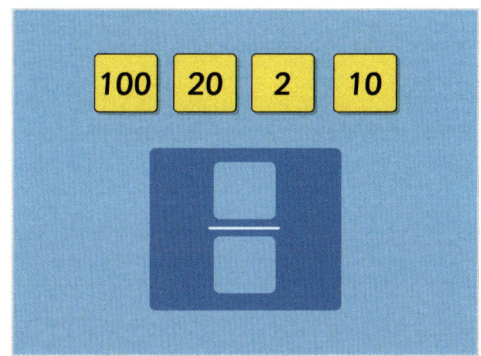

25. What is the same and what is different about these sequences?

- **Open question, no set answer**

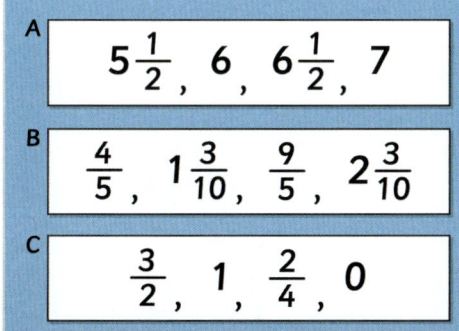

Level 4: Problem solving with greater depth

✱ **Required:** 5/5 ✱ **Pupil Navigation:** on
✱ **Randomised:** off

26. A box contains 120 crayons. 25 crayons are pink and 1/5 of the crayons are red. How many crayons are **green**?

- **71** ■ **49** ■ **24**

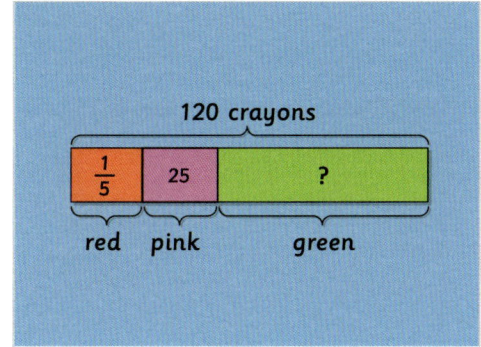

27. Julie makes nine portions of pizza dough. The total weight of the ingredients that Julie uses is _____ kilograms (kg).

Enter the missing mixed number fraction. Don't include the units in your answer.

- **1 44/100** - **1 11/25kilograms** - **1 22/50**
- **1 11/25** - **1 11/25 kilograms** - **1 11/25kg**
- **1 11/25 kg** ■ **1.44** ■ **144/100**

28. Norah is thinking of an equivalent fraction to 3/15.

The numerator is an even number between 20 and 30.

The sum of the denominator's digits is 4.

What is Norah's fraction?
Put a forward slash (/) between the numerator and denominator. For example, one-half is 1/2.

- **26/130** ■ **22/110** ■ **26/15** ■ **24/120** ■ **21/105**
- **30/150** ■ **28/140**

Level 4 *continued*

29. An animal rescue centre has 4 full bags of

hay. They use 1/4 of a bag of hay for the
hamsters and 9/8 bags of hay for the rabbits.
If there are 3/2 bags of hay left at the end of
the day, how much hay do they give to the
donkey?
Give your answer as an improper fraction.
For example, 3/2.

- 9/8 ▪ 21/8

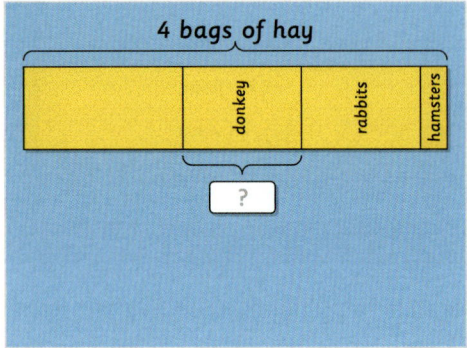

30. Dan and his friends take part in a relay race.
a
b
c
April runs 1 2/5 kilometres (km).
Charlie runs 1 7/20 km.
Dan runs 1 7/10 km further than April and
Charlie run in total.

All of the children run a total of _____ km.
Enter the missing mixed number fraction.
Don't include the units in your answer.

- 7 1/5 ▪ 7 2/10 km ▪ 7 1/5 kilometres ▪ 7 1/5 km
- 7 4/20 ▪ 7 2/10 kilometres ▪ 7 2/10 ▪ 7 4/20 km
- 7 4/20 kilometres

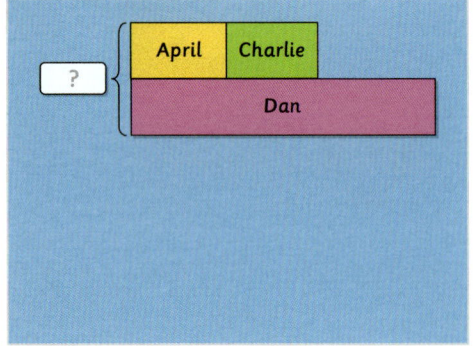

Mathematics Y5

Decimals and Percentages

Decimals
Percentages
Equivalence

Read, Write, Compare and Order Numbers with up to Three Decimal Places

Objective: I can read, write, order and compare numbers with up to 3 decimal places.

Quick Search Ref: 10036

Level 1: Understanding: Read, write, compare and order numbers with up to three decimal places with image support.

Required: 7/10 **Pupil Navigation:** on **Randomised:** off

1. In the number 45.17, what digit is in the tenths column?

- **1**
- 4
- 0.1

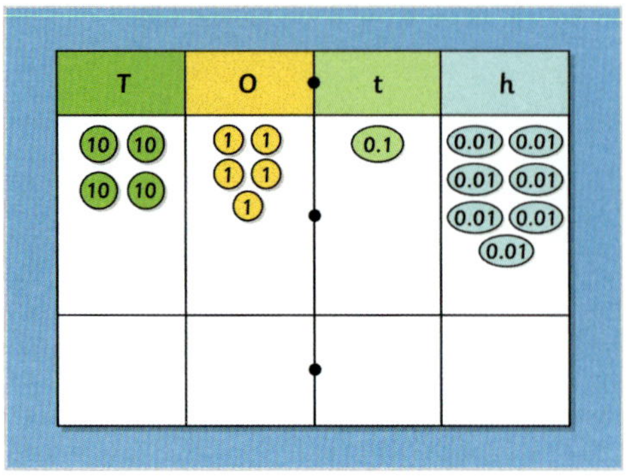

2. What digit is in the **hundredths column** in the number 5.37?

- **7**
- 0.07

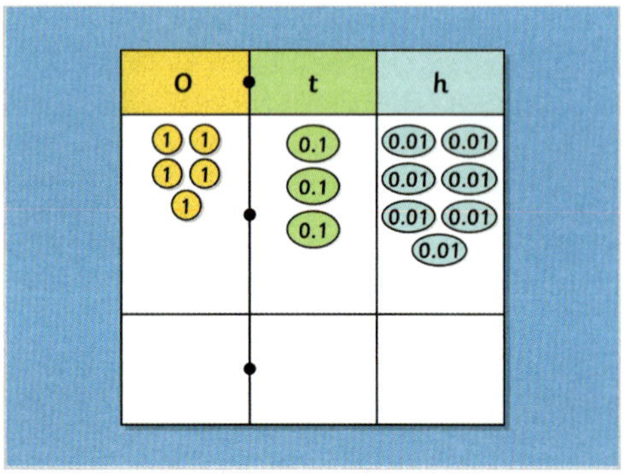

3. What is seventy-two and eight tenths in **digits**?

- **72.8**
- 728

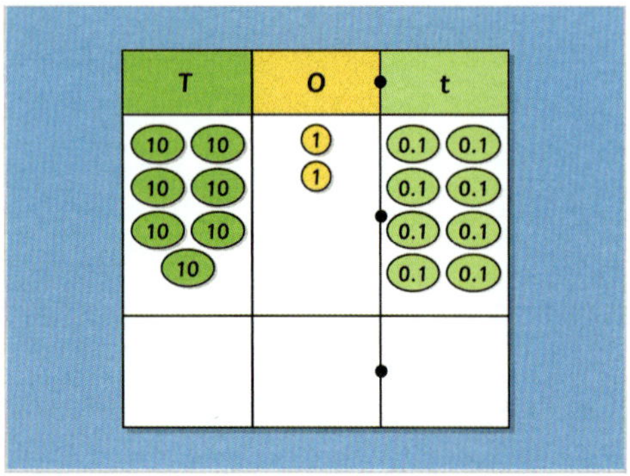

4. Which is the largest number?

1/4

- 1.31
- 2.32
- **2.39**
- 1.99

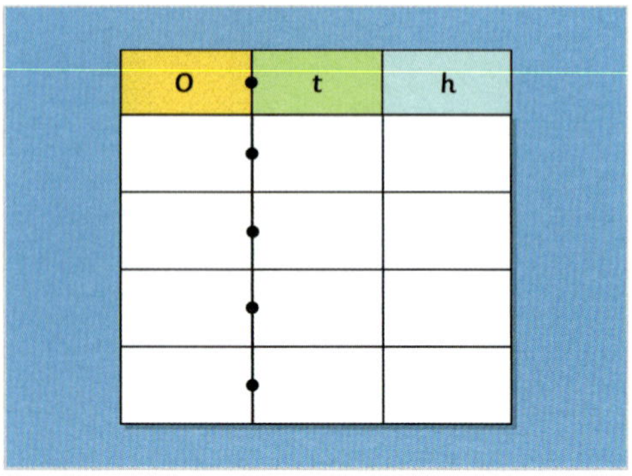

5. What is the value of the 2 in the number 83.012?
Give your answer as a decimal number.

- **0.002**
- 0.2
- 0.02

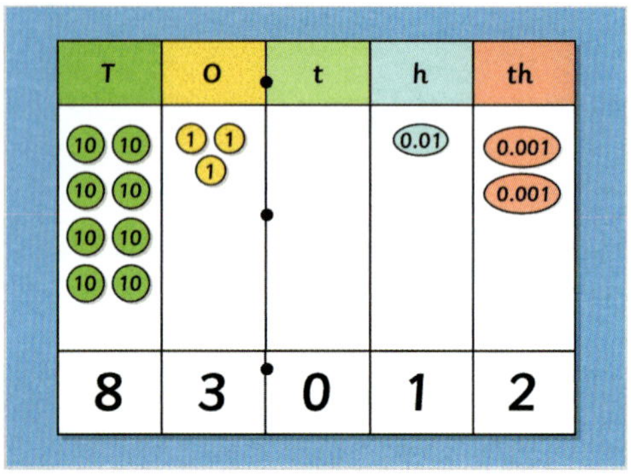

6. What decimal number is made using these arrow cards?

- **8.402**
- 8402
- 8.42

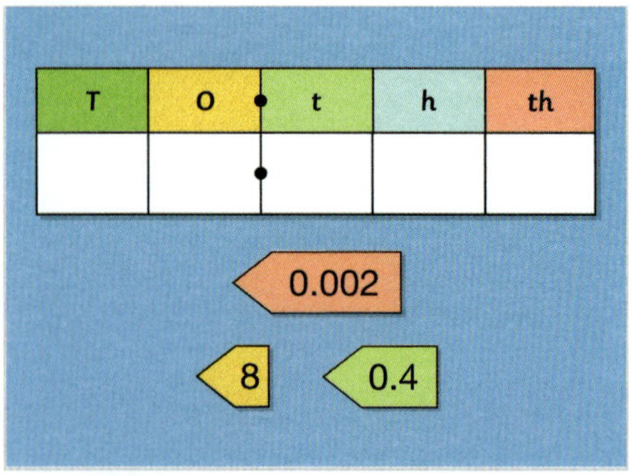

Level 1 *continued*

7. Sort the decimal numbers starting with the smallest value first.

- 5.612 ■ 7.246 ■ 7.249

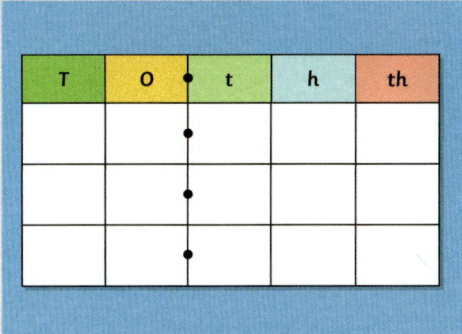

8. Which is the smallest number?

- 2.32 ■ 1.99 ■ 2.39 ■ 1.31

1/4

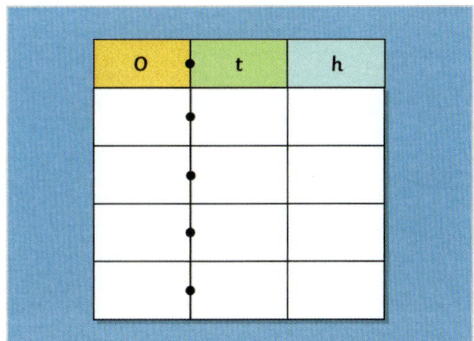

9. What decimal number is made using these arrow cards?

- 3.501 ■ 3501 ■ 3.51

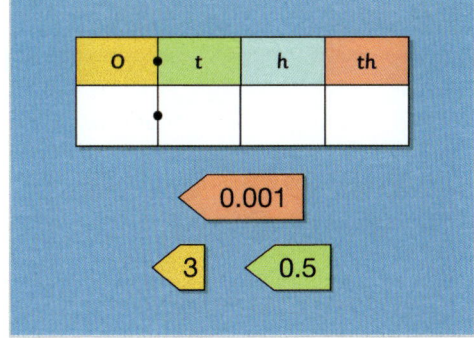

10. What is the value of the 3 in the number 57.123?
Give your answer as a decimal number.

- 0.003 ■ 0.3 ■ 0.03

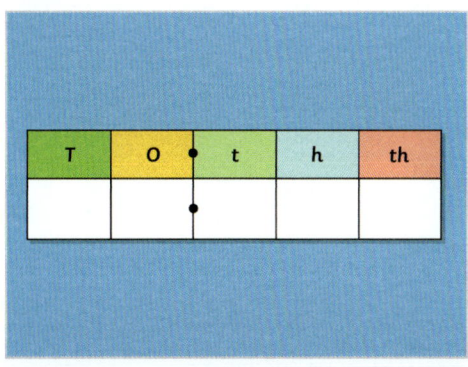

Level 2: Fluency: Read, write, compare and order numbers with up to three decimal places.

❋ **Required:** 7/10   ❋ **Pupil Navigation:** on
❋ **Randomised:** off

11. In the image, a decimal number is represented using base ten blocks. What decimal number is represented?

- 2.43 ■ 243

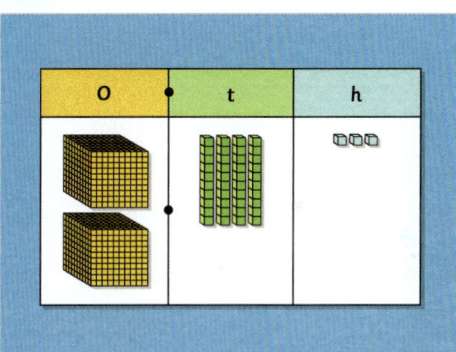

12. What is the value of the 7 in the number 836.537?
Give your answer as a decimal number.

- 0.7 ■ 0.007 ■ 0.07

836.537

Level 2 *continued*

13. Which is the largest number?

 ■ 7.85

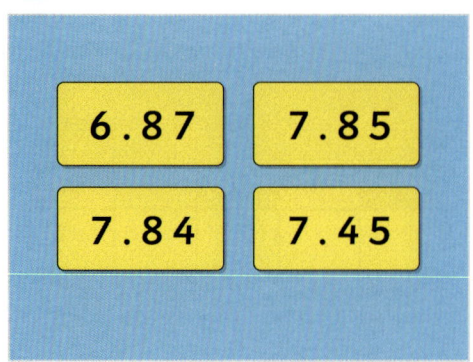

14. What is the value of 3.461 written in words?

1/4

- ■ three point four hundred and sixty-one
- ■ three ones, four tenths, six hundredths and one thousandth
- ■ three thousand, four hundred and sixty-one
- ■ three ones, six tenths, six hundredths and one thousandth

15. Which is the smallest number?

  ■ 4.35

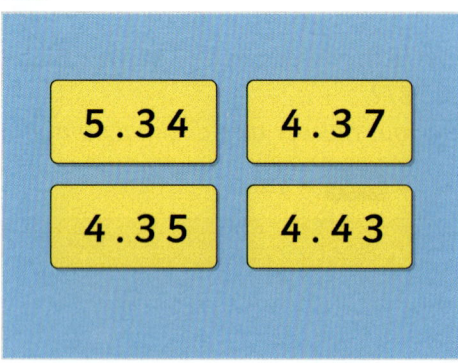

16. Sort the numbers into descending order (largest first).

■ 27.736 ■ 27.735 ■ 26.010 ■ 24.764

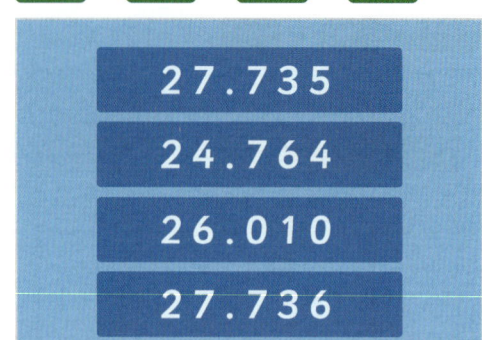

17. The table shows the daily number of visitors to a museum for each year from 2010 to 2015. Which **year** had the most visitors?

■ 2012

Museum visitors 2010 to 2015

year	average number of daily visitors
2010	321.569
2011	315.713
2012	321.801
2013	314.998
2014	318.004
2015	321.089

18. Frankie represents a decimal number using place value counters. What is Frankie's decimal number?

■ 3.521

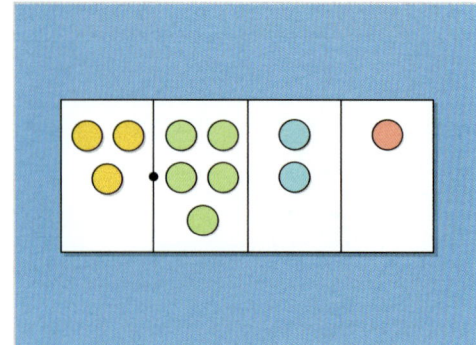

Level 2 *continued*

19. Sort the numbers into ascending order (smallest first).

↑↓

- 52.515 ■ 52.731 ■ 52.736 ■ 54.761

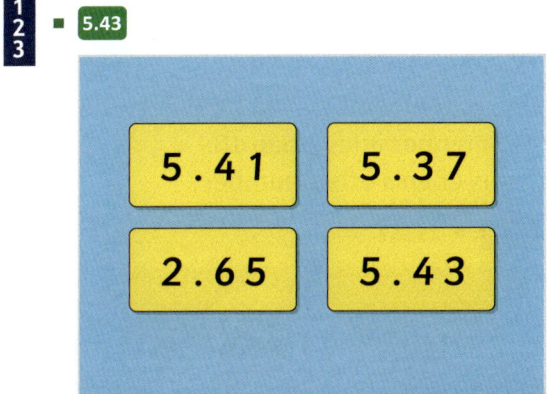

20. Which is the largest number?

123

- 5.43

Level 3: Reasoning: Reason about numbers with up to 3 decimal places.

❋ **Required:** 5/5   ❋ **Pupil Navigation:** on
❋ **Randomised:** off

21. Which symbol makes the statement true?

☐☒☐
1/3

27.13 ____ 27.099

- < ■ = ■ >

27.13 ☐ 27.099

22. Using every digit card once, make the smallest possible number.

123

- 0.579 ■ 5.79 ■ 5.079

23. Amira is trying to find the number halfway between 2.5 and 3.7. Explain how she can do this and find the correct answer for her.

abc

- **Open question, no set answer**

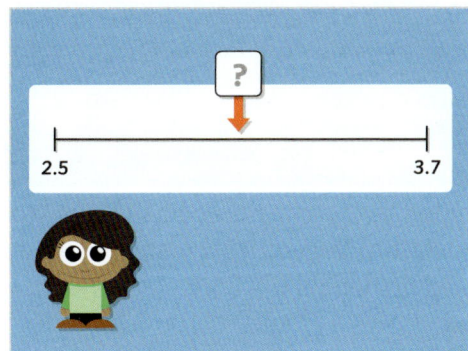

24. Which of the following statements is **false**?

☐☒☐
1/4

- 1.009 < 1.09 ■ 21.249 > 21.25 ■ 4.35 > 4.053
- 85.023 < 85.2

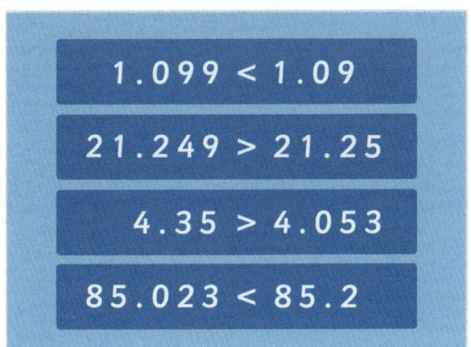

Level 3 *continued*

25. Using the decimal numbers shown in the
a image, Orla and Brad are trying to find the
b closest number to 7.5. Brad says, "7.6 is
c closest because it is only 1 away." Do you
agree with Brad? Explain your answer.

- Open question, no set answer

Level 4: Problem solving with greater depth: Solve
multi-step problems involving numbers with
up to three decimal places.

❋ **Required:** 5/5   ❋ **Pupil Navigation:** on
❋ **Randomised:** off

26. Ahmed puts some numbers in order from
1 smallest to largest. How many possible digits
2 could be behind the ink?
3

■ **6**   ■ 5

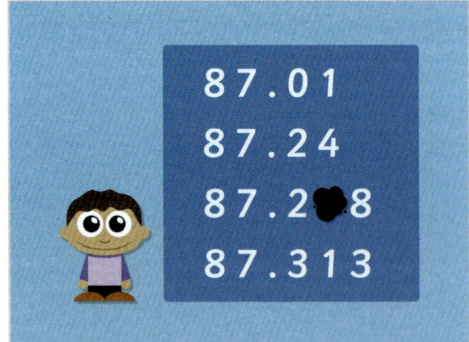

27. Use the following clues to find the missing
1 decimal number:
2
3 • The number has 3 digits.
• The digits have a sum of 13.
• All of the digits are different.

■ **6.25**   ■ 6.16   ■ 6.52

28. Chen is thinking of two numbers.
1
2 • The numbers have a difference of 3.487.
3 • One of the numbers is 11.665 and the
other number is less than 10.
What is Chen's **other number**?

■ **8.178**   ■ 15.152

29. Ellen buys the second most expensive pizza
a on the menu. In a special deal, she gets a
b discount of one-tenth off the original price.
c How much does she pay?
Include the £ sign in your answer.

■ £495   ■ 4.95   ■ **£4.95**   ■ £0.55   ■ £4.95p

■ £5.50   ■ 495p

Level 4 *continued*

30. Arabella is thinking of a number.

• The number is greater than 6.34 and smaller than 7.56.

• The number has **3 digits** and all of the digits are odd.

• The same digit can be used more than once.

How many possible numbers could Arabella be thinking of?

▪ 13

Round Decimals With Two Decimal Places to the Nearest Whole Number and to One Decimal Place

Objective: I can round decimals with 2 decimal places to the nearest whole number and to 1 decimal place.

Quick Search Ref: 10133

Level 1: Understanding: Round decimals with two decimal places to the nearest whole number and to one decimal place with support.

✱ Required: 7/10   ✱ Pupil Navigation: on   ✱ Randomised: off

1. Select the whole number that comes **before** 8.2 and the whole number that comes **after** 8.2.

2/6 ▪ 2 ▪ 3 ▪ **8** ▪ **9** ▪ 8.1 ▪ 8.3

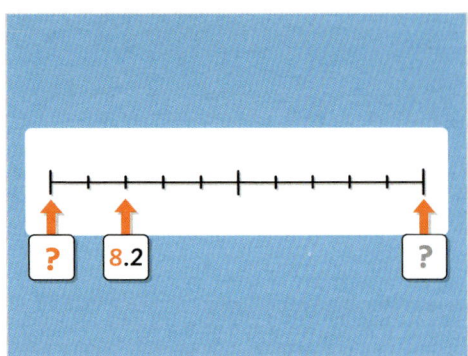

2. What is 8.2 rounded to the nearest whole number?

▪ **8** ▪ 9 ▪ 7

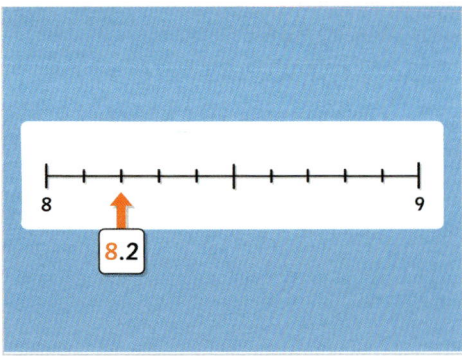

3. The number 3.16 is between the decimal numbers 3.1 and ____.

1/4 ▪ 3 ▪ 3.10 ▪ 3.11 ▪ **3.2**

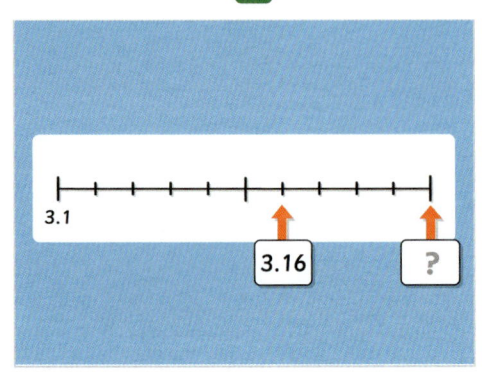

4. What is 3.16 to the nearest tenth?

▪ **3.2** ▪ 3 ▪ 3.1

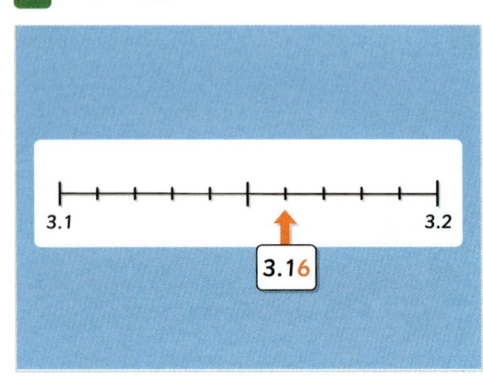

5. What is 6.81 rounded to the nearest tenth?

▪ 7 ▪ **6.8** ▪ 6.9

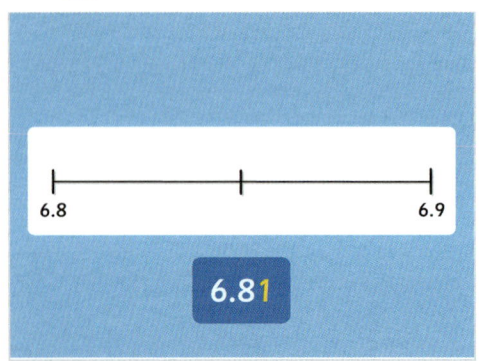

6. Round 67.42 to the nearest whole number.

▪ **67** ▪ 66 ▪ 67.4 ▪ 68

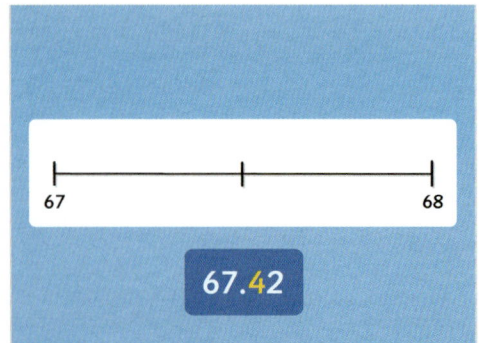

Level 1 *continued*

7. Round 94.61 to the nearest whole number.

1 2 3 ▪ 95 ▪ 94.6 ▪ 94

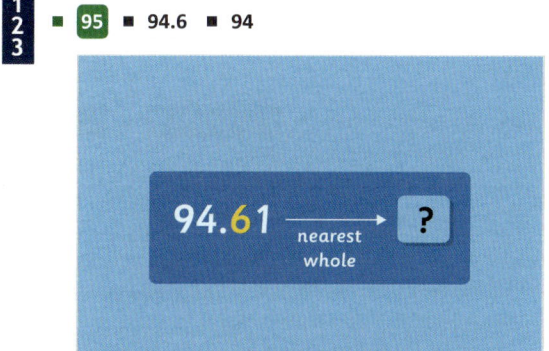

8. What is 45.29 to the nearest whole number?

1 2 3 ▪ 46 ▪ 45 ▪ 44 ▪ 45.3

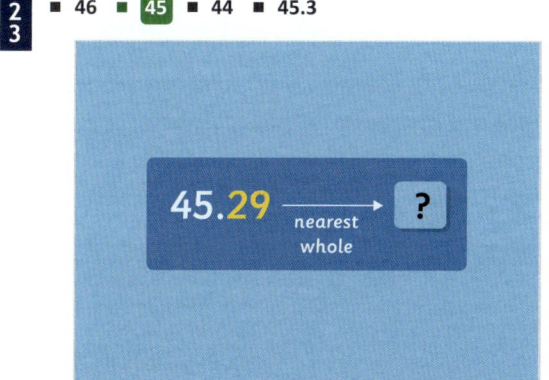

9. What is 6.89 to the nearest tenth?

1 2 3 ▪ 6.9 ▪ 6.8 ▪ 7

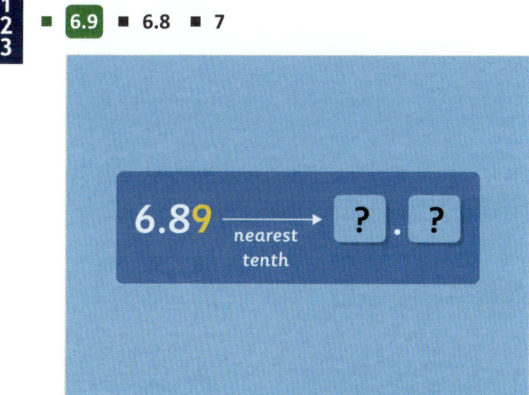

10. What is 12.4 rounded to the nearest whole number?

1 2 3 ▪ 12 ▪ 13 ▪ 11

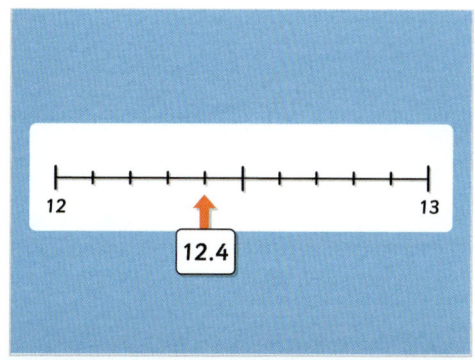

Level 2: Fluency: Round decimals with 2 decimal places to the nearest whole number and to one decimal place.

✸ **Required:** 7/10 ✸ **Pupil Navigation:** on
✸ **Randomised:** off

11. 28.5 rounded to the nearest whole number is _____.
Enter the missing number.

1 2 3 ▪ 29 ▪ 28

12. What is 34.62 rounded to the nearest whole number?

1 2 3 ▪ 34.6 ▪ 35 ▪ 34

Level 2 *continued*

13. What is 5.49 rounded to **1 decimal place**?

- 5.4   ■ **5.5**   ■ 5

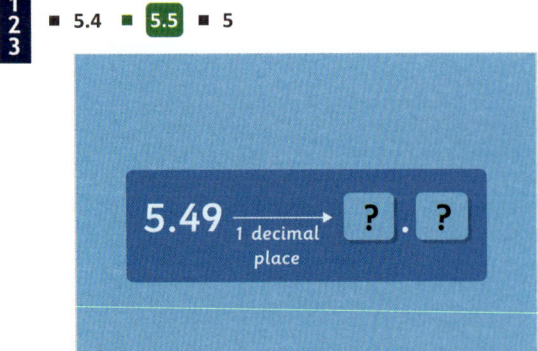

14. What is 1,534.42 rounded to the nearest **tenth**?

- **1534.4**   ■ 1534.5   ■ 1534

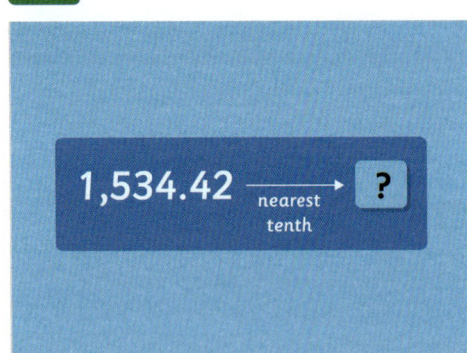

15. When rounded to the nearest whole number, which **four** numbers round to 10?

4/6   ■ **10.12**   ■ 10.5   ■ **9.70**   ■ **10.39**   ■ 9.49   ■ **10.04**

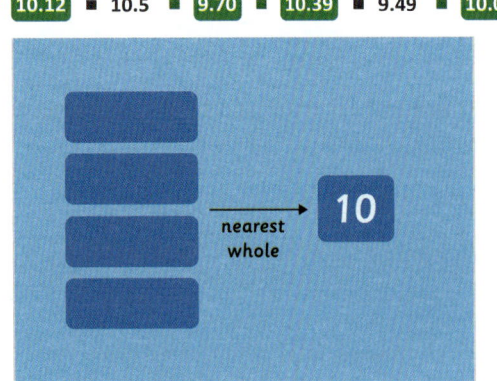

16. A pencil is 8.27 centimetres (cm) long. What is the length of the pencil rounded to the **nearest centimetre**?
Include the units cm (centimetres) in your answer.

- 9 cm   ■ 9 centimetres   ■ **8 centimetres**   ■ **8 cm**
- 8.3 cm   ■ 8   ■ 8.3 centimetres

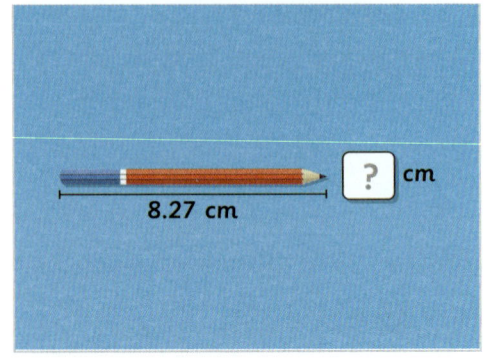

17. 52.78% of Bluewater's population are 60 years old or above. What is this percentage to one decimal place?

- 53   ■ **52.8**   ■ 52.7

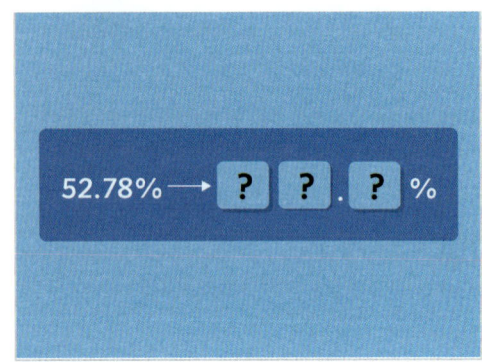

18. Robert scores 87.26% on a test. Rounded to one decimal place, Robert's score is _____%.
Enter the missing number.

- **87.3**   ■ 87.2   ■ 87

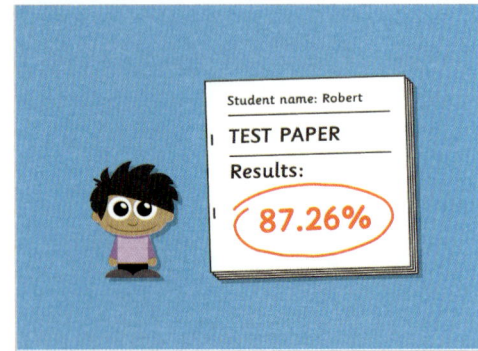

Level 2 *continued*

19. When rounded to the nearest whole number, which **four** numbers round to 3?

4/6

- 3.09 ■ 3.81 ■ 3.23 ■ 2.46 ■ 2.61 ■ 2.57

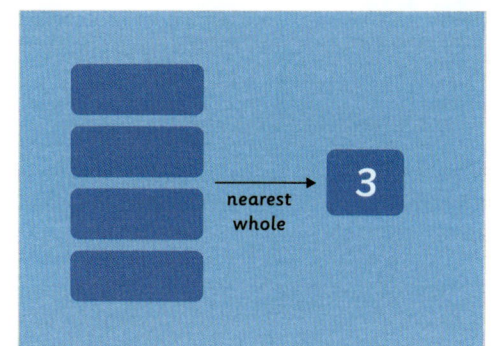

20. What is 99.52 rounded to the nearest whole number?

1
2
3

- 100 ■ 99.5 ■ 99

Level 3: Reasoning: Reason about rounding decimal numbers.

✴ **Required:** 5/5   ✴ **Pupil Navigation:** on
✴ **Randomised:** off

21. Farrah is thinking of a number with 2 decimal places. She rounds it to the nearest tenth and she gets the answer 75.4. What is the **largest** possible number that Farrah could be thinking of?

1
2
3

- 75.4 ■ 75.44 ■ 75.49 ■ 75.45

22. Ryan says, "I'm thinking of a number. When I round it to the nearest tenth, I get the same answer as when I round it to the nearest whole number."
Is this possible? Explain your answer and give at least one example.

a
b
c

- Open question, no set answer

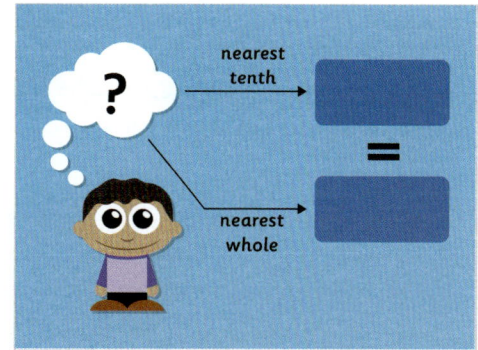

23. Claudia is ordering a new laptop case. She measures the length of the laptop as 24 centimetres (cm) to the nearest centimetre. What is the maximum length the laptop can be to **2 decimal places**?
Include the units cm (centimetres) in your answer.

a
b
c

- 24.49 ■ 24.49 cm ■ 24.49 centimetres ■ 24.4 cm
- 24.4 centimetres

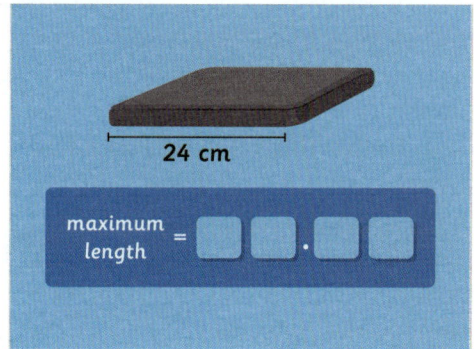

24. A runner sets off on a 1-mile run. After 0.47 miles she stops for a rest. Explain how to find out how much further she has to run to be closer to the finish line than the start line.

a
b
c

- Open question, no set answer

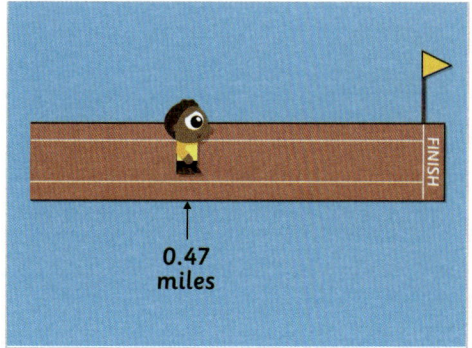

Level 3 *continued*

25. Isla has exactly £10.00 to spend and has put
a several items in her shopping basket. She
b rounds the prices to the nearest pound and
c estimates that they will cost £10.00. When
Isla gets to the checkout, she doesn't have
enough money.

Explain a better way to estimate the prices
so that Isla definitely knows that she has
enough money to buy all of the items.

- Open question, no set answer

Level 4: Problem solving with greater depth: Use
rounding to solve multi-step problems.

❋ **Required:** 5/5   ❋ **Pupil Navigation:** on
❋ **Randomised:** off

26. Jed is thinking of a number with 2 decimal
1 places. Use the following clues to find Jed's
2 number:
3
• The number has a different digit in each
column.
• All of the digits are even.
• The number rounds to 85 to the nearest
whole number.
• None of the digits are 0.

■ **84.62**   ■ 86.42

27. Rounded to the nearest tenth, *x* is 8.1 and *y*
1 is 4.9. What is the **largest possible difference**
2 between *x* and *y*?
3
Give your answer to 2 decimal places.

■ **3.29**   ■ 3.2

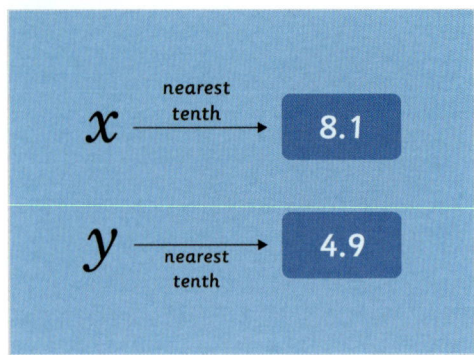

28. Using every digit card, make two different
1 numbers which round to 39 to the nearest
2 whole number. Find the **sum** of the two
3 numbers.

■ 39.48   ■ **78.42**   ■ 38.94

29. Pablo is thinking of a number with 2 decimal
1 places. Use the following clues to find
2 Pablo's number:
3
• The number rounds to 2.9 to the nearest
tenth.
• All of the digits in his number are even.
• The sum of the digits is 18.

■ **2.88**

Level 4 *continued*

30. Kane is buying a new carpet for downstairs in his house. He needs to make sure that he has more than enough carpet, so he rounds each measurement **up** to the next whole number. How many square metres (m²) of carpet does he need to buy in total?
Don't include the units in your answer.

■ 74   ■ 81   ■ 65.69

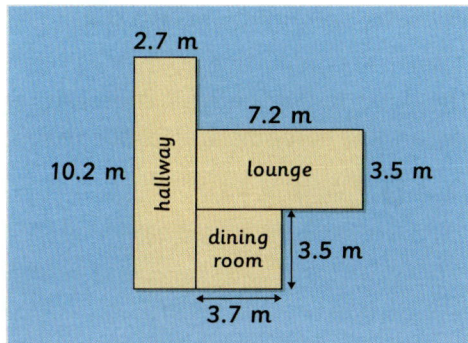

Solve Problems Involving Numbers up to Three Decimal Places

Objective: I can solve problems involving numbers up to 3 decimal places.

Quick Search Ref: 10038

Level 1: Problem solving with greater depth: Solve multi-step problems involving numbers with up to three decimal places.

❄ Required: 10/10   ❄ Pupil Navigation: on   ❄ Randomised: off

1. Using each digit card **once**, make the closest possible number to 5.

▪ **4.951** ▪ 5.149

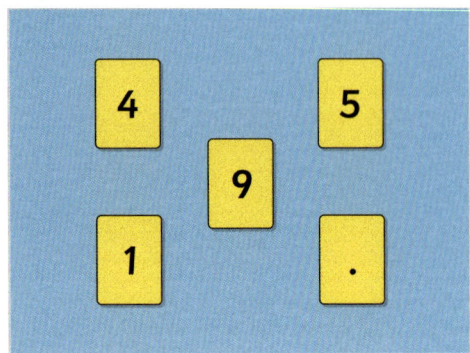

2. Add two side-by-side decimals to find the answer to the box above. What decimal number goes in box A?

▪ **10.604**

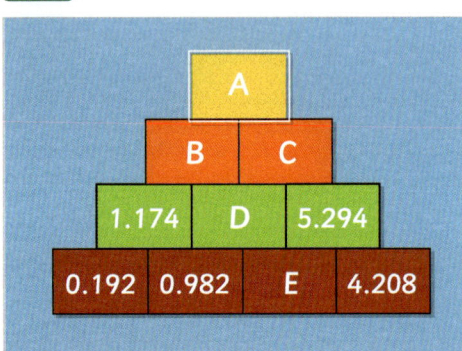

3. Five children take part in a 100 metre race and their times are recorded in a table. If Gary finished 3rd, what is his fastest possible finishing time?
Give your answer to 2 decimal places.

▪ **10.88** ▪ 11.04 ▪ 10.871 ▪ 11.05

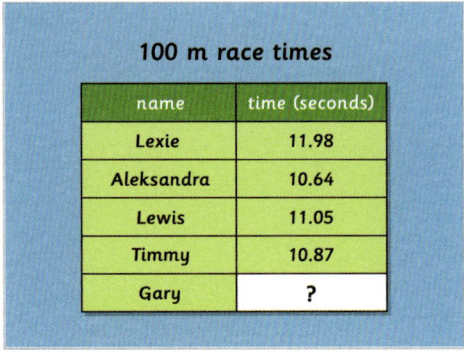

100 m race times

name	time (seconds)
Lexie	11.98
Aleksandra	10.64
Lewis	11.05
Timmy	10.87
Gary	?

4. Libby is thinking of a number. Use the following clues to find how many possible numbers Libby could be thinking of:
• The number is between 4 and 5.
• It has 3 decimal places.
• The digits in the tenths, hundredths and thousandths columns are **consecutive**.

▪ 10 ▪ **8** ▪ 4.123 ▪ 4.012

5. Zoe buys one bottle of milk, one loaf of bread and one block of cheese. How much does she pay in total?
Include the £ sign in your answer.

▪ **£2.20** ▪ 2.2 ▪ £2.2 ▪ 2.20

Level 1 *continued*

6. Jeni has written four decimal numbers containing only the digits 0, 1, 3 and 5. She has represented each number as a letter. Use the following clues to find the value of **A.BCD**:

• **D.AAA** is a whole number.

• **B.DAC** is the number with the greatest value.

• **C.ADB** is the number closest to 1.

■ 0.513

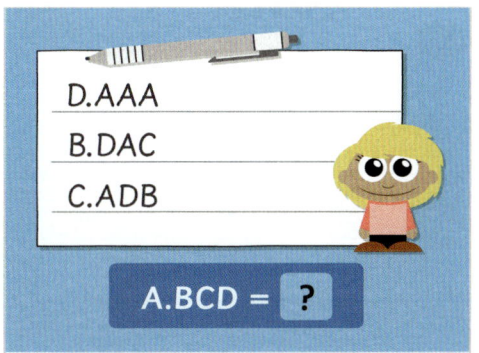

A.BCD = ?

7. Five children take part in a sponsored swim and their distances are recorded in a table. How many **metres** does Jared swim?
Include the units m (metres) in your answer.

■ 1,452 metres ■ 1452 m ■ 1452 metres ■ 5705 m

■ 1,452 m ■ 5,705 metres ■ 1452

■ 1.452 kilometres ■ 1.452 km

Sponsored swim

name	distance (km)
Chloe	2.171
Abdul	1.145
Pedro	0.931
Jared	?
Jack	1.458
total	7.157

8. Two numbers with 2 decimal places are represented with the letters *x* and *y*. When rounded to the nearest tenth, *x* is 6.4 and *y* is 9.9. What is the **smallest possible difference** between *x* and *y*?
Give your answer to 2 decimal places.

■ 3.41

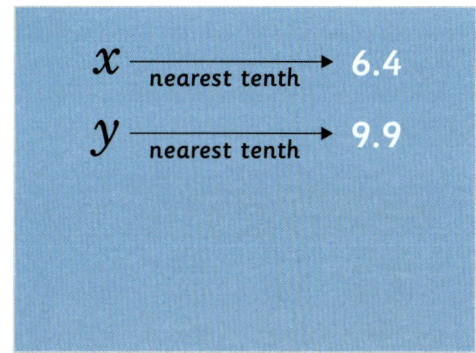

$x \xrightarrow{\text{nearest tenth}} 6.4$

$y \xrightarrow{\text{nearest tenth}} 9.9$

9. The ages of the people at a show are presented in a table. What proportion of the audience are male **and** under 12?
Give your answer as a decimal number.

■ 0.2 ■ 20

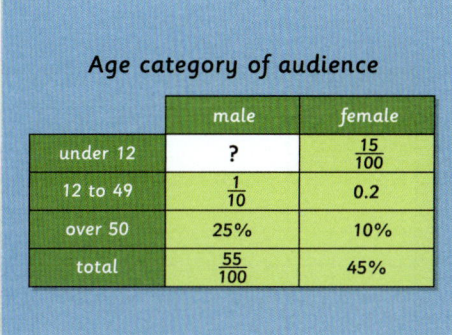

Age category of audience

	male	female
under 12	?	$\frac{15}{100}$
12 to 49	$\frac{1}{10}$	0.2
over 50	25%	10%
total	$\frac{55}{100}$	45%

10. Each type of shape represents a decimal number. What is the value of the triangle?

■ 1.5 ■ 0.17 ■ 0.32 ■ 0.75

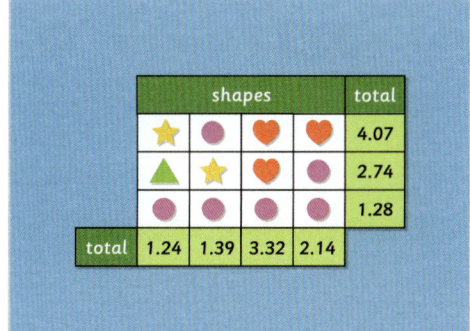

	shapes			total
	★	●	♥	♥	4.07
	▲	★	♥	●	2.74
	●	●	●	●	1.28
total	1.24	1.39	3.32	2.14

Understand Thousandths in Decimal Numbers

Objective: I can recognise and use thousandths and relate them to tenths, hundredths and decimal equivalents.

Quick Search Ref: **10184**

Level 1: Understanding: Recognise place value in decimals and relate decimal place value to fractions. Link thousandths to hundredths and tenths.

✿ **Required:** 7/10   ✿ **Pupil Navigation:** on   ✿ **Randomised:** off

---

1. In the number 59.34, which digit is in the tenths column?

▪ **3** ▪ 4 ▪ 9 ▪ 0.3 ▪ 5

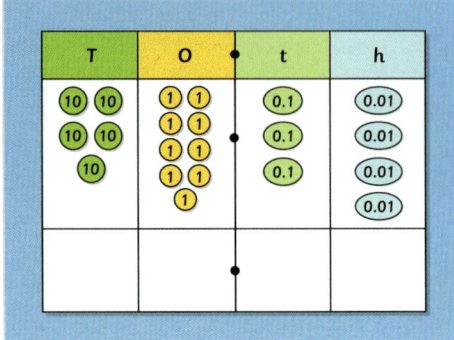

2. What is 7/100 as a decimal?

▪ **0.07** ▪ 7.10 ▪ 0.7 ▪ 7.1

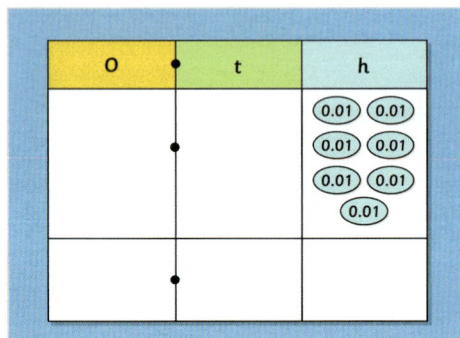

3. What is the value of the 9 in the number 3.96?
Select the two correct answers.

2/5

▪ 9 ones ▪ **9 tenths** ▪ 9 hundredths ▪ **0.9** ▪ 0.09

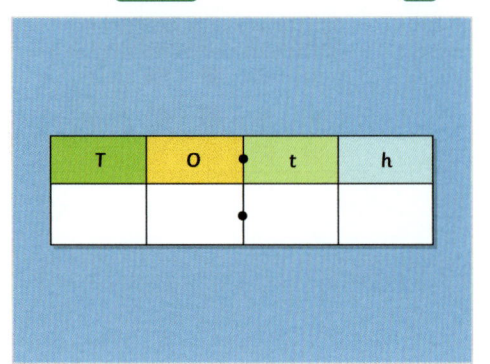

4. What is the value of the 8 in the number 47.18?
Select the two correct answers.

2/5

▪ 8 tenths ▪ **8 hundredths** ▪ 8 thousandths ▪ 0.8
▪ **0.08**

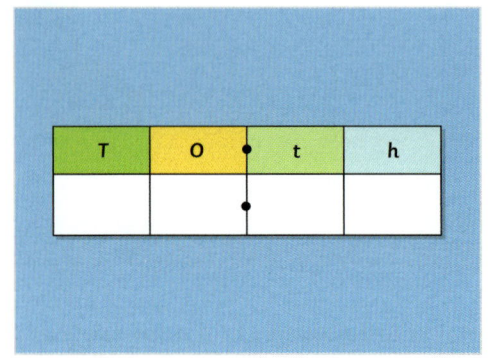

5. In the number 0.145, which digit is in the thousandths column?

▪ **5** ▪ 1 ▪ 4 ▪ 0.005

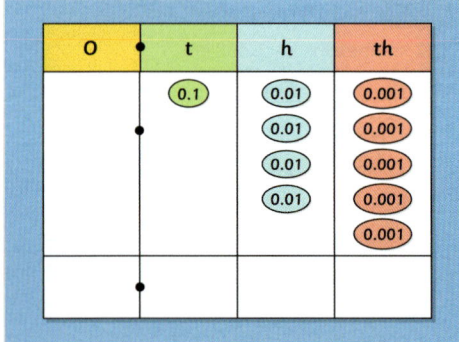

6. Jack partitions the number 0.782 into fractions. In the sum of the fractions, what is the value of the missing numerator?

▪ **782** ▪ 2

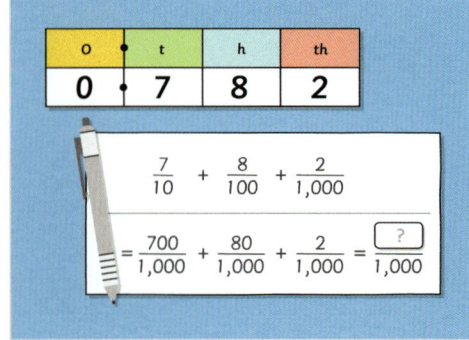

Level 1 *continued*

7. 2 hundredths are the same as ____ thousandths.
Enter the missing number.

- **20** - **0.2**

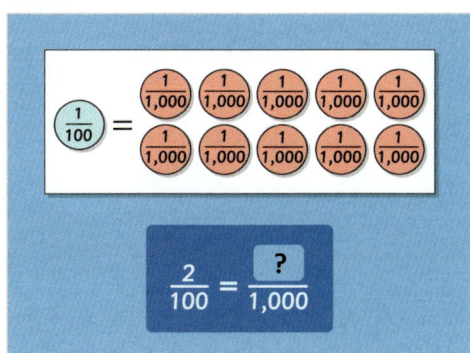

8. Ola partitions the number 0.427. In the sum of the fractions, what is the value of the missing numerator?

- **427** - **7**

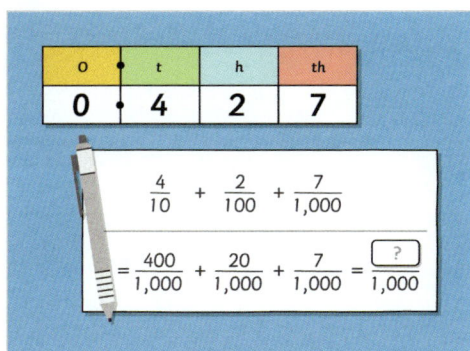

9. In the number 4.863, which digit is in the thousandths column?

- **3** - **0.003** - **8** - **6** - **4**

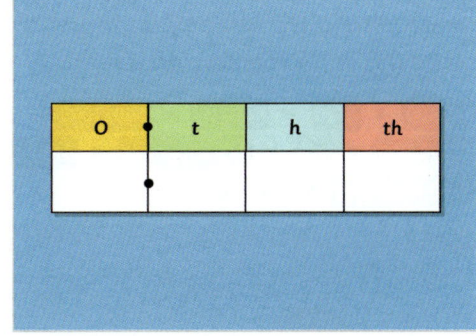

10. What is the value of the 4 in the number 27.142?
Select the two correct answers.

2/5

- 4 tenths - **4 hundredths** - 4 thousandths - 0.4
- **0.04**

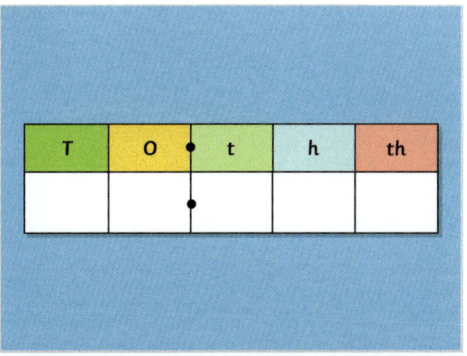

Level 2: Fluency: Identify decimal place value up to thousandths and relate thousandths to tenths and hundredths.

✱ **Required:** 7/10   ✱ **Pupil Navigation:** on
✱ **Randomised:** off

11. What is one thousandth as a fraction and a decimal?
Select the two correct answers.

2/6

- 1/10 - 1/100 - **1/1,000** - 0.1 - 0.01 - **0.001**

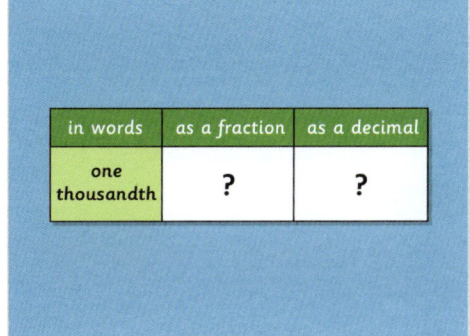

12. What is the value of the 6 in the number 18.256?
Select the three correct answers.

3/7

- **6 thousandths** - 6/100 - **6/1,000** - 6 ones
- 0.6 - **0.006** - 6 hundredths

Level 2 *continued*

13. What is the 5 worth in the number 41.252?
Give your answer as a decimal.

■ 0.05

14. What is the missing decimal number?
64.852 = 64 + 0.8 + 0.05 + _____

■ 0.002   ■ 2

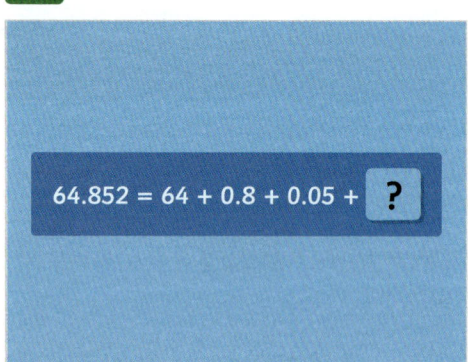

15. Josh partitions a decimal number. What is the missing decimal number?

■ 41.862

16. 6 tenths is equal to _____ thousandths.
Enter the missing number.

■ 600   ■ 6

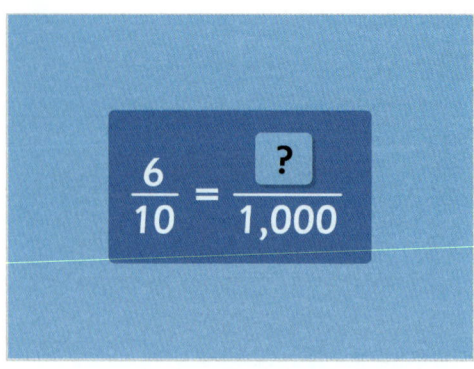

17. A snail travels 9/1,000 m (metres) in one minute. How many **metres** does it travel in one minute?
Include the units m (metres) in your answer.

■ 9/1000   ■ 0.009 metres   ■ 0.009 m   ■ 9/1,000 m
■ 0.009   ■ 9/1,000   ■ 9/1,000 metres   ■ 9/1000 m
■ 9/1000 metres

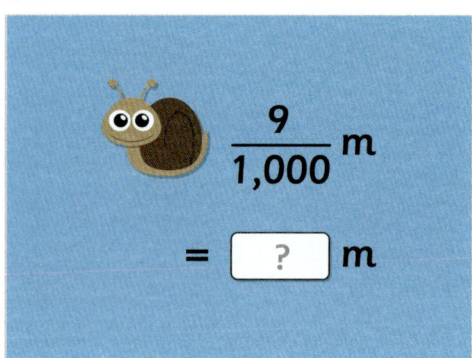

18. A basketball weighs 0.6 kg (kilograms). What does the basketball weigh as a **fraction** of a kilogram?
Put a forward slash (/) between the numerator and the denominator and don't include the units in your answer.

■ 60/100   ■ 6/10   ■ 6/10 kg   ■ 600/1000
■ 600/1,000   ■ 3/5   ■ 3/5 kilograms   ■ 3/5 kg
■ 6/10 kilograms   ■ 6 tenths

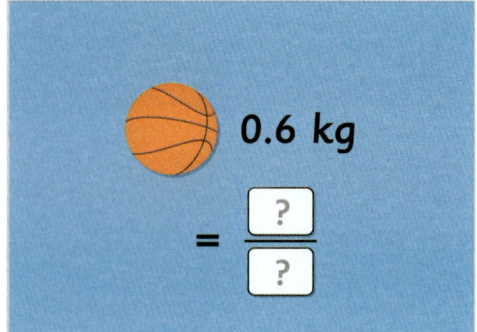

Level 2 *continued*

19. What is the missing decimal number?

1
2
3

$3 + 0.7 + 0.09 +$ _____ $= 3.791$

- ■ 0.001 ■ 1

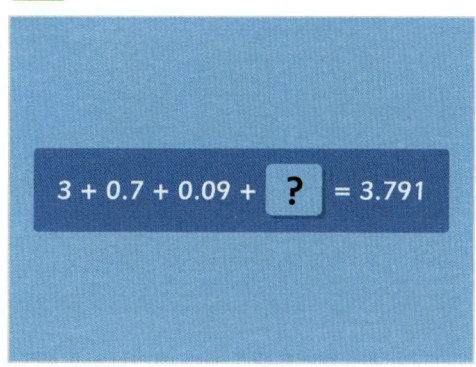

20. What does the 4 represent in 92.481?
Select the three correct answers.

3/7

- ■ 4/10 ■ 4/1,000 ■ 0.4 ■ 0.004 ■ 4 ones
- ■ 4 tenths ■ 4 hundredths

Level 3: Reasoning: Reason about place value up to thousandths, including comparing decimals and fractions.

❋ Required: 5/5 ❋ Pupil Navigation: on
❋ Randomised: off

21. Tamara is trying to solve a maths problem.

a
b
c

Explain how she can find the missing denominator.

- Open question, no set answer

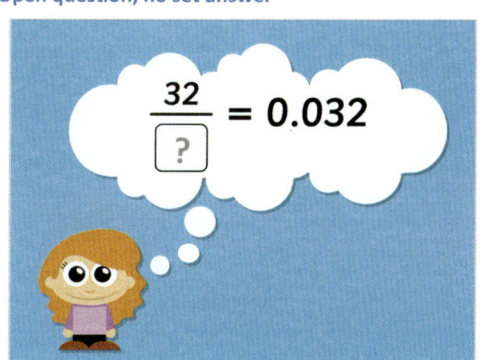

22. Which symbol makes the statement true?

1/3

- ■ < ■ = ■ >

23. Using only four of the digits on the cards, what is the second smallest possible number you can make?

1
2
3

- ■ 0.148 ■ 0.146 ■ 9.864

24. Milly says, "4.009 is larger than 4.1 because 9 is larger than 1." Do you agree with Milly?

a
b
c

Explain your answer.

- Open question, no set answer

4.009 > 4.1

Level 3 *continued*

25. What do you need to add to 5.34 to get 5.7?
Give your answer as a decimal number.

■ **0.36**　■ **11.04**

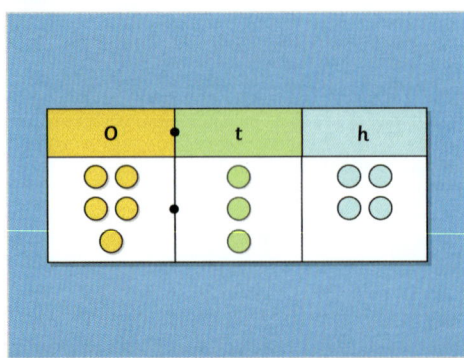

Level 4: Problem solving with greater depth: Use knowledge of place value in decimals and their equivalent fractions to solve multi-step problems.

❋ **Required:** 5/5　❋ **Pupil Navigation:** on
❋ **Randomised:** off

26. Jemma has used some symbols in place of decimal numbers. The symbols are worth the following as fractions:
• Circle = 4/10
• Star = 3/100
• Triangle = 2/1,000
What decimal number is represented in the image?

■ **0.862**　■ **862/1000**

27. Using every digit card once, make the closest possible number to 65.

■ **64.951**　■ **65.149**

28. Timmy is thinking of a decimal number with three decimal places. Use the following clues to find Timmy's number:
• The digit in the thousandths column is equal to 2^2.
• The hundredths digit is a prime number between 6 and 10.
• The ones digit is the smallest possible odd digit.
• The digit in the tenths column is a quarter of 12.

■ **1.374**　■ **1.394**

Level 4 *continued*

29. The table shows how many seconds it takes athletes from different countries to run 100 metres (m). Use the following clues to complete the table and find the time taken by the **Canadian** athlete:

• USA is 4/100 of a second slower than Jamaica.

• France is 0.1 seconds slower than the USA.

• The USA time is exactly halfway between the Canada and France times.

• The Canada athlete has the fastest time.

■ 9.75 ■ 9.95

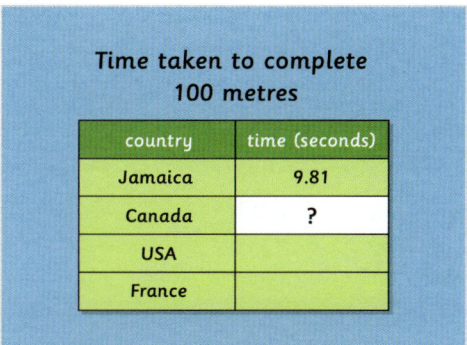

Time taken to complete 100 metres

country	time (seconds)
Jamaica	9.81
Canada	?
USA	
France	

30. Pablo has written down four numbers containing only the digits 0, 1, 3 and 5. He has represented each digit as a letter. Use the clues to find what number is represented by **A.BCD**:

• **D.AAA** is a whole number.

• **B.DAC** is the number with the greatest value.

• **C.ADB** is the number closest to 1.

■ 0.513

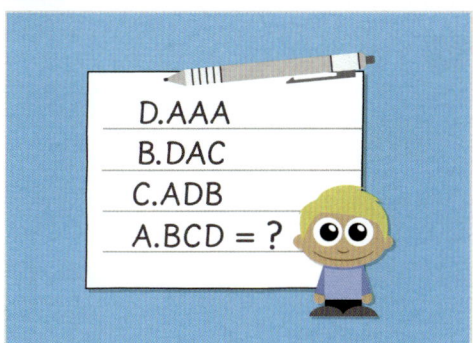

D.AAA
B.DAC
C.ADB
A.BCD = ?

Decimals and Percentages Topic Review

Objective:	I can answer questions involving decimals and percentages from the Year 5 curriculum.

Quick Search Ref:	**10727**

Level 1: Understanding

❋ **Required:** 7/10   ❋ **Pupil Navigation:** off   ❋ **Randomised:** off

---

1. What is the value of the 3 in the number 57.123?
Give your answer as a decimal number.

■ **0.003**　■ 0.03　■ 0.3

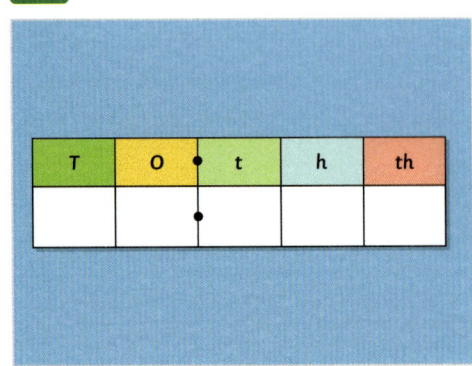

2. Round 94.61 to the nearest whole number.

■ **95**　■ 94　■ 94.6

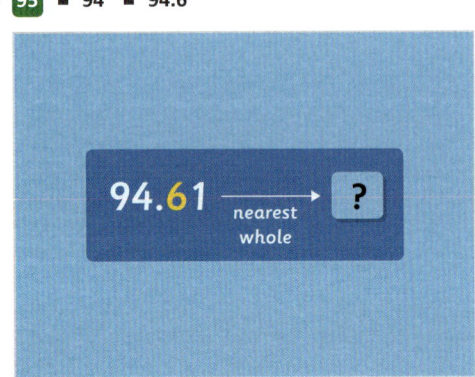

3. What is 1.2 divided by 100?

■ **0.012**　■ 0.12　■ 120

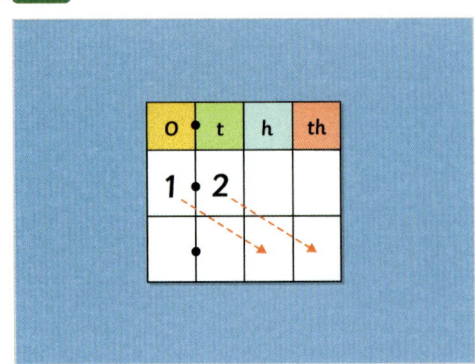

4. What is 0.1 as a fraction?

■ **1/10**　■ 1/100　■ 1/1,000

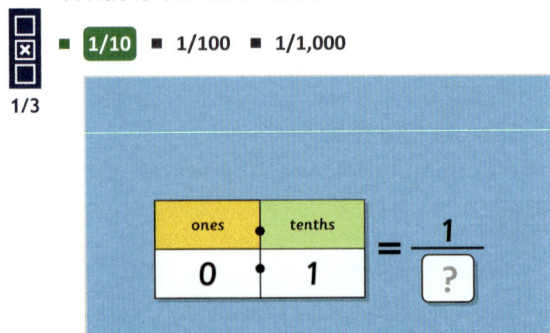

5. Subtract 6.2 from 9.5.

■ **3.3**　■ 33

6. Ola partitions the number 0.427. In the sum of the fractions, what is the value of the missing numerator?

■ **427**　■ 7

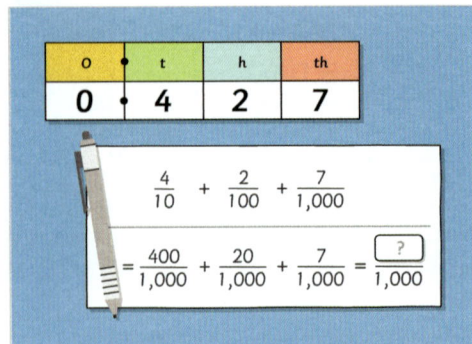

Level 1 *continued*

7. What percentage of the rectangle is **not** shaded?

a b c

- ■ 75 ■ 75% ■ 25% ■ 3/4 ■ 3% ■ 3 ■ 25

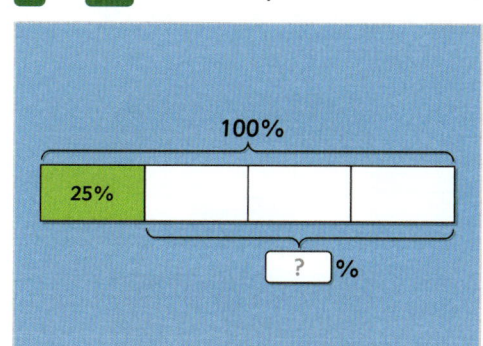

8. What percentage of the hundred square is shaded?

a b c

- ■ 46% ■ 46 ■ 54 ■ 46/100 ■ 54%

9. What is 12.4 rounded to the nearest whole number?

1 2 3

- ■ 12 ■ 11 ■ 13

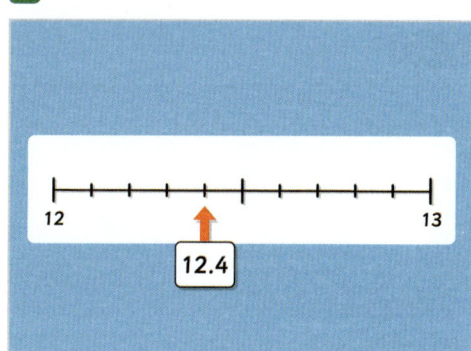

10. In the number 4.863, which digit is in the thousandths column?

1 2 3

- ■ 3 ■ 8 ■ 4 ■ 0.003 ■ 6

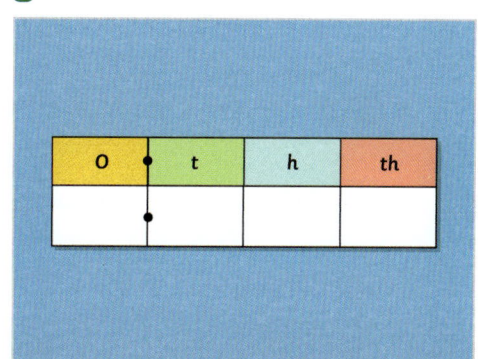

Level 2: Fluency

✱ **Required:** 7/10   ✱ **Pupil Navigation:** off
✱ **Randomised:** off

11. What is 7.3 divided by 100?

1 2 3

- ■ 0.073

$$7.3 \div 100 = \boxed{?}$$

12. Sort the numbers into ascending order (smallest first).

↑ ↓

- ■ 52.515 ■ 52.731 ■ 52.736 ■ 54.761

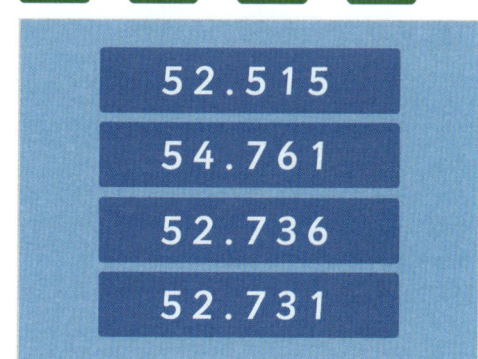

52.515

54.761

52.736

52.731

Level 2 *continued*

13. What is 0.08 as a fraction?

 Put a forward slash (/) between the numerator and denominator. For example, one-half is 1/2.

- 8/10 - **8/100** - 8/1,000 - 08/100 - 8/1000

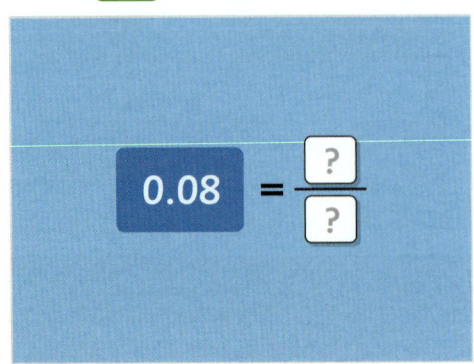

$$0.08 = \frac{?}{?}$$

14. What is the missing decimal number?

123 3 + 0.7 + 0.09 + _____ = 3.791

- **0.001** - 1

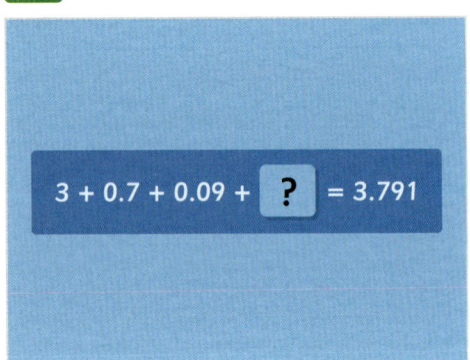

3 + 0.7 + 0.09 + **?** = 3.791

15. Robert scores 87.26% on a test. Rounded to one decimal place, Robert's score is _____%.
123 *Enter the missing number.*

- **87.3** - 87 - 87.2

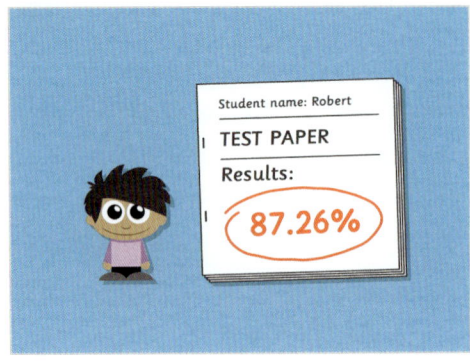

Student name: Robert

TEST PAPER

Results:

87.26%

16. Dale uses 100 tiles in his bathroom. The plan
abc shows the numbers of blue and white tiles that he uses. What percentage of the tiles are blue?

- **60%** - **60** - 40% - 40

17. Chelsea adds 25.6 millilitres of cordial to
123 134.8 millilitres of water. What volume of liquid does Chelsea have altogether?
Don't include the units in your answer.

- **160.4** - 1604 - 150.4 - 159.4

18. What is 521.854 × 100?
123

- **52185.4**

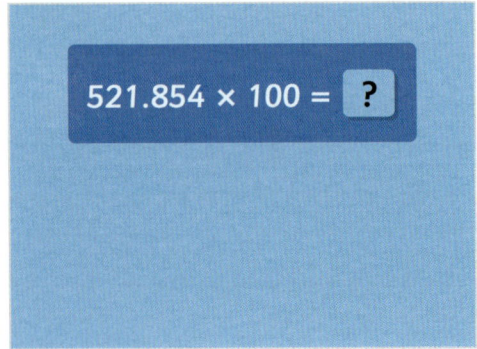

521.854 × 100 = **?**

Level 2 *continued*

19. Which is the largest number?

 ▪ 5.43

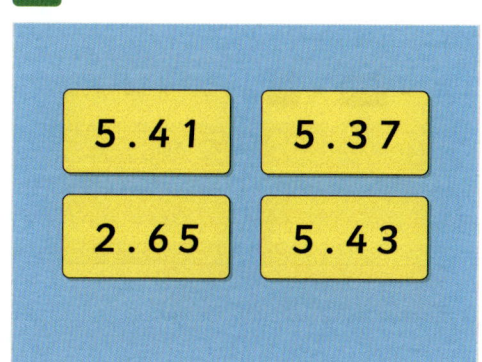

5.41   5.37
2.65   5.43

20. What is 0.193 as a fraction?

 ▪ 193/10 ▪ 193/1,000 ▪ 193/100

1/3

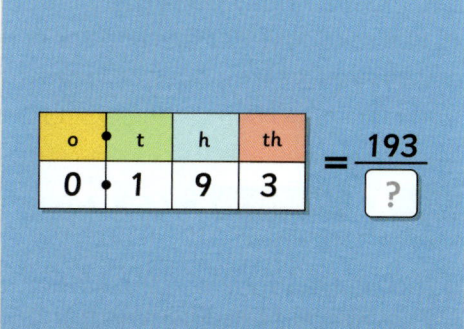

o	t	h	th
0	1	9	3

= 193/?

Level 3: Reasoning

✦ **Required:** 5/5   ✦ **Pupil Navigation:** off
✦ **Randomised:** off

21. Tobey says, "To multiply a number by 100, I can just add two zeros on the end of the number." Is Tobey correct? Explain your answer.

- Open question, no set answer

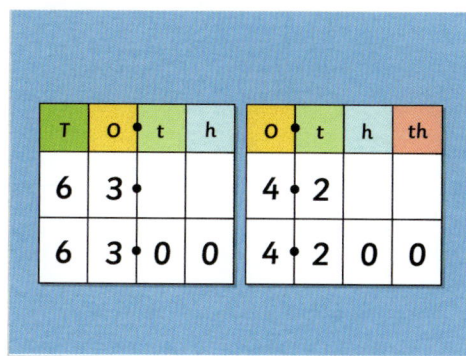

T	O	t	h
6	3		
6	3	0	0

O	t	h	th
4	2		
4	2	0	0

22. Claudia is ordering a new laptop case. She measures the length of the laptop as 24 centimetres (cm) to the nearest centimetre. What is the maximum length the laptop can be to **2 decimal places**?
Include the units cm (centimetres) in your answer.

▪ 24.49 cm ▪ 24.49 centimetres ▪ 24.4 centimetres
▪ 24.49 ▪ 24.4 cm

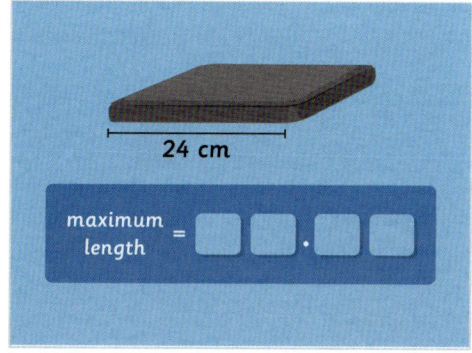

24 cm

maximum length = ☐☐.☐☐

23. Milly says, "4.009 is larger than 4.1 because 9 is larger than 1." Do you agree with Milly? Explain your answer.

- Open question, no set answer

4.009 > 4.1

24. Which symbol makes the following statement true?

1/4 ____ 0.23

1/3 ▪ < ▪ = ▪ >

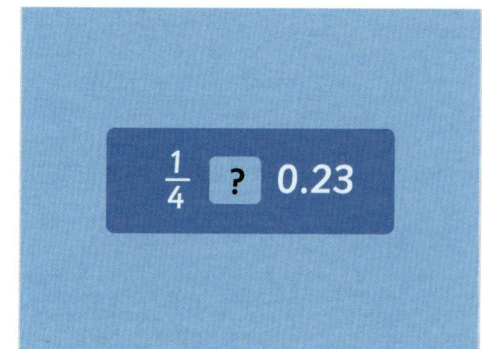

¼ ? 0.23

25. Josh says, *"Three parts of the shape are shaded. This means that 3% of the whole shape is shaded."*

Explain why Josh is incorrect and find the correct percentage of the shape that is shaded.

- Open question, no set answer

Level 4: Problem solving with greater depth

✱ **Required:** 5/5   ✱ **Pupil Navigation:** off
✱ **Randomised:** off

26. Use the following clues to find the missing decimal number:
- The number has 3 digits.
- The digits have a sum of 13.
- All of the digits are different.

■ 6.25   ■ 6.52   ■ 6.16

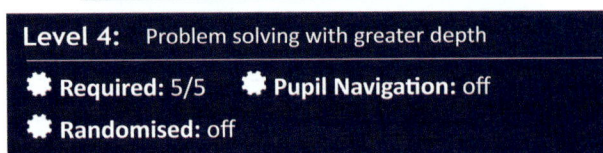

$$6.15 < \boxed{} \cdot \boxed{} \boxed{} > 6.27$$

27. Will buys a pack of 100 sheets of paper. He uses half of the pack to print a school project and 2/10 of the pack to make paper aeroplanes. What **percentage of the pack** does he have left?

■ 3/10   ■ 30%   ■ 30   ■ 70%   ■ 50   ■ 50%   ■ 70

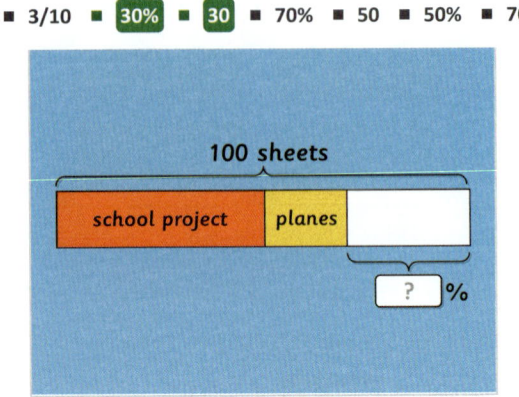

28. At an athletics track, Henry completes 10 laps of track 2 and 100 laps of track 4. How many **kilometres** (km) does he run in total? *Include the units km (kilometres) in your answer.*

■ 7.7 km   ■ 7.7 kilometres   ■ 7.7   ■ 7,700 m
■ 7,700   ■ 7,700 metres

Length of running tracks

track	distance (metres)
track 1	400
track 2	352
track 3	75.4
track 4	41.8

Level 4 *continued*

29. Dale is thinking of a decimal number.

**a
b
c**

The ones digit is less than 1.
The tenths digit is an even number between 0 and 5, but it is **not a prime number**.
The hundredths digit is double the tenths digit.
The thousandths digit is a square number between 5 and 10.

He converts his decimal number to a fraction. What is Dale's **fraction**?
Put a forward slash (/) between the numerator and denominator. For example, one-half is 1/2.

- 489/1000   - 489/1,000   - 0.489   - 249/1,000
- 249/1000

30. Each type of shape represents a decimal number. What is the value of the triangle?

**a
b
c**

- 0.17   - 0.75   - 1.5   - 0.32

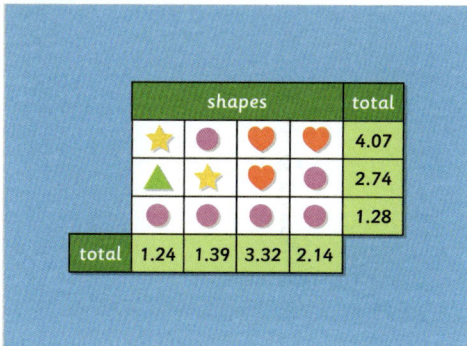

shapes				total
⭐	🟣	🧡	🧡	4.07
🔺	⭐	🧡	🟣	2.74
🟣	🟣	🟣	🟣	1.28
total	1.24	1.39	3.32	2.14

Mathematics Y5

Properties of Number

Multiples and Factors
Prime Numbers
Squares and Cubes

Identify Factors

Objective: I can identify factors, including finding all factor pairs of a number and common factors of two numbers.

Quick Search Ref: **11667**

Level 1: Understanding: Identify factors, factor pairs and common factors with support.

✿ **Required:** 7/10   ✿ **Pupil Navigation:** on   ✿ **Randomised:** off

1. A factor of 6 is a whole number that...

1/3
- ■ is the product of 6 multiplied by itself.
- ■ divides exactly into 6 without leaving a remainder.
- ■ can be divided by 6 without leaving a remainder.

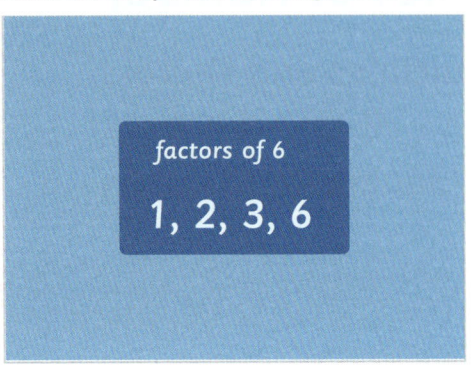

2. Six counters can be arranged in two different ways to show the factors of 6. Which number is one of the factors of 6?

1/4   ■ **2**   ■ 4   ■ 12   ■ 36

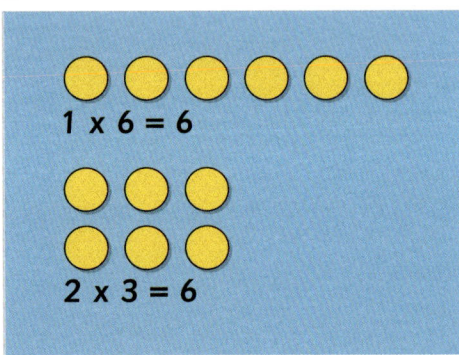

3. 1 and 12 are a factor pair of 12 because 1 × 12 = 12. Which **two** of these numbers are another factor pair of 12?

2/5   ■ **2**   ■ 4   ■ **6**   ■ 7   ■ 9

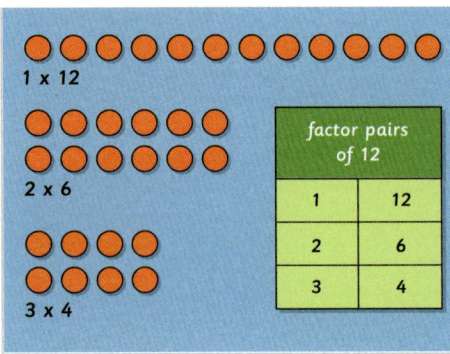

4. Which **two pairs** of numbers are factor pairs of 20?

1/4
- ■ **1 and 20**   ■ 2 and 5   ■ **4 and 5**   ■ 4 and 10

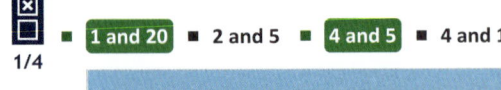
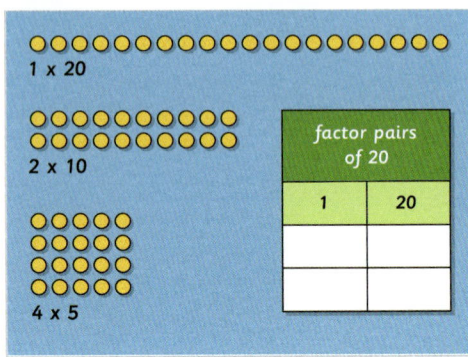

5. Select **all three** factor pairs of 18.

3/6
- ■ **1 and 18**   ■ 2 and 6   ■ 3 and 3   ■ **3 and 6**
- ■ 4 and 6   ■ **2 and 9**

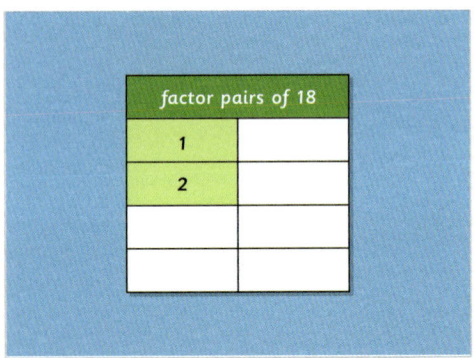

6. 1 is a **common factor** of 15 and 21 because it is a factor of both 15 and 21. Which **other** number is a common factor of 15 and 21?

- ■ **3**   ■ 1   ■ 7   ■ 5

Level 1 *continued*

7. Give one of the **common factors** of 24 and 42.

■ **2** ■ **3** ■ **6** ■ **1**

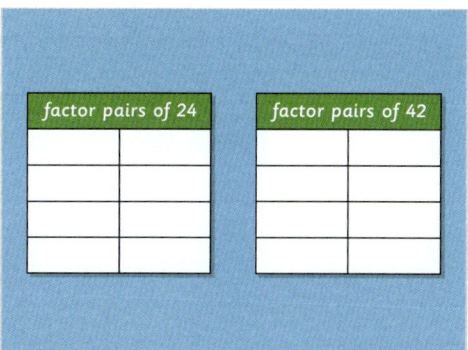

10. Eight counters can be arranged in two different ways to show the factors of 8. Which number is one of the factors of 8?

1/4 ■ 3 ■ **4** ■ 16 ■ 64

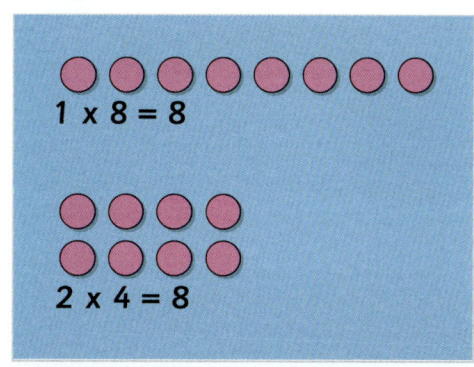

$1 \times 8 = 8$

$2 \times 4 = 8$

Level 2: Fluency: Identify factors, factor pairs and common factors. Solve one-step problems involving factors.

❋ **Required:** 7/10   ❋ **Pupil Navigation:** on
❋ **Randomised:** off

8. 5 is a factor of both 10 and 15. Which **other** number is a common factor of 10 and 15?

■ 5 ■ **1** ■ 3 ■ 2

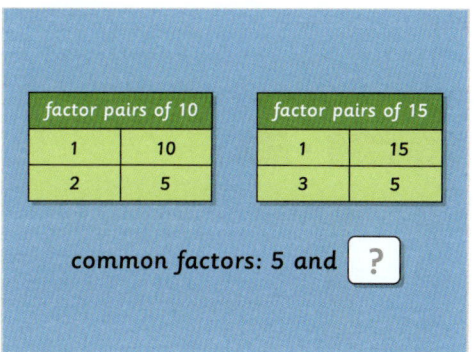

common factors: 5 and [?]

11. Select the **four** factors of 21.

4/6 ■ **1** ■ 2 ■ **3** ■ **7** ■ 11 ■ **21**

9. Select **all three** factor pairs of 32.

3/5 ■ **1 and 32** ■ **2 and 16** ■ 3 and 12 ■ **4 and 8**
■ 6 and 7

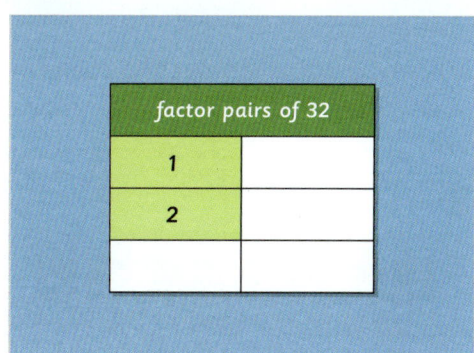

12. Which factor of 18 is missing from the table?

■ **6**

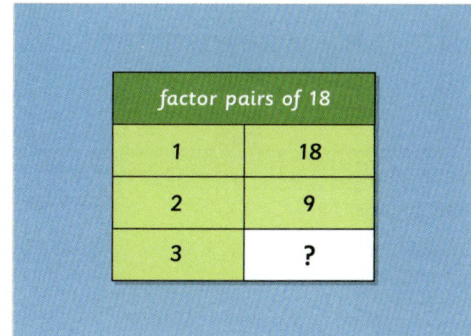

Level 2 *continued*

13. How many factors does 40 have?

 ▪ **8** ▪ 4

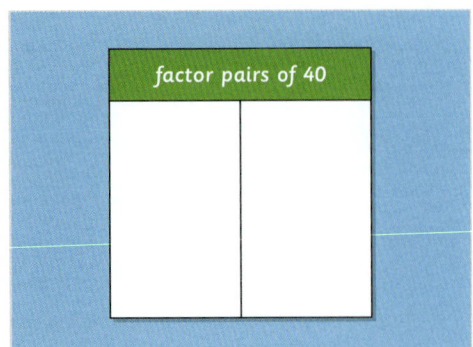

factor pairs of 40

14. Which number (other than 1) is a common factor of 14 and 35?

▪ 1 ▪ **7** ▪ 2 ▪ 5

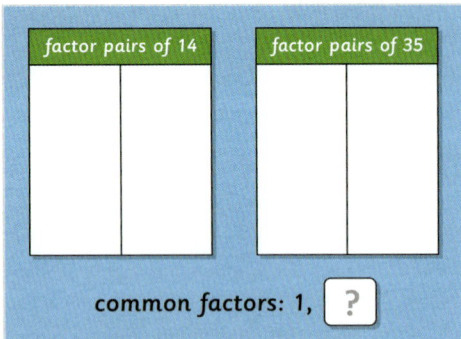

factor pairs of 14

factor pairs of 35

common factors: 1, ?

15. What is the highest number that is a common factor of 6 and 15?

▪ **3** ▪ 1 ▪ 2 ▪ 5

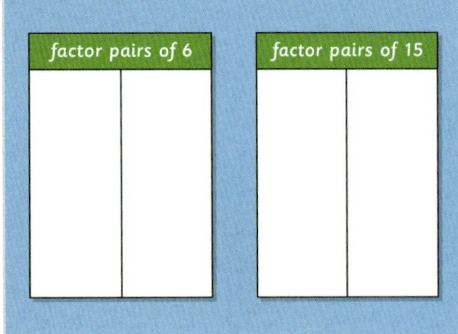

factor pairs of 6

factor pairs of 15

16. 15 sweets are shared equally between a group of children. If each child gets 3 sweets and there are none left over, how many children are there?

▪ **5**

15 sweets

17. Matthew builds a tower using nine identical bricks. If the tower is 63 centimetres (cm) tall, what is the height of each brick in the tower?
Include the units cm (centimetres) in your answer.

▪ **7 centimetres** ▪ **7 cm** ▪ 7

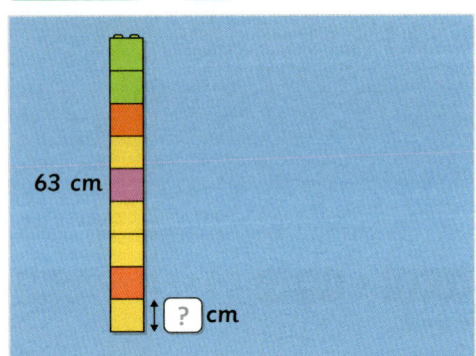

63 cm

? cm

18. A school caretaker arranges 56 chairs in rows of eight. How many rows of chairs does he make?

▪ **7**

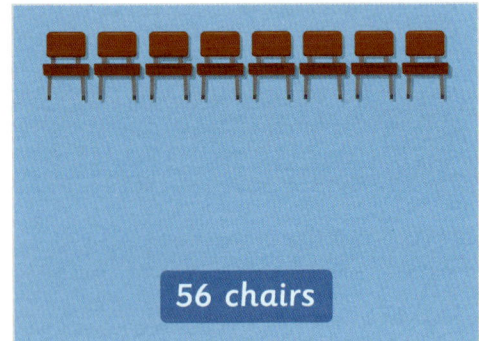

56 chairs

Level 2 *continued*

19. Which number (other than 1) is a common factor of 22 and 32?

[1][2][3]

■ 4 ■ 1 ■ **2** ■ 11 ■ 8 ■ 16

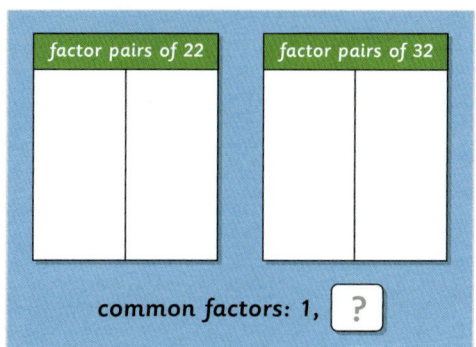

common factors: 1, **?**

20. Which factor of 28 is missing from the table?

[1][2][3]

■ **14**

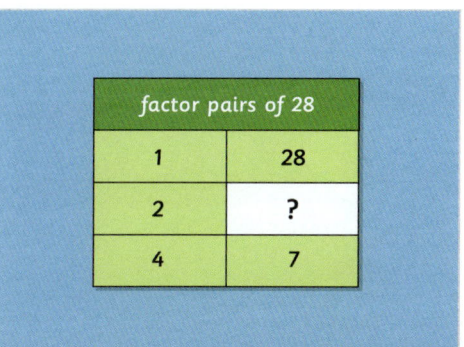

Level 3: Reasoning: Reason about factors and common factors.

✸ **Required:** 5/5   ✸ **Pupil Navigation:** on
✸ **Randomised:** off

21. Which number is a common factor of **all** numbers?

[1][2][3]

■ **1**

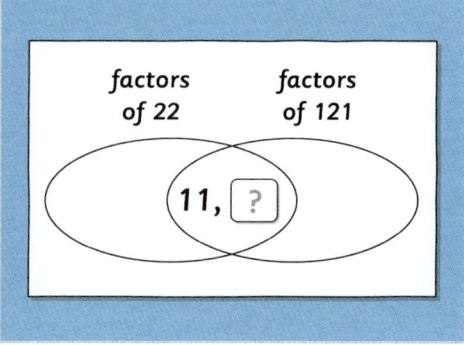

22. Explain whether a number **always**, **sometimes** or **never** has an even number of factors.

[a][b][c]

- Open question, no set answer

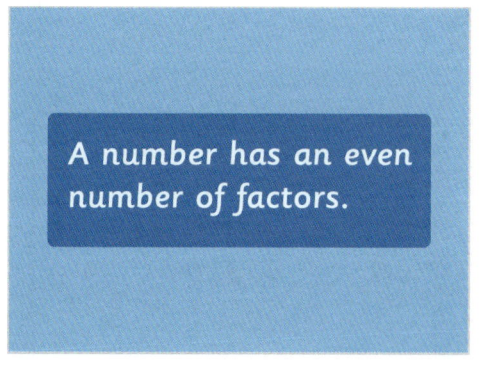

A number has an even number of factors.

23. Nathan says that 6 is a factor of 24, so 2 is also a factor of 24.
Zack says that 6 is a factor of 24, so 3 is also a factor of 24.
Explain who is correct.

[a][b][c]

- Open question, no set answer

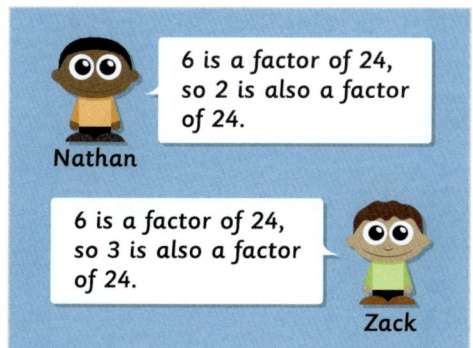

6 is a factor of 24, so 2 is also a factor of 24.

Nathan

6 is a factor of 24, so 3 is also a factor of 24.

Zack

24. Anya has made a table to find all the factors of 64. Explain whether Anya has found all the factors of 64 and how you know.

[a][b][c]

- Open question, no set answer

factor pairs of 64	
1	64
2	32
3	✕
4	16
5	✕
6	✕
7	✕
8	8

Level 3 *continued*

25. The Venn diagram shows the factors of 9 and 12. Which number is in the wrong part of the diagram?

■ 6 ■ 1

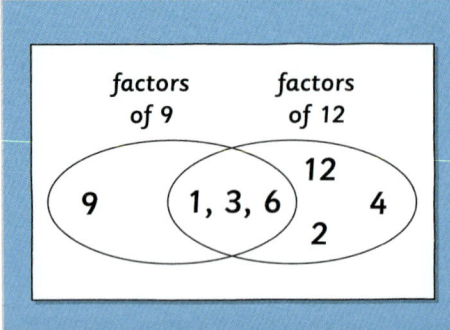

Level 4: Problem solving with greater depth: Solve multi-step problems involving factors.

❋ **Required:** 5/5 ❋ **Pupil Navigation:** on
❋ **Randomised:** off

26. What is the total value of all the factors of 63 added together?

■ 101

27. What is the highest number that is a common factor of 72 and 84?

■ 12 ■ 2 ■ 6 ■ 1 ■ 3 ■ 4

28. The Venn diagram shows the factors of 33 and the factors of which other number?

■ 7 ■ 21 ■ 1

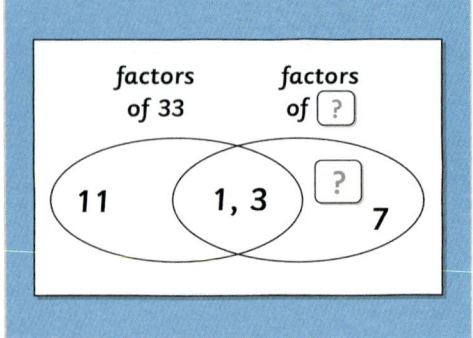

29. Jamie is thinking of a number. His number is four less than a factor of 28 and one more than a factor of 10. What is the product of Jamie's's number multiplied by itself?

■ 9

30. Suzy and her father have ages with a common factor of 6. Suzy's father is older than 40 and younger than 70. If Suzy is one-third of the age of her father, how old is Suzy?

■ 18 ■ 54

Identify Multiples

Objective:	I can identify multiples of whole numbers.
Quick Search Ref:	10216

Level 1: Understanding: Identify multiples and numbers that are multiples of more than one number with support.

✿ Required: 7/10　　　✿ Pupil Navigation: on　　　✿ Randomised: off

1. A multiple of 6 is a whole number that . . .

- divides exactly into 6 without a remainder.
- is the product of 6 multiplied by itself.

1/3
- can be divided by 6 without leaving a remainder.

multiples of 6

6, 12, 18, 24, 30, 36

2. You can use number patterns to find some multiples. Which **three** number patterns are true?

3/5
- All multiples of 2 are even numbers.
- All multiples of 3 are odd numbers.
- All multiples of 4 are even numbers that halve to give an even number.
- All multiples of 5 end in 0.
- All multiples of 9 have digits that add to a total of 9.

numbers	multiples
2	2, 4, 6, ...
3	3, 6, 9, ...
4	4, 8, 12, ...
5	5, 10, 15, ...
9	9, 18, 27, ...

3. Which number is **not** a multiple of 4?

- 4　- 24　- 42

1/3

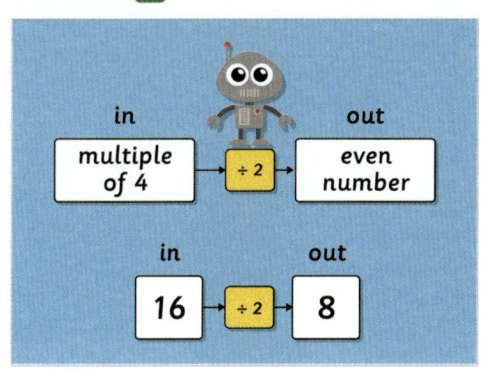

4. Which of the following numbers is a multiple of 8?

- 28　- 38　- 44　- 56

1/4

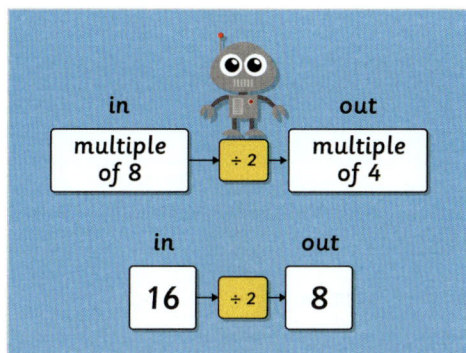

5. Which of these numbers is a multiple of 7?

- 17　- 24　- 35　- 47

1/4

multiples of 7
7
14

6. Which number is a multiple of both 3 and 5?

- 6　- 10　- 9　- 15　- 25

1/5

multiples of	
3	5
3	5
6	10

Level 1 *continued*

7. What is the **lowest number** that is a multiple of both 4 and 6?

■ 12 ■ 24

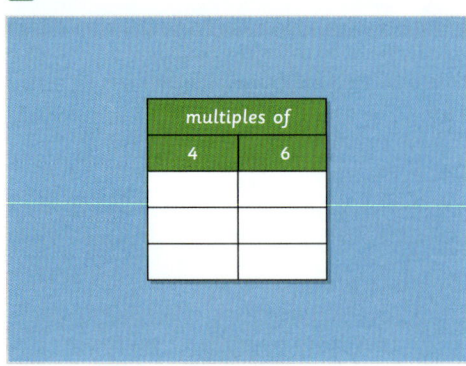

multiples of	
4	6

8. Which of the following numbers is a multiple of both 6 and 9?

■ 6 ■ 9 ■ 12 ■ 18 ■ 27

1/5

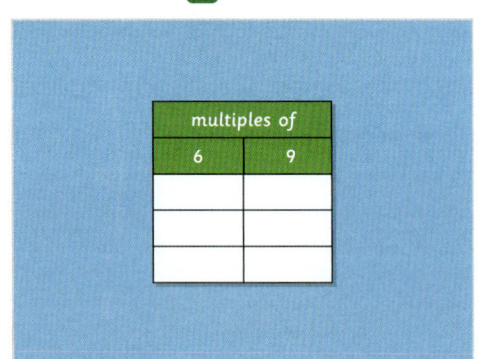

multiples of	
6	9

9. Which of the following numbers is a multiple of 6?

■ 9 ■ 15 ■ 16 ■ 24 ■ 26

1/5

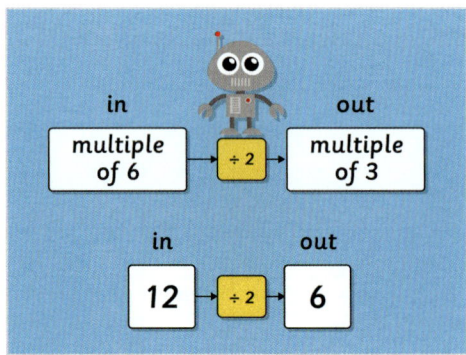

10. Which number is **not** a multiple of 9?

■ 18 ■ 45 ■ 59 ■ 72 ■ 108

1/5

Digits in a multiple of 9 have a total of 9.

numbers	digit total
9	9
18	1 + 8 = 9
27	2 + 7 = 9

Level 2: Fluency: Identify multiples and numbers that are multiples of more than one number. Identify the highest multiple within a range. Solve one-step problems involving multiples.

✳ **Required:** 7/10   ✳ **Pupil Navigation:** on
✳ **Randomised:** off

11. Select **three** multiples of 4.

■ 12 ■ 14 ■ 16 ■ 26 ■ 34 ■ 48

3/6

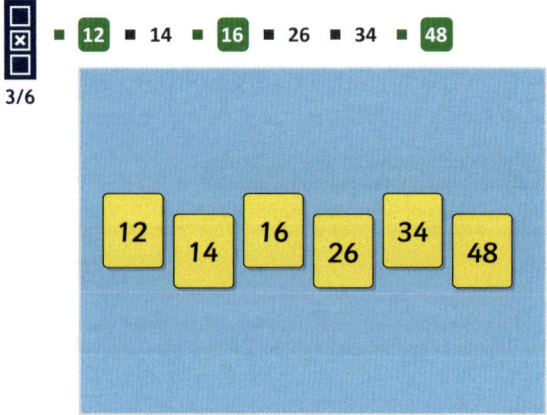

12. Which **three** numbers are multiples of 9?

■ 59 ■ 63 ■ 109 ■ 81 ■ 126

3/5

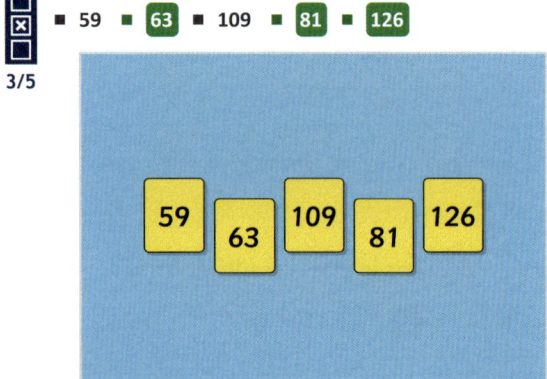

Level 2 *continued*

13. Which **three** numbers are multiples of 3?

3/5

■ 109 ■ **135** ■ 149 ■ **180** ■ **330**

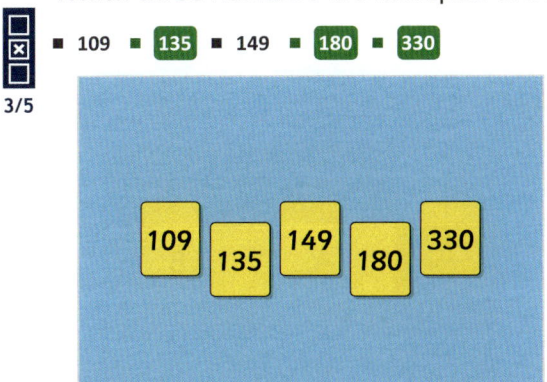

14. What is the highest multiple of 3 that is less than 50?

■ 50 ■ **48** ■ 51 ■ 49

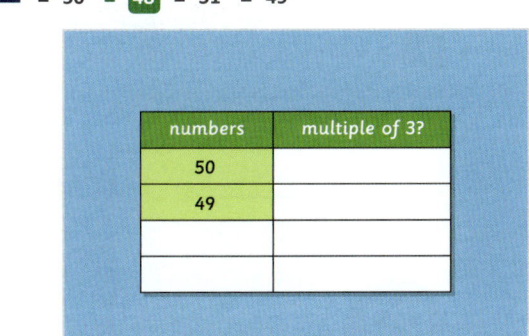

numbers	multiple of 3?
50	
49	

15. What is the lowest number that is a multiple of both 6 and 8?

■ **24** ■ 48

16. Davey saves £2 coins until he has enough to change into a note. A £____ note is the lowest value note he can exchange for his coins.
Enter the missing number.

■ **10** ■ 5 ■ 20

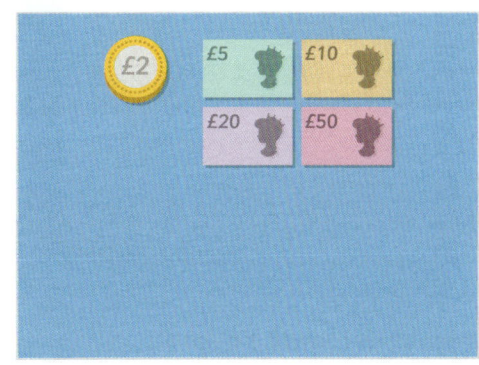

17. Sophie has 120 beads. If she makes a necklace using groups of nine beads, what is the largest number of beads Sophie can use?

■ **117** ■ 120 ■ 108

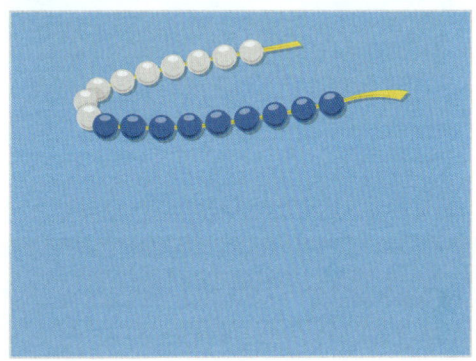

18. The maximum number of people allowed in a room for a concert is 50. If Mr Charlton arranges chairs in rows of 8, what is the largest number of people that can be seated?

■ **48** ■ 50

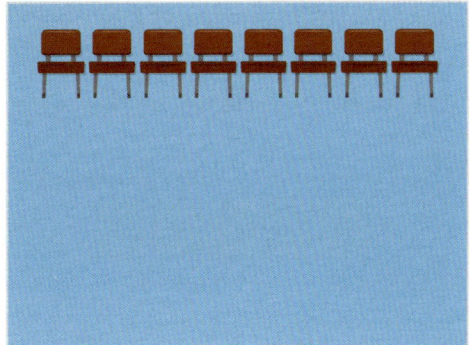

Level 2 continued

19. What is the lowest number that is a multiple of both 4 and 5?

 ■ 20

20. Select all four multiples of 6.

■ 12 ■ 16 ■ 18 ■ 32 ■ 36 ■ 48

3/6

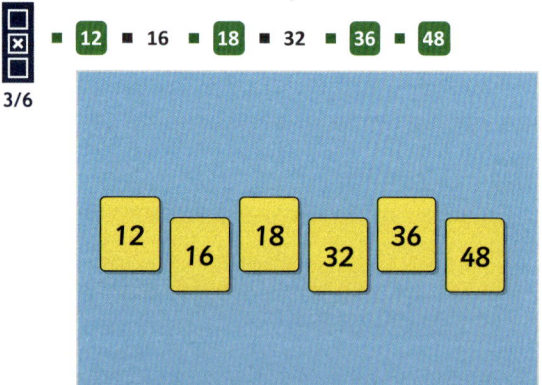

Level 3: Reasoning: Reason about multiples.

❋ **Required:** 5/5   ❋ **Pupil Navigation:** on
❋ **Randomised:** off

21. Noah is sorting multiples of 4 and 5 on the Venn diagram. Which number should he put in **section B**?

1/5 ■ 55 ■ 56 ■ 60 ■ 64 ■ 65

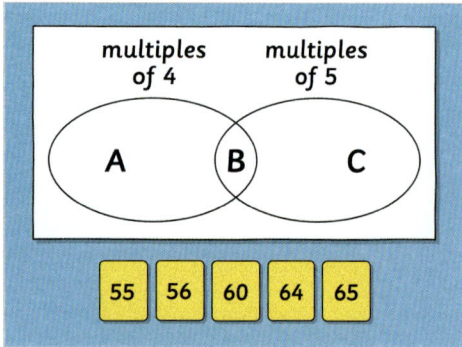

22. William says, "In my diagram, there will never be a number in box B." Is William correct? Explain your answer.

- Open question, no set answer

	multiples of 50	not a multiple of 50
multiple of 100	A	B
not multiple of 100	C	D

23. Ahmed uses the calculation 200 ÷ 7 to find the highest multiple of 7 that is less than 200. What is his answer?

■ 200 ■ 196 ■ 203

24. When you multiply an even number by another even number, explain whether the product is **always**, **sometimes** or **never** a multiple of an odd number.

- Open question, no set answer

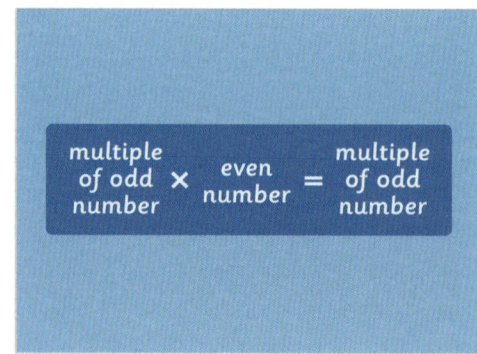

multiple of odd number × even number = multiple of odd number

Level 3 *continued*

25. Mrs Wilson arranges chairs in four groups of
a
b
c
8. She asks Sammy to rearrange the chairs into groups of 6. Explain why Sammy won't be able to make complete groups of 6.

- Open question, no set answer

Level 4: Problem solving with greater depth: Solve multi-step problems involving multiples.

❋ **Required:** 5/5   ❋ **Pupil Navigation:** on
❋ **Randomised:** off

26. Suzy's age is 4 less than a number that is a
1
2
3
multiple of both 6 and 9. If Suzy is younger than 30 years old, what is her age?

- **14** ▪ 18 ▪ 32

27. Bananas are sold in packs of 5, and apples
1
2
3
are sold in packs of 6. If Mary buys enough packs of fruit to give 43 children one banana and one apple each, how many pieces of fruit will she have left over?

- **7** ▪ 5 ▪ 2

pack of 5 bananas   pack of 6 apples

28. Sort the numbers between 1 and 20 into the
1
2
3
Venn diagram. Which number goes in section A of the diagram?

- **12**

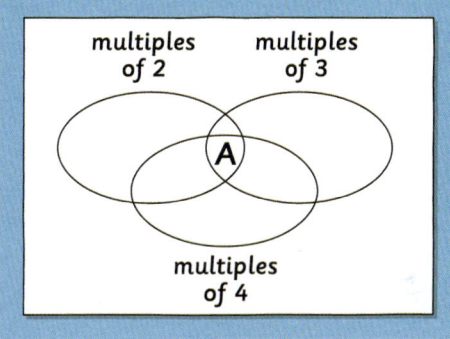

multiples of 2   multiples of 3

A

multiples of 4

29. Ben is thinking of a number. Use the
1
2
3
following clues to find his number. The number is:
• an odd number
• less than 100
• a multiple of 7
• 1 less than a multiple of 5.

- **49**

Level 4 continued

30. Max's magic multiple machine only allows a ball to pass through a sorting stage if the number on the ball is a multiple of one of the magic numbers. Only one ball goes all the way through the sorting machine. What is the number on the ball?

■ 81   ■ 12

Identify Multiples and Factors

Objective: I can identify multiples, factors and factor pairs of whole numbers as well as common factors of two numbers.

Quick Search Ref: 11715

Level 1: Understanding: Identify multiples, factors, factor pairs and common factors with support.

✿ **Required:** 7/10   ✿ **Pupil Navigation:** on   ✿ **Randomised:** off

---

1. A multiple of 6 is a whole number that . . .

1/3
- ■ divides exactly into 6 without a remainder.
- ■ is the product of 6 multiplied by itself.
- ■ can be divided by 6 without leaving a remainder.

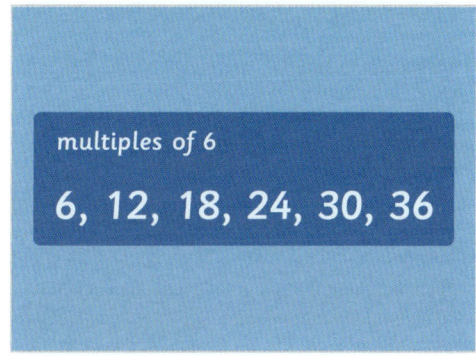

multiples of 6

6, 12, 18, 24, 30, 36

2. You can use number patterns to find some multiples. Which **three** number patterns are true?

3/5
- ■ All multiples of 2 are even numbers.
- ■ All multiples of 3 are odd numbers.
- ■ All multiples of 4 are even numbers that halve to give an even number.
- ■ All multiples of 5 end in 0.
- ■ All multiples of 9 have digits that add to a total of 9.

numbers	multiples
2	2, 4, 6, ...
3	3, 6, 9, ...
4	4, 8, 12, ...
5	5, 10, 15, ...
9	9, 18, 27, ...

3. A factor of 6 is a whole number that...

1/3
- ■ is the product of 6 multiplied by itself.
- ■ divides exactly into 6 without leaving a remainder.
- ■ can be divided by 6 without leaving a remainder.

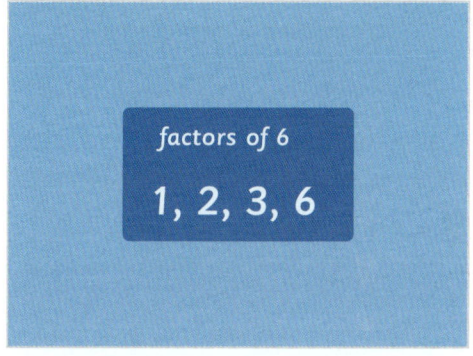

factors of 6

1, 2, 3, 6

4. 1 and 12 are a factor pair of 12 because 1 × 12 = 12. Which **two** of these numbers are another factor pair of 12?

2/5
■ 2   ■ 4   ■ 6   ■ 7   ■ 9

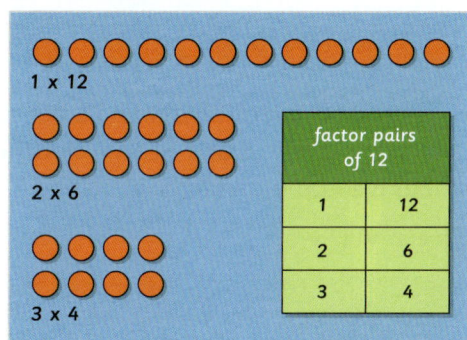

1 x 12

2 x 6

3 x 4

factor pairs of 12	
1	12
2	6
3	4

5. Which of the following numbers is a multiple of 6?

1/5
■ 9   ■ 15   ■ 16   ■ 24   ■ 26

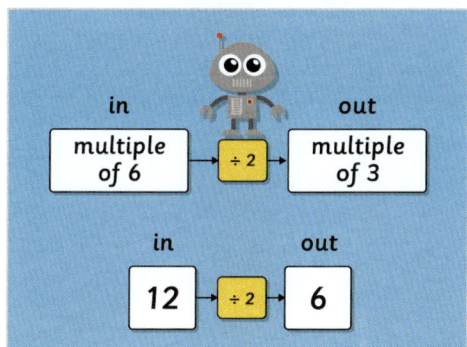

in   out

multiple of 6 → ÷ 2 → multiple of 3

in   out

12 → ÷ 2 → 6

---

6. Select **all three** factor pairs of 32.

- **1 and 32** - **2 and 16** - 3 and 12 - **4 and 8**
- 6 and 7

3/5

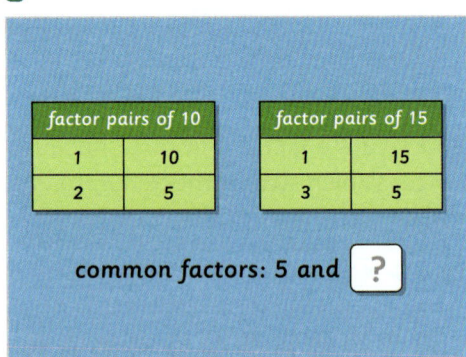

7. 5 is a factor of both 10 and 15. Which **other** number is a common factor of 10 and 15?

- **1** - 2 - 5 - 3

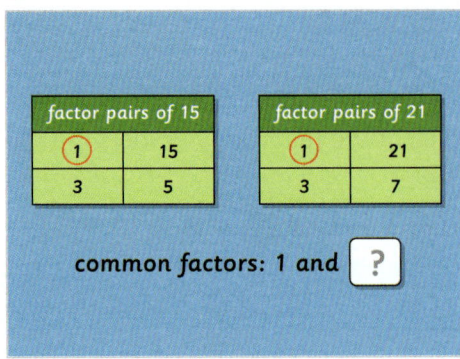

8. 1 is a **common factor** of 15 and 21 because it is a factor of both 15 and 21. Which **other** number is a common factor of 15 and 21?

- **3** - 7 - 1 - 5

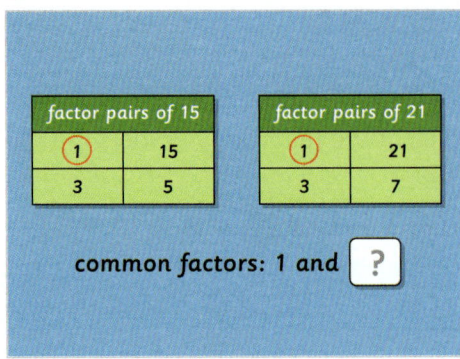

9. Which number is **not** a multiple of 9?

- 18 - 45 - **59** - 72 - 108

1/5

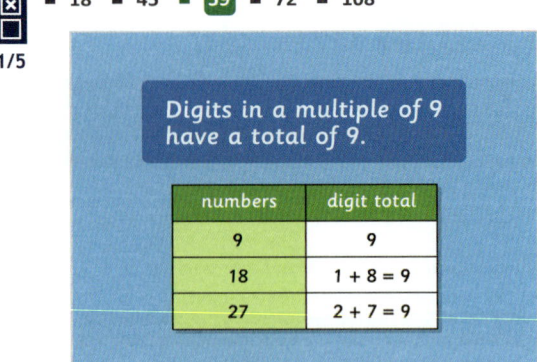

10. Which **two pairs** of numbers are factor pairs of 20?

- **1 and 20** - 2 and 5 - **4 and 5** - 4 and 10

1/4

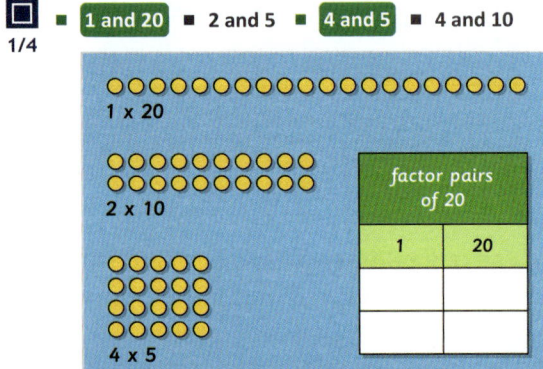

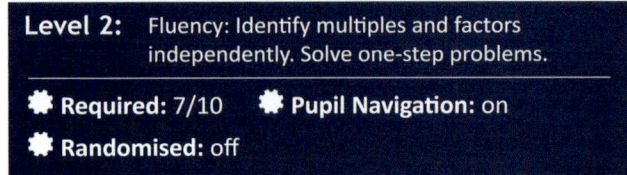

Level 2: Fluency: Identify multiples and factors independently. Solve one-step problems.

✱ **Required:** 7/10   ✱ **Pupil Navigation:** on
✱ **Randomised:** off

11. Which **three** numbers are multiples of 3?

- 109 - **135** - 149 - **180** - **330**

3/5

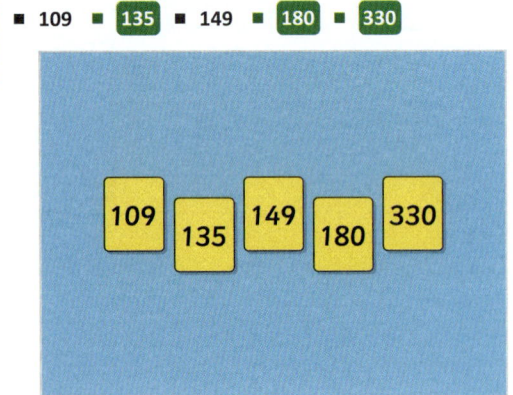

12. Which factor of 28 is missing from the table?

 ▪ **14**

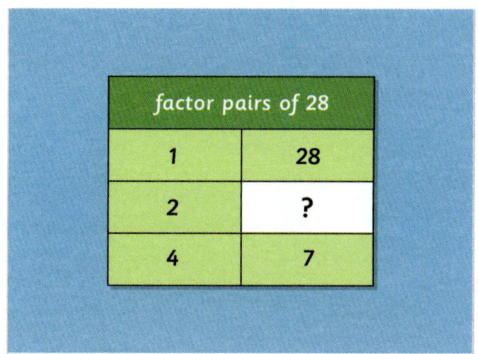

factor pairs of 28	
1	28
2	**?**
4	7

13. Which number (other than 1) is a common factor of 22 and 32?

▪ 1 ▪ **2** ▪ 8 ▪ 4 ▪ 11 ▪ 16

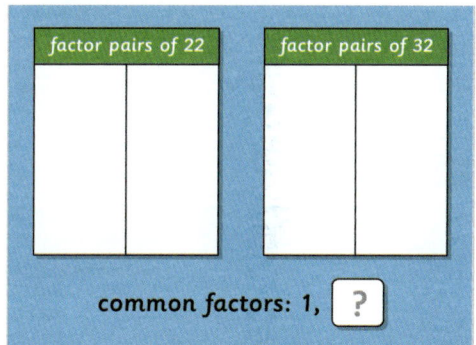

factor pairs of 22		factor pairs of 32	

common factors: 1, **?**

14. What is the highest multiple of 3 that is less than 50?

▪ **48** ▪ 49 ▪ 50 ▪ 51

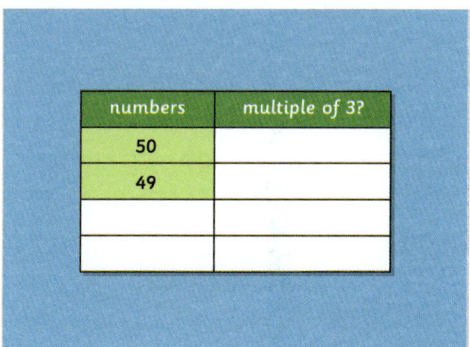

numbers	multiple of 3?
50	
49	

15. What is the lowest number that is a multiple of both 4 and 5?

 ▪ **20**

16. A school caretaker arranges 56 chairs in rows of eight. How many rows of chairs does he make?

▪ **7**

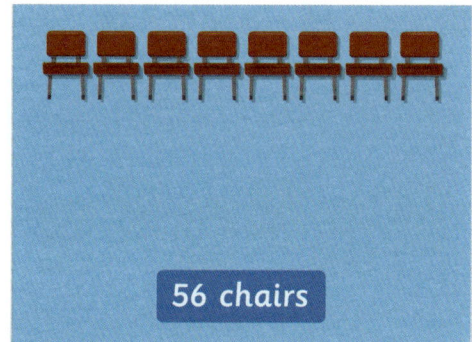

56 chairs

17. Sophie has 120 beads. If she makes a necklace using groups of nine beads, what is the largest number of beads Sophie can use?

▪ **117** ▪ 108 ▪ 120

18. 15 sweets are shared equally between a group of children. If each child gets 3 sweets and there are none left over, how many children are there?

▪ **5**

15 sweets

19. Select **three** multiples of 4.

▪ **12** ▪ 14 ▪ **16** ▪ 26 ▪ 34 ▪ **48**

3/6

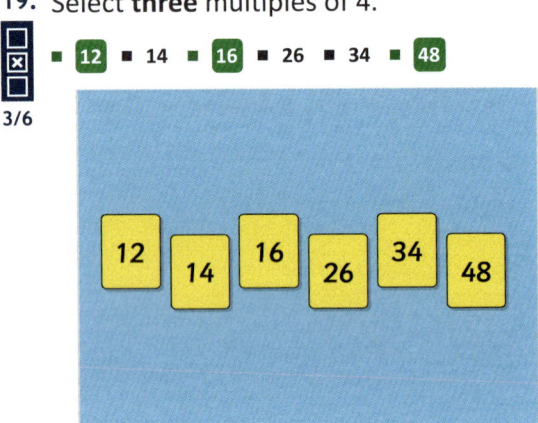

20. Select the **four** factors of 21.

▪ **1** ▪ 2 ▪ **3** ▪ **7** ▪ 11 ▪ **21**

4/6

factor pairs of 21	

Level 3: Reasoning: Reason about multiples or factors.

❋ **Required:** 5/5   ❋ **Pupil Navigation:** on
❋ **Randomised:** off

21. William says, "In my diagram, there will never be a number in box B." Is William correct? Explain your answer.

- Open question, no set answer

	multiples of 50	not a multiple of 50
multiple of 100	A	B
not multiple of 100	C	D

22. Ahmed uses the calculation 200 ÷ 7 to find the highest multiple of 7 that is less than 200. What is his answer?

▪ **196** ▪ 200 ▪ 203

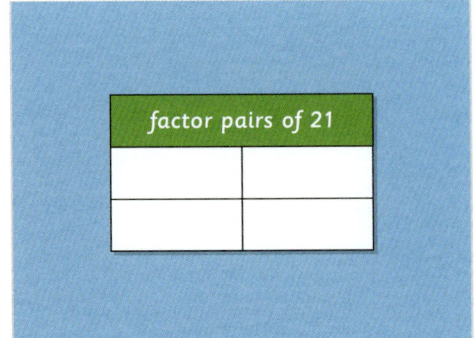

23. Anya has made a table to find all the factors of 64. Explain whether Anya has found all the factors of 64 and how you know.

- Open question, no set answer

factor pairs of 64	
1	64
2	32
3	✕
4	16
5	✕
6	✕
7	✕
8	8

Level 3 *continued*

24. Mrs Wilson arranges chairs in four groups of 8. She asks Sammy to rearrange the chairs into groups of 6. Explain why Sammy won't be able to make complete groups of 6.

a b c

- Open question, no set answer

25. The Venn diagram shows the factors of 9 and 12. Which number is in the wrong part of the diagram?

1 2 3

▪ **6** ▪ 1

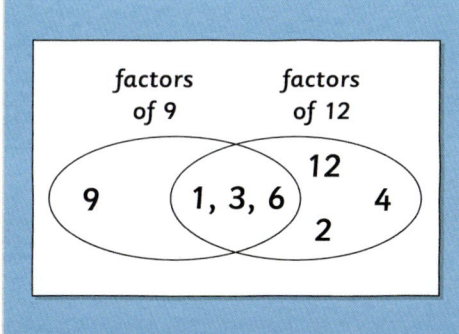

Level 4: Problem solving with greater depth: Solve multi-step problems involving multiples or factors.

❋ **Required:** 5/5   ❋ **Pupil Navigation:** on
❋ **Randomised:** off

26. Suzy's age is 4 less than a number that is a multiple of both 6 and 9. If Suzy is younger than 30 years old, what is her age?

1 2 3

▪ **14** ▪ 32 ▪ 18

27. What is the highest number that is a common factor of 72 and 84?

a b c

▪ **12** ▪ 6 ▪ 3 ▪ 2 ▪ 1 ▪ 4

28. Sort the numbers between 1 and 20 into the Venn diagram. Which number goes in section A of the diagram?

1 2 3

▪ **12**

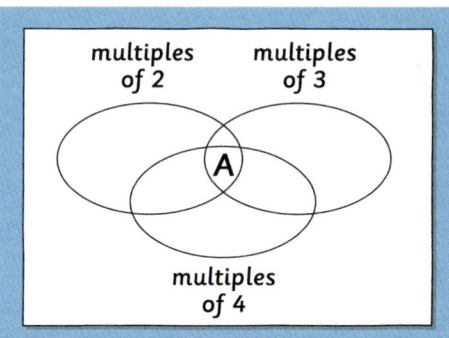

29. Jamie is thinking of a number. His number is four less than a factor of 28 and one more than a factor of 10. What is the product of Jamie's's number multiplied by itself?

a b c

▪ **9**

Level 4 *continued*

30. Max's magic multiple machine only allows a ball to pass through a sorting stage if the number on the ball is a multiple of one of the magic numbers. Only one ball goes all the way through the sorting machine. What is the number on the ball?

- 81   - 12

Understand Prime Factors

Objective: I can identify prime numbers and prime factors.

Quick Search Ref: 10062

Level 1: Understanding: Identify prime factors using prime factor trees with support.

✱ **Required:** 7/10   ✱ **Pupil Navigation:** on   ✱ **Randomised:** off

1. A prime factor is a factor that is also a prime number. Which **two** factors of 12 are prime factors?

2/6   ▪ 1 ▪ **2** ▪ **3** ▪ 4 ▪ 6 ▪ 12

factors of 12	prime number	prime factor
1	✗	✗
2	✓	✓
3		
4		
6		
12		

2. One of the prime factors of 6 is 2. What is the **other** prime factor of 6?

▪ 2 ▪ **3** ▪ 6 ▪ 1

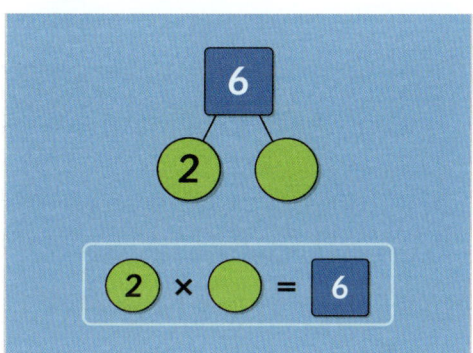

3. Composite numbers can be broken down into prime factors. What is the missing prime factor of 14?

▪ 1 ▪ **7** ▪ 2 ▪ 14

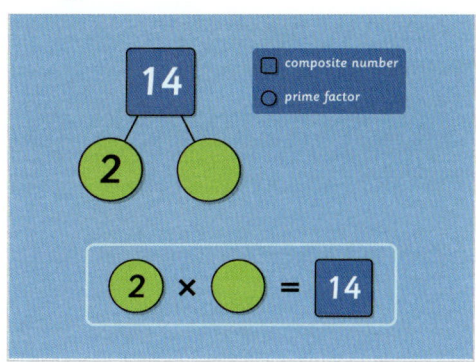

4. Which number is the product of prime factors 2 and 5?

▪ **10** ▪ 2 ▪ 5

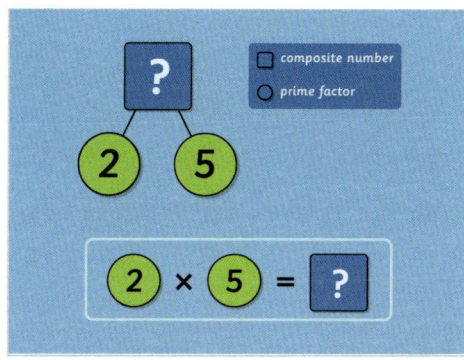

5. What are the **two** prime factors of 15?

▪ 1 ▪ **3** ▪ **5** ▪ 10 ▪ 15

2/5

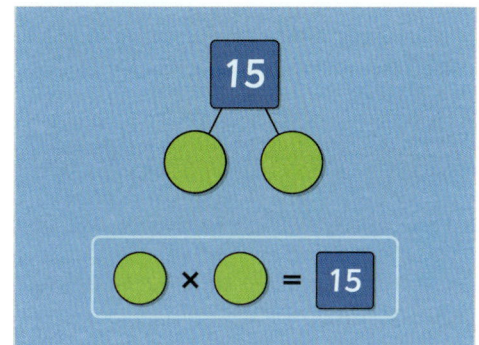

6. Factors that are composite numbers can be broken down into prime factors. What is the missing prime factor of 12?

▪ **3** ▪ 2

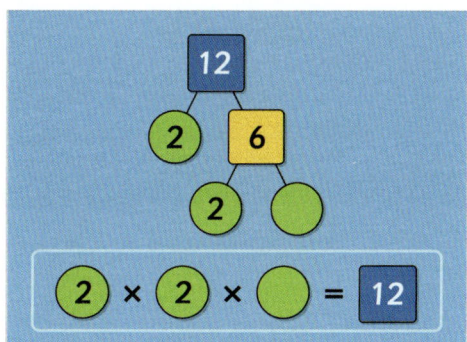

7. 2 is one of the prime factors of 18. What is the **other** prime factor of 18?

1
2
3

- **3** - 9

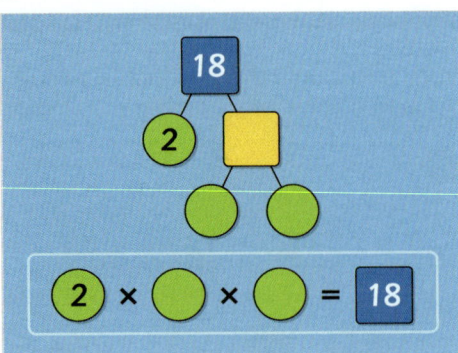

8. Composite numbers can be broken down into prime factors. What is the missing prime factor of 20?

1
2
3

- **5** - 2

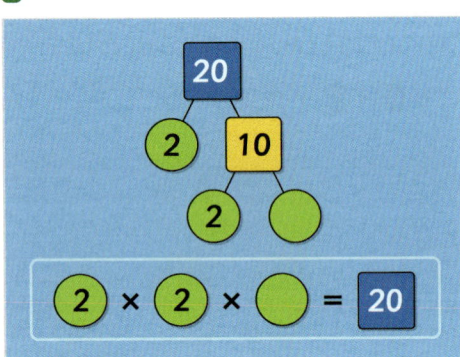

9. Which number is the product of prime factors 3 and 7?

1
2
3

- **21** - 7 - 3

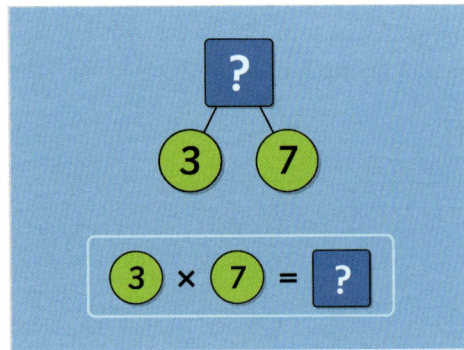

10. Composite numbers can be broken down into prime factors. What is the missing prime factor of 22?

1
2
3

- 2 - **11** - 22 - 1

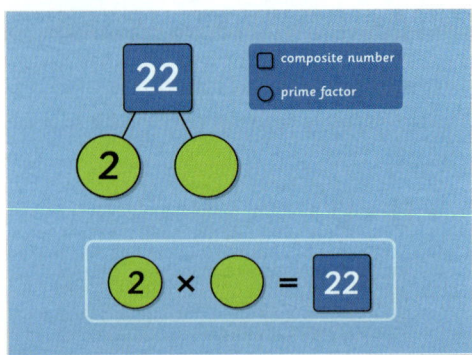

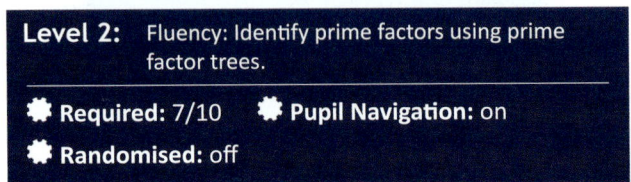

Level 2: Fluency: Identify prime factors using prime factor trees.

✲ **Required:** 7/10   ✲ **Pupil Navigation:** on
✲ **Randomised:** off

11. Select the **two** prime factors of 35.

☐
☒
☐

- 3 - **5** - **7** - 15 - 30

2/5

12. What number is a prime factor of 7?

1
2
3

- 14 - **7** - 1

13. What is the only prime factor of 9?

■ **3** ■ 9 ■ 1

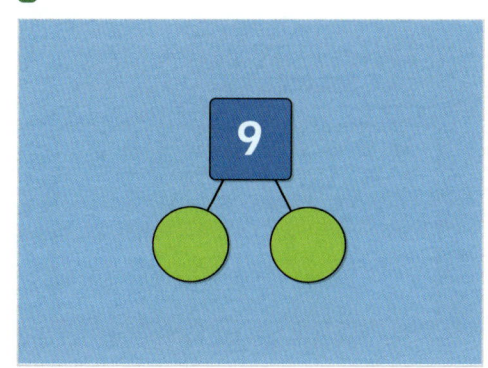

14. Which **three** numbers are prime factors of 30?

3/6

■ **2** ■ 15 ■ **3** ■ 6 ■ 10 ■ **5**

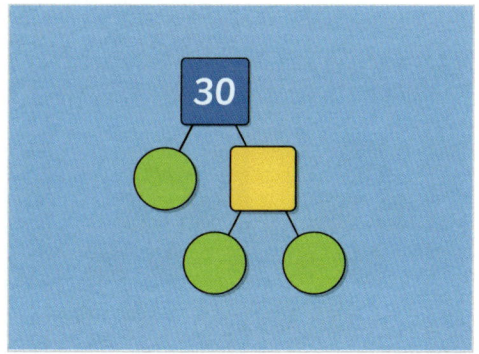

15. The prime factors of 56 are 2 and ___.
Enter the missing number.

■ 28 ■ **7** ■ 2 ■ 14

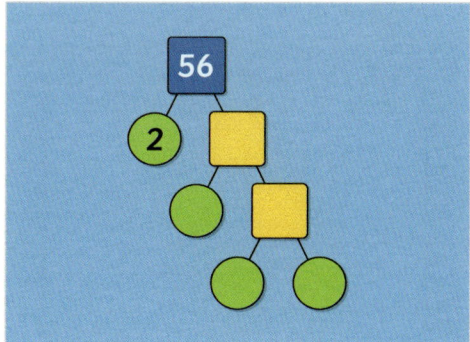

16. What number is represented by the prime factor tree?

■ 9 ■ **27** ■ 3

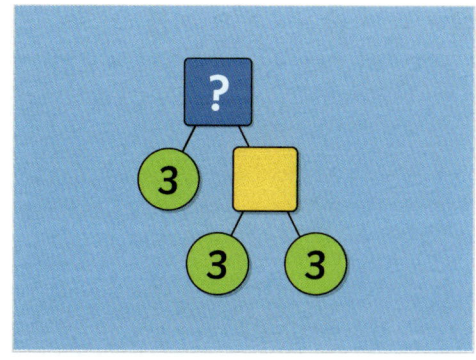

17. What number is represented by the prime factor tree?

■ 6 ■ **36** ■ 9 ■ 18

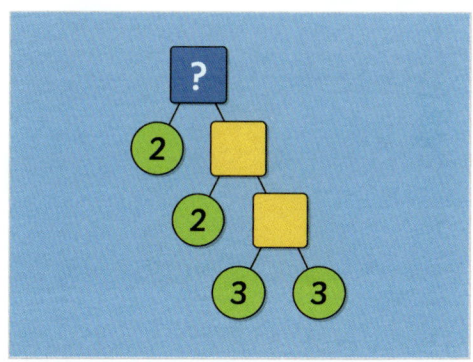

18. What number is represented by the prime factor tree?

■ 2 ■ **8** ■ 4

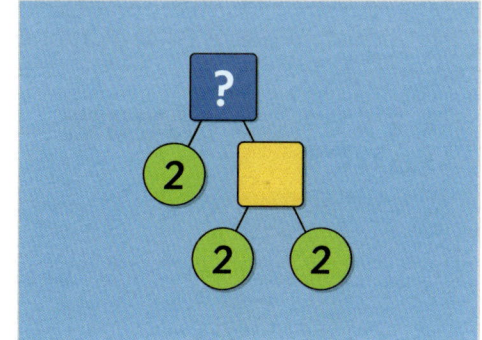

19. Which **three** numbers are the prime factors of 42?

3/6

■ **2** ■ 14 ■ **3** ■ 6 ■ 21 ■ **7**

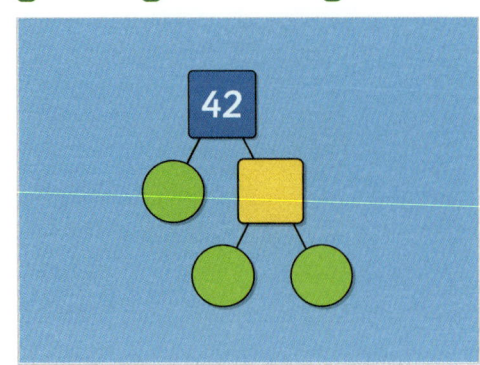

20. Select the two prime factors of 26.

2/4

■ 1 ■ **2** ■ **13** ■ 26

Level 3: Reasoning: Reason about prime factors.

❖ **Required:** 5/5 ❖ **Pupil Navigation:** on
❖ **Randomised:** off

21. Which calculation shows 24 as a product of its **prime factors**?

1/4

■ 24 = 3 × 8 ■ 24 = 4 × 6 ■ **24 = 2 × 2 × 2 × 3**
■ 24 = 2 × 3 × 4

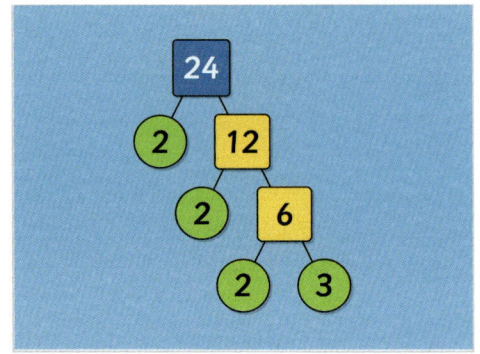

22. Explain why all multiples of 6 have at least two prime factors.

a b c

- Open question, no set answer

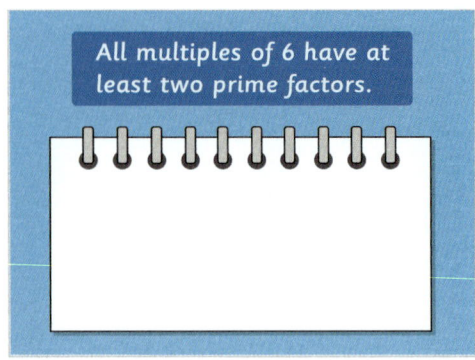

23. Chloe and Dylan have both drawn prime factor trees for 40. Whose diagram is correct: Chloe's, Dylan's or both? Explain your answer.

a b c

- Open question, no set answer

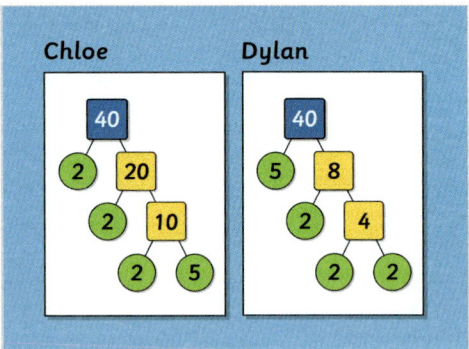

24. What is the **smallest** number that has the prime factors 2, 5 and 7?

1 2 3

■ 35 ■ **70** ■ 10 ■ 14

25. Which number has more than one prime factor?

☐
☒
☐

1/4

■ 4 ■ **12** ■ 16 ■ 64

Level 4: Problem solving with greater depth: Solve multi-step problems involving prime factors.

✱ **Required:** 5/5 ✱ **Pupil Navigation:** on
✱ **Randomised:** off

26. Which square number less than 100 has more than one prime factor?

1
2
3

■ **36**

27. Identify the prime number that is a common factor of the following numbers: 63, 70, 105.

1
2
3

■ 2 ■ **7** ■ 5 ■ 3

28. What is the **highest** prime factor of 156?

1
2
3

■ **13** ■ 2 ■ 3

29. What is the difference between the number with the most prime factors and the number with the fewest prime factors?

1
2
3

■ **13**

30. Sammy folds a piece of square paper in half several times. When he completely unfolds the paper, there are creases making 32 equal rectangles. How many times does Sammy fold the paper?

1
2
3

■ **5**

Understand Prime Numbers, Composite Numbers and Prime Factors

Objective: I can identify prime numbers, composite numbers and prime factors.

Quick Search Ref: 10231

Level 1: Understanding: Prime numbers, composite numbers and prime factors.

✿ **Required:** 7/10   ✿ **Pupil Navigation:** on   ✿ **Randomised:** off

---

1. What is a prime number?

1/4
- ■ A whole number which exactly divides into another whole number.
- ■ A number that can only be divided by itself and 1.
- ■ A number that can be divided evenly by numbers other than itself and 1.
- ■ A number which is multiplied by itself.

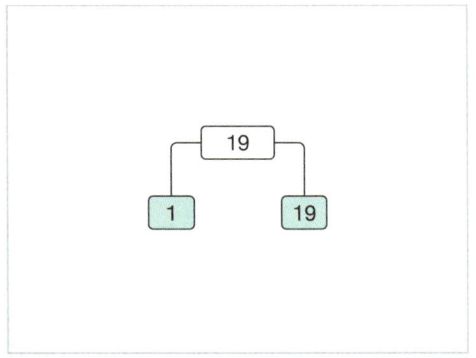

2. What is a composite number?

1/4
- ■ A whole number which exactly divides into another whole number.
- ■ A number that has exactly 2 factors.
- ■ A number that can be divided evenly by numbers other than itself and 1.
- ■ A number which is multiplied by itself.

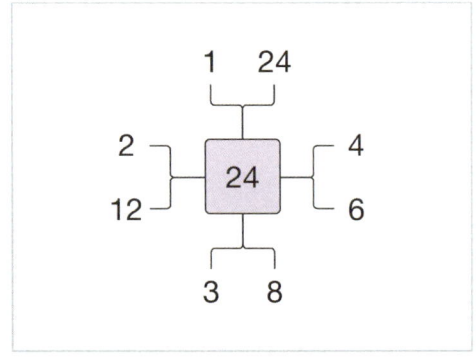

3. What is a factor?

1/4
- ■ A whole number which exactly divides into another whole number.
- ■ A number that can only be divided by itself and 1.
- ■ A number that can be divided evenly by numbers other than itself and 1.
- ■ A number which is multiplied by itself.

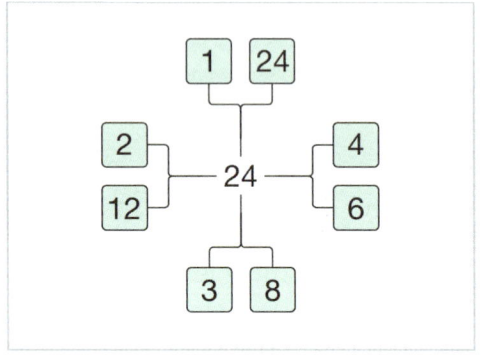

4. Is 19 a prime or composite number?

1/2
- ■ composite number   ■ 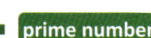 prime number

5. 49 is a _____ number.

1/2
- ■ composite   ■ prime

Level 1 *continued*

6. What is a prime factor?

1/4

- ■ A whole number which exactly divides into another whole number.
- ■ A number that can only be divided by itself and 1.
- ■ A number that can be divided evenly by numbers other than itself and 1.
- ■ **A factor that is a prime number.**

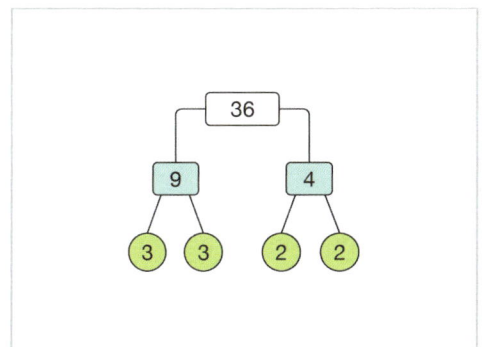

7. Which number is neither prime or composite?

1/4

- ■ 15 ■ 41 ■ **1** ■ 18

8. Select the two composite numbers.

2/5

- ■ 11 ■ **12** ■ 13 ■ **15** ■ 19

9. Select the **two** prime numbers.

2/5

- ■ **5** ■ 6 ■ **7** ■ 8 ■ 9

10. Is 7 a prime or composite number?

1/2

- ■ composite number ■ **prime number**

Level 2:
Fluency: Identifying prime numbers, composite numbers and prime factors.

✱ **Required:** 6/8   ✱ **Pupil Navigation:** on
✱ **Randomised:** off

11. Which number between 12 and 16 is a prime number?

- ■ 16 ■ 14 ■ **13** ■ 12 ■ 15

12. Use the factor tree to find the **two prime factors** of 36.

2/6

- ■ **3** ■ 4 ■ 6 ■ **2** ■ 12 ■ 9

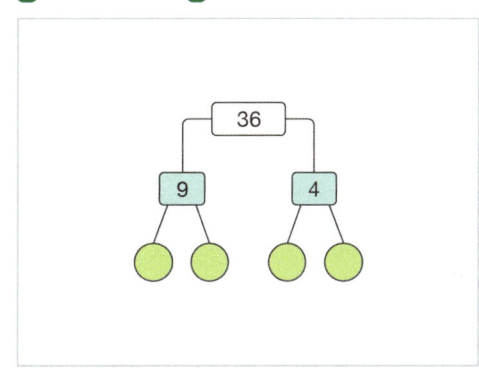

13. What is the missing number in the sequence?

- ■ **5** ■ 4 ■ 6

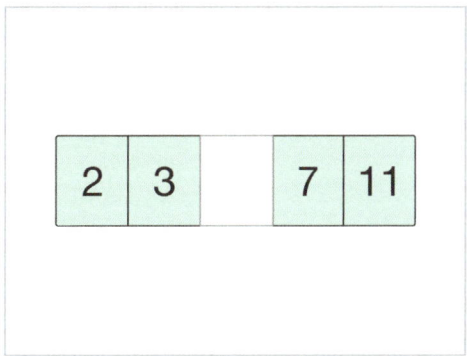

Level 2 *continued*

14. Use the factor tree to find the **three prime factors** of 30.

■ 15 ■ **3** ■ **2** ■ 6 ■ 10 ■ **5**

3/6

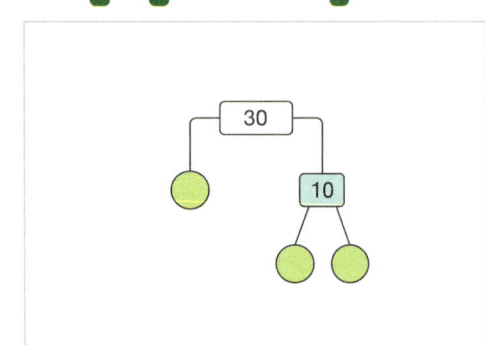

15. Complete the number sequence.

■ **10**

16. The prime factors of 48 are 2 and ___.

■ **3**

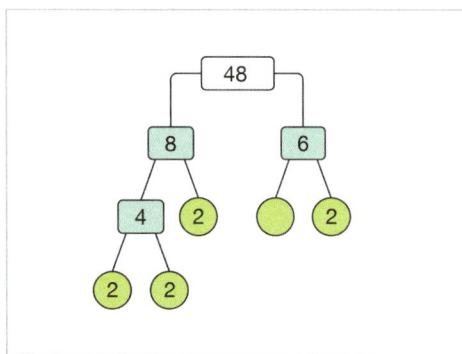

17. What is the missing number in the sequence?

1
2
3

■ **13** ■ 12 ■ 14 ■ 15 ■ 16

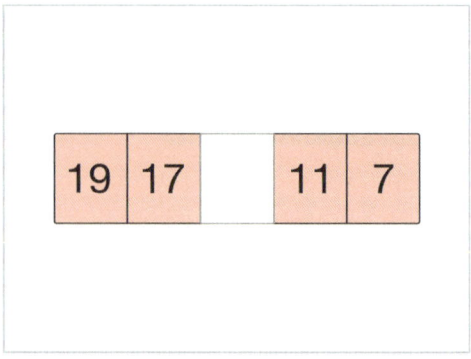

18. Which two numbers are prime factors of 20?

■ 1 ■ **2** ■ 4 ■ **5** ■ 10 ■ 20

2/6

Level 3: Reasoning: Use reasoning to identify prime numbers, composite numbers and prime factors.

✿ **Required:** 6/8   ✿ **Pupil Navigation:** on
✿ **Randomised:** off

19. How many prime numbers are there between 1 and 20?

1
2
3

■ **8** ■ 9

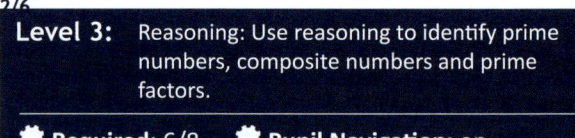

20. The sum of three prime numbers is 20. One of the numbers is 13. What are the other **two** numbers?

2/6   ■ 1 ■ **2** ■ 3 ■ 4 ■ **5** ■ 6

21. Other than 2 and 3, are there any other consecutive prime numbers between 1 and 100?
Explain your answer.

a
b
c

- Open question, no set answer

Level 3 *continued*

22.  The number 24 is written as a product of its **prime factors**. Which one of the following is correct?

1/4
- 24 = 3 × 8   ■ 24 = 4 × 6   ■ **24 = 2 × 2 × 2 × 3**
- 24 = 2 × 3 × 4

23. What is the **smallest** number which has **two** different prime factors?

- 3   ■ 2   ■ **6**   ■ 4   ■ 5

24. Jamila says, "If you multiply two prime numbers, the answer will always be a prime number." Is she correct? Explain your answer.

- Open question, no set answer

25. Which of these numbers is the odd one out?

- 13   ■ **10**   ■ 19   ■ 3   ■ 7

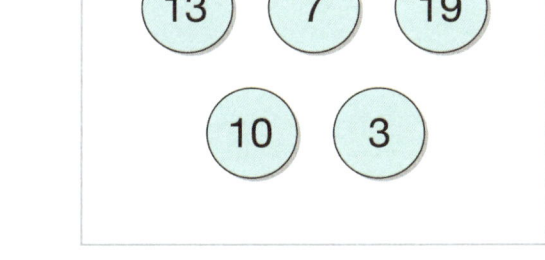

26. What is the only **prime** number between 90 and 100?

- **97**

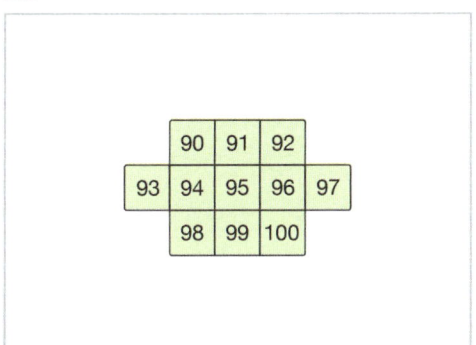

Level 4: Problem Solving: Calculate prime numbers up to 100.

❋ **Required:** 5/5   ❋ **Pupil Navigation:** on
❋ **Randomised:** off

27. Derek collects football cards. The number of cards he has in the morning is a **prime** number **between 90 and 100**. On the way to school he buys 3 packets which each contain 6 cards.
When Derek gets to school, he gives Keon some cards (the **greatest prime factor of 33**). How many cards does Derek have left?

- **104**   ■ 112   ■ 115

28. I am a **prime** number between 70 and 100. My **tens** digit is **greater** than my **ones** digit. The **sum** of my digits is 10. What number am I?

- **73**

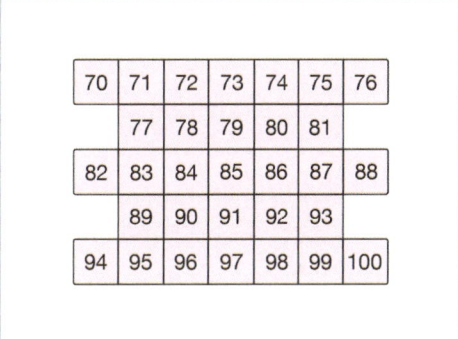

29. The two missing numbers in the equation are **prime** numbers between 2 and 60. What number goes in the **answer** box?

- **2**

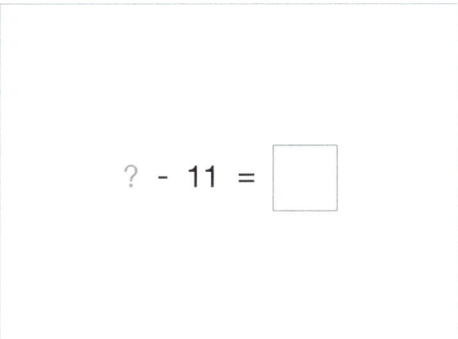

Level 4 *continued*

30. Sam adds together two consecutive **composite** numbers between 40 and 45 (inclusive). He then adds a number which only has the **prime factors** 5, 3 and 2. What answer does Sam get?

■ **119** ■ 117 ■ 115 ■ 111 ■ 113

31. Which number has the **most prime factors**?

■ 21 ■ 45 ■ **66** ■ 63 ■ 85 ■ 91

1/6

Identify Prime Numbers and Composite Numbers

Objective: I can identify prime numbers and composite numbers.

Quick Search Ref: 10118

Level 1: Understanding: Identify prime and composite numbers with support.

✿ **Required:** 7/10 ✿ **Pupil Navigation:** on ✿ **Randomised:** off

---

1. Which **two** statements are true for every prime number?

2/4

- ■ A prime number can only be divided by itself and 1 without leaving a remainder.
- ■ A prime number can be divided by other numbers as well as by itself and 1 without leaving a remainder.
- ■ A prime number has more than two factors.
- ■ A prime number has only two factors.

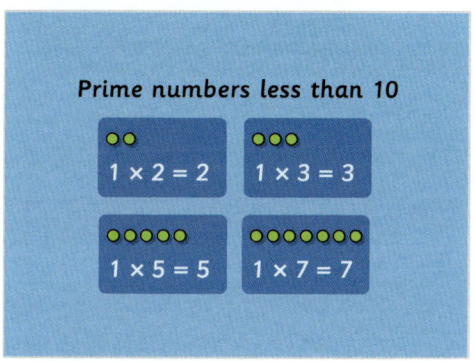

2. Which **two** statements are true for all composite numbers?

2/4

- ■ A composite number has only two factors.
- ■ A composite number can be divided by other numbers as well as by itself and 1 without leaving a remainder.
- ■ A composite number has more than two factors.
- ■ A composite number can only be divided by itself and 1 without leaving a remainder.

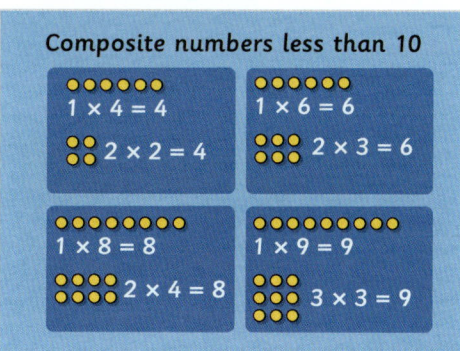

3. Which number is a **prime number**?

1/4

■ 1 ■ 6 ■ 9 ■ 13

number	arrays	factors
1	● 1 × 1	1
6	●●●●●● 1 × 6 2 × 3	1, 6, 2, 3
9	●●●●●●●●● 1 × 9 3 × 3	1, 9, 3
13	●●●●●●●●●●●●● 1 × 13	1, 13

4. Which number is a **composite number**?

1/4

■ 7 ■ 11 ■ 15 ■ 19

number	factors
7	1, 7
11	
15	
19	

5. What are the first **three** prime numbers?

3/5

■ 1 ■ 2 ■ 3 ■ 4 ■ 5

number	factors
1	
2	
3	
4	
5	

Level 1 continued

6. Select the statement that is **true**.

1/3

- All odd numbers are prime numbers.
- All numbers are either prime numbers or composite numbers.
- The only even prime number is 2.

number	factors	prime	composite
1	1	✗	✗
2	1, 2	✓	✗
4	1, 2, 4	✗	✓
6	1, 2, 3, 6	✗	✓
9	1, 3, 9	✗	✓

7. Which number is a **prime number**?

1/4

- 14 ■ 15 ■ **17** ■ 18

Testing for prime numbers between 10 and 100

A prime number is **not**:
- an even number
- a multiple of 3
- a multiple of 5
- a multiple of 7

8. Which number is a **prime number**?

1/3

- 9 ■ 10 ■ **11**

Testing for prime numbers between 10 and 100

A prime number is **not**:
- an even number
- a multiple of 3
- a multiple of 5
- a multiple of 7

9. What is the smallest **composite number**?

- **4** ■ 3 ■ 2

number	factors
1	
2	
3	
4	
5	

10. The number 1 . . .

1/3

- is a composite number. ■ is a prime number.
- isn't a prime number or a composite number.

Prime numbers have exactly two factors.

Composite numbers have more than two factors.

Level 2: Fluency: Identify prime numbers and composite numbers.

✦ **Required:** 7/10   ✦ **Pupil Navigation:** on
✦ **Randomised:** off

11. What are the **four** prime numbers less than 10?

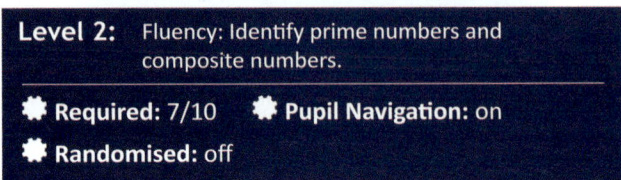

4/7

- 1 ■ **2** ■ **3** ■ 4 ■ **5** ■ **7** ■ 9

Level 2 *continued*

12. Which **four** numbers between 10 and 20 are prime numbers?

4/6

- 11 - 13 - 15 - 16 - 17 - 19

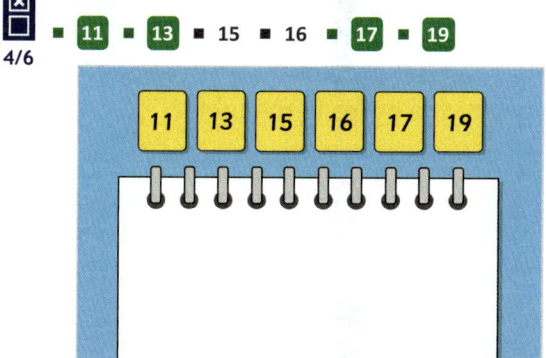

13. What is the first prime number greater than 20?

- 29 - 23 - 21

14. What is the only prime number between 89 and 101?

- 93 - 97 - 91

15. Which number is not a composite number?

1/5

- 80 - 81 - 83 - 85 - 86

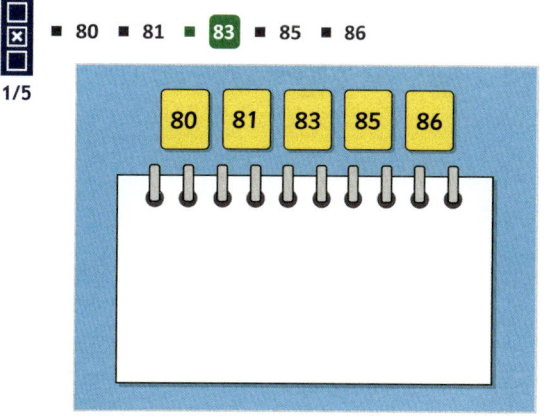

16. What is the next prime number?

41, 43, 47, 53, _____

- 59 - 61

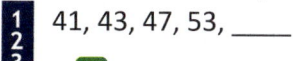

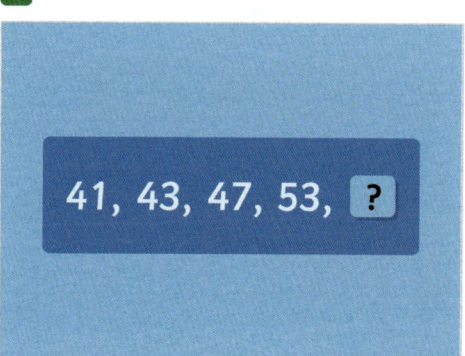

17. What is the missing prime number?

79, 73, 71, _____, 61

- 67

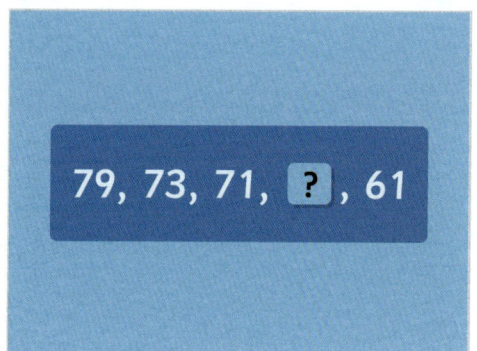

Level 2 *continued*

18. Select the **three** prime numbers.

▪ 41 ▪ 43 ▪ 45 ▪ 47 ▪ 49

3/5

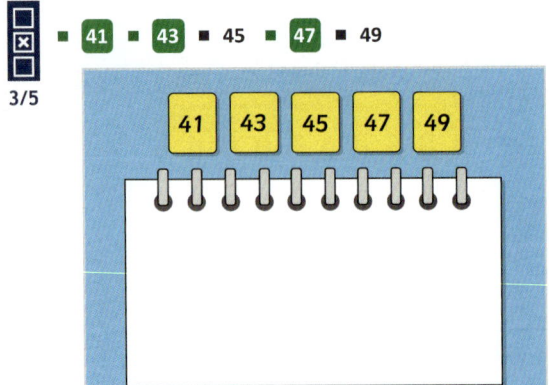

19. Which number is not a composite number?

▪ 30 ▪ 32 ▪ 34 ▪ 37 ▪ 39

1/5

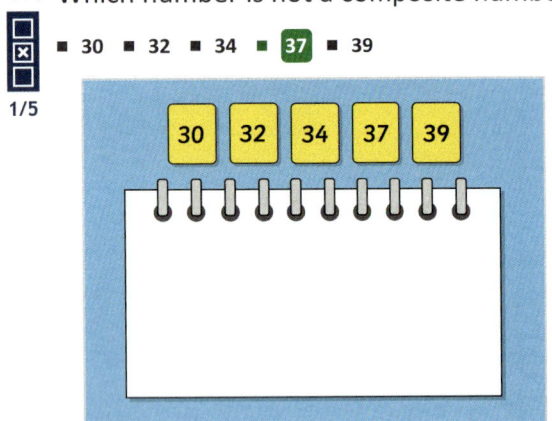

20. What is the missing prime number that is less than 20?

▪ 13

2, 3, 5, 7, 11, ? 17, 19

Level 3: Reasoning: Reason about prime numbers and composite numbers.

❋ Required: 5/5   ❋ Pupil Navigation: on
❋ Randomised: off

21. What number greater than zero, does not go in any of the boxes in the table?

▪ 1 ▪ 0

	even numbers	odd numbers
prime numbers		
composite numbers		

22. Is the following statement true? Explain your answer.
If you multiply two prime numbers together, the answer is always a composite number.

- Open question, no set answer

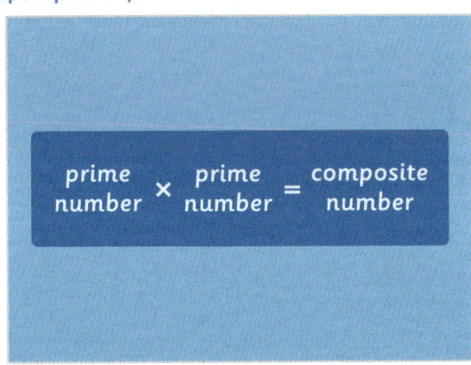

prime number × prime number = composite number

23. Explain whether odd numbers are **always**, **sometimes** or **never** prime numbers.

- Open question, no set answer

24. The sum of three prime numbers is 20. If one
of the numbers is 5, what is the largest of
the **three** numbers?

- 2 - **13** - 5

25. Explain why Mr Brown can't split his class of
37 pupils into smaller equal groups.

- Open question, no set answer

Level 4: Problem solving with greater depth: Solve
multi-step problems involving prime numbers
and composite numbers.

❋ **Required:** 5/5   ❋ **Pupil Navigation:** on
❋ **Randomised:** off

26. Simon thinks of a prime number. Use the
following clues to find Simon's number:
- The number is greater than 70 and less
than 100.
- The tens digit is greater than the ones
digit.
- The sum of the digits is 10.

- **73**

27. Faizan adds together **two consecutive
composite numbers** shown on the cards.
What prime number does Faizan make?

- **89** - 45 - 44

Level 4 *continued*

28. There are five **square numbers** less than 100 that are two more than a prime number. What is the total of these five square numbers?

   ■ 168   ■ 284   ■ 285

29. What is the total of the prime numbers between 30 and 60 that are one more than a multiple of 6?

   ■ 111   ■ 311

30. Which prime number between 70 and 80 is 30 more than one prime number, and 10 less than another prime number?

   ■ 73   ■ 79   ■ 71

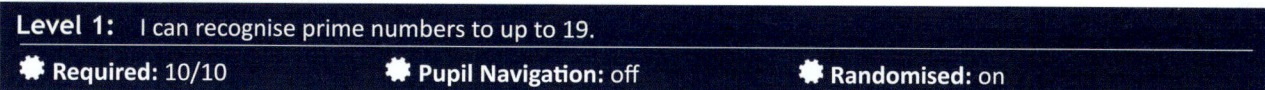
Practise Recognising Prime Numbers

Objective:	I can recognise prime numbers up to 100.
Quick Search Ref:	**10026**

Level 1: I can recognise prime numbers to up to 19.

✿ **Required:** 10/10   ✿ **Pupil Navigation:** off   ✿ **Randomised:** on

1. Which number is a prime number?

1 2 3 · ■ **2**

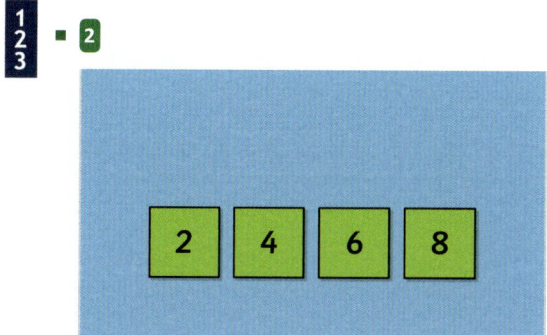

2. Which number is a prime number?

1 2 3 · ■ 6 ■ **3** ■ 9 ■ 1

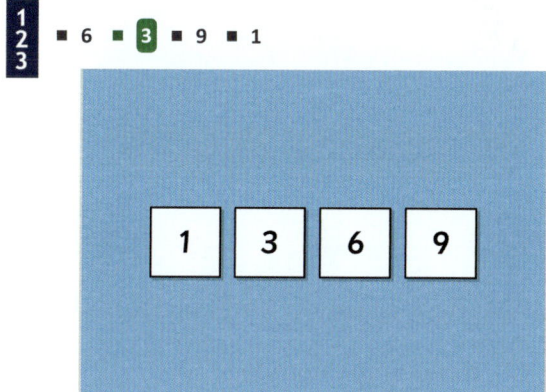

3. Which number is a prime number?

1 2 3 · ■ **5** ■ 4 ■ 9 ■ 8

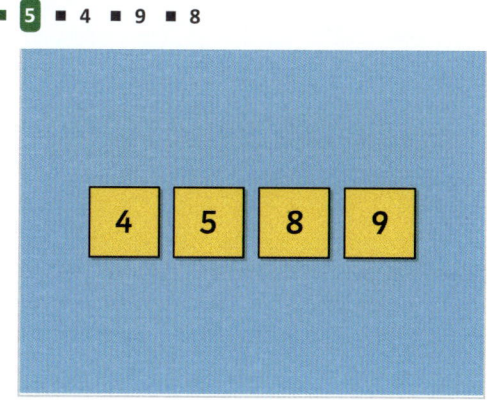

4. Which number is a prime number?

1 2 3 · ■ **7** ■ 1 ■ 4 ■ 9

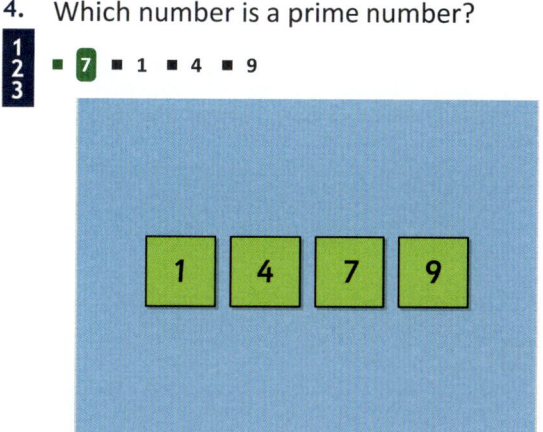

5. Which number is a prime number?

1 2 3 · ■ 9 ■ **11** ■ 12 ■ 10

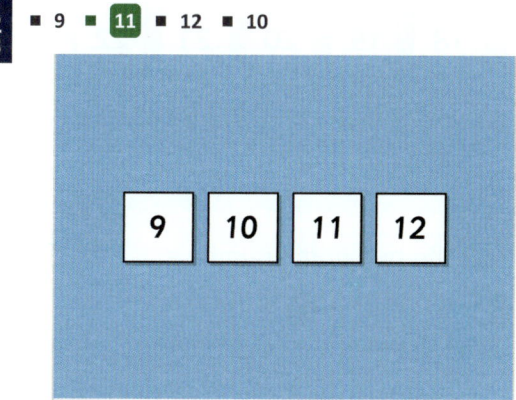

6. Which number is a prime number?

1 2 3 · ■ **13**

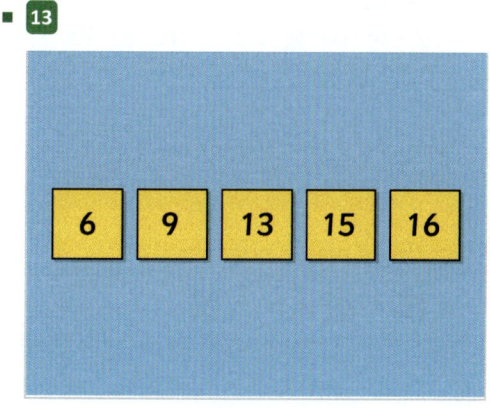

Level 1 *continued*

7. Which number is a prime number?

 ▪ **17**

8. Which number is a prime number?

 ▪ **19**

9. Select the **two** prime numbers.

 ▪ **7** ▪ 10 ▪ 12 ▪ **13** ▪ 15

2/5

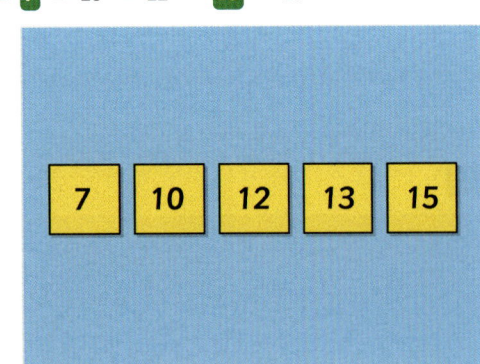

10. Select **three** prime numbers.

 ▪ 9 ▪ **11** ▪ **13** ▪ 15 ▪ **17**

3/5

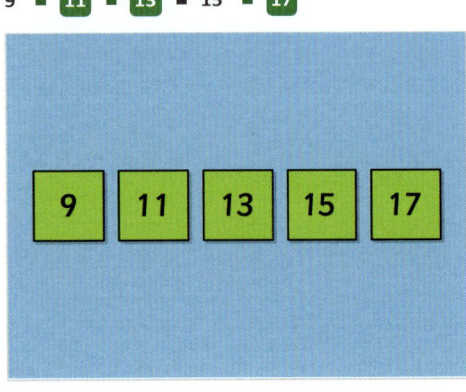

Level 2: I can recognise prime numbers beyond 19.
- ✿ **Required:** 10/10   ✿ **Pupil Navigation:** off
- ✿ **Randomised:** on

11. Select **two** prime numbers.

▪ 21 ▪ 22 ▪ **23** ▪ 24 ▪ 25 ▪ 27 ▪ **29**

2/7

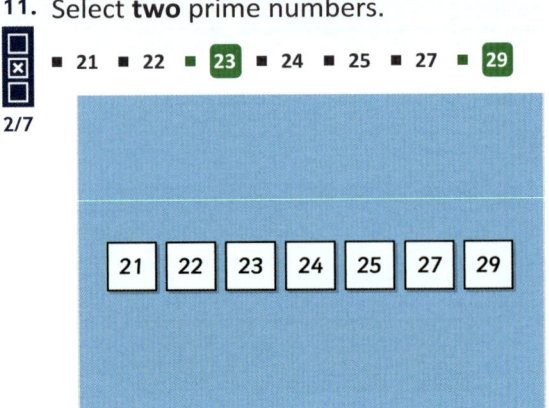

12. Select **two** prime numbers.

▪ **31** ▪ 32 ▪ 33 ▪ 34 ▪ 35 ▪ **37** ▪ 39

2/7

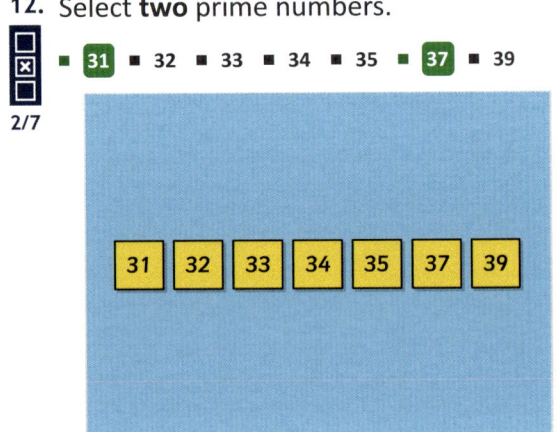

13. Select **three** prime numbers.

▪ **41** ▪ 42 ▪ **43** ▪ 45 ▪ 46 ▪ **47** ▪ 49

3/7

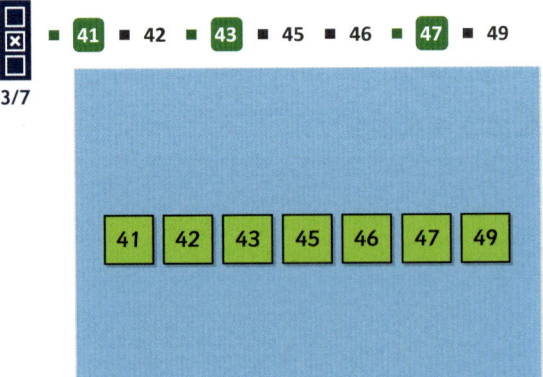

Level 2 *continued*

14. Select **two** prime numbers.

■ 51 ■ 52 ■ **53** ■ 55 ■ 56 ■ 57 ■ **59**

2/7

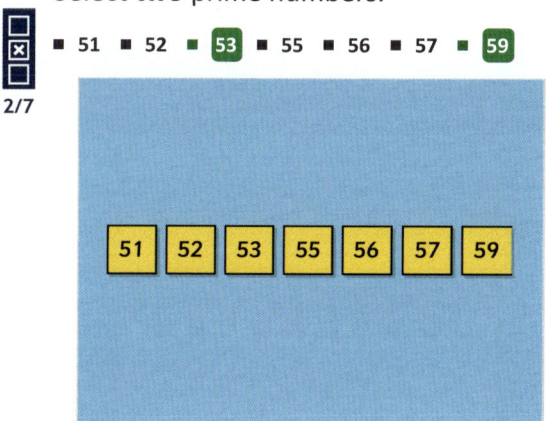

15. Select **two** prime numbers.

■ **61** ■ 62 ■ 63 ■ 64 ■ 65 ■ 66 ■ **67**

2/7

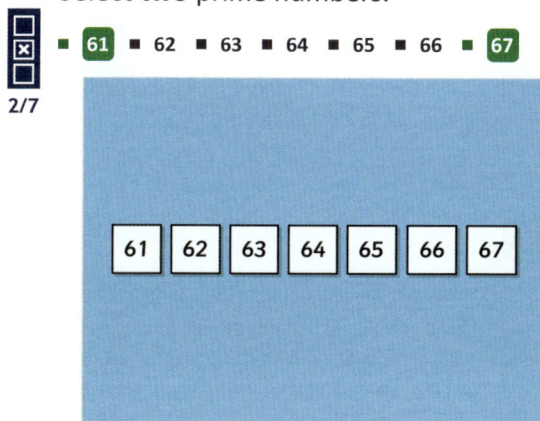

16. Select **three** prime numbers.

■ **71** ■ 72 ■ **73** ■ 75 ■ 76 ■ 77 ■ **79**

3/7

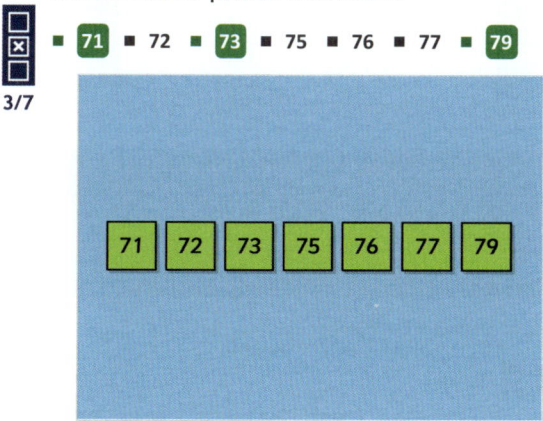

17. Select **two** prime numbers.

■ 81 ■ 82 ■ **83** ■ 85 ■ 86 ■ 87 ■ **89**

2/7

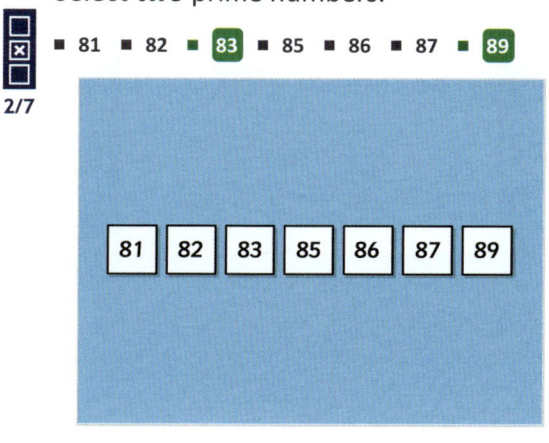

18. Which number is a prime number?

■ **97** ■ 91

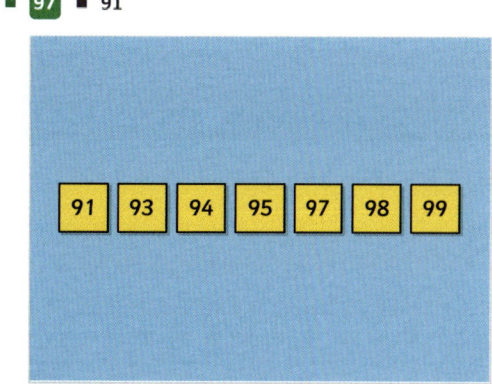

19. Select **two** prime numbers.

■ 27 ■ **37** ■ 49 ■ 57 ■ **61** ■ 75 ■ 87

2/7

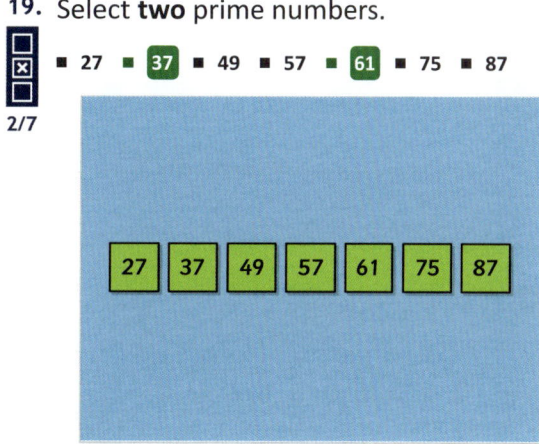

20. Select **three** prime numbers.

■ **11** ■ 24 ■ 39 ■ **43** ■ 57 ■ **89** ■ 91

3/7

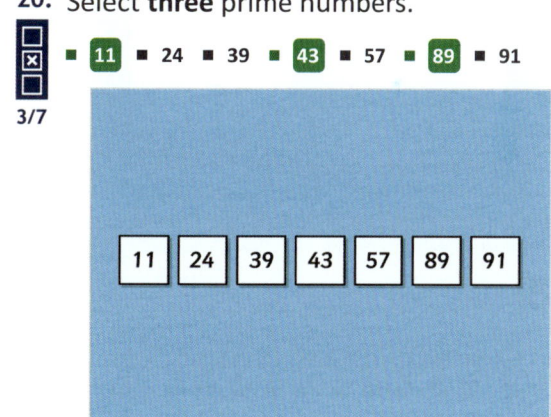

Practise Cube Numbers

Objective: I can recognise and find cube numbers and use the notation for cubed (^{3}).

Quick Search Ref: 10430

Level 1: I can find cube numbers up to 5^3.

✻ **Required:** 5/5   ✻ **Pupil Navigation:** off   ✻ **Randomised:** on

1. What is 1^3?

1 2 3 ▪ 1

2. What is 2^3?

1 2 3 ▪ 8

3. What is 3^3?

1 2 3 ▪ 27

4. What is 4^3?

1 2 3 ▪ 64

5. What is 5^3?

1 2 3 ▪ 125

Level 2: I can recognise cube numbers.

✻ **Required:** 10/20   ✻ **Pupil Navigation:** off
✻ **Randomised:** on

6. Is 1 a cube number?

☐ ☒ ☐ ▪ yes ▪ no
1/2

Level 2 *continued*

7. Is 8 a cube number?

1/2
■ yes ■ no

8. Is 27 a cube number?

1/2
■ yes ■ no

9. Is 64 a cube number?

1/2
■ yes ■ no

10. Is 125 a cube number?

1/2
■ yes ■ no

11. Is 1,000 a cube number?

1/2
■ yes ■ no

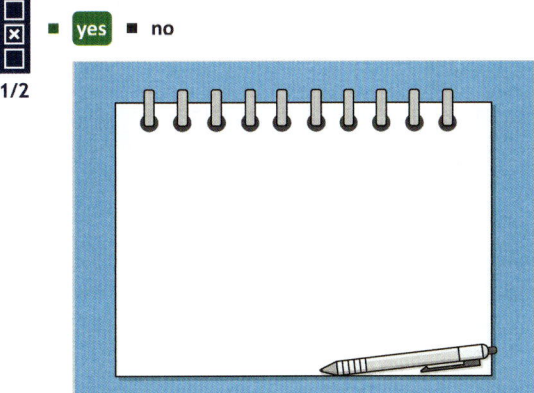

12. Is 25 a cube number?

1/2
■ yes ■ no

13. Is 2 a cube number?

1/2
■ yes ■ no

14. Is 4 a cube number?

1/2
■ yes ■ no

Ref:10430
Practise Cube Numbers

Level 2 *continued*

15. Is 100 a cube number?

☒ ▪ yes ▪ no

1/2

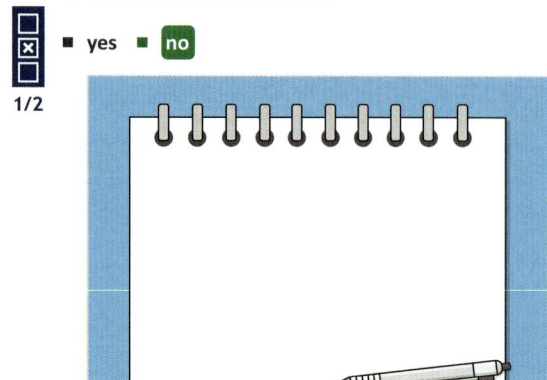

16. Is 9 a cube number?

☒ ▪ yes ▪ no

1/2

17. Is 16 a cube number?

☒ ▪ yes ▪ no

1/2

18. Is 50 a cube number?

☒ ▪ yes ▪ no

1/2

19. Is 99 a cube number?

☒ ▪ yes ▪ no

1/2

20. Is 3 a cube number?

☒ ▪ yes ▪ no

1/2

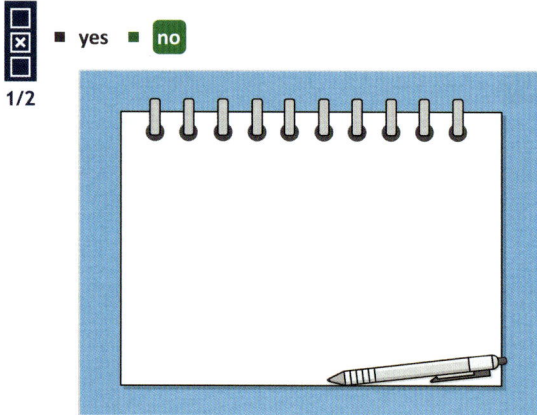

21. Is 88 a cube number?

☒ ▪ yes ▪ no

1/2

22. Select **all** the cube numbers.

☒ ▪ 8 ▪ 10 ▪ 16 ▪ 27 ▪ 48

2/5

Level 2 continued

23. Select **all** the cube numbers.

■ **1** ■ 2 ■ 4 ■ **8** ■ 16 ■ 32 ■ **64**

3/7

24. Select **all** the cube numbers.

■ **1** ■ 3 ■ 9 ■ **27** ■ 81 ■ **125** ■ 144

3/7

25. Select **all** the cube numbers.

■ 12 ■ **64** ■ 65 ■ 100 ■ **125** ■ **1,000**

3/6

Level 3: I can solve simple problems involving cube numbers.

✿ **Required:** 5/8 ✿ **Pupil Navigation:** off

✿ **Randomised:** on

26. Which cube number is between 0 and 5?

■ **1**

27. Which cube number is between 120 and 130?

■ **125**

28. Which cube number is between 6 and 20?

■ **8**

29. Which cube number is between 20 and 40?

 ▪ **27**

30. Which cube number is between 50 and 70?

▪ **64**

31. $1^3 + 2^3 =$ _____

▪ **9**

32. $2^3 + 10 =$ _____

▪ **18**

33. $3^3 + 1^3 =$ _____

▪ **28**

Practise Square Numbers

Objective:	I can recognise and find square numbers to 144 and use the notation for squared (²).
Quick Search Ref:	**10477**

1. What is 1^2?

123 ▪ 1

2. What is 2^2?

123 ▪ 4

3. What is 3^2?

123 ▪ 9

4. What is 4^2?

123 ▪ 16

5. What is 5^2?

123 ▪ 25

6. What is 6^2?

123 ▪ 36

Level 1 *continued*

7. What is 7²?

1 2 3 ▪ 49

8. What is 8²?

1 2 3 ▪ 64

9. What is 9²?

1 2 3 ▪ 81

10. What is 11²?

1 2 3 ▪ 121

11. What is 10²?

1 2 3 ▪ 100

12. What is 12²?

1 2 3 ▪ 144

Level 2: I can recognise square numbers.

✿ **Required:** 10/23   ✿ **Pupil Navigation:** off
✿ **Randomised:** on

13. Is 25 a square number?

☐ ☒ ☐ ▪ yes ▪ no
1/2

Level 2 continued

14. Is 49 a square number?

☐ ☒ ☐ 1/2   ▪ yes   ▪ no

15. Is 1 a square number?

☐ ☒ ☐ 1/2   ▪ yes   ▪ no

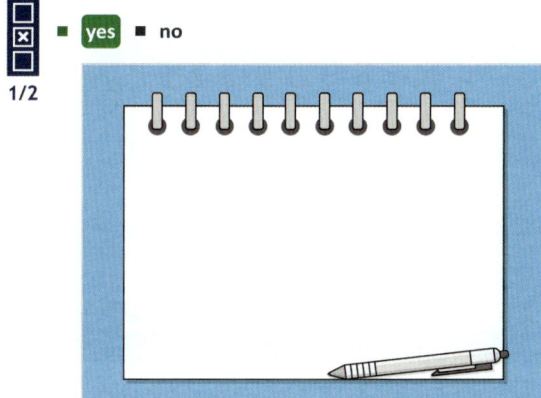

16. Is 64 a square number?

☐ ☒ ☐ 1/2   ▪ yes   ▪ no

17. Is 36 a square number?

☐ ☒ ☐ 1/2   ▪ yes   ▪ no

18. Is 121 a square number?

☐ ☒ ☐ 1/2   ▪ yes   ▪ no

19. Is 100 a square number?

☐ ☒ ☐ 1/2   ▪ yes   ▪ no

20. Is 16 a square number?

☐ ☒ ☐ 1/2   ▪ yes   ▪ no

21. Is 4 a square number?

☐ ☒ ☐ 1/2   ▪ yes   ▪ no

Level 2 *continued*

22. Is 9 a square number?

■ yes ■ no

1/2

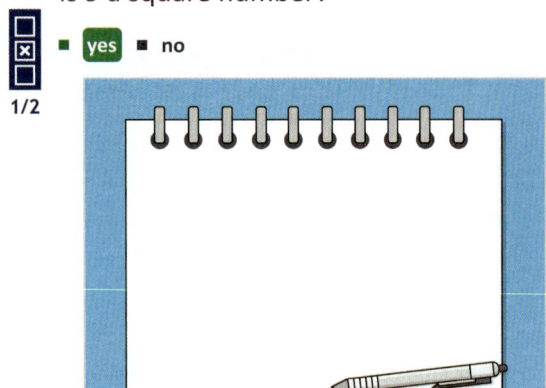

23. Is 81 a square number?

■ yes ■ no

1/2

24. Is 5 a square number?

■ yes ■ no

1/2

25. Is 12 a square number?

■ yes ■ no

1/2

26. Is 10 a square number?

■ yes ■ no

1/2

27. Is 20 a square number?

■ yes ■ no

1/2

28. Is 55 a square number?

■ yes ■ no

1/2

29. Is 78 a square number?

■ yes ■ no

1/2

Ref:10477   Practise Square Numbers

Level 2 *continued*

30. Is 99 a square number?

▪ yes ▪ no

1/2

31. Is 110 a square number?

▪ yes ▪ no

1/2

32. Is 140 a square number?

▪ yes ▪ no

1/2

33. Select **all** the square numbers.

2/7

▪ 1 ▪ 2 ▪ 3 ▪ 4 ▪ 5 ▪ 6 ▪ 7

34. Select **all** the square numbers.

4/7

▪ 30 ▪ 36 ▪ 45 ▪ 49 ▪ 64 ▪ 65 ▪ 81

35. Select **all** the square numbers.

3/7

▪ 100 ▪ 110 ▪ 121 ▪ 125 ▪ 130 ▪ 140 ▪ 144

Recognise and Use Square Numbers and Cube Numbers

Objective: I can recognise and use square numbers and cube numbers and the notation for squared (^{2}) and cubed (^{3}).

Quick Search Ref: 10041

Level 1: Understanding: Recognise and find square and cube numbers with image support.

✿ Required: 7/10 ✿ Pupil Navigation: on ✿ Randomised: off

1. Select the **two** square numbers.

■ **1** ■ 2 ■ 3 ■ **4** ■ 5

2/5

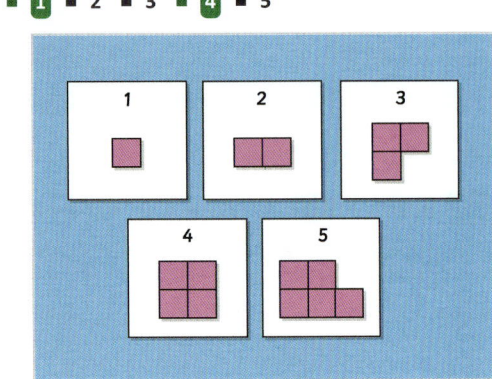

2. Which image shows how you can calculate 3 squared (3^2)?

■ image A ■ **image B** ■ image C ■ image D

1/4

A
$3^2 = 3 \times 2 = 6$

B
$3^2 = 3 \times 3 = 9$

C
$3^2 = 3 + 3 = 6$

D
$3^2 = 3 \times 3 \times 3 = 27$

3. What is 4^2?

■ **16** ■ 8 ■ 64

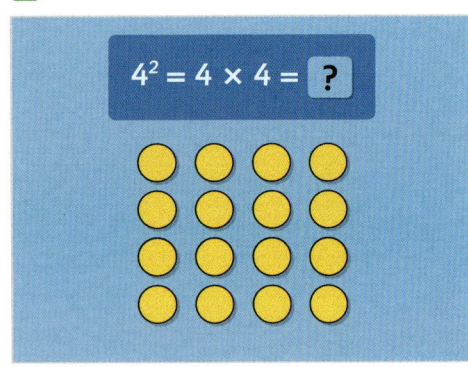

$4^2 = 4 \times 4 = \boxed{?}$

4. Calculate the value of 6^2.

■ **36** ■ 12 ■ 216

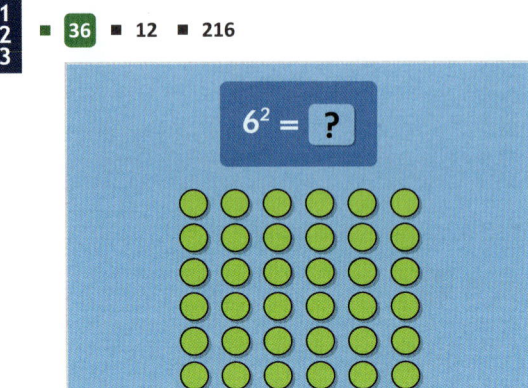

$6^2 = \boxed{?}$

5. Select the **two** cube numbers.

■ **1** ■ 2 ■ 4 ■ 5 ■ **8** ■ 6

2/6

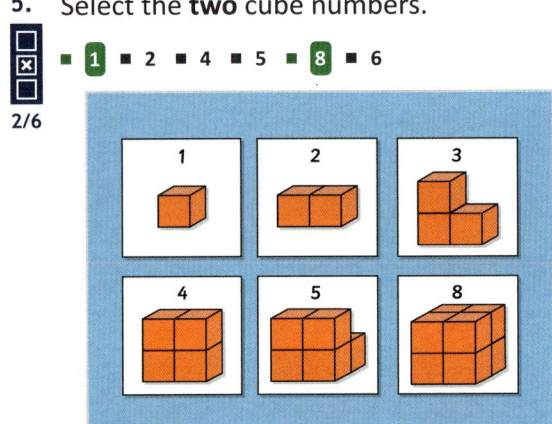

6. Which image shows how you can calculate 4 cubed (4^3)?

■ **image A** ■ image B ■ image C

1/3

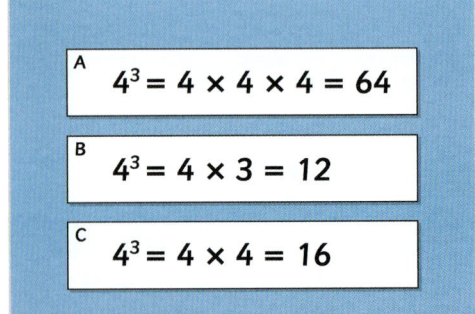

A
$4^3 = 4 \times 4 \times 4 = 64$

B
$4^3 = 4 \times 3 = 12$

C
$4^3 = 4 \times 4 = 16$

Level 1 *continued*

7. Calculate the value of 2^3.

■ **8** ■ 4 ■ 6

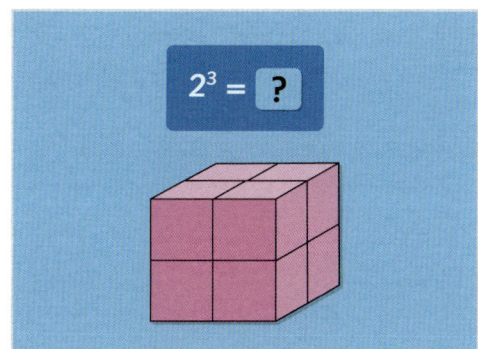

8. What is 3 cubed?

■ **27** ■ 9

9. Calculate the value of 10^2.

■ **100** ■ 20 ■ 1000

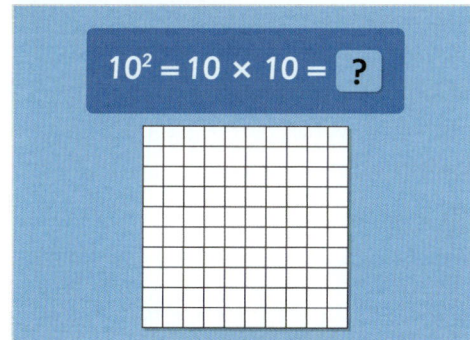

10. Select the correct option to complete the following statement:
5^2 means that you need to calculate _____.

1/4   ■ $5 \times 5 \times 5$ ■ 5×2 ■ **5×5** ■ $5 + 5$

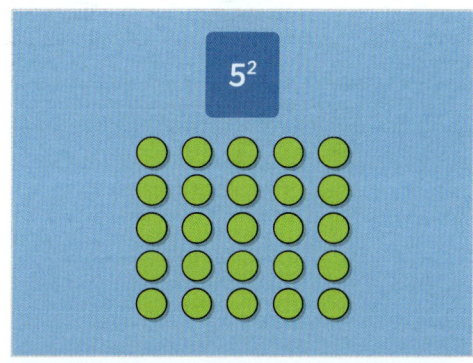

Level 2: Fluency: Recognise and find square and cube numbers.

✱ **Required:** 7/10   ✱ **Pupil Navigation:** on
✱ **Randomised:** off

11. What is 8^2?

■ 16 ■ **64** ■ 512

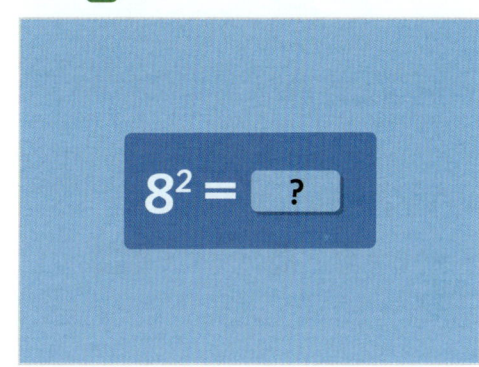

12. Select the **three** square numbers.

3/6   ■ **9** ■ **16** ■ 20 ■ **25** ■ 30 ■ 10

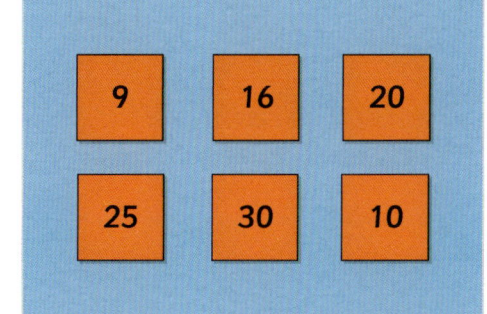

Level 2 *continued*

13. What is 4 cubed?

■ 16 ■ **64** ■ 12

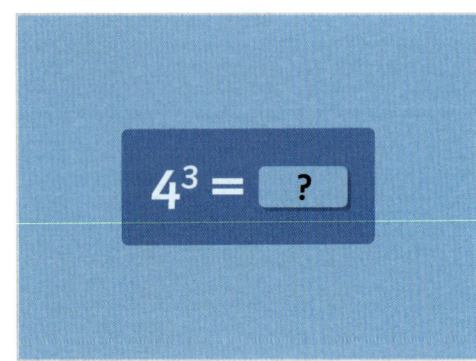

$$4^3 = \boxed{?}$$

14. What is 5^3?

■ **125** ■ 25 ■ 15

$$5^3 = \boxed{?}$$

15. Select the **three** cube numbers.

3/6

■ 9 ■ 16 ■ **8** ■ **27** ■ 100 ■ **1,000**

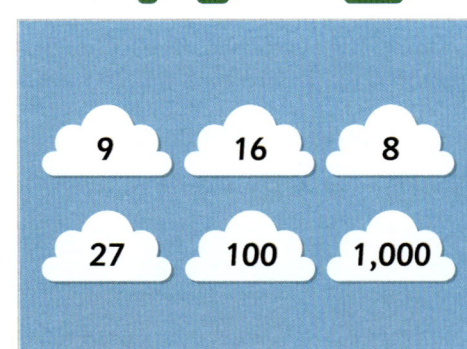

9 16 8

27 100 1,000

16. One side of a square playground measures 20 metres (m). What is the area of the playground in square metres (m²)?
Don't include the units in your answer.

■ **400** ■ 80 ■ 4 ■ 8000 ■ 40

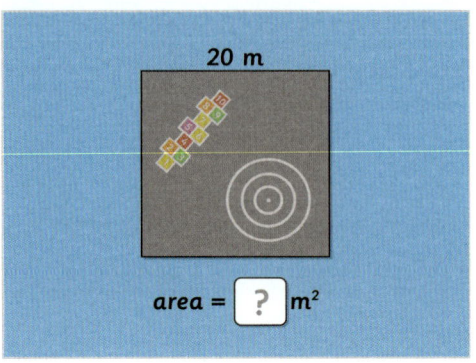

20 m

area = $\boxed{?}$ m²

17. Each edge of a small cube measures 1 centimetre (cm). How many small cubes do you need to make a large cube where each edge measures 10 cm?

■ **1000** ■ 100

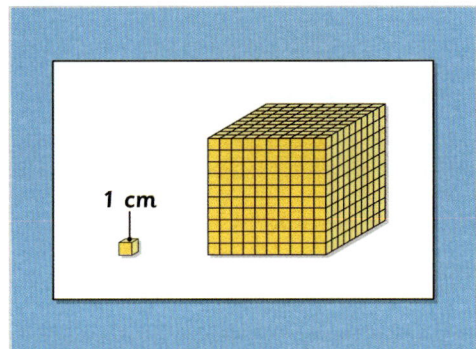

1 cm

18. One side of a square garden measures 9 metres (m). What is the area of the garden in square metres (m²)?

■ **81** ■ 36 ■ 729

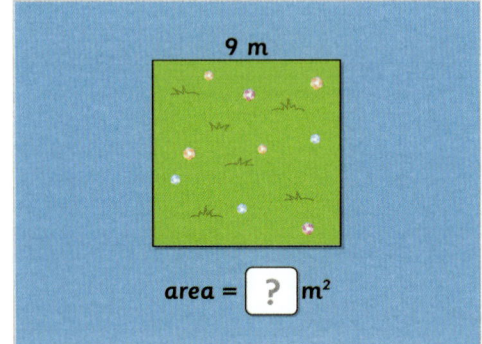

9 m

area = $\boxed{?}$ m²

Level 2 continued

19. What is 1³?

■ 1   ■ **1**   ■ 3

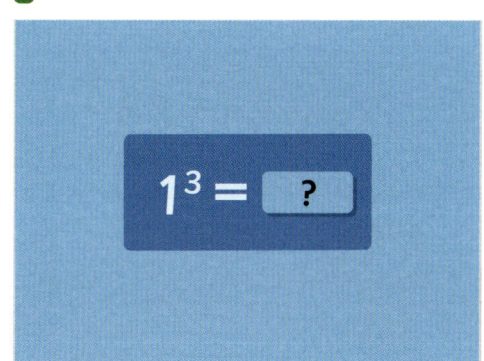

20. 7² = _____

■ 14   ■ **49**   ■ 343

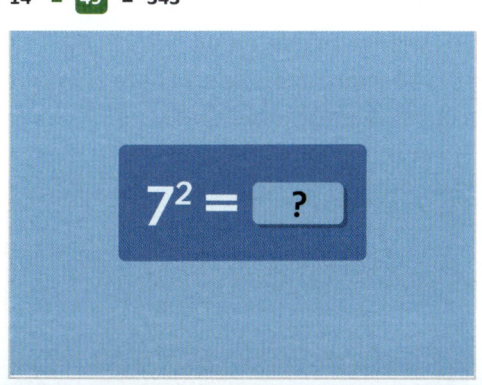

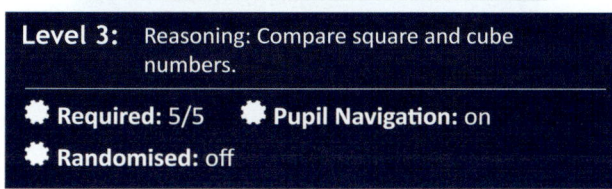

Level 3: Reasoning: Compare square and cube numbers.

🔧 **Required:** 5/5   🔧 **Pupil Navigation:** on

🔧 **Randomised:** off

21. Josh completes some questions on square numbers for his homework but gets all of the answers wrong. Explain why Josh's answers are wrong and give the correct answer for each question.

- Open question, no set answer

22. Select the symbol that makes the following statement true.

7² _____ 4³

1/3   ■ **<**   ■ >   ■ =

23. Which of these numbers is a square number **and** a cube number?

■ 27   ■ **64**   ■ 8   ■ 81   ■ 1000   ■ 49   ■ 125

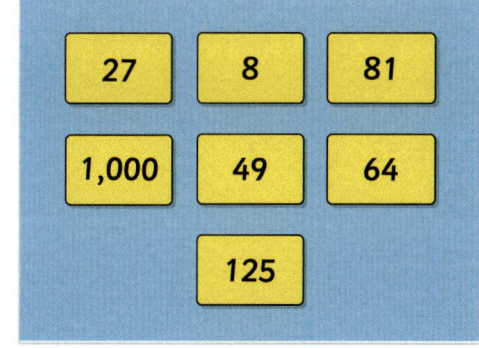

24. Which two calculations are equivalent to 9³?

■ **9 × 9²**   ■ 9 × 9   ■ **9 × 81**   ■ 9 + 99   ■ 9² + 1

2/5

Level 3 *continued*

25. Michael says, "The answer to $3^2 + 7^2$ is the same as the answer to 10^2."

a b c Do you agree with Michael? Explain your answer and prove that you are correct.

- Open question, no set answer

$3^2 + 7^2 = 10^2$

Level 4: Problem solving with greater depth: Solve multi-step problems involving cube and square numbers.

❋ **Required:** 5/5   ❋ **Pupil Navigation:** on
❋ **Randomised:** off

26. What is $52 + 6^2 - 2^3$?

1 2 3   ▪ 80

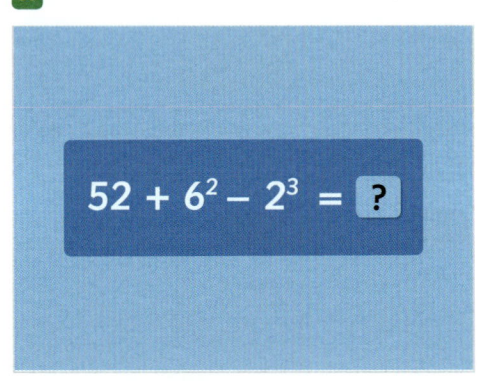

$52 + 6^2 - 2^3 = $?

27. What is the missing digit?

1 2 3 $3^2 + \_\_^2 = 25$

▪ 4   ▪ 16

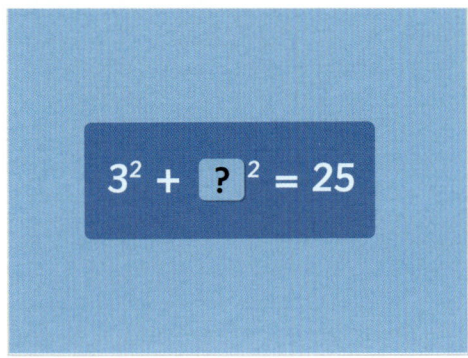

$3^2 + ?^2 = 25$

28. Fabrice's age is a **cube number**. Next year, his age will be a square number. How old is he now?

1 2 3

▪ 9   ▪ 8   ▪ 27

29. Pat is thinking of a number. Use the following clues to find Pat's number:

1 2 3
• His number is between 1 and 100.
• It is **10 more** than one square number and **7 less** than another.
• His number is **not** a square number.

▪ 74   ▪ 64   ▪ 81

30. Kit puts a whole number into the function machine and gets the answer 10^2. What number did he start with?

1 2 3

▪ 8   ▪ 64

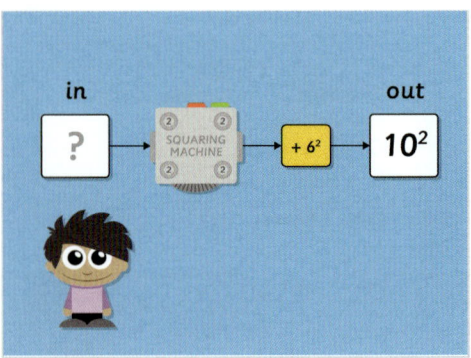

in → ? → SQUARING MACHINE → $+ 6^2$ → out 10^2

Mathematics **Y5**

Measurement

Area and Perimeter
Length
Mass
Volume and Capacity
Time

Find and Compare the Area of Rectangles and Squares

Objective: I can calculate and compare the area of rectangles, squares and composite shapes, using standard units.

Quick Search Ref: 10117

Level 1: Understanding: Finding the area of squares and rectangles.

✿ **Required:** 7/10   ✿ **Pupil Navigation:** on   ✿ **Randomised:** off

1. Which of the following statements describes **area**?

1/3

■ Area is the total distance around a 2D shape.
■ **Area is a measure of how much space there is on the surface of a 2D shape.**
■ Area is the amount that something can hold.

2. To find the area of a rectangle or square you need to perform which of the following calculations?

1/4

■ length + width   ■ length ÷ width   ■ **length x width**
■ length - width

3. **Area** is measured in:

■ Centimetres (cm)   ■ Cubic units (^{3})   ■ Metres (m)
■ **Square units (^{2})**

1/4

4. What is the area of the rectangle?

■ 16 cm   ■ **15 cm²**   ■ 15 cm   ■ 8 cm²

1/4

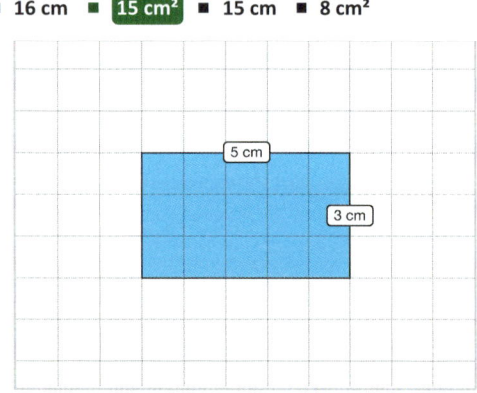

5. What is the area of the square?

■ 12 cm   ■ **9 cm²**   ■ 6 cm²

1/3

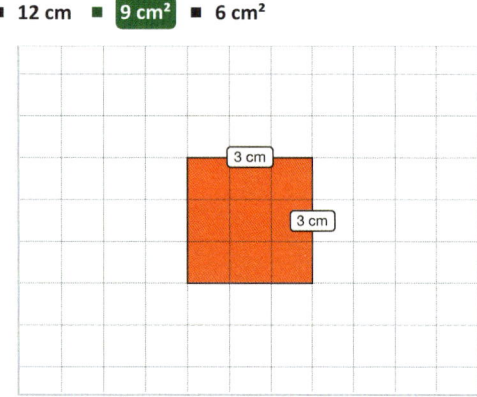

6. The area of the rectangle is _____ cm².

1 2 3

■ **70**   ■ 17   ■ 34

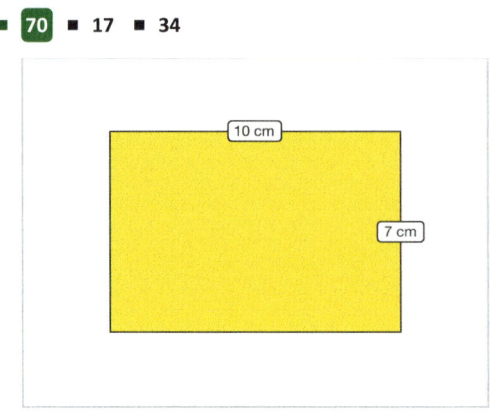

7. The area of the square is _____ cm².

1 2 3

■ **36**   ■ 24   ■ 12

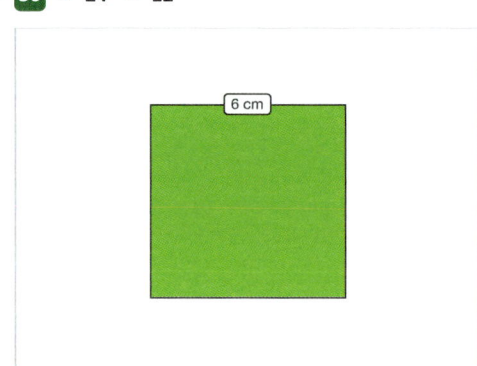

8. The area of the rectangle is _____ m².

1 2 3

■ **36**   ■ 30   ■ 15

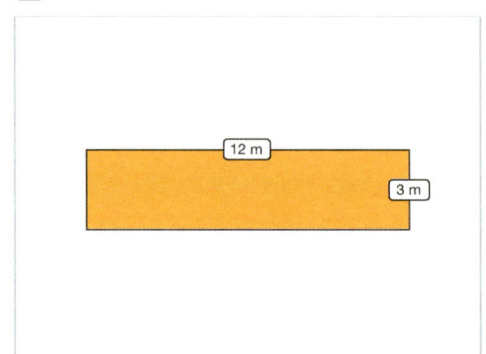

Level 1 continued

9. The area of the rectangle is _____ m².

■ 28 ■ **48** ■ 14

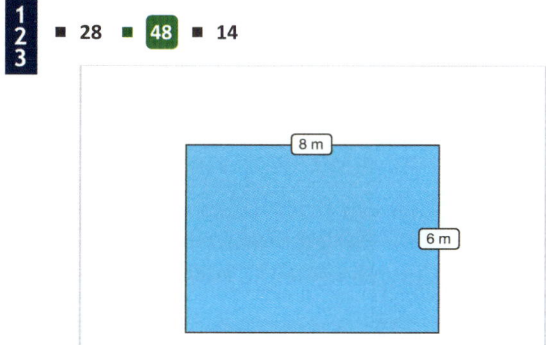

10. The area of the rectangle is _____ cm².

■ **54** ■ 30 ■ 15

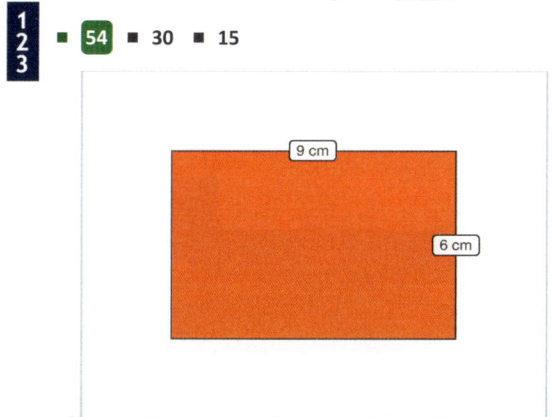

Level 2: Fluency: Comparing and estimating the area of squares, rectangles and composite shapes.

✿ **Required:** 7/10 ✿ **Pupil Navigation:** on
✿ **Randomised:** off

11. Select **three** shapes on the centimetre squared paper which have an area of 12 square centimetres.

3/5

■ **Shape A** ■ Shape B ■ Shape C ■ **Shape D**
■ **Shape E**

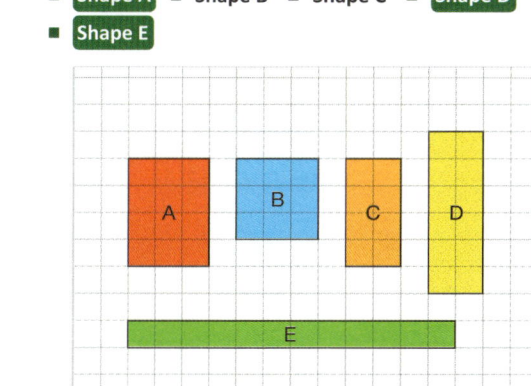

12. Find the area of each shape, then order them, **smallest** area first.

■ Shape B ■ Shape E ■ Shape A ■ Shape D
■ Shape C

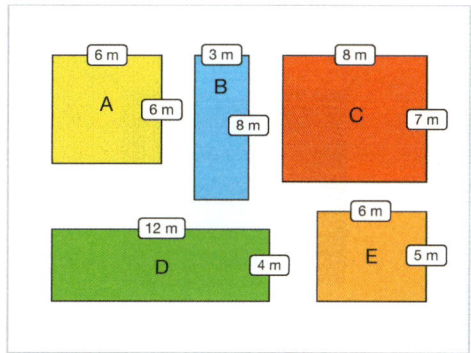

13. **Estimate** the area of the rectangle in square centimetres.

■ **18** ■ 12 ■ 17.01 ■ 14 ■ 9 ■ 21

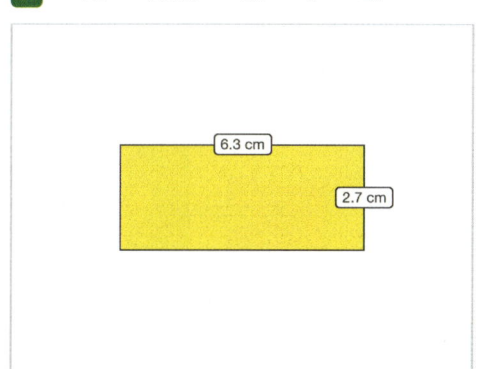

14. Calculate the **total area** of the composite rectilinear shape in square metres.

■ **28** ■ 16 ■ 12 ■ 22

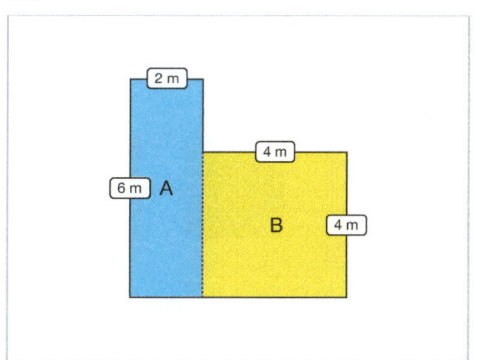

15. **Estimate** the area of the rectangle in square
metres.

- 54 ▪ **56** ▪ 63 ▪ 55.89 ▪ 48

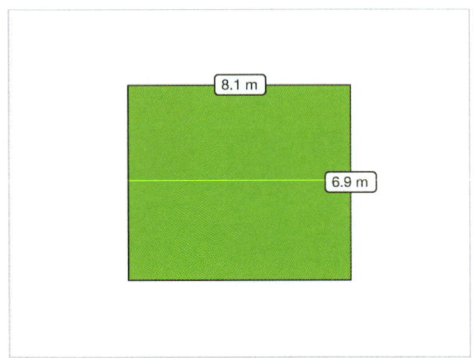

16. Select **two** shapes which have an area of 18
square centimetres.

2/5

- Shape A ▪ Shape B ▪ **Shape C** ▪ **Shape D**
- Shape E

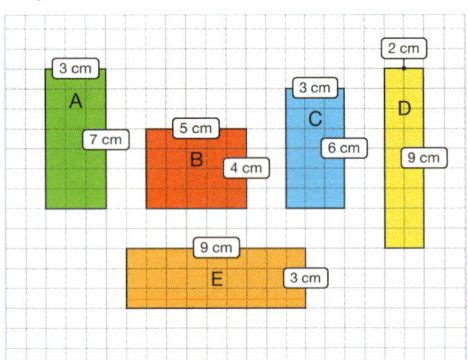

17. Find the area of each shape, then order
them, **smallest** area first.

- **Shape B** ▪ **Shape C** ▪ **Shape A** ▪ **Shape D**
- **Shape E**

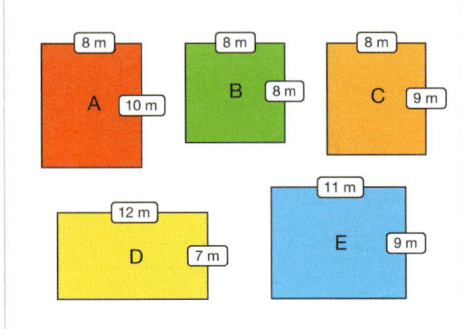

18. Calculate the **total area** of the composite
rectilinear shape in square metres.

- **122** ▪ 45 ▪ 50 ▪ 77

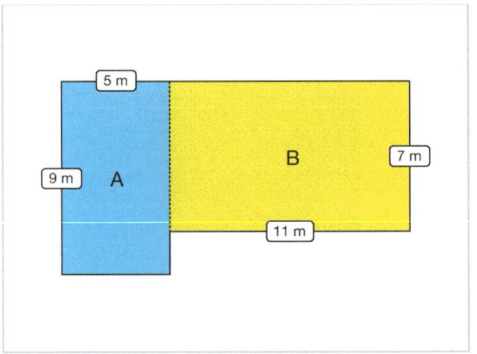

19. Calculate the **total area** of the composite
rectilinear shape in square centimetres.

- **49** ▪ 32

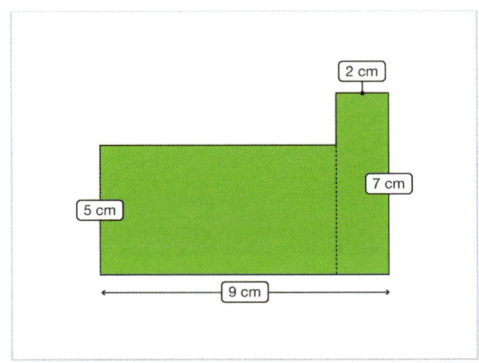

20. Find the **total area** of the composite
rectilinear shape in square metres.

- **40** ▪ 30

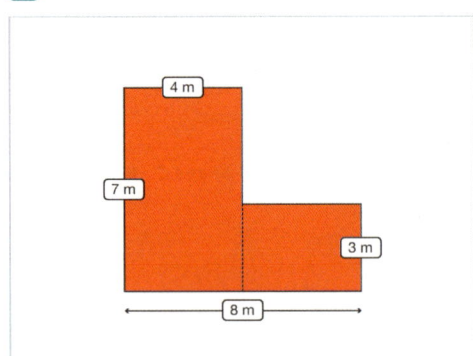

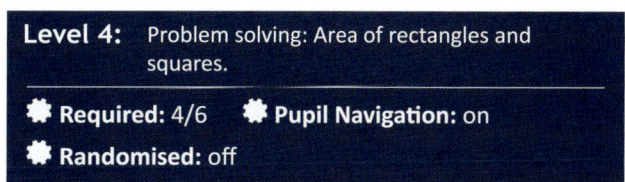

Level 3: Reasoning: Area of rectangles and squares.

❋ **Required:** 4/6   ❋ **Pupil Navigation:** on
❋ **Randomised:** off

21. The rectangle has a perimeter of 4,200
centimetres and width of seven metres.

What is the **surface area** of the rectangle in
square metres?

■ 294   ■ 98   ■ 196

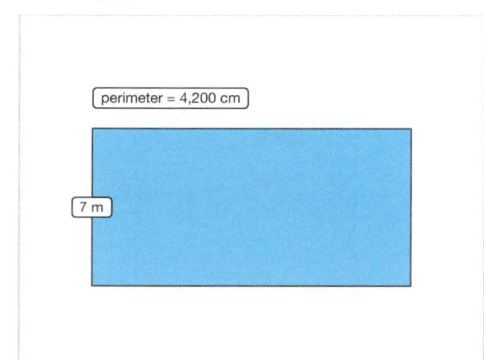

perimeter = 4,200 cm

7 m

22. Sam says, "The area of a square will always
be a square number."

Is he correct? Explain your answer.

- Open question, no set answer

23. A square has an area of 4 square metres.
What is its area in **square centimetres**?

■ 100,000   ■ 40,000   ■ 200   ■ 800   ■ 400

1/5

24. A page in a sticker book has a length and
width of 12 centimetres and each sticker has
an area of four square centimetres. What is
the maximum number of stickers you could
stick on one page?

■ 36   ■ 576

25. Erica says, "The area of a rectangle will
never be a square number."

Is she correct? Explain your answer.

- Open question, no set answer

26. A square has an area of 25 square metres.
What is its area in **square centimetres**?

■ 250,000   ■ 500   ■ 2,000   ■ 2,500

1/4

Level 4: Problem solving: Area of rectangles and
squares.

❋ **Required:** 4/6   ❋ **Pupil Navigation:** on
❋ **Randomised:** off

27. Highbury Primary School is having the
netball court resurfaced. The court is 30
metres long and 15 metres wide.

It costs **£20.00** to resurface **one square
metre** of the court. How much, in pounds,
will it cost to resurface the full court?

■ 9000   ■ 900   ■ 450   ■ 1800

30 m

15 m

28. Diagram (a) shows a bathroom tile,
which has the design of an **equilateral
triangle** within a square. The tile has an area
of **81 square centimetres**.

Diagram (b) shows four of the tiles laid out
on the floor.

In **centimetres**, what is the **perimeter of the
red shape** in diagram (b)?

■ 72   ■ 27

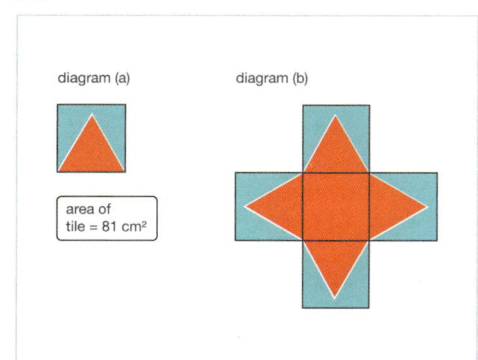

diagram (a)   diagram (b)

area of
tile = 81 cm²

29. Salma's mum is decorating her living room and needs to buy some wallpaper for one wall. The wall measures five metres by three metres.

The wallpaper is sold in rolls of **two square metres**. How many **full rolls** will she need to buy to decorate the wall?

■ 15 ■ **8** ■ 16

30. Highbury Primary School is having a gymnastics competition and needs to buy new mats to cover the gym floor.

What is the **total cost** of the gym mats in pounds if each mat costs £2.50?

■ 32 ■ **80** ■ 128

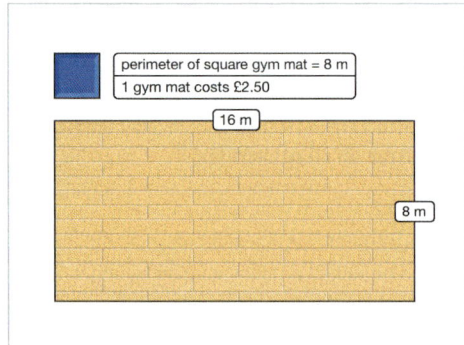

31. Jack puts a poster for his new band in the window of the local post office.

The window measures eight metres wide and has a length of six metres. The poster covers one quarter of the window.

What is the **area** of the poster in **square metres**?

■ **12** ■ 14 ■ 48

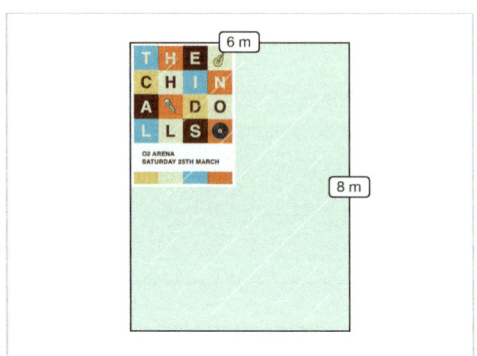

32. Bacup Cricket Club is having its field returfed for the new season.

The field is 60 metres wide and 140 metres long. The wicket area, which is **3 metres** wide and **20 metres** long, **does not** need to be returfed

If one square metre of turf costs **£2.00**, what will be the **total cost** in pounds to returf the cricket field excluding the wicket area?

■ **16680** ■ 8400 ■ 8340 ■ 60

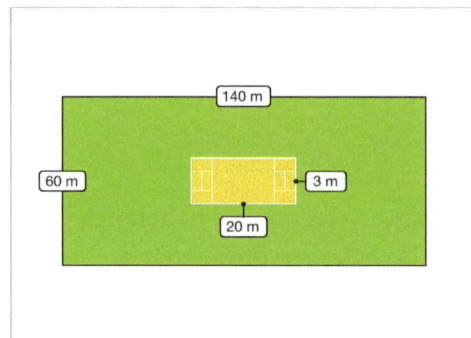

Convert Between Different Units of Length (cm and mm)

Objective: I can convert between centimetres and millimetres.

Quick Search Ref: 10697

Level 1: Understanding: Convert centimetres and millimetres.

✱ **Required:** 7/10   ✱ **Pupil Navigation:** on   ✱ **Randomised:** off

---

1. How many millimetres are there in one centimetre?

1/3

- ■ **10 millimetres** ■ 100 millimetres
- ■ 1,000 millimetres

2. How do you convert a measurement from centimetres to millimetres?

1/4

- ■ divide by 10   ■ divide by 100   ■ **multiply by 10**
- ■ multiply by 100

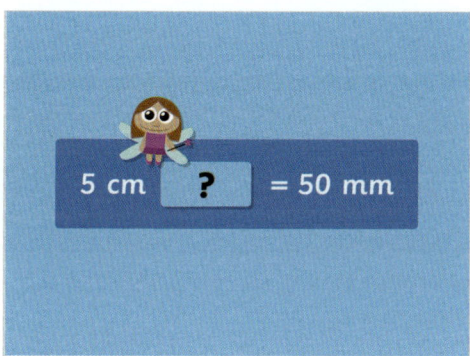

5 cm **?** = 50 mm

3. 3 centimetres = _____ millimetres
Don't include the units in your answer.

- ■ 300 ■ **30** ■ 0.3

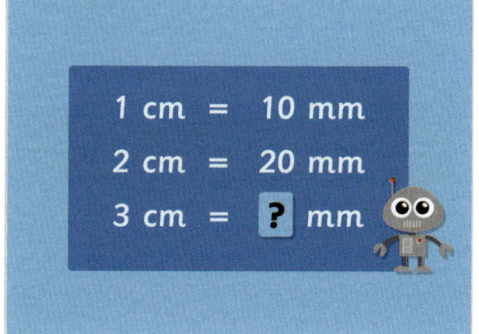

1 cm = 10 mm
2 cm = 20 mm
3 cm = **?** mm

4. How many centimetres are there in 10 millimetres?
Include the units cm (centimetres) in your answer.

- ■ **1 centimetres** ■ 100 ■ **1 centimetre** ■ **1 cm**
- ■ 100 cm ■ 1 ■ 100 centimetres

5. How do you convert a measurement from millimetres to centimetres?

1/4

- ■ multiply by 10   ■ multiply by 100   ■ **divide by 10**
- ■ divide by 100

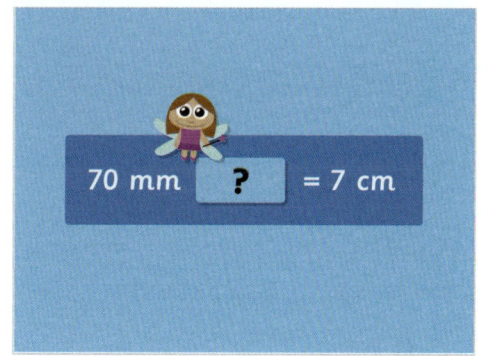

70 mm **?** = 7 cm

Level 1 continued

6. How many centimetres are there in 50 millimetres?

a b c

Include the units cm (centimetres) in your answer.

- **5 cm** - 5 - **5 centimetres** - 500 - 500 cm
- 500 centimetres

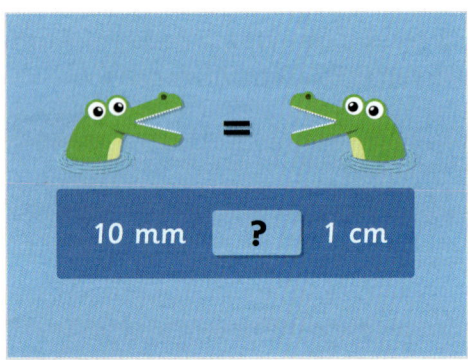

7. What is the missing symbol in this number sentence?

10 millimetres ____ 1 centimetre

1/3 - > - < - **=**

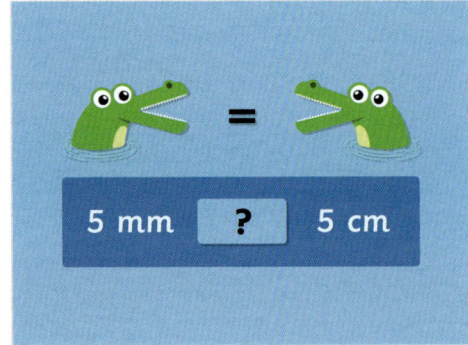

8. What is the missing symbol in this number sentence?

5 millimetres ____ 5 centimetres

1/3 - > - **<** - =

9. 6 centimetres = ____ millimetres

1 2 3

Don't include the units in your answer.

- 600 - **60** - 0.6 - 6000

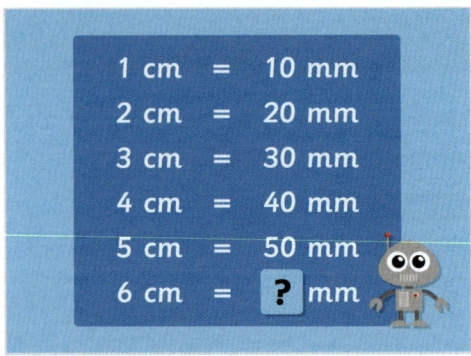

10. 40 millimetres = ____ centimetres

1 2 3

Don't include the units in your answer.

- **4** - 400

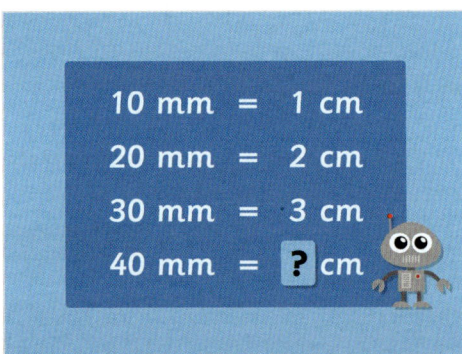

Level 2: Fluency:- Convert centimetres with 1 decimal place to millimetres and vice versa.
___
✱ **Required:** 7/10   ✱ **Pupil Navigation:** on
✱ **Randomised:** off

11. How many millimetres are there in 6.5 centimetres?

a b c

Include the units mm (millimetres) in your answer.

- 0.65 millimetres - **65 mm** - **65 millimetres** - 65
- 0.65 mm - 0.65

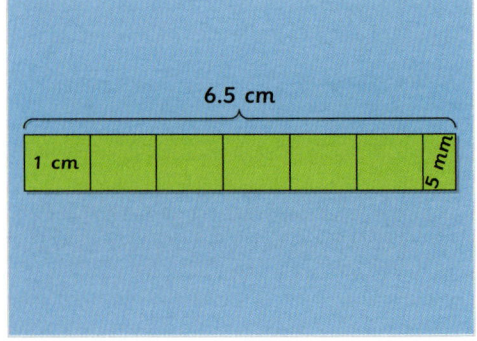

Level 2 *continued*

12. 9.7 centimetres = ____ millimetres

Don't include the units in your answer.

- **97**   ▪ **0.97**

13. 58 millimetres is the same as how many centimetres?

Include the units cm (centimetres) in your answer.

- **5.8 cm**   ▪ **5.8 centimetres**   ▪ 580 cm   ▪ 5.8   ▪ 580
- 580 centimetres

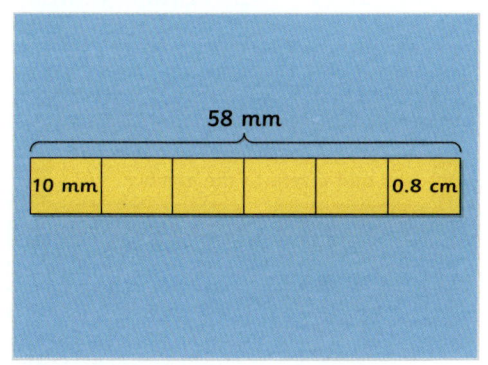

14. What is the missing symbol in this number sentence?

6.7 centimetres ____ 76 millimetres

1/3

▪ >   ▪ **<**   ▪ =

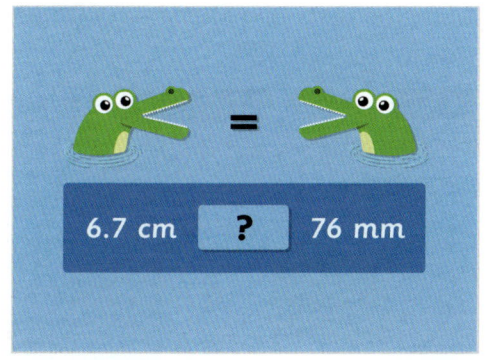

15. Sort the lengths from shortest to longest. Put the shortest length first.

- **29 millimetres**   ▪ **3.5 centimetres**   ▪ **37 millimetres**
- **4.2 centimetres**

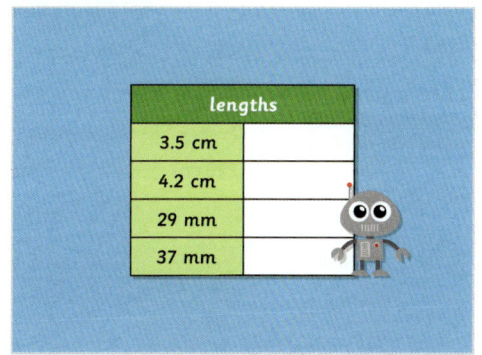

16. A book is 145 millimetres tall. How tall is the book in centimetres?

Include the units cm (centimetres) in your answer.

- 14.5   ▪ 1,450   ▪ **14.5 cm**   ▪ **14.5 centimetres**
- 1,450 centimetres   ▪ 1,450 cm   ▪ 1450 cm   ▪ 1450
- 1450 centimetres

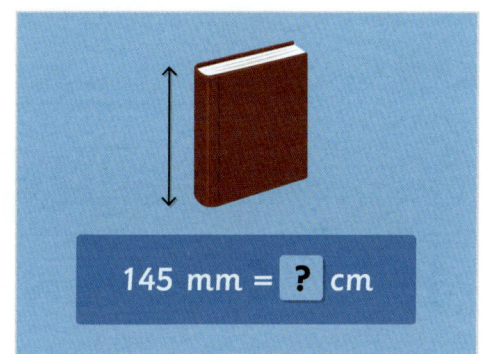

17. Pete's watch is 13.4 centimetres long. How long is Pete's watch in millimetres?

Include the units mm (millimetres) in your answer.

- 134   ▪ 1.34 millimetres   ▪ **134 mm**
- **134 millimetres**   ▪ 1.34 mm   ▪ 1.34

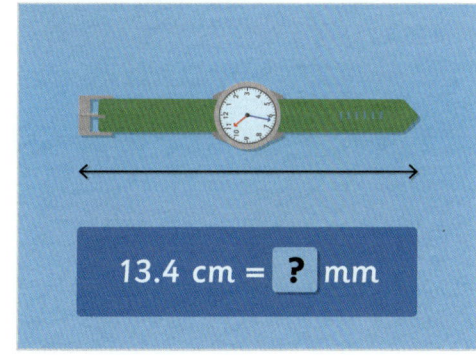

Level 2 *continued*

18. Alexa makes a paper aeroplane that is 227 millimetres long. How long is the paper aeroplane in centimetres?
Include the units cm (centimetres) in your answer.

- 2,270 centimetres - **22.7 cm** - 2270
- **22.7 centimetres** - 22.7 - 2270 centimetres
- 2270 cm - 2,270 - 2,270 cm

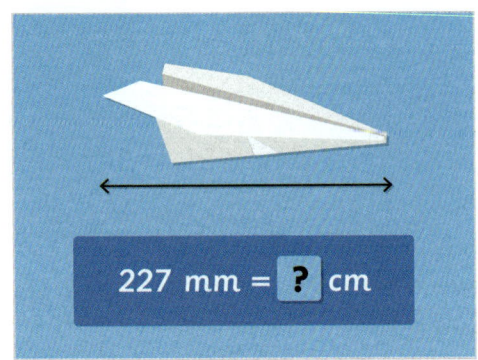

227 mm = **?** cm

19. What is the missing symbol in this number sentence?
5.7 centimetres ___ 78 millimetres

1/3 - > - **<** - =

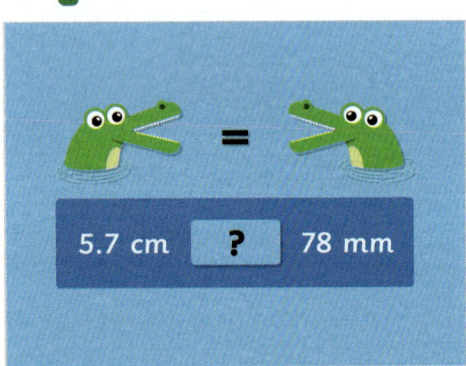

5.7 cm **?** 78 mm

20. Sort the lengths from longest to shortest. Put the longest length first.

- **11.2 centimetres** - **101 millimetres**
- **8.9 centimetres** - **72 millimetres**

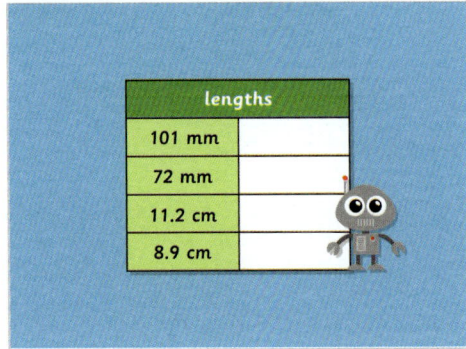

lengths	
101 mm	
72 mm	
11.2 cm	
8.9 cm	

Level 3: Reasoning: Use reasoning to convert lengths.

- **Required:** 5/5 - **Pupil Navigation:** on
- **Randomised:** off

21. What is the missing symbol in this number sentence?
37 mm ___ 5 cm – 12 mm

- **<** - > - =

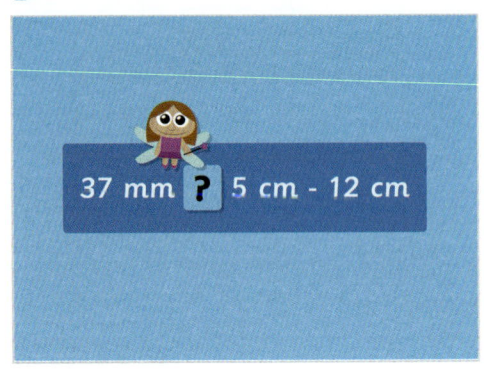

37 mm **?** 5 cm - 12 cm

22. Ariana says, "To convert centimetres to millimetres, you add a zero to the number." Is Ariana correct? Explain your answer.

- Open question, no set answer

To convert centimetres to millimetres you add a zero to the number.

23. Each row in the box adds up to 8 centimetres. What is the total value of A, B and C in millimetres?
Include the units mm (millimetres) in your answer.

- **107 mm** - 107 - **107 millimetres** - 27 mm - 27
- 27 millimetres

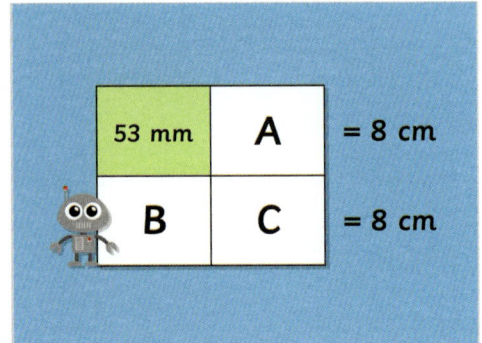

53 mm	A	= 8 cm
B	C	= 8 cm

Level 3 *continued*

24. Mark says, "5 centimetres is 10 times longer than 5 millimetres." Is Mark correct? Explain your answer.

a b c

- Open question, no set answer

25. Name one thing that is the same, and one thing that is different, about the measurements 1/2 centimetre and 5 millimetres.

a b c

- Open question, no set answer

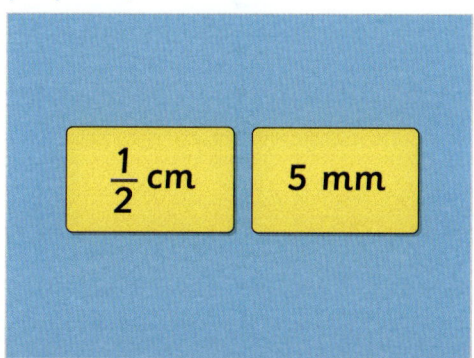

Level 4: Problem Solving: Solve multi-step problems.

✱ **Required:** 5/5 ✱ **Pupil Navigation:** on
✱ **Randomised:** off

26. A snail crawls 250 millimetres every 2 minutes. How many minutes does the snail take to crawl 100 centimetres?
Don't include the units in your answer.

1 2 3

▪ **8** ▪ 4

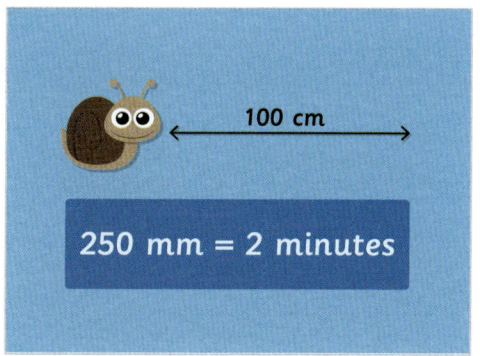

27. Alex and Bert each throw a dart to see who can get closest to the bullseye. Alex's dart is 10.8 centimetres from the bullseye. Bert's dart is 23 millimetres closer to the bullseye than Alex's dart. How far from the bullseye is Bert's dart in millimetres?
Include the units mm (millimetres) in your answer.

a b c

▪ 85 ▪ **85 mm** ▪ **85 millimetres** ▪ 8.5 cm ▪ 8.5
▪ 8.5 centimetres

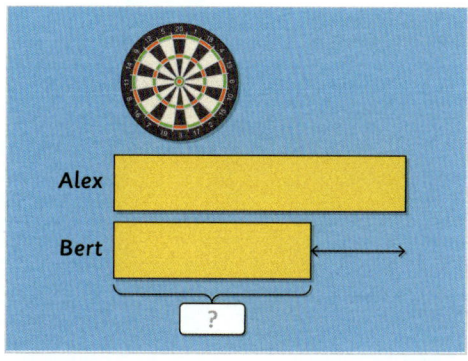

28. Georgina has two pieces of card. Card A is 13 centimetres (cm) wide and card B is 270 millimetres wide. Georgina joins card A and B together to make a piece of card that is 30 cm wide. What is the length of the overlap in centimetres?
Include the units (cm) in your answer.

a b c

▪ 100 centimetres ▪ 10 ▪ 100 ▪ **10 cm**
▪ **10 centimetres** ▪ 40 cm ▪ 40 centimetres
▪ 100 cm ▪ 40

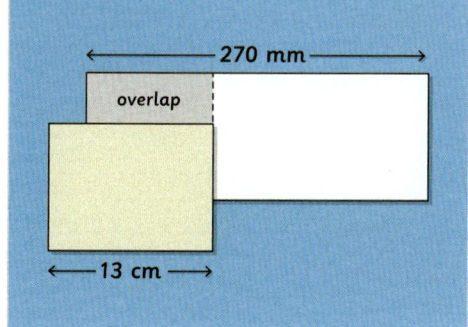

Level 4 *continued*

29. Serena is painting a beach and the sea on a piece of card for an art project. The piece of card is 24 centimetres tall. The sea is 3 times bigger than the beach. What is the length of the section of the card that is painted with the sea in millimetres?

a
b
c

Include the units mm (millimetres) in your answer.

- 18 centimetres   - 180   - 180 mm
- 180 millimetres   - 18 cm   - 18

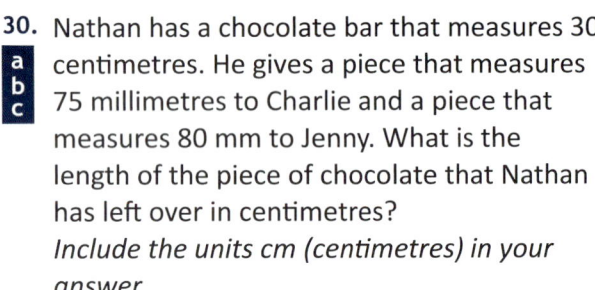

30. Nathan has a chocolate bar that measures 30 centimetres. He gives a piece that measures 75 millimetres to Charlie and a piece that measures 80 mm to Jenny. What is the length of the piece of chocolate that Nathan has left over in centimetres?

a
b
c

Include the units cm (centimetres) in your answer.

- 14.5 centimetres   - 14.5 cm   - 14.5
- 145 millimetres   - 145 mm   - 145

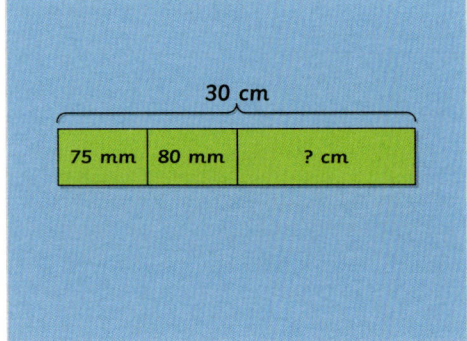

Convert Between Different Units of Length (km and m)

Objective: I can convert between kilometres and metres.

Quick Search Ref: 10637

1. How many metres are there in one kilometre?

1/3

- 10 metres   - 100 metres   - **1,000 metres**

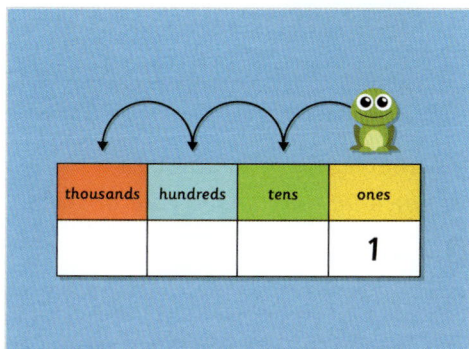

2. How do you convert a measurement from kilometres to metres?

1/4

- divide by 100   - divide by 1,000   - multiply by 100
- **multiply by 1,000**

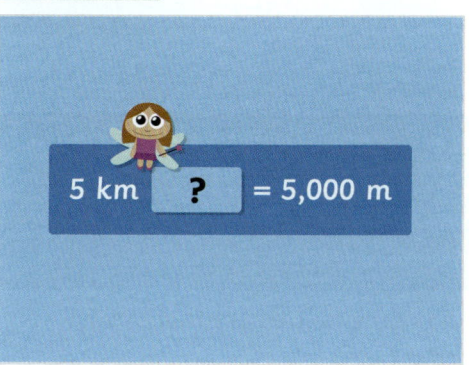

3. 3 kilometres = _____ metres
Don't include the units in your answer.

- 0.003   - **3000**   - 300

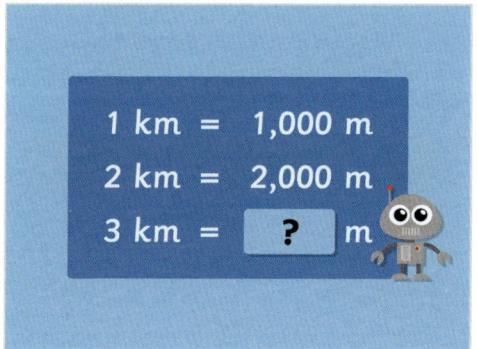

4. How many kilometres are there in 1,000 metres?

- **1**   - 10

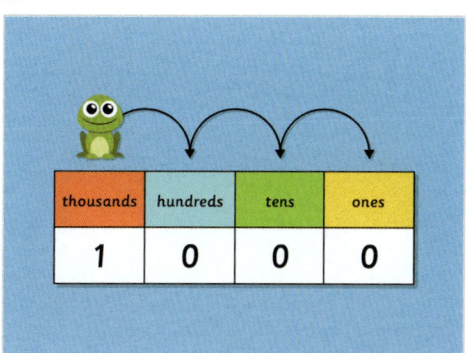

5. How do you convert a measurement from metres to kilometres?

1/4

- divide by 100   - **divide by 1,000**   - multiply by 100
- multiply by 1,000

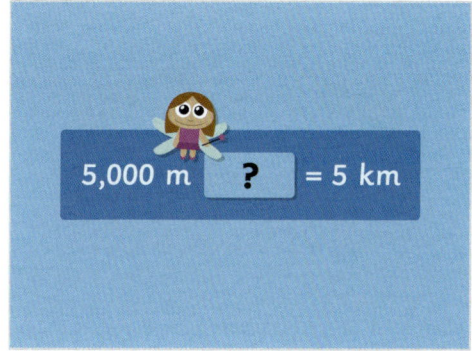

Level 1 *continued*

6. How many kilometres are there in 5,000 metres?

Include the units km (kilometres) in your answer.

- 5000000 ▪ **5 kilometres** ▪ 5000000 km ▪ **5 km**
- 50 km ▪ 5 ▪ 5,000,000 ▪ 5,000,000 km
- 50 kilometres ▪ 5,000,000 kilometres
- 5000000 kilometres ▪ 50

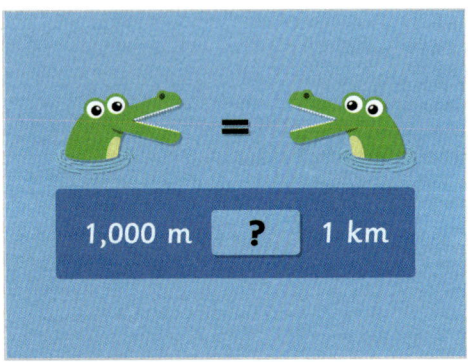

```
1,000 m  =  1 km
2,000 m  =  2 km
3,000 m  =  3 km
4,000 m  =  4 km
5,000 m  =  ? km
```

7. What is the missing symbol in this number sentence?

1,000 metres ___ 1 kilometre

1/3 ▪ > ▪ < ▪ **=**

1,000 m **?** 1 km

8. What is the missing symbol in this number sentence?

5 metres ___ 5 kilometres

1/3 ▪ > ▪ **<** ▪ =

5 m **?** 5 km

9. 6 kilometres = ___ metres

Don't include the units in your answer.

- 600 ▪ **6000** ▪ 0.006

```
1 km  =  1,000 m
2 km  =  2,000 m
3 km  =  3,000 m
4 km  =  4,000 m
5 km  =  5,000 m
6 km  =  ? m
```

10. 4,000 metres = ___ kilometres

Don't include the units in your answer.

- **4** ▪ 40

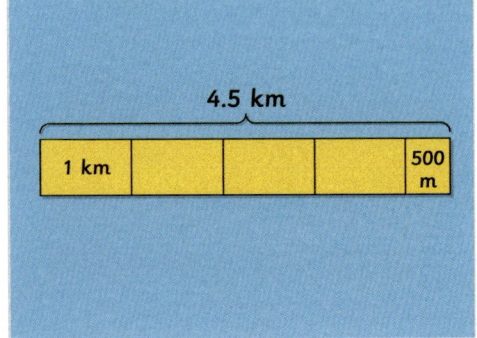

```
1,000 m  =  1 km
2,000 m  =  2 km
3,000 m  =  3 km
4,000 m  =  ? km
```

Level 2: Fluency: Convert kilometres with 1 decimal place to metres and vice versa.

❋ **Required:** 7/10   ❋ **Pupil Navigation:** on
❋ **Randomised:** off

11. How many metres are there in 4.5 kilometres?

Include the units m (metres) in your answer.

- 450 ▪ **4,500 metres** ▪ **4500 metres** ▪ **4,500 m**
- 4,500 ▪ **4500 m** ▪ 4500 ▪ 45

4.5 km

| 1 km | | | 500 m |

Level 2 continued

12. 8.2 kilometres = ___ metres
Don't include the units in your answer.

■ 820   ■ **8200**   ■ 82

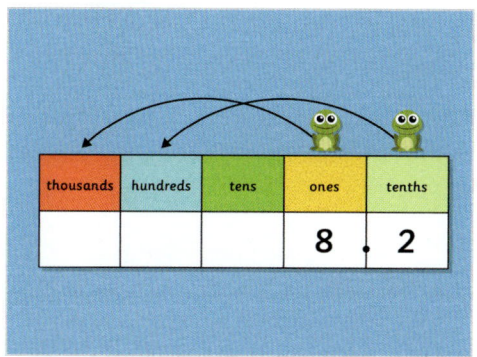

13. How many kilometres is 7,800 metres the same as?
Include the units km (kilometres) in your answer.

■ 7.8   ■ 78   ■ 780   ■ **7.8 kilometre**   ■ **7.8 km**
■ 780 kilometres   ■ 780 km   ■ 78 km   ■ 78 kilometres

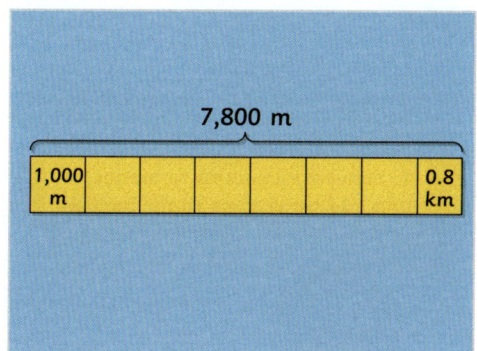

14. What is the missing symbol in this number sentence?
4.5 kilometres ___ 5,400 metres

1/3

■ >   ■ **<**   ■ =

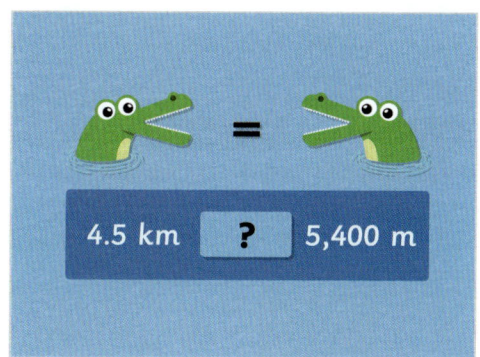

15. Sort the distances from shortest to longest. Put the shortest distance first.

■ **500 metres**   ■ **1.2 kilometres**   ■ **1,800 metres**
■ **10 kilometres**

distances
1.2 km
10 km
500 m
1,800 m

16. Chris rides 9,600 metres on his bike. How many kilometres does he ride?
Include the units km (kilometres) in your answer.

■ 960 kilometres   ■ 96 km   ■ 960   ■ **9.6 kilometres**
■ **9.6 km**   ■ 9.6   ■ 96 kilometres   ■ 96   ■ 960 km

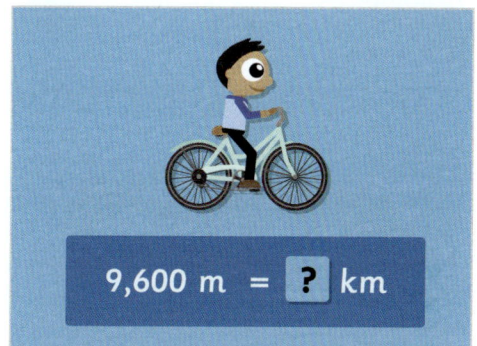

9,600 m = [?] km

17. A race is 4.8 kilometres long. How many metres is this?
Include the units m (metres) in your answer.

■ **4,800 metres**   ■ **4,800 m**   ■ 480 m   ■ **4800 metres**
■ **4800 m**   ■ 48 metres   ■ 48 m   ■ 480 metres   ■ 480
■ 48

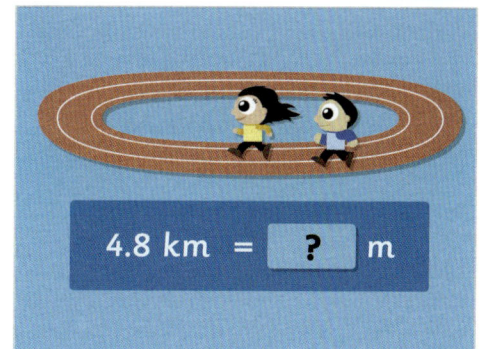

4.8 km = [?] m

Level 2 *continued*

18. The distance from Downtown to Tipsville is
11,300 metres. How many kilometres is this?
Include the units km (kilometres) in your answer.

- 1,130 ▪ 11.3 ▪ 113 km ▪ **11.3 km** ▪ 1,130 km
- 1130 ▪ **11.3 kilometres** ▪ 113 ▪ 1,130 kilometres
- 113 kilometres ▪ 1130 kilometres ▪ 1130 km

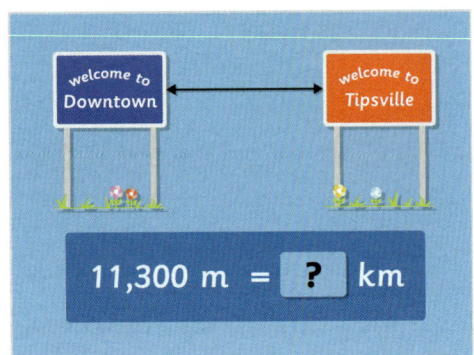

19. What is the missing symbol in this number sentence?

8.7 kilometres ____ 7,800 metres

1/3 ▪ **>** ▪ < ▪ =

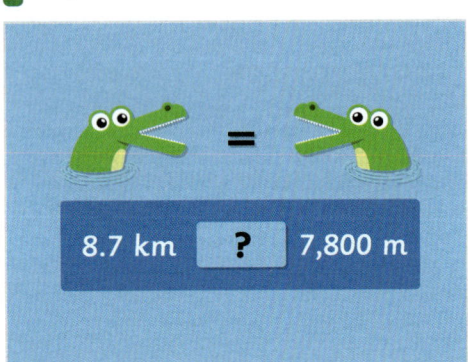

20. Sort the distances from longest to shortest. Put the longest distance first.

- **8 kilometres** ▪ **6,200 metres** ▪ **4.8 kilometres**
- **480 metres**

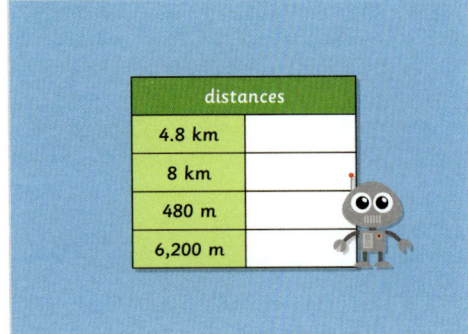

Level 3: Reasoning: Use reasoning to convert lengths.

❋ **Required:** 5/5 ❋ **Pupil Navigation:** on
❋ **Randomised:** off

21. Saffron says, "10.2 kilometres is the same as the total of 10 kilometres and 2 metres." Is Saffron correct? Explain your answer.

- Open question, no set answer

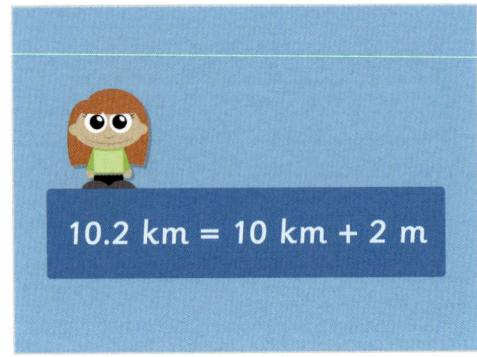

22. Marsha says, "To convert kilometres to metres, you add three zeroes to the number." Is Marsha correct? Explain your answer.

- Open question, no set answer

23. Each row in the box adds up to 5 kilometres. What is the total value of A, B and C in kilometres?
Include the units km (kilometres) in your answer.

- **6.5 km** ▪ 1.5 ▪ **6.5 kilometres** ▪ 5 km
- 1.5 kilometres ▪ 6.5 ▪ 5 kilometres ▪ 5 ▪ 1.5 km

Level 3 continued

24. Reggie says, "5 kilometres is 1,000 times longer than 5 metres." Is Reggie correct? Explain your answer.

a b c

- Open question, no set answer

5 kilometres is 1,000 times longer than 5 metres.

25. A taxi journey costs £2 for every 500 metres. Rick spends £6 on his journey. How far did he travel in kilometres?

a b c

Include the units km (kilometres) in your answer.

- **1.5 kilometres** ∎ 1500 metres ∎ 1,500 m ∎ **1.5 km**
- 1500 ∎ 1,500 metres ∎ 1,500 ∎ 1.5 ∎ 1500 m

price: £2 every 500 m

26. Gemma can run 250 metres in 2 minutes. How many minutes does it take Gemma to run 2 kilometres?

1 2 3

Don't include the units in your answer.

- **16** ∎ 8

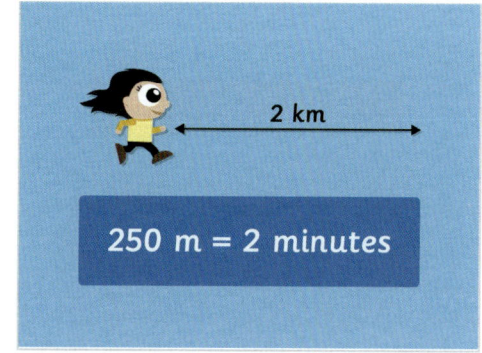

250 m = 2 minutes

27. Hillside is 2.8 kilometres away from the beach. Waverton is 900 metres closer to the beach than Hillside. How many kilometres is Waverton from the beach?

a b c

Include the units km (kilometres) in your answer.

- **1.9 kilometres** ∎ 1.9 ∎ **1.9 km** ∎ 1900 m
- 1,900 metres ∎ 1,900 ∎ 1900 ∎ 1900 metres
- 1,900 m

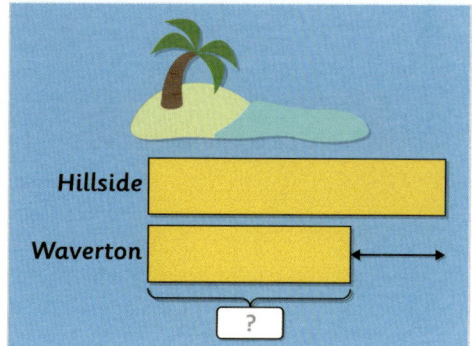

Hillside

Waverton

?

Level 4 *continued*

28. Luka and Sara walk 12 kilometres in total for
    a b c charity. Luka walks 3 times further than Sara.
    If they get £2 for every 1,000 metres they
    walk, how much money does Sara make for
    charity?
    Include the £ sign in your answer.

    - 6.00   - £6.00   - £6   - 6

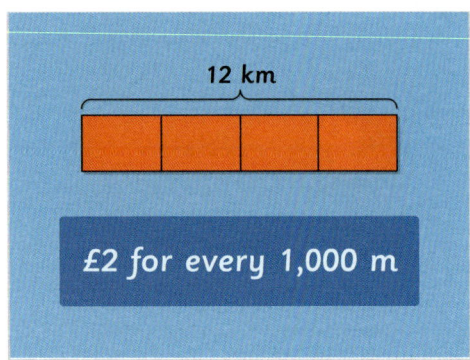

29. Luka and Sara walk 12 kilometres in total for
    a b c charity. Luka walks 3 times further than Sara.
    How many metres does Luka walk?
    Include the units m (metres) in your answer.

    - 9000 metres   - 9,000 metres   - 9,000 m
    - 9000 m   - 9 km   - 9 kilometres   - 3,000 metres
    - 9   - 3000 metres   - 3000 m   - 3,000   - 9,000
    - 9000   - 3,000 m   - 3000

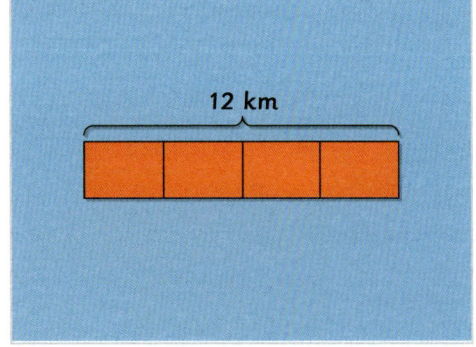

30. Eric walks 1,250 metres from his house to
    a b c the newsagent. He then walks to the sweet
    shop, which is half a kilometre back towards
    Eric's house. How far is the sweet shop from
    Eric's house in metres?
    Include the units m (metres) in your answer.

    - 750 metres   - 750   - 500   - 750 m   - 500 m
    - 500 metres

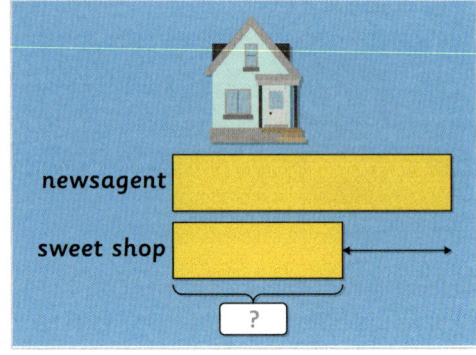

Convert Units of Measure: g and kg, l and ml

Objective:	I can convert between grams and kilograms and betweeen litres and millilitres.

Quick Search Ref:	**10144**

Level 1: Understanding: Methods for converting and identifying units of measure.

✳ **Required:** 7/10   ✳ **Pupil Navigation:** on   ✳ **Randomised:** off

---

1. How many grams are in one kilogram?

■ 10   ■ 100   ■ **1,000**

1/3

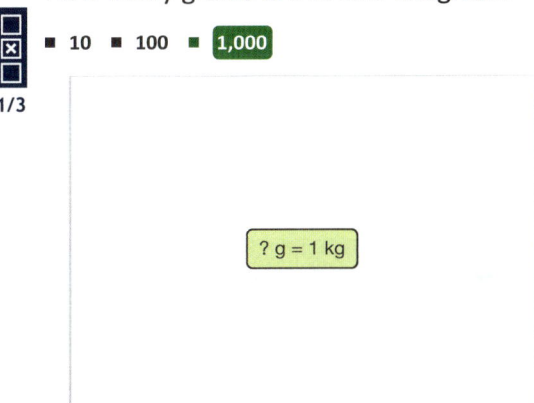

2. How many millilitres are in one litre?

■ 10   ■ 100   ■ **1,000**

1/3

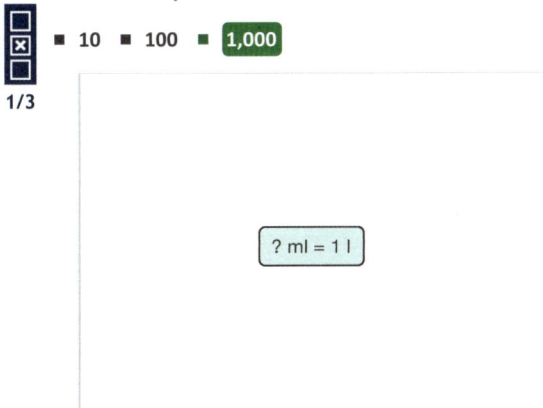

3. To convert from kilograms to grams, . . .

■ multiply by 100.   ■ divide by 100.
■ **multiply by 1,000.**   ■ divide by 1,000.

1/4

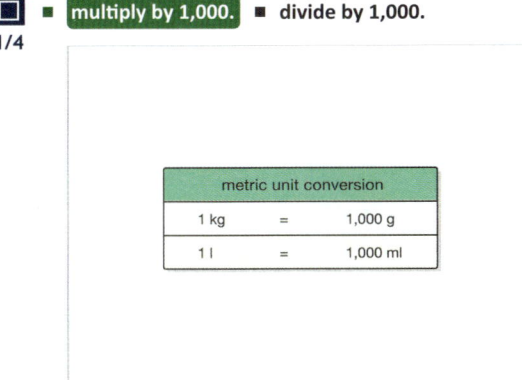

4. To convert from litres to millilitres, . . .

■ **multiply by 1,000.**   ■ divide by 1,000.
■ multiply by 100.   ■ divide by 100.

1/4

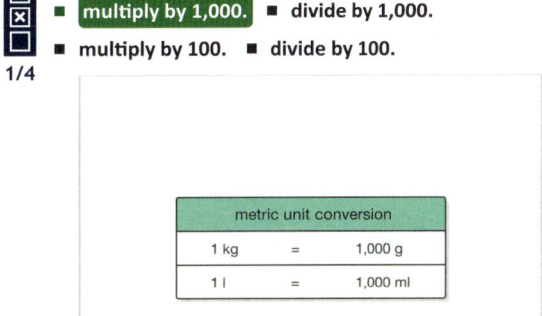

5. To convert from grams to kilograms, . . .

■ multiply by 100.   ■ divide by 100.
■ multiply by 1,000.   ■ **divide by 1,000.**

1/4

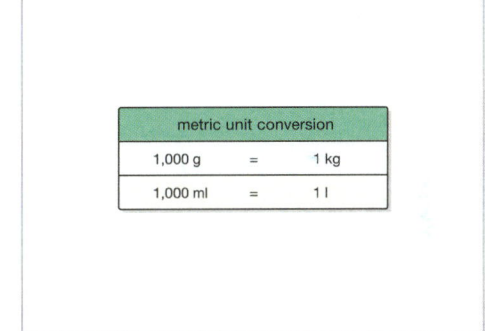

6. To convert from millilitres to litres, . . .

■ **divide by 1,000.**   ■ multiply by 1,000.
■ multiply by 100.   ■ divide by 100.

1/4

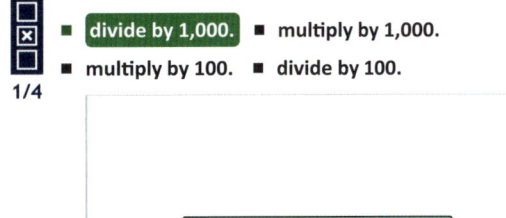

Level 1 *continued*

7. Which unit of measure are you likely to see on a small carton of apple juice?

☐☒☐ 1/4

- litres ■ kilograms ■ **millilitres** ■ grams

8. Which unit of measure are you likely to see on a bag of sugar?

☐☒☐ 1/4

- litres ■ **kilograms** ■ millilitres ■ grams

9. Which unit of measure are you likely to see on a small piece of cheese?

☐☒☐ 1/4

- litres ■ kilograms ■ millilitres ■ **grams**

10. Which unit of measure are you likely to see on a large bottle of water?

☐☒☐ 1/4

- **litres** ■ kilograms ■ millilitres ■ grams

Level 2: Fluency: Converting and ordering units of measure.

✸ **Required:** 7/10   ✸ **Pupil Navigation:** on
✸ **Randomised:** off

11. What is the amount shown in litres?

ⓐⓑⓒ *Include the units l (litres) in your answer.*

- **0.350 l** ■ **0.35 l** ■ 3,500 l ■ 0.35 ■ 35,000 l
- **0.35 litres** ■ **0.350 litres** ■ 350,000 l ■ 35 l
- 3500 l ■ 3.5 l ■ 350000 l ■ 35000 l

350 millilitres

12. What is the weight shown converted to

ⓐⓑⓒ **grams**?
Include the units g (grams) in your answer.

- **1,500 grams** ■ **1,500 g** ■ 0.15 g ■ 150 g
- **1500 grams** ■ **1500 g** ■ 0.015 g ■ 0.0015 g
- 1,500 ■ 1500 ■ 15 g

1.5 kg

Level 2 *continued*

13. 6.425 litres = _____ millilitres

a b c

- **6,425 ml** - **6,425** - **6,425 millilitres** - 64.25
- 642.5 - **6425 millilitres** - **6425** - **6425 ml**
- 0.00645 - 0.645 - 0.0645

14. 835 grams = _____ kilograms

a b c

- 835,000 - 83.5 - **0.835 kilograms** - 835000
- **0.835** - **0.835 kg** - 8350 - 8,350 - 8.35
- 83,500 - 83500

15. Arrange the quantities into ascending order according to size (smallest first).

↑↓

- **half a litre** - **5,300 ml** - **8.5 l** - **8,750 ml**

16. Choose the correct symbol to complete this statement.

☐ ☒ ☐

three-quarters of a kilogram ___ 586 grams

1/3 - < - **>** - =

17. What is the total amount of liquid shown in litres?

a b c

Don't include the units in your answer.

- **3.220** - **3.22** - 3,220 - 3220

18. Arrange the masses into descending order according to size (largest first).

↑↓

- **5 kg** - **3,450 g** - **3.42 kg** - **2,500 g**

19. 3.250 kg + 4,100 g = _____ kg

a b c

- **7.35** - **7.350** - **7.350 kilograms** - **7.35 kilograms**
- **7.350 kg** - **7.35 kg** - 7,350 - 7350

20. 3 litres - 2,250 millilitres = _____ millilitres

a b c

- **750 milliltres** - 2,247 - **750** - **750 ml** - 2247
- 5,250 - 0.75 - 5250

Level 3:
Reasoning: Converting values to the same unit of measure and using four operations.

✿ **Required:** 7/8   ✿ **Pupil Navigation:** on
✿ **Randomised:** off

21. What is the combined weight of the three objects in kilograms?

a b c

Include the units in your answer.

- 5,795 kg - **5.795 kilograms** - **5.795 kg** - 5.795
- 5795 kg

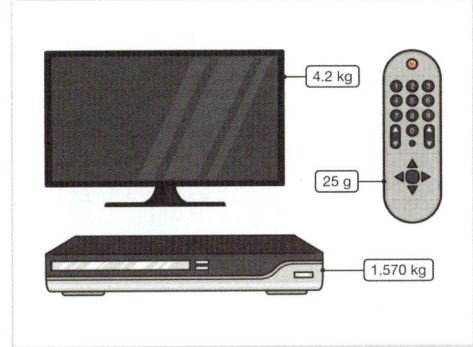

22. Marcel says, "Three-quarters of a kilogram is equal to 75 grams."

a b c

Is he correct? Explain your answer.

- Open question, no set answer

23. Kane has a jug containing 1.64 litres of juice. Lara has a jug containing 2,120 millilitres of juice. How many more millilitres of juice does Lara have than Kane?

1 2 3

Don't include the units in your answer.

- **480** - 0.48

Level 3 *continued*

24. Jodie buys 2,500 grams of bird feed at 80 pence per kilogram, and 1,500 grams of fish food at £1.24 per kilogram. She pays with a £10 note. How much change does she receive?

Include the £ sign in your answer.

- 3.86   - **£6.14**   - 6.14   - £6.14p   - £3.86

25. Choose **two** calculations that equal 0.684 kg.

- **342 g × 2**   - 2 kg - 1.556 kg   - 0.5 kg + 174 g
- **1 kg - 316 g**   - 0.5 kg + 0.084 g

2/5

26. A basket weighs 1.2 kilograms.
A basket containing three equal-sized apples weighs 1,440 grams. How much does each apple weigh in **grams**?

Include the units in your answer.

- **80 grams**   - **80 g**   - 0.08 g   - 80

27. A plastic cup holds 200 millilitres of liquid. Laura fills it with water and pours it into a large jug. She repeats this process eight times. How many litres of water will be in the large jug when she has finished?

Don't include the units in your answer.

- **1.6**   - 1600

28. 3/4 of a litre - 0.03 litres = . . .

- 500 millilitres + 0.25 litres   - half of 14,440 millilitres
- 753 millilitres   - **half of 1,440 millilitres**

1/7

- 7.200 litres   - 2/10 of a litre   - 50% of a litre

Level 4: Problem Solving: Converting between different units of measure (including fractions and decimals).

✿ **Required:** 5/5   ✿ **Pupil Navigation:** on
✿ **Randomised:** off

29. Thomas and Ella each have a bag of seashells.
Thomas has 4/10 of a kilogram of shells. Ella has 1/5 of a kilogram of shells.
How much more does Thomas' bag of shells weigh?

Include the units g (grams) in your answer.

- **200 grams**   - **200 g**   - 200

Level 4 *continued*

30. Dennis, Tom and Adam are having a water fight. They fill a bucket with 3,580 millilitres of water. Dennis throws 1.05 litres of water. Tom throws twice as much water as Dennis. Adam throws 0.12 litres of water. How much water is left in the bucket in millilitres?
Include the units in your answer.

- 0.31 - 3,270 - 3270 millilitres - **310 millilitres**
- 310 - **310 ml** - 3270 ml - 0.31 litres - 0.31 l
- 3270 - 3,270 ml - 3,270 millilitres

31. A family buys five 1 kilogram bags of pasta. On Monday, dad uses 1/10 of the pasta in a salad. On Wednesday, the twins each have a 750 gram bowl of pasta for lunch. On Thursday, friends come over and they eat three times the amount of pasta used on Monday. How many kilograms of pasta are left?
Include the units in your answer.

- **1.5 kilograms** - 1,500 - 1500 grams - **1.5 kg**
- 1500 - 1.5 - 1,500 grams

32. Carol's car has a fuel tank capacity of 52 litres.
She fills the tank **three-quarters full**. She uses half of the fuel travelling to a football match, 15 litres travelling to an art fair, and another 1500 millilitres travelling to the supermarket. How many litres of fuel does the car have left in its tank?
Don't include the units in your answer.

- **3** - 3000

Fuel tank capacity 52 litres

33. Three equal-sized cakes have a combined weight of 1.2 kilograms. Jay eats 2/10 of one cake. His sister takes half of a cake to school. His grandma eats 250 grams of one cake. How many kilograms of cake are left?
Include the units in your answer.

- 0.67 - **0.670 kilograms** - **0.670 kg** - **0.67 kg**
- **0.67 kilograms** - 670 g

1.2 kg

Estimate Volume and Capacity

Objective: I can estimate volume and capacity.

Quick Search Ref: 10154

Level 1: Understanding: Units of measure for volume and capacity.

✹ **Required:** 6/7   ✹ **Pupil Navigation:** on   ✹ **Randomised:** off

1. Capacity is:
- ☐ The amount of space that an object fills.
- ☒ How heavy or light an object is.
- ☐ 1/3 ■ The amount of space in a container or the amount of liquid that it can hold.

2. Volume is:
- ☒ The amount of space that an object fills.
- ☐ How heavy or light an object is.
- ☐ 1/3 The amount of space in a container or the amount of liquid that it can hold.

3. Which unit of measure would you use to measure water in a glass?

1/4 ■ Litres ■ Kilograms ■ **Millilitres** ■ Metres

4. Which unit of measure would you use to measure the water in a swimming pool?

1/4 ■ millilitres ■ grams ■ centimetres ■ **litres**

5. Which unit of measure would you use to measure the volume of a cube?

1/4 ■ square centimetres (cm²) ■ millilitres
■ **cubic centimetres (cm³)** ■ kilometres

6. In cubic centimetres (cm³), what is the volume of the model shown?

1 2 3 ■ **5** ■ 6

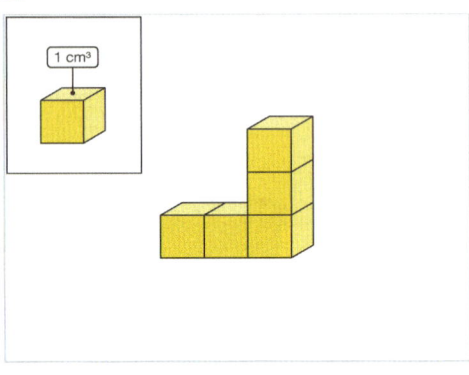

1 cm³

7. In cubic centimetres (cm³), what is the volume of the model shown?

1 2 3 ■ **8** ■ 9

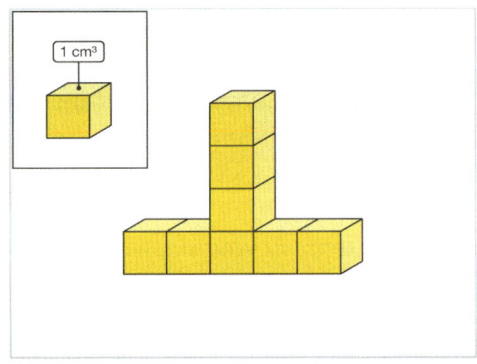

1 cm³

Level 2: Fluency: Estimating volume and capacity, using pictorial representation.

✹ **Required:** 7/10   ✹ **Pupil Navigation:** on
✹ **Randomised:** off

8. What is the most likely capacity of a small carton of fruit juice?

1/3 ■ 1 litre ■ **300 millilitres** ■ 26 millilitres

Level 2 *continued*

9. In cubic centimetres (cm³), estimate the volume of the cuboid.

123

- 20 ▪ **40** ▪ 10 ▪ 6

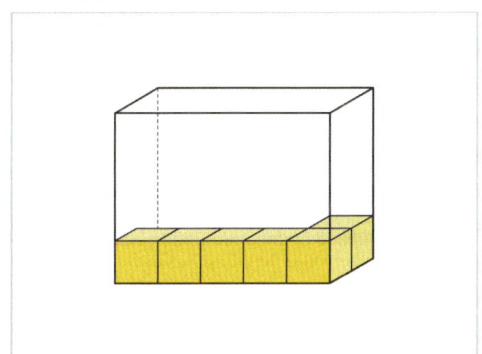

10. Estimate the total amount of liquid in both containers. Give your answer in **millilitres** (ml).

123

- **625** ▪ 1500 ▪ 500 ▪ 0.625 ▪ 125

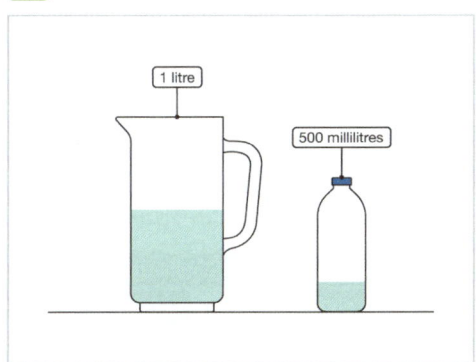

11. The water has been poured from the small bottle into an empty six litre cylinder. Estimate the capacity of the small bottle to the **nearest litre**.

123

- **3** ▪ 1

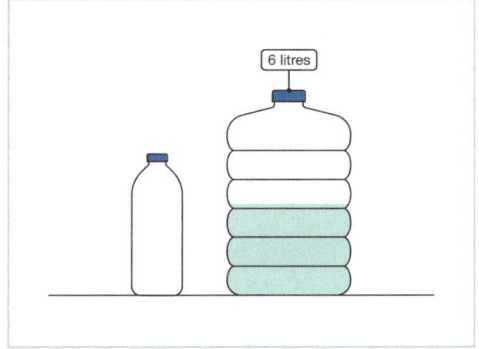

12. Which of the models are equal in volume?

- (b) and (c) ▪ (a) and (d) ▪ **(a), (b), (c) and (d)**
- (a), (b), (c), (d) and (e)

1/4

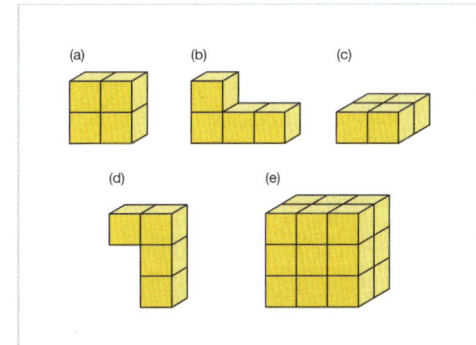

13. Estimate the **capacity** of the flask.

- 500 millilitres ▪ 1.5 litres ▪ **1 litre**

1/3

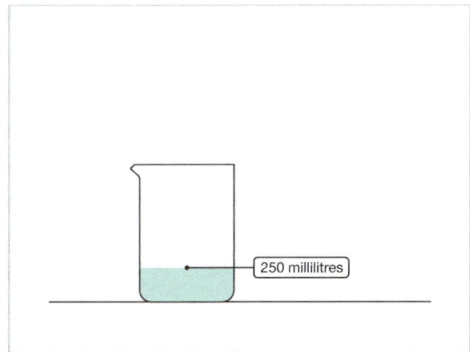

14. What is the most likely capacity of an average-sized ketchup bottle?

- more than one litre ▪ **less than one litre**

1/2

15. The capacity of the container is two litres. What is the best estimate for the volume of liquid it is holding?

1/3

- 1,500 millilitres ▪ 250 millilitres ▪ **500 millilitres**

Level 2 continued

16. Which measurement is closest to the volume of **two jars** of honey?

1/3

- **0.7 litres** ■ 500 mililitres ■ 7 litres

17. Gary has four identical glasses of water, each containing 0.315 litres of water. If he rounds each glass of water to the nearest 100 mililitre, approximately how much water will he have in total?

1/4

■ **1,200 ml** ■ 1,260 ml ■ 1,300 ml ■ 1,600 ml

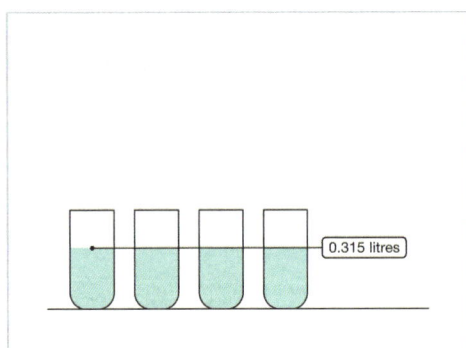

Level 3: Reasoning: Use rounding to estimate volume and capacity in various contexts.

✻ **Required:** 3/3   ✻ **Pupil Navigation:** on

✻ **Randomised:** off

18. Maya and Samuel pour water from a bottle into an empty plastic container. Who do you agree with?

1/2

- **Maya** "I think the volume of water in the bottle is larger than the volume of water in the container."
- **Samuel** "I think the volume of the water in the bottle is the same as the volume of water in the container."

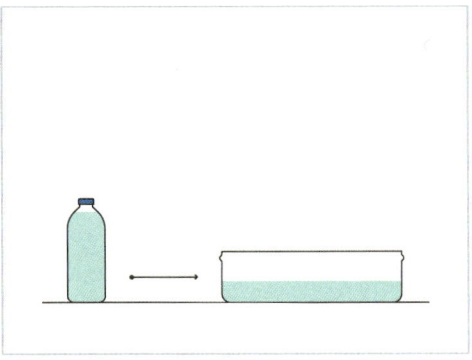

19. In millilitres, estimate how much more oil there is than water. Give your answer to the **nearest 100 millilitres**.
Don't include the units in your answer.

- **1,600** ■ **1600ml** ■ **1600millilitres** ■ 2,450
- **1600** ■ **1,600millilitres** ■ **1,600ml** ■ 1,550
- 2450 ■ 1550

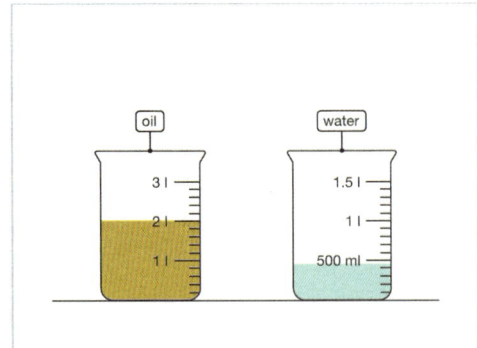

Level 3 continued

20. Round the capacities to the nearest 100 millilitres and estimate the **total capacity** of the items shown. Give your answer to the **nearest litre**.

- **6** ■ **5551** ■ **5700**

Level 4: Problem Solving: Estimating volume and capacity.

- ✱ **Required:** 3/5   ✱ **Pupil Navigation:** on
- ✱ **Randomised:** off

21. Mr Lee is filling a **six-litre bucket** with water for his gardening project. **Round to the nearest hundred** and find out approximately how many jugs of water he will need to fill his bucket if the capacity of one jug is **206 millilitres**.

- **30**

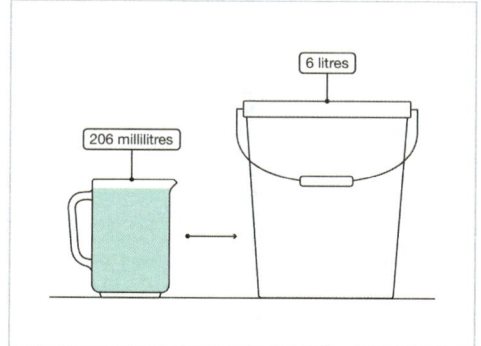

22. A can of juice has a capacity of **316 millilitres**. A bottle of juice has a capacity of **562 millilitres**. Round both to the **nearest 100 millilitres** and estimate in litres how much juice Hamad will have altogether if he buys **sixteen cans** and **four bottles** of juice.

- ■ **7304** ■ **7.2** ■ **7200**

23. A tennis ball has a **volume** of 130 cubic centimetres. The capacity of a tennis ball holder is 4,221 cubic centimetres. Round both to the **nearest 100 cubic centimetres** and estimate how many tennis balls can fit into **three** tennis ball holders.

- **126** ■ **129** ■ **120** ■ **42**

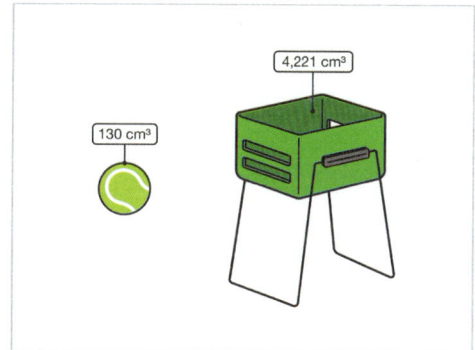

Level 4 *continued*

24.
1
2
3
The capacity of a lemonade bottle is 1.5 litres. The capacity of a glass is 229 millilitres. Round the capacity of the glass to the **nearest 100 millilitres** and estimate how many **full glasses** can be filled by **three bottles** of lemonade.

- ■ **22** ■ 22.5

25.
1
2
3
A cereal box has a capacity of 1,872 cubic centimetres. A box of tea bags has a capacity of 1,238 cubic centimetres. Round both to the **nearest 100 cubic centimetres** and estimate the total capacity of **nine** cereal boxes and **five** tea bag boxes.

- ■ **23100** ■ 3100 ■ 6000 ■ 17100 ■ 23.1 ■ 23038

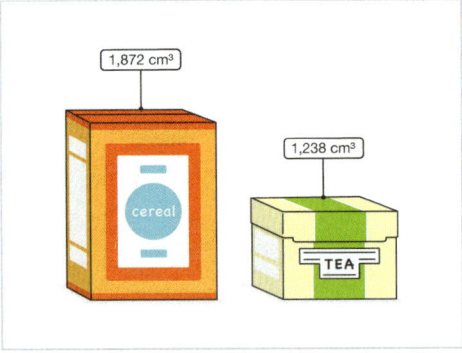

Solve Problems Involving Converting Between Units of Time

Objective: I can solve problems involving converting between units of time.

Quick Search Ref: 10280

Level 1: Problem Solving: Converting between different units of time (seconds, minutes, hours, days, weeks, months and years).

❄ **Required:** 7/10   ❄ **Pupil Navigation:** on   ❄ **Randomised:** off

1. Harriet is 8 years, 45 weeks, 46 days and 72 hours old. How many years old is Harriet?
▪ 9

2. Farhana has a cat that is 62 months old. Jade has a dog that is 4 years and 8 months old. Safeera has a rabbit that is 45 months old. Who has the oldest pet?

▪ Farhana ▪ Jade ▪ Safeera

62 months   4 years 8 months   45 months

3. Two movies have a total running time of 3 hours and 30 minutes. The first film starts at 6.25 p.m. and finishes at 7.45 p.m. How long is the second film in hours and minutes?

▪ 1 hour 20 minutes ▪ Two hours
▪ 2 hours 10 minutes ▪ 2 hours 20 minutes

4. Jane takes 3 minutes and 14 seconds to plant a flower. She then takes three times as long to water her garden. How long in **seconds** does she spend watering her garden?
Don't include units in your answer.

▪ 582

5. It takes Mercury 88 days to orbit the Sun. It takes Venus 225 days to orbit the Sun. What is the total duration of the two orbits to the **nearest week**?
Don't include units in your answer.

▪ 313 ▪ 45 ▪ 44

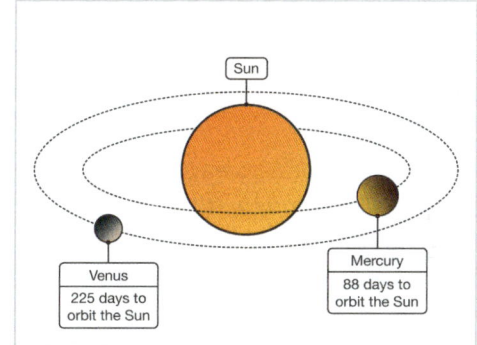

Sun

Venus
225 days to orbit the Sun

Mercury
88 days to orbit the Sun

6. The table shows the gestation periods for different animals. What is the difference between the gestation periods of the red deer and the red fox?

▪ 25 weeks ▪ 25 weeks and 3 days
▪ 25 weeks and 5 days ▪ 41 weeks and 5 days

animal	gestation period
mouse	20 days
red deer	236 days
sheep	147 days
chipmunk	31 days
red fox	58 days

Level 1 *continued*

7. The table shows the gestation periods for different animals. Which two animals have a combined gestation period of 25 weeks and 3 days?

1/4

- mouse and red deer
- red deer and sheep
- sheep and chipmunk
- chipmunk and red fox

animal	gestation period
mouse	20 days
red deer	236 days
sheep	147 days
chipmunk	31 days
red fox	58 days

8. A team of six builders each work 12 hours a day to repair a bridge. It takes them a total of five days to complete the work. How many hours does it take to repair the bridge in total?

- 360
- 72
- 60

9. The duration of a luxury cruise is 210 hours. It departs on the 14th of January at 9.00 a.m. On which date will the journey be completed?

1/4

- 20th January
- 21st January
- 22nd January
- 23rd January

10. Jodie and Steph did a sponsored silence. Jodie raised 20 pence for every minute of silence and Steph raised 8 pence for every 20 seconds of silence. They raised a total of £2.32. Jodie was silent for 8 minutes. For how many seconds was Steph silent?
Don't include units in your answer.

- 180
- 72

Measurement Topic Review

Objective: I can answer questions on different types of measurement and units.

Quick Search Ref: 10739

Level 1: Understanding

✱ **Required:** 7/10 ✱ **Pupil Navigation:** on ✱ **Randomised:** off

---

1. The area of the rectangle is _____ m².

1
2
3

▪ **48** ▪ 28 ▪ 14

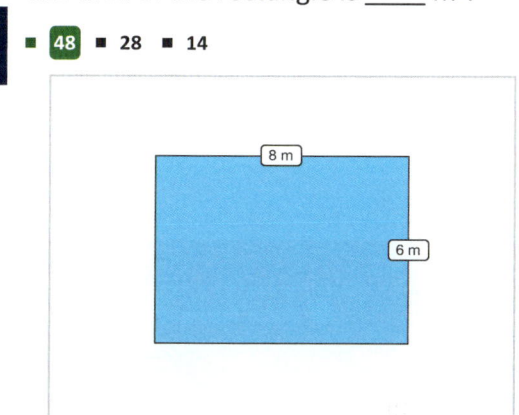

2. What is the perimeter of the composite shape?
Don't include the units in your answer.

1
2
3

▪ 56 ▪ **44** ▪ 96

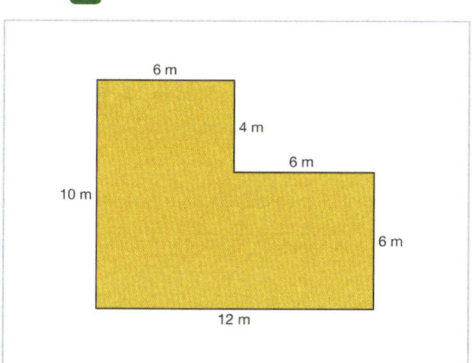

3. To convert from millimetres to centimetres . . .

☐
☒
☐

1/4

▪ multiply by 10. ▪ **divide by 10.** ▪ multiply by 100.
▪ divide by 100.

metric unit conversion		
10 mm	=	1 cm
100 cm	=	1 m
1,000 m	=	1 km

4. To convert from centimetres to metres . . .

☐
☒
☐

1/4

▪ **divide by 100.** ▪ multiply by 100. ▪ divide by 1,000.
▪ multiply by 1,000.

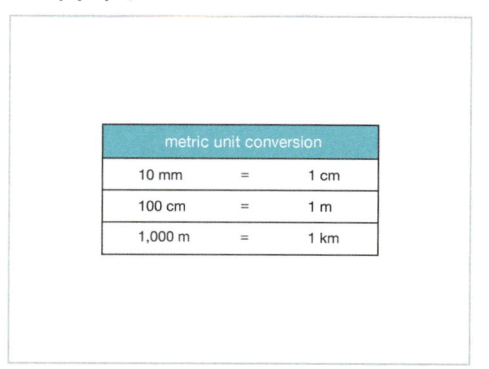

metric unit conversion		
10 mm	=	1 cm
100 cm	=	1 m
1,000 m	=	1 km

5. Which unit of measure are you likely to see on a small piece of cheese?

☐
☒
☐

1/4

▪ litres ▪ kilograms ▪ millilitres ▪ **grams**

6. One pound is approximately how many kilograms?

☐
☒
☐

1/4

▪ **0.5 kg** ▪ 1 kg ▪ 5 kg ▪ 7 kg

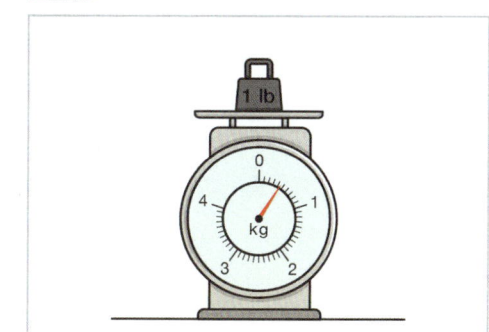

Level 1 *continued*

7. In cubic centimetres (cm³), what is the volume of the model shown?

- **8** ▪ **9**

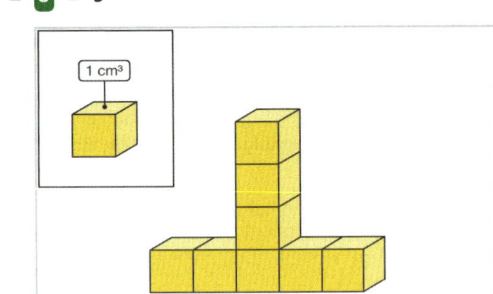
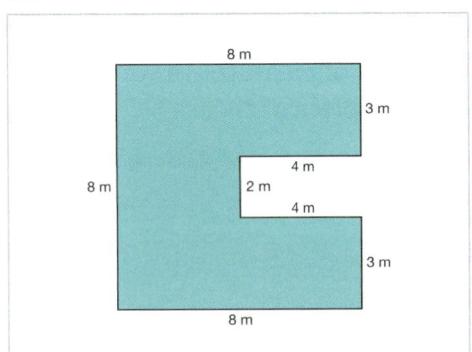

8. The perimeter of the **composite shape** is ____ metres.

▪ **40**

9. Which unit of measure are you likely to see on a large bottle of water?

1/4

▪ **litres** ▪ **kilograms** ▪ **millilitres** ▪ **grams**

10. How many metres are there in one kilometre?

1/3

▪ **10** ▪ **100** ▪ **1,000**

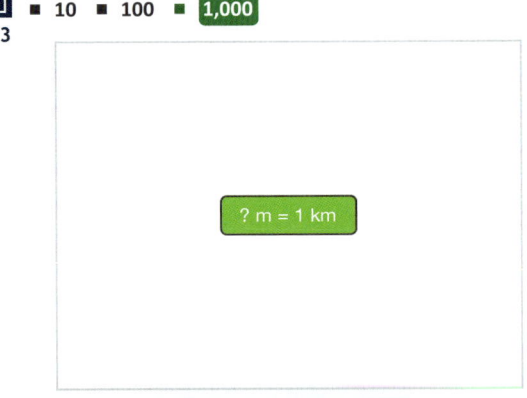

Level 2: Fluency

✿ **Required:** 7/10   ✿ **Pupil Navigation:** on
✿ **Randomised:** off

11. 3 litres - 2,250 millilitres = _____ millilitres

▪ **750 milliltres** ▪ **750** ▪ **750 ml** ▪ **5,250** ▪ **5250**
▪ **2,247** ▪ **2247** ▪ **0.75**

12. The capacity of the container is two litres. What is the best estimate for the volume of liquid it is holding?

1/3

▪ **1,500 millilitres** ▪ **250 millilitres** ▪ **500 millilitres**

13. 3.250 kg + 4,100 g = _____ kg

▪ **7.35** ▪ **7.350** ▪ **7.350 kilograms** ▪ **7.35 kilograms**
▪ **7.350 kg** ▪ **7.35 kg** ▪ **7350** ▪ **7,350**

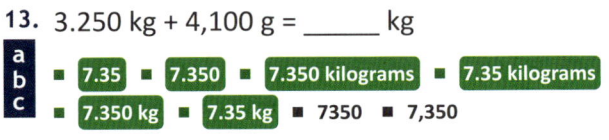

Level 2 *continued*

14. Calculate the **total area** of the composite rectilinear shape in square metres.

1 2 3

- ■ **122** ■ 50 ■ 45 ■ 77

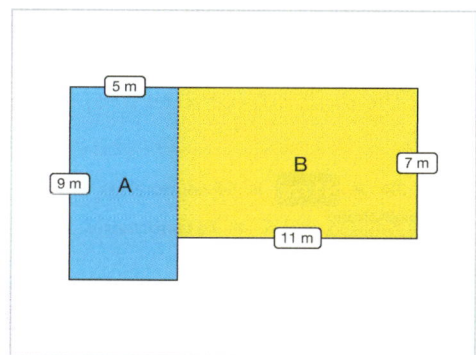

15. A large mat measures 1.2 m and a small mat measures 78 cm. When they are joined together, what is the combined length of the mats in centimetres?

a b c

Include the unit cm (centimetres) in your answer.

- ■ **198 centimetres** ■ 198 ■ **198 cm** ■ 79.2 ■ 1.98

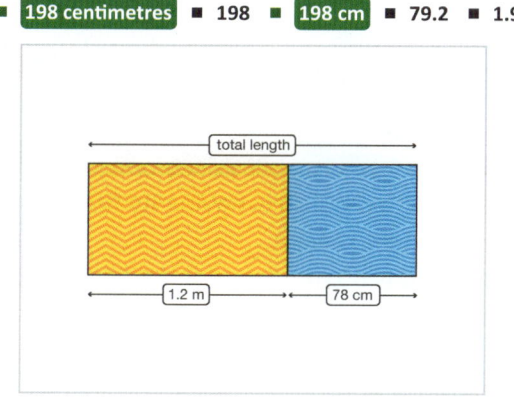

16. The net is made up of squares that each have a side length of nine centimetres. What is the perimeter of the net?

a b c

Include the units cm (centimetres) in your answer.

- ■ 216 ■ 216 centimetres ■ **126 centimetres**
- ■ **126 cm** ■ 216 cm ■ 486 centimetres ■ 486 cm
- ■ 36 cm ■ 36 centimetres ■ 486 ■ 36 ■ 126

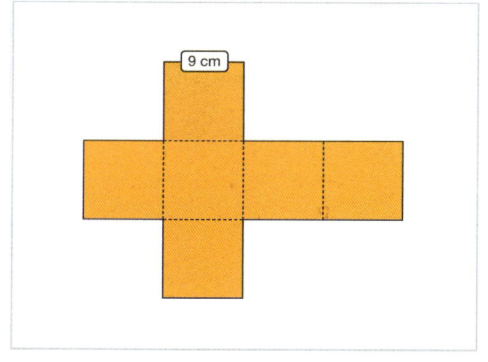

17. Using the estimate 1 litre = 2 pints, approximately how many pints of water are there in the fish tank?

a b c

Include the units l (litres) in your answer.

- ■ **20 l** ■ 5 litres ■ **20 litres** ■ 5 ■ 20 ■ 5 l

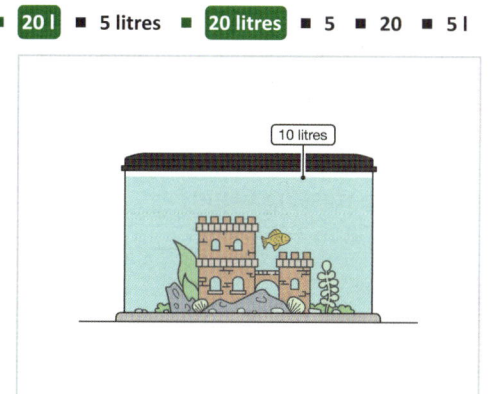

18. Arrange the masses into descending order according to size (largest first).

↑ ↓

- ■ 5 kg ■ 3,450 g ■ 3.42 kg ■ 2,500 g

19. Calculate the **total area** of the composite rectilinear shape in square centimetres.

1 2 3

- ■ **49** ■ 32

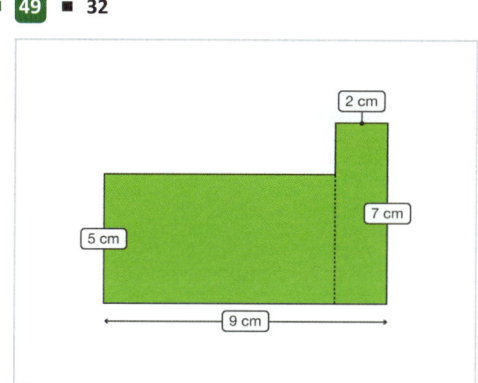

20. What is the perimeter of the composite shape?

a b c

Include the units m (metres) in your answer.

- ■ **104 m** ■ 76 m ■ 118 m ■ **104 metres** ■ 76
- ■ 104 ■ 118 ■ 76 metres ■ 118 metres ■ 345
- ■ 345 metres ■ 345 m

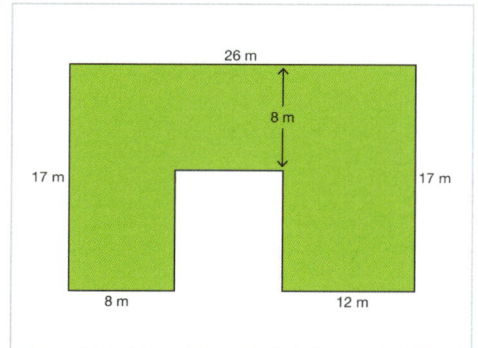

Level 3: Reasoning
_____
✳ **Required:** 5/5   ✳ **Pupil Navigation:** on
✳ **Randomised:** off

21. Marcel says, "Three-quarters of a kilogram is
a equal to 75 grams."
b Is he correct? Explain your answer.
c

- Open question, no set answer

22. What is the missing value?
1 43 cm + 1.08 m = 23 cm + _____ m
2
3   ■ **1.28** ■ **128**

23. Sam says, "The area of a square will always
a be a square number."
b
c Is he correct? Explain your answer.

- Open question, no set answer

24. Maya and Samuel pour water from a bottle
☐ into an empty plastic container. Who do you
☒ agree with?
☐

1/2
■ **Maya "I think the volume of water in the bottle is larger
  than the volume of water in the container."**
■ **Samuel "I think the volume of the water in the bottle is
  the same as the volume of water in the container."**

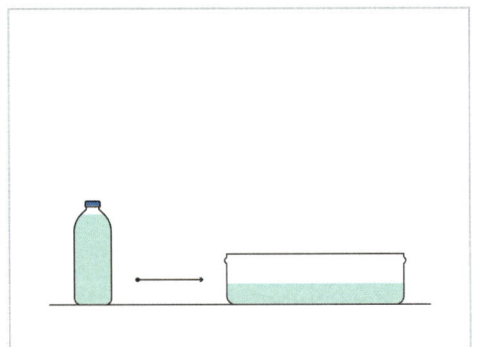

25. Shapes A, B and C are squares. Shape C has a
a perimeter of 32 centimetres and its length is
b twice the size of shape B. The length of
c shape B is twice the size of shape A.
All three shapes are joined together to make
a new shape. What is the perimeter of the
composite shape?
*Include the units cm (centimetres) in your
answer.*

■ **44** ■ **38** ■ **44 cm** ■ **56 centimetres**
■ **44 centimetres** ■ **56** ■ **38 centimetres** ■ **38 cm**
■ **56 cm**

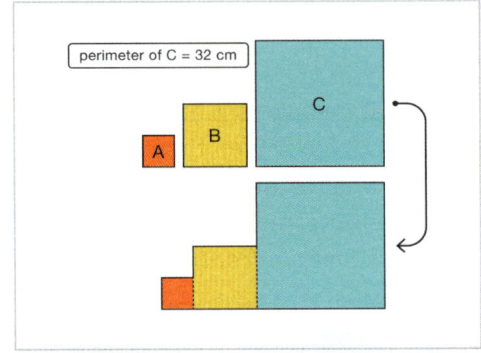

perimeter of C = 32 cm

Level 4: Problem Solving
_____
✳ **Required:** 5/5   ✳ **Pupil Navigation:** on
✳ **Randomised:** off

26. Bacup Cricket Club is having its field returfed
a for the new season.
b
c The field is 60 metres wide and 140 metres
long. The wicket area, which is **3 metres**
wide and **20 metres** long, **does not** need to
be returfed
If one square metre of turf costs **£2.00**, what
will be the **total cost** in pounds to returf the
cricket field excluding the wicket area?

■ **8340** ■ **16680** ■ **60** ■ **16,680** ■ **8,400** ■ **8,340**
■ **8400**

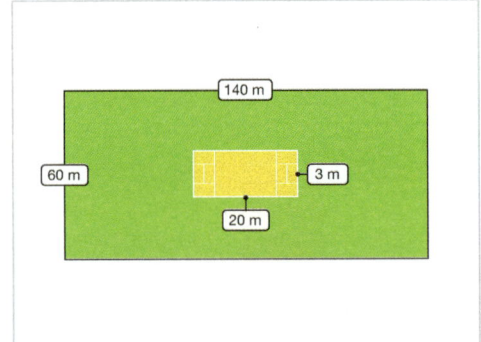

Level 4 *continued*

27. Dennis, Tom and Adam are having a water fight. They fill a bucket with 3,580 millilitres of water. Dennis throws 1.05 litres of water. Tom throws twice as much water as Dennis. Adam throws 0.12 litres of water. How much water is left in the bucket in millilitres?
Include the units in your answer.

- 3,270   - **310 millilitres**   - **310 ml**   - 0.31 litres
- 3270   - 3,270 millilitres   - 0.31   - 3270 millilitres
- 310   - 3270 ml   - 0.31 l   - 3,270 ml

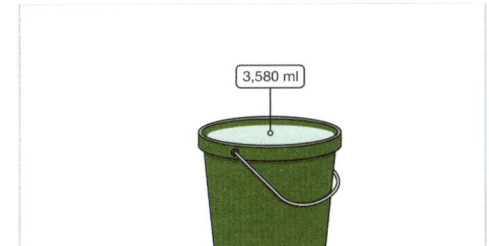

28. A family buys five 1 kilogram bags of pasta. On Monday, dad uses 1/10 of the pasta in a salad. On Wednesday, the twins each have a 750 gram bowl of pasta for lunch. On Thursday, friends come over and they eat three times the amount of pasta used on Monday. How many kilograms of pasta are left?
Include the units in your answer.

- **1.5 kilograms**   - 1500 grams   - **1.5 kg**   - 1.5
- 1,500   - 1500   - 1,500 grams

29. A team of six builders each work 12 hours a day to repair a bridge. It takes them a total of five days to complete the work. How many hours does it take to repair the bridge in total?

- **360**   - 60   - 72

30. A bakery uses 4.5 kilograms of butter and 6 pounds of sugar a day. In **pounds**, what is the approximate combined mass of butter and sugar used in a 3 day period?
Don't include the units in your answer.

- **45**   - 22.5

Solve Problems Involving Measurements: All Four Operations

Objective:	REVIEW IN PROGRESS 9/8/19 I can use all four operations to solve problems involving length, mass, volume and money.

Quick Search Ref: 10022

Level 1: Understanding: Use all four operations to solve problems involving length, distance and height.

✱ Required: 4/6	✱ Pupil Navigation: on	✱ Randomised: off

1. What is the most suitable unit of measure for recording the length of a football pitch?

1/4

■ Kilometres ■ Litres ■ **Metres** ■ Kilograms

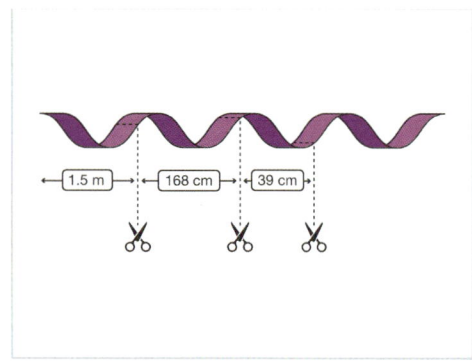

2. Kevin walked 43.8 kilometres on Saturday and 32.2 kilometres on Sunday. How many kilometres did he walk **in total** over the weekend?

■ **76**

3. A ribbon is 5.2 metres long. Three lengths are cut from the ribbon.
- 1.5 metres
- 168 centimetres
- 39 centimetres
How much ribbon is left over in **metres**?

■ 8.77 ■ **1.63** ■ 163 ■ 3.57

4. Shazad makes a tower from 70 pence worth of ten pence coins. If a ten pence coin is 1.9 millimetres thick, what is the height of the tower in **centimetres**?

■ 19 ■ **1.33** ■ 13.3

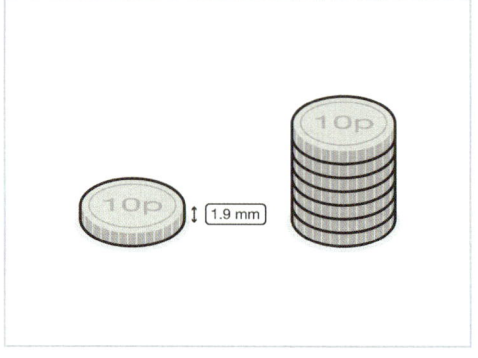

5. Usain Bolt is the only sprinter to win the **100 metre** and **200 metre** titles at **three** consecutive Olympics. What is the total distance in **kilometres** covered by Usain over the three Olympics?

■ **0.9** ■ 0.6 ■ 90 ■ 900 ■ 9

6. Two long jumpers are practising their jumps for the athletic championships. They each take two attempts.

Jade jumps 6.64 metres and 6.52 metres. Grace jumps 7.52 metres and 814 centimetres.

In **centimetres**, how much further does Grace jump than Jade over the two attempts?

■ **250** ■ 2.5 ■ 2882 ■ 80836

Level 2: Fluency: Use all four operations to solve problems involving weight.

✱ **Required:** 5/7 ✱ **Pupil Navigation:** on

✱ **Randomised:** off

7. What is the most suitable unit of measure for weighing a bag of cement?

1/5

- **Kilograms** ▪ Kilometres ▪ Grams ▪ Litres
- Metres

8. A box of 17 dried apricots weighs 184 grams. The weight of the empty box is 14 grams. How much does **one dried apricot** weigh in **grams**?

- **10** ▪ 2890 ▪ 10.8

9. Mr. Brown asked the children in year 5 to keep a diary of how much sweets and chocolate they ate over the weekend. 8 children each ate 50 grams of sweets. 15 children each ate 45 grams of chocolate. How much **chocolate**, to the **nearest kilogram**, did the children eat in total?

- **1** ▪ 675 ▪ 0.675 ▪ 0.4

10. Maria has a recipe for 12 pancakes. She wants to make 30 pancakes. How many grams of flour will she use?

- **275** ▪ 220 ▪ 3300

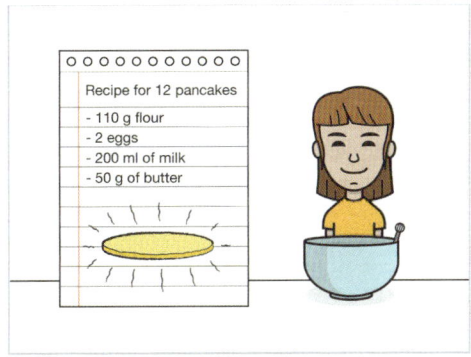

11. Mrs. Jolly the baker uses 125 grams of caster sugar to make thirty biscuits. A bag of caster sugar weighs one kilogram. How many biscuits can Mrs. Jolly make from **two bags** of caster sugar?

- **480** ▪ 60 ▪ 240

12. Yasir places a 0.5 kilogram weight and a 150 gram weight on one side of the balance scales. Rashid places a 0.75 kilogram weight and a 350 gram weight on the other side. How much more weight, in **grams**, does Yasir need to add to make the scales balance?

- 650 ▪ **450** ▪ 1100 ▪ 0.45

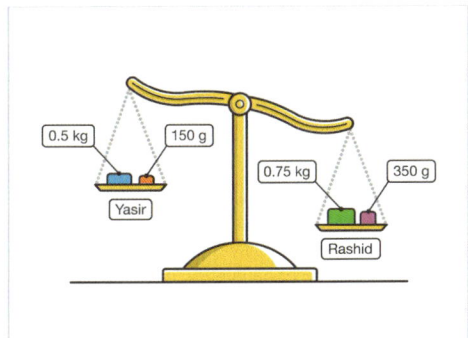

Level 2 *continued*

13. A basketball weighs 0.6 kilograms. **Two basketballs** weigh the same as **eight hockey balls**. How many **grams** does one hockey ball weigh?

- 0.15 ▪ **150** ▪ 75 ▪ 1200 ▪ 1.2

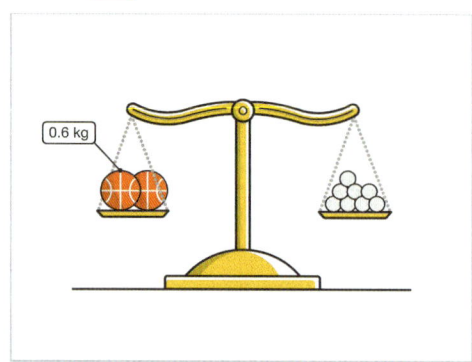

Level 3: Reasoning: Use all four operations to solve problems involving volume.

✳ **Required:** 4/5   ✳ **Pupil Navigation:** off
✳ **Randomised:** off

14. What is the most suitable unit of measure for recording the **volume** of water in the bottle?

1/4 ▪ Metres ▪ Kilograms ▪ Millimetres ▪ **Millilitres**

15. Anna keeps a log of how much water she drinks in a week. How many **litres** of water did Anna drink in total on Monday, Tuesday and Wednesday?

- **2.45** ▪ 6.388 ▪ 4.25

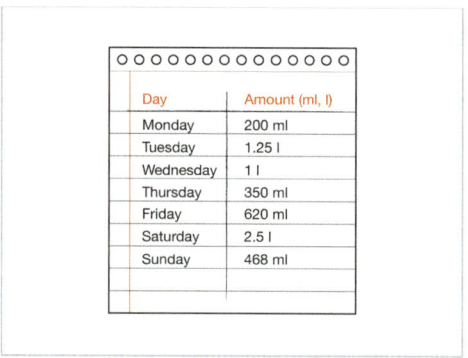

Day	Amount (ml, l)
Monday	200 ml
Tuesday	1.25 l
Wednesday	1 l
Thursday	350 ml
Friday	620 ml
Saturday	2.5 l
Sunday	468 ml

16. Sharon has to take 7.5 millilitres of medicine three times a day for a 30 days. How many millilitres of medicine will she consume in this time?

- **675** ▪ 22.5 ▪ 225

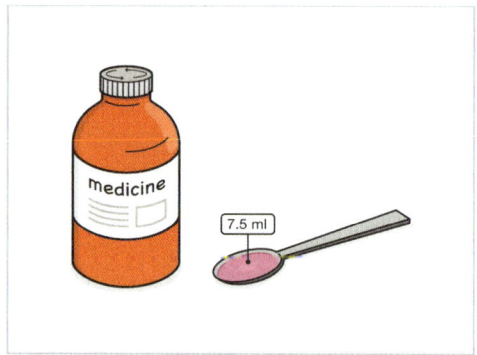

17. Tony makes a tropical fruit drink for breakfast. It contains the following:
One-tenth of a litre of pineapple juice.
1/3 Three-fifths of a litre of orange juice.
One-eighth of a litre of mango juice.
Which is the most suitable jug for Tony to serve his drink in?

- jug 1 ▪ jug 2 ▪ **jug 3**

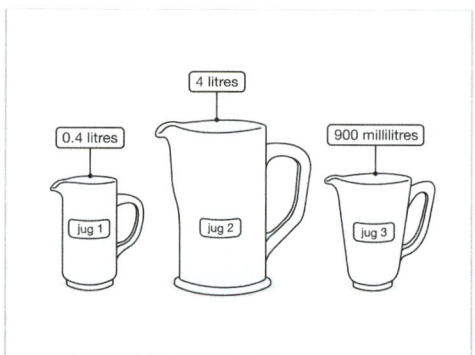

18. One can holds 355 millilitres of pop. There are **24 cans** in one case. How many **litres** of pop are in **three cases**?

- 8520 ▪ **25.56** ▪ 25560 ▪ 8.52

19. How many pennies are equivalent to £8.00?

- ■ **800** ■ **80** ■ **100** ■ **0.8**

20. Shazia buys six books costing £5.99 each. She pays with a £50.00 note. How much change does she receive?

- ■ **£35.94** ■ **£14.06p** ■ **£14.06** ■ **1406p** ■ **£44.01**

21. A family cinema ticket costs £8.67. Mr. Finch buys one family ticket and a drink for each of his two children. He spends a total of £15.07. How much does **one drink** cost?

- ■ **£3.20** ■ **£12.80** ■ **£3.20p** ■ **320p** ■ **£6.40**

22. Joseph was given £120 for his birthday. He spent one-quarter of the amount on a computer game, **and from the money left over** he spent two-fifths on gifts for his family. How much money did Joseph have left after buying the game and the gifts?

- ■ **£54.00** ■ **£66.00** ■ **£42** ■ **£54** ■ **£90.00**
- ■ **540p** ■ **£72.00** ■ **£42.00** ■ **£90** ■ **£66** ■ **£72**

23. Danny's dad works nine hours a day from Monday to Friday. He is paid £7.50 per hour. How much does Danny's dad earn in one week?

- ■ **£337.50** ■ **£337.50p** ■ **£67.50**

24. There is a sale at a department store offering a 25% discount on clothes. Sarah buys a coat that originally cost £58. How much does she pay?

- ■ **£14.00** ■ **£72.50** ■ **£43.50** ■ **£14** ■ **£43.50p**
- ■ **£14.50**

Mathematics **Y5**

Geometry

2D Shape
3D Shape
Angles
Transformation

Recognise Regular and Irregular Polygons

Objective: I can identify polygons and use reasoning about equal sides and angles to recognise regular and irregular polygons.

Quick Search Ref: 10251

Level 1: Understanding: Recognise the properties of regular and irregular polygons.

❂ **Required:** 7/10 ❂ **Pupil Navigation:** on ❂ **Randomised:** off

1. Which **two** shapes are polygons?

☐☒☐
2/6

■ shape A ■ shape B ■ shape C ■ shape D
■ shape E ■ shape F

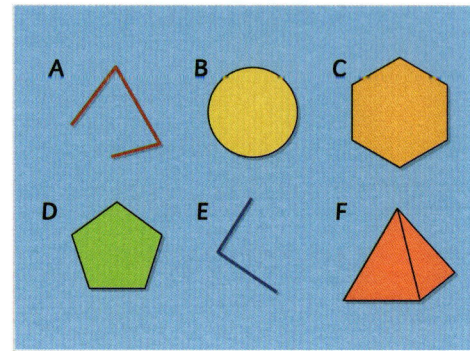

2. Select the shape which is **not** a polygon.

☐☒☐
1/4

■ hexagon ■ pentagon ■ circle ■ nonagon

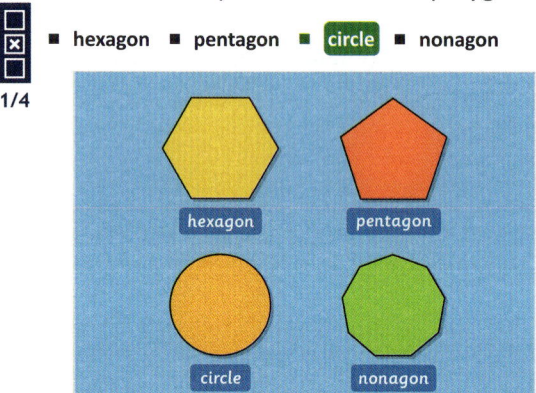

3. How many angles are there in a pentagon?

123
■ 5

Number of sides and angles in polygons		
shape	number of sides	number of angles
triangle	3	3
square	4	4
pentagon	5	?

4. Select **two** properties of a **regular** polygon.

☐☒☐
2/4

■ Two of the sides are the same length.
■ All of the sides are the same length.
■ The total of all the angles is 180°.
■ All of the angles are equal.

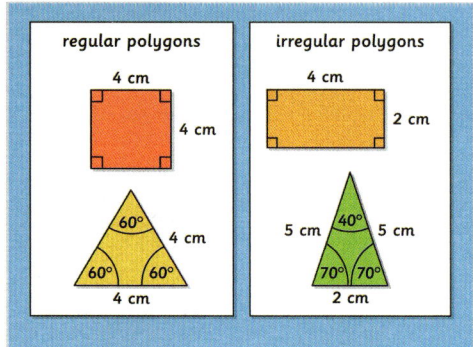

5. Which shape is a regular polygon?

☐☒☐
1/4

■ parallelogram ■ kite ■ rectangle ■ square

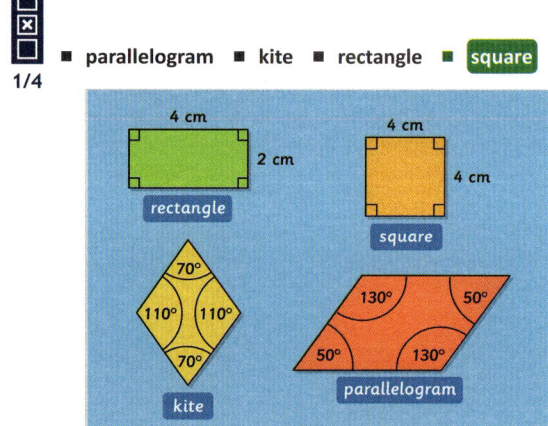

6. How many irregular **hexagons** are shown in the image?

123
■ 3 ■ 2

Level 1 *continued*

7. Select the **three** shapes that are not regular polygons.

☐
☒
☐

3/5

■ shape A ■ shape B ■ shape C ■ shape D
■ shape E

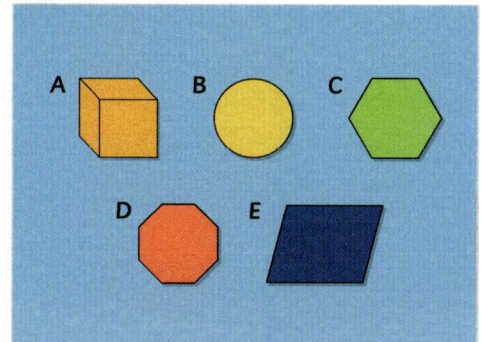

8. Which shape is an irregular polygon?

☐
☒
☐

1/4

■ pentagon ■ square ■ triangle ■ hexagon

9. This shape is a polygon.

☐
☒
☐

1/2

■ True ■ False

10. How many sides does an irregular quadrilateral have?

1
2
3

■ 4

Level 2: Fluency: Identify regular and irregular polygons.

✸ **Required:** 7/10   ✸ **Pupil Navigation:** on
✸ **Randomised:** off

11. A _____ has ten sides and ten angles.
Enter the name of a 2D shape.

a
b
c

■ decagon

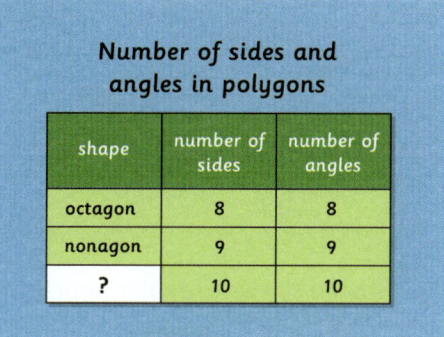

Number of sides and angles in polygons

shape	number of sides	number of angles
octagon	8	8
nonagon	9	9
?	10	10

12. Which triangle is a regular polygon?

☐
☒
☐

1/3

■ isosceles ■ equilateral ■ scalene

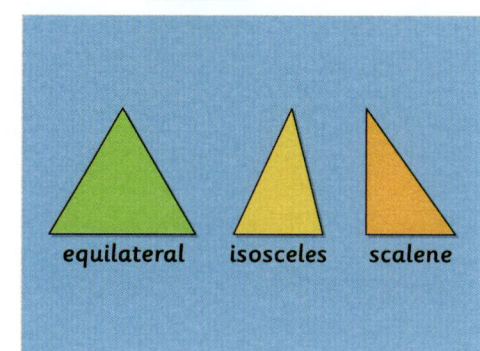

Level 2 continued

13. What is a regular polygon called?

1/5

- kite - trapezium - rhombus - **square**
- rectangle

Properties of quadrilaterals

quadrilateral	always has equal sides	always has equal angles
kite	✗	✗
rhombus	✓	✗
square	✓	✓
trapezium	✗	✗
rectangle	✗	✓

14. Select **two** options that describe the shape in the image.

2/5

- a quadrilateral - **an irregular polygon** - a pentagon
- **a hexagon** - a regular polygon

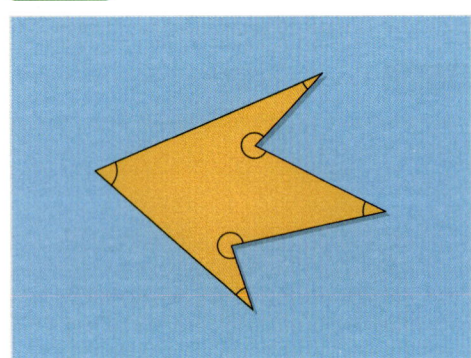

15. Which image shows irregular pentagons?

1/5

- image A - image B - **image C** - image D
- image E

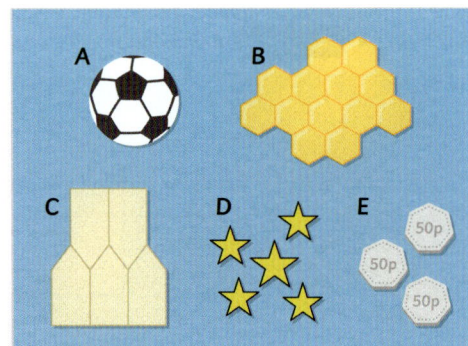

16. Belle cuts a regular decagon exactly in half to make two irregular polygons. How many sides does each of the new polygons have?

1
2
3

- 10 - **6** - 5

17. A regular octagon has a perimeter of 72 centimetres (cm). How long is one side of the octagon?
Include the units cm (centimetres) in your answer.

a
b
c

- **9 cm** - **9 centimetres** - 9

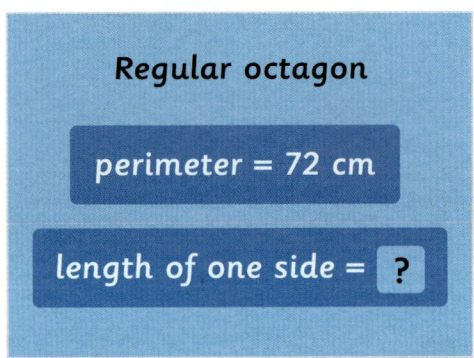

Regular octagon

perimeter = 72 cm

length of one side = ?

18. A regular shape has a perimeter of 24 centimetres (cm). If the length of one side of the shape is 4 cm, how many sides does the shape have?

1
2
3

- **6**

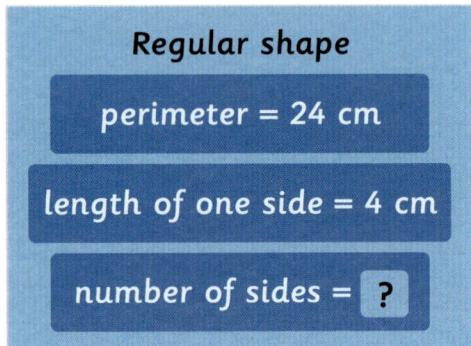

Regular shape

perimeter = 24 cm

length of one side = 4 cm

number of sides = ?

Level 2 *continued*

19. Select **two** options that describe the shape in the image.

- a quadrilateral   ■ **a regular polygon**   ■ a hexagon
- an irregular polygon   ■ **an equilateral triangle**

1/5

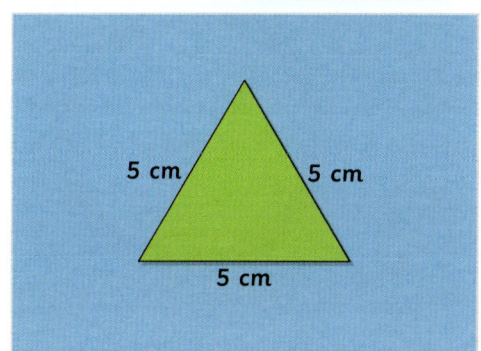

20. The triangle is half of which shape?

- shape A   ■ shape B   ■ **shape C**

1/3

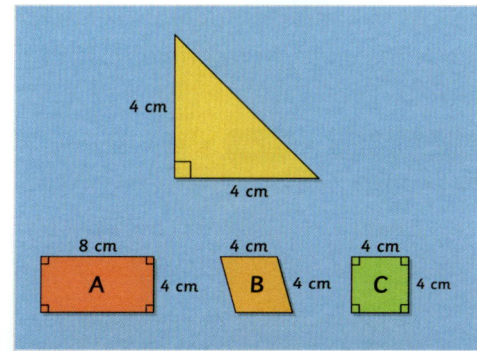

Level 3: Reasoning: Reason about the properties of regular and irregular polygons.

✱ **Required:** 5/5   ✱ **Pupil Navigation:** on
✱ **Randomised:** off

21. Hamza says, "I am thinking of a polygon which has three short sides and two longer sides. It has two obtuse angles and three acute angles." What polygon is Hamza thinking of? Explain how you know whether it is regular or irregular.

- Open question, no set answer

22. Maisie says that a rectangle is a regular polygon because all of the angles are the same. Do you agree with Maisie? Explain your answer.

- Open question, no set answer

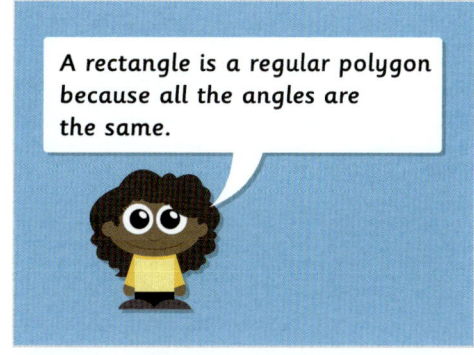

23. Dan has sorted some polygons into a Venn diagram. Where in the Venn diagram should he put the trapezium?

1/4

- **section A**   ■ section B   ■ section C   ■ section D

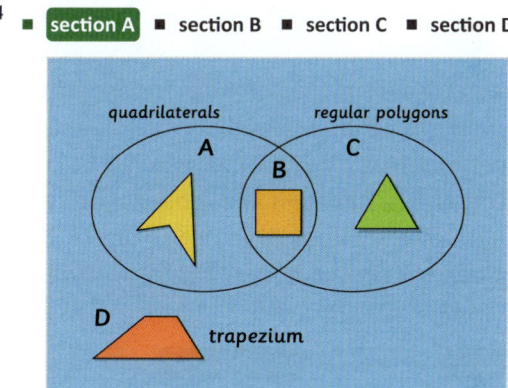

24. All of the angles in this regular polygon are **acute**. Prove that the shape is an equilateral triangle.

- Open question, no set answer

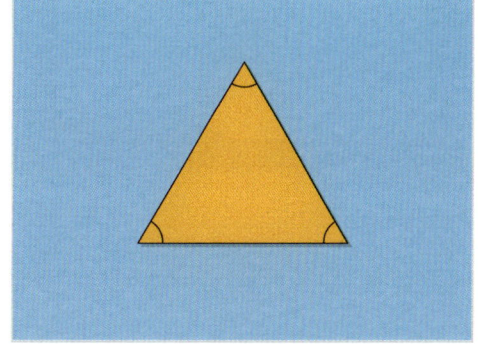

Level 3 *continued*

25. Which shape is the **only regular quadrilateral**?

1/4

- rhombus ▪ parallelogram ▪ rectangle ▪ **square**

Properties of quadrilaterals

	opposite sides equal	opposite angles equal	all sides equal	all angles equal
rhombus	✓	✓	✓	✗
parallelogram	✓	✓	✗	✗
rectangle	✓	✓	✗	✓
square	✓	✓	✓	✓

Level 4: Problem Solving in greater depth: Solve multi-step problems involving the properties of regular and irregular polygons.

✿ **Required:** 5/5   ✿ **Pupil Navigation:** on
✿ **Randomised:** off

26. Brad arranges pieces of wool to form three regular polygons with sides each measuring 5 centimetres (cm). If Brad makes a **pentagon**, a **hexagon** and a **heptagon**, what length of wool does Brad use in total? *Include the units cm (centimetres) in your answer.*

- **90 cm** ▪ 18 cm ▪ **90 centimetres** ▪ 18 centimetres
- 18 ▪ 90

27. Jamal has drawn the first two sides of a **regular polygon** onto a clock face. How many sides will the completed polygon have?

1
2
3

- **6**

28. The composite shape is made up of **regular polygons** joined to a **rectangle**. The perimeter of each square is 26 centimetres (cm) and the perimeter of each triangle is 10.5 cm. What is the perimeter of rectangle? *Include the units cm (centimetres) in your answer.*

a
b
c

- **20 centimetres** ▪ **20 cm** ▪ 10 centimetres ▪ 10
- 20 ▪ 10 cm

29. How many irregular polygons are there inside the large square?

1
2
3

- 6 ▪ **9** ▪ 4 ▪ 2 ▪ 11

Level 4 *continued*

30. Eric is making a play area from rectangular
pieces of grass. The play area is a four-sided
regular polygon which is 15 metres (m) wide.
How many pieces of grass will Eric need to
cover the whole play area?

■ 20 ■ 10

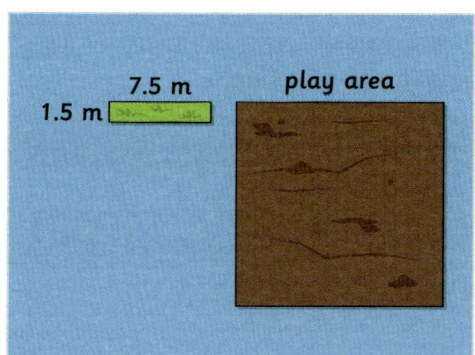

Use the Properties of Rectangles to Find Missing Lengths and Angles

Objective: I can use the properties of rectangles to find missing lengths and angles.

Quick Search Ref: 10116

Level 1: Understanding: Recognise the properties of squares and rectangles.

✹ Required: 7/10 **✹ Pupil Navigation:** on **✹ Randomised:** off

1. Select **three** properties of a rectangle.

   3/5
   - All the angles in a rectangle are 90°.
   - The angles in a rectangle add up to 180°.
   - All the sides of a rectangle are equal in length.
   - The opposite sides of a rectangle are parallel.
   - The opposite sides of a rectangle are equal in length.

   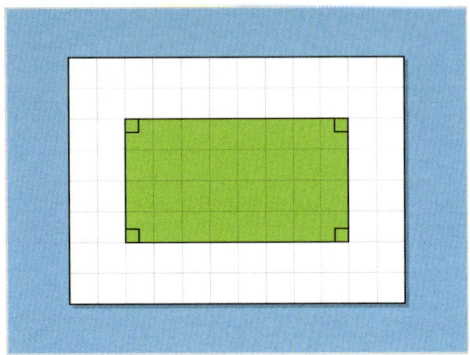

2. Shape *ABCD* is a rectangle. How long is side *DC*?

   Include the units m (metres) in your answer.

   - 8 metres - 8 m - 8

   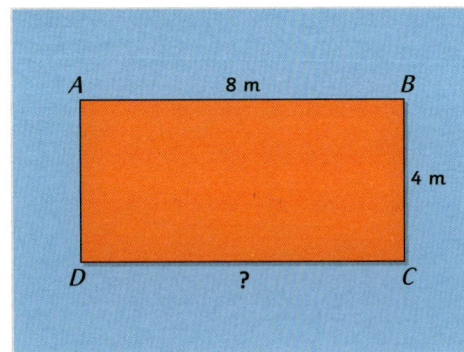

3. A diagonal of a quadrilateral is a straight line joining two opposite corners.
   Select **three** correct statements about the diagonals of a rectangle.
   3/4
   - The diagonals of a rectangle are the same length as each other.
   - A rectangle has four diagonals.
   - The diagonals of a rectangle cross at the midpoint of each other.
   - A rectangle has two diagonals.

   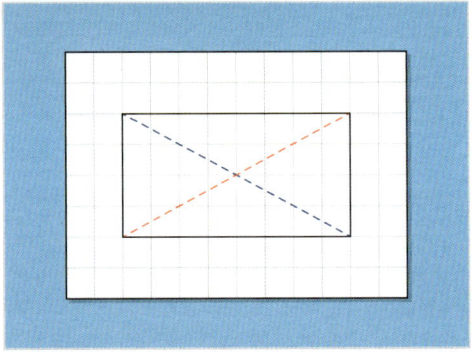

4. Line *AC* is a diagonal of the rectangle *ABCD* and measures 7 centimetres (cm). How long is diagonal *BD*?
   Include the units cm (centimetres) in your answer.

   - 7 cm - 7 centimetres - 7

   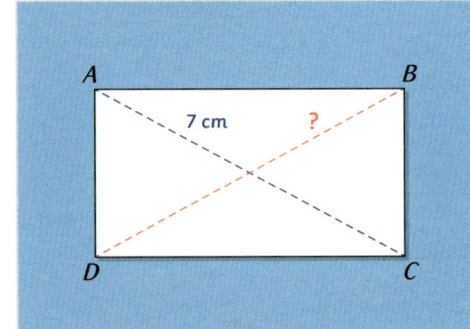

Level 1 *continued*

5. Select **two** statements about the diagonals of a square that are **not true** for the diagonals of a **rectangle**.

2/4
- The diagonals split each corner angle into two angles of 45°.
- The shape has two diagonals.
- The diagonals meet at 90°.
- The diagonals cross at the midpoint of each other.

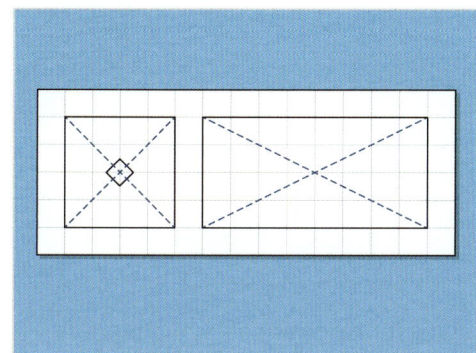

6. What is the size of angle *x* in the square?

1/3
- 90° ■ 45° ■ 60°

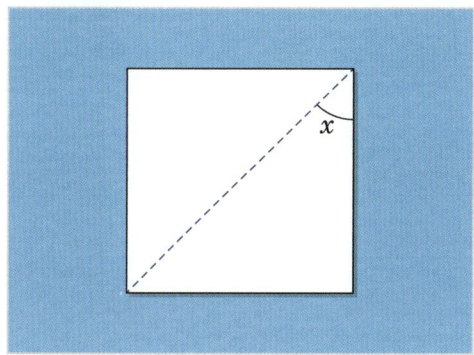

7. Select **three** statements to describe the triangles created when the square *ABCD* is split by diagonal *AC*.

3/5
- The triangles are equilateral triangles.
- The triangles are the same size.
- The triangles are scalene triangles.
- The triangles are right-angled triangles.
- The triangles are isosceles triangles.

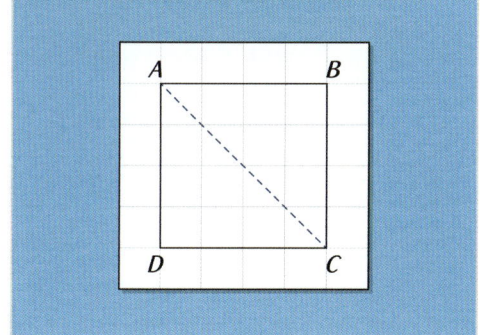

8. Select **two** statements that are **not true** about the diagonals of the rectangle.

2/4
- The diagonals split each corner angle into two angles of 45°.
- The shape has two diagonals.
- The diagonals are at right angles to each other.
- The diagonals cross at the midpoint of each other.

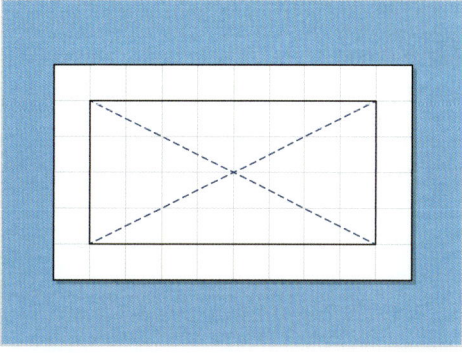

9. What is the size of angle *x* in the square?

1/3
- 60° ■ 90° ■ 45°

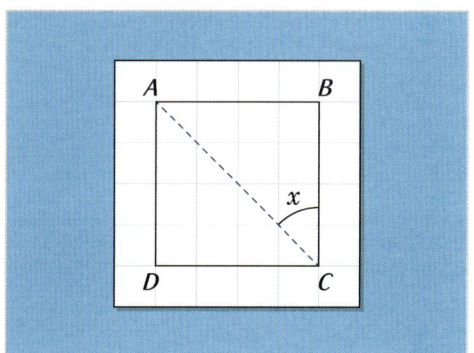

10. Line *BD* is a diagonal of the square *ABCD* and measures 10 centimetres (cm). How long is diagonal *AC*?
Include the units cm (centimetres) in your answer.

- 10 centimetres ■ 10 cm ■ 10

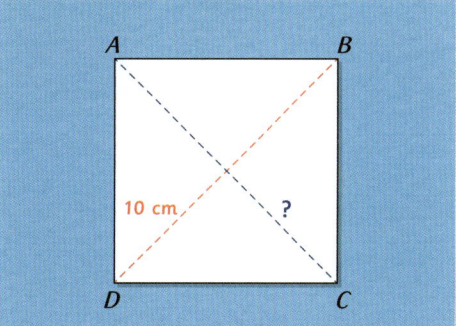

Level 2: Fluency: Use the properties of squares and rectangles to find missing lengths and angles.

❋ **Required:** 7/10   ❋ **Pupil Navigation:** on
❋ **Randomised:** off

11. In the rectangle, the total of angles *a*, *b* and *c*
is _____ degrees (°).
Enter the missing angle.

- **90** ▪ 30

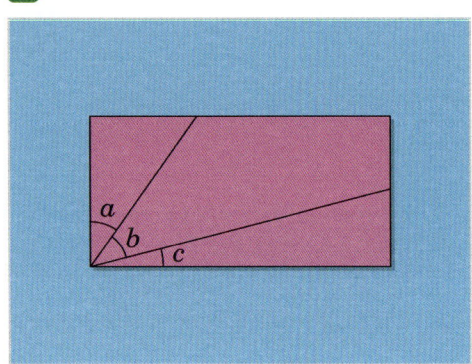

12. *ABCD* is a rectangle. Angle *x* is _____ degrees
(°).
Enter the missing angle.

- **60**

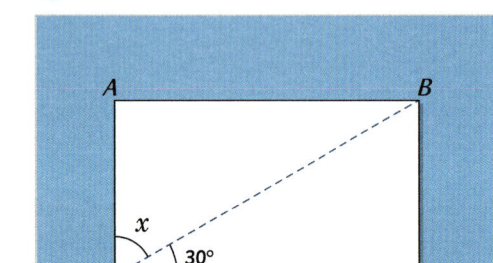

13. Line *AC* is a diagonal of the square *ABCD* and
measures 15 metres (m). *E* is the point
where the diagonals of the square cross each
other. How long is line *EB*?
Include the units m (metres) in your answer.

- 7.5 ▪ **7.5 metres** ▪ **7.5 m** ▪ 15 m ▪ 15 metres
- 15

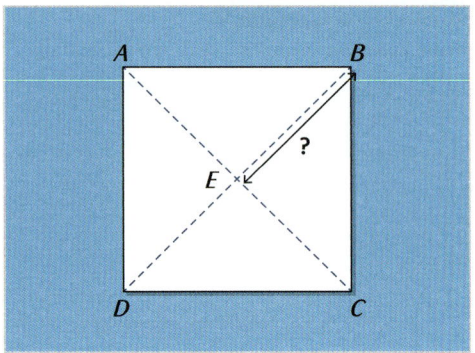

14. Angle *x* in the shape is _____ degrees (°).
Enter the missing angle.

- **135**

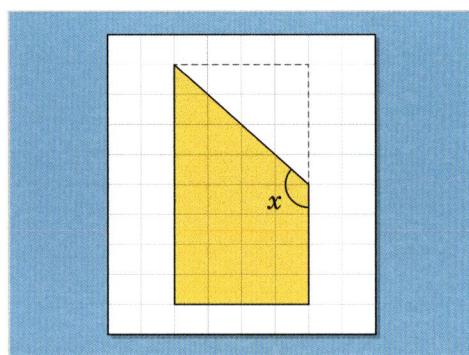

15. The **perimeter** of the tennis court is 64
metres (m). What is the missing side length?
Include the units m (metres) in your answer.

- 24 ▪ **24 metres** ▪ **24 m** ▪ 48 ▪ 48 metres
- 48 m

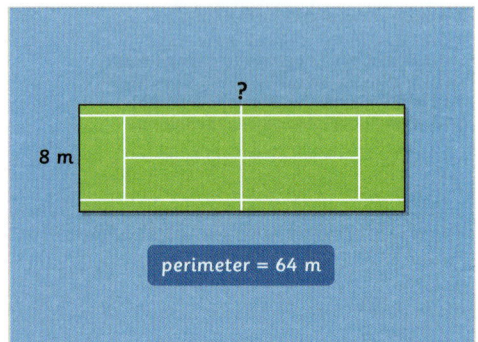

16. The shape has been made from two squares and a triangle. Angle *x* is _____ degrees (°).
1
2
3
Enter the missing angle.

- **225** ▪ 180

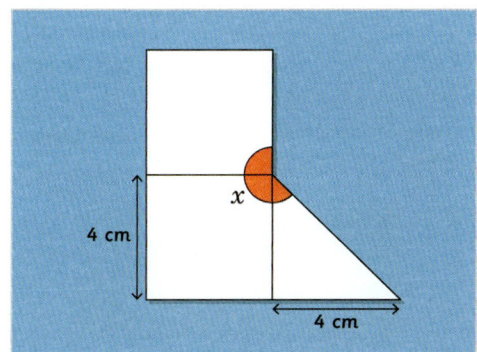

17. Peter is drawing a rectangle with a **perimeter**
a
b
c of 16 centimetres (cm). He has drawn one side measuring 3 cm. What length must he draw the next side?
Include the units cm (centimetres) in your answer.

- **5 centimetres** ▪ 10 ▪ 5 ▪ 13 cm ▪ **5 cm**
- 10 cm ▪ 10 centimetres ▪ 13 centimetres ▪ 13

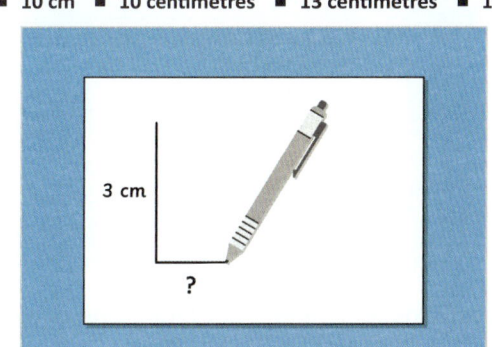

18. Line *AC* is a diagonal of the rectangle *ABCD*
a
b
c and measures 10 centimetres (cm). How long is the part of the diagonal marked on the diagram?
Include the units cm (centimetres) in your answer.

- 10 centimetres ▪ **5 centimetres** ▪ 5 ▪ **5 cm**
- 10 ▪ 10 cm

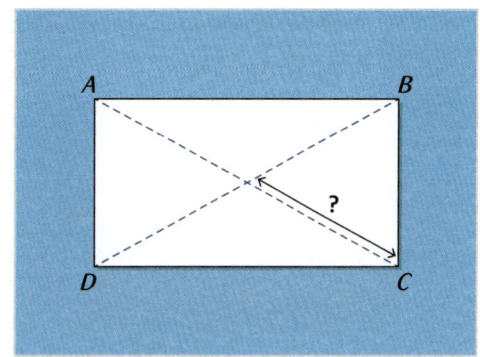

19. A square has been cut in half diagonally to
a
b
c make two triangles. What is the missing side length?
Include the units cm (centimetres) in your answer.

- **6 centimetres** ▪ **6 cm** ▪ 6

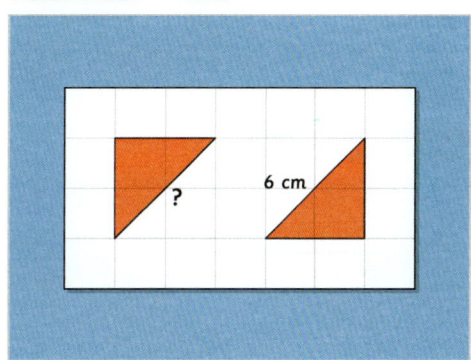

20. *ABCD* is a rectangle. Angle *x* is _____ degrees
1
2
3 (°).
Enter the missing angle.

- **65**

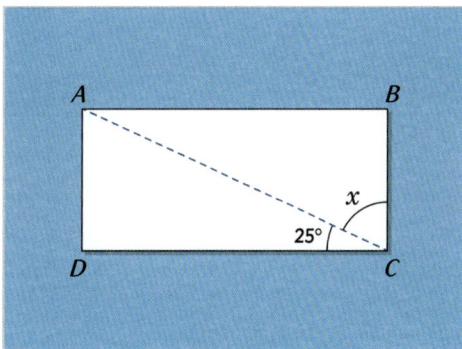

21. Bobby joins two rectangles to make one larger rectangle. He says that the total of the angles in the new rectangle is 720°. Do you agree with Bobby? Explain your answer.

- Open question, no set answer

22. One of the four triangles created by the diagonals of the rectangle has been shaded. If each diagonal measures 5 centimetres (cm), what is the perimeter of the shaded triangle?
Include the units cm (centimetres) in your answer.

▪ 8 centimetres ▪ 8 cm ▪ 8

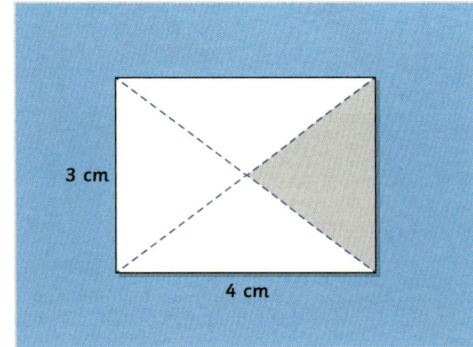

23. Starting from one of the corners, Alfie walks in a straight line to the centre of a rectangular field. Starting from a different corner, Beatrice walks in a straight line to the centre of the same field. Who walks further? Explain your answer.

- Open question, no set answer

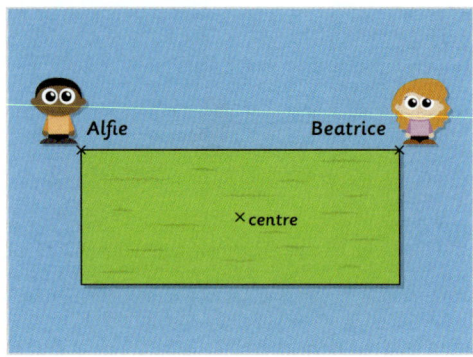

24. *AB* and *BC* are diagonals of two squares. Prove that the total of angle *x* and angle *y* is 90°.

- Open question, no set answer

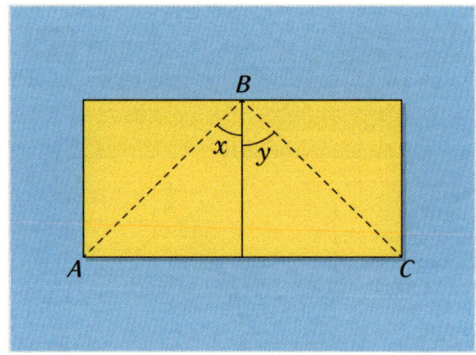

25. A rectangle has been cut along one diagonal to form triangles A and B. The triangles are joined with shape C to form a composite shape. What is the perimeter of the composite shape?
Include the units cm (centimetres) in your answer.

▪ 22 centimetres ▪ 22 cm ▪ 22

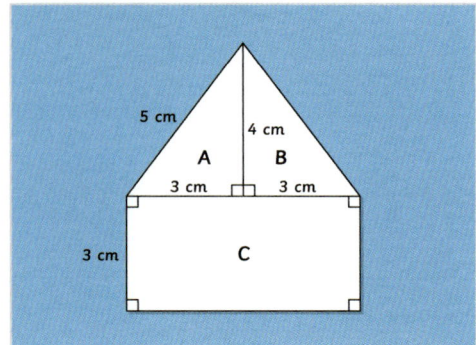

26. A rectangular piece of card has a perimeter
 of 40 centimetres (cm) and width of 8 cm. The area of the card is _____cm².
Enter the missing number.

- ■ **96** ■ **12**

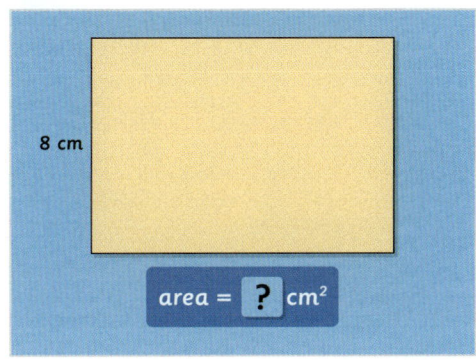

27. Clemmie has four pieces of square card with
1 2 3 a side length of 5 centimetres (cm). Two pieces of card are black and two pieces are white. Clemmie cuts along one diagonal of each piece of card and makes the pattern in the image.
The total area of the pattern Clemmie makes is _____ cm².
Enter the missing number.

- ■ **100** ■ **25** ■ **10**

28. A rectangle has an area of 84 cm² and a
a b c perimeter of 38 cm. The lengths of the sides of the rectangle are all whole numbers. What is the length of the longer side of the rectangle?
Include the units cm (centimetres) in your answer.

- ■ **12 centimetres** ■ **12 cm** ■ **7 cm** ■ **7 centimetres**
- ■ **12** ■ **7**

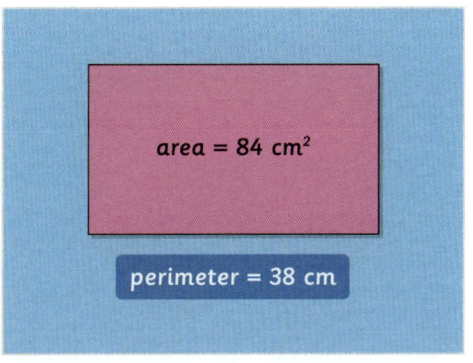

29. Two rectangles are joined together to make
1 2 3 the letter T. The total of all the angles inside the new shape is _____°.
Enter the missing number.

- ■ **1080**

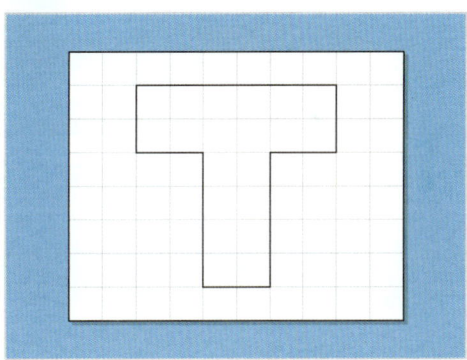

30. A rectangle is made up from three identical
1 2 3 squares and has a perimeter of 96 millimetres (mm). The area of one of the squares is _____ mm².

- ■ **144** ■ **12** ■ **432**

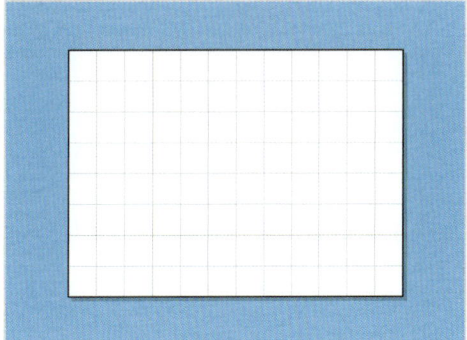

Identify 3D Shapes from 2D Representations

Objective: I can identify 3D shapes from 2D representations.

Quick Search Ref: 10193

Level 1: Understanding: Recognise the properties of 3D shapes and their 2D faces.

⚙ **Required:** 7/10   ⚙ **Pupil Navigation:** on   ⚙ **Randomised:** off

1. Select the **two** 3D shapes.

2/5

- pyramid ▪ square ▪ rectangle ▪ pentagon
- cylinder

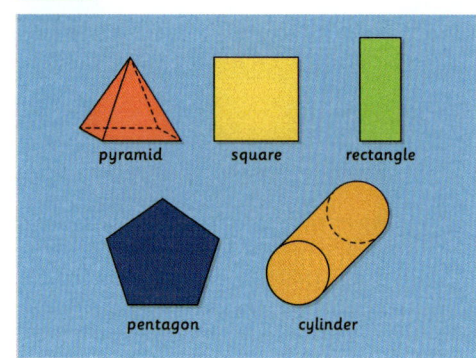

2. Sort the labels for the 3D shape into order, starting with the label in position 1.

- vertex ▪ face ▪ edge

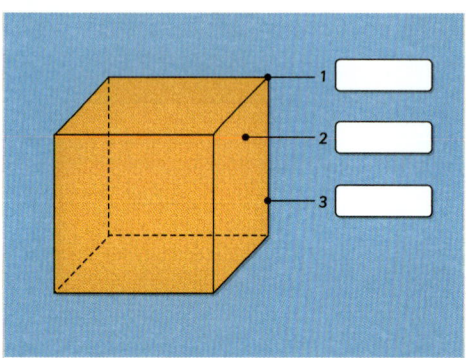

3. A **net** of a shape is . . .

1/3

- ▪ one of the faces of a 3D shape.
- ▪ a 2D shape that can be folded to form a 3D shape.
- ▪ the view of a 3D shape from one side.

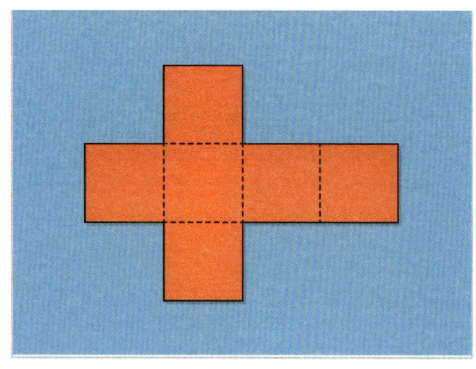

4. Which **two** 2D shapes are faces of the prism?

2/4

- triangle ▪ hexagon ▪ rectangle ▪ circle

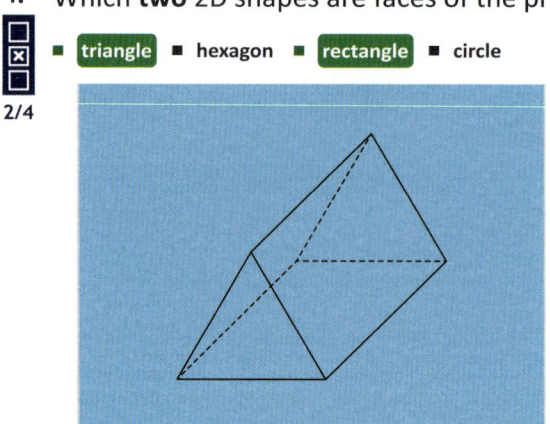

5. Which diagram shows the six faces of the cuboid?

1/3

- diagram A ▪ diagram B ▪ diagram C

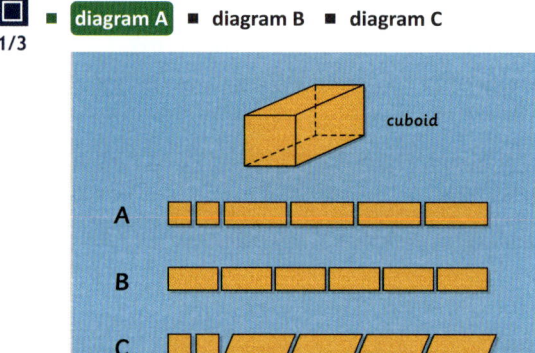

6. How many triangular faces does a tetrahedron have?

1 2 3

- 4

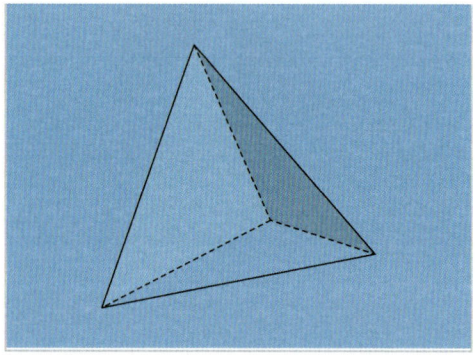

Level 1 *continued*

7. The dashed lines on the 2D net show some of the _____ of the 3D shape.

■ vertices ■ faces ■ **edges**

1/3

8. The corner points on the 2D net show some of the _____ of the 3D shape.

■ **vertices** ■ faces ■ edges

1/3

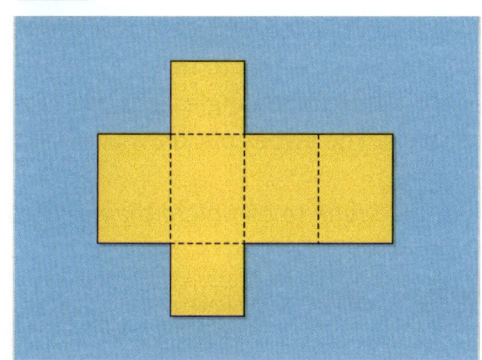

9. Select **two** statements that do **not** describe a hexagonal prism.

■ **six hexagonal faces** ■ six rectangular faces
■ **four rectangular faces** ■ two hexagonal faces

2/4

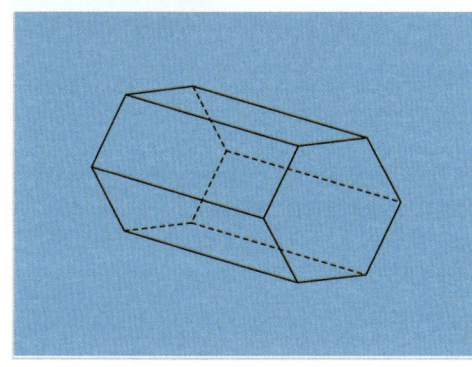

10. Select the **two** 3D shapes.

■ circle ■ **cube** ■ triangle ■ **cuboid** ■ rectangle

2/5

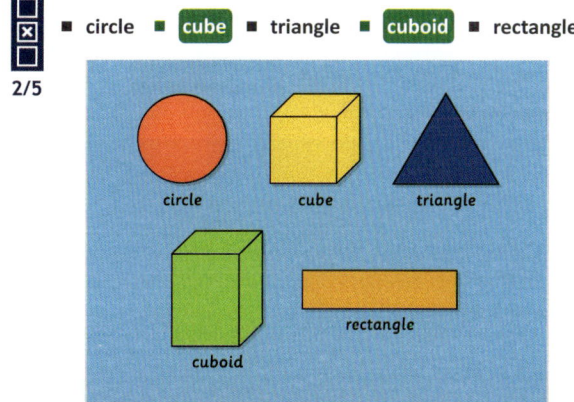

Level 2: Fluency: Identify 3D shapes from 2D nets.

✱ **Required:** 7/10 ✱ **Pupil Navigation:** on
✱ **Randomised:** off

11. The net is made from six squares and makes a _____ when folded.
Enter the name of a 3D shape.

■ square ■ **cube** ■ cuboid

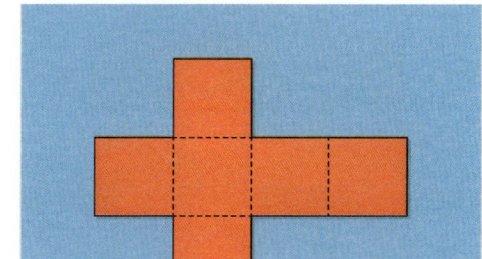

12. How many of the diagrams show the net of a cuboid?

■ **2**

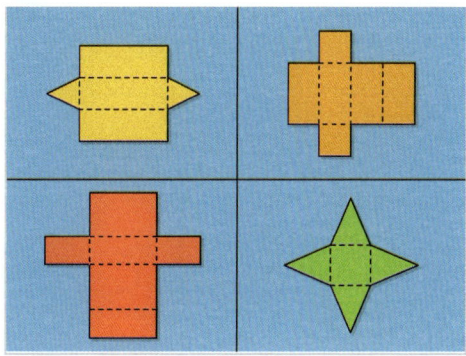

13. This net folds into which 3D shape?

1/4

- cuboid ■ tetrahedron ■ **triangular prism**
- square-based pyramid

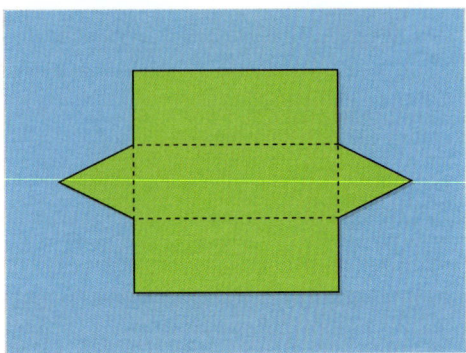

14. The net forms a 3D shape. How many vertices does the 3D shape have?

■ **8** ■ 6 ■ 12

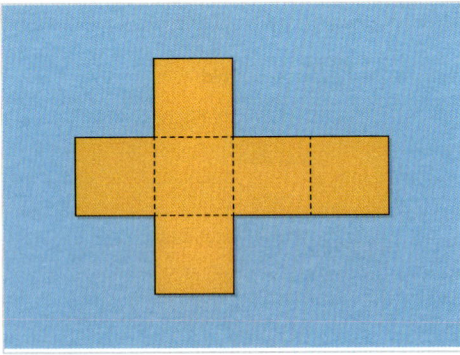

15. Sam unfolds a cereal box to make a net. The net shows that Sam's cereal box is a _____ when folded into its 3D form.
Fill in the blank with the name of a 3D shape.

- **cuboid** ■ cube

16. Janie has stacked three 10 centimetre (cm) cubes to make a tower. She uses the net to make a box that is exactly the same size as her tower. What is the missing side length on the net?
Include the units cm (centimetres) in your answer.

- **30 cm** ■ **30 centimetres** ■ **30**

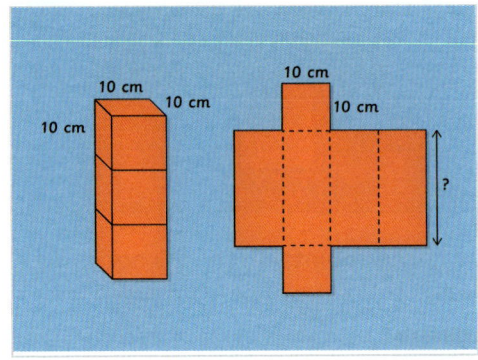

17. Chloe has a packet of sweets. She unfolds the cardboard packet to make a net which has six identical rectangles and two regular hexagons. The sweet packet is a hexagonal _____ .

Fill in the blank to complete the name of the 3D shape.

- **prism** ■ cuboid ■ pyramid

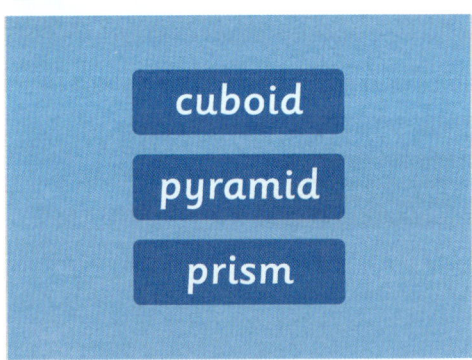

18. The net forms a 3D shape. How many edges does the 3D shape have?

■ **9** ■ 10 ■ 5 ■ 6

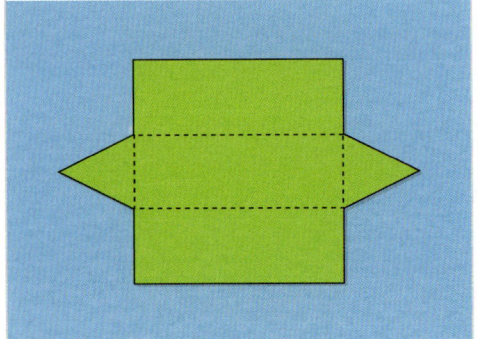

Level 2 *continued*

19. The net folds into which 3D shape?

☐
☒
☐

1/4

■ cube ■ **cuboid** ■ square-based pyramid

■ triangular prism

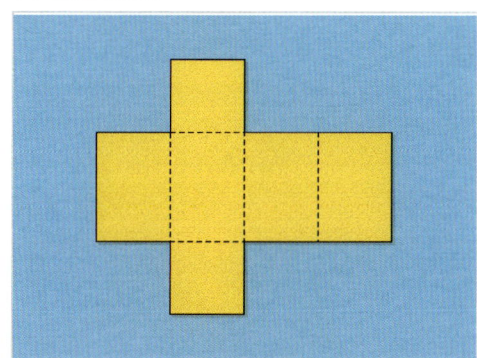

20. How many squares are there in the net of a cube?

1
2
3

■ **6**

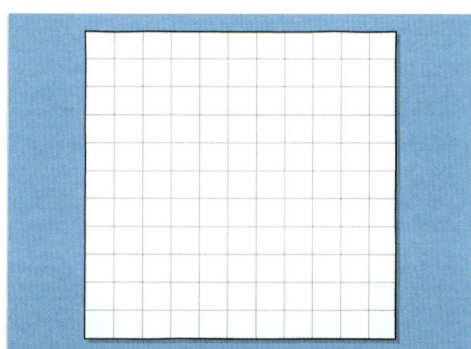

Level 3: Reasoning: Reason about 3D shapes and their 2D representations.

✱ **Required:** 5/5 ✱ **Pupil Navigation:** on

✱ **Randomised:** off

21. The image shows the net of a dice. Jen folds the net to make a 3D shape. When she rolls the dice, the letter B is showing on top. Which letter is on the **bottom** of the dice?

☐
☒
☐

1/5

■ A ■ C ■ D ■ **E** ■ F

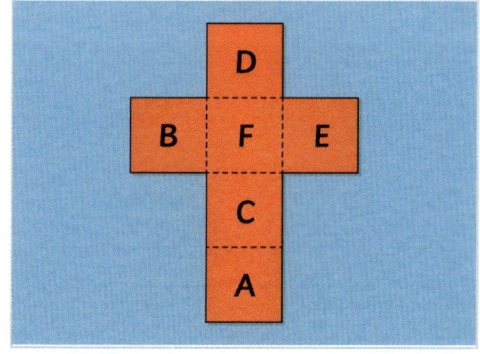

22. If two identical cubes are joined face-to-face to form a cuboid, the new shape will have 12 faces, 16 vertices and 24 edges. Is this statement correct? Explain your answer.

a
b
c

- Open question, no set answer

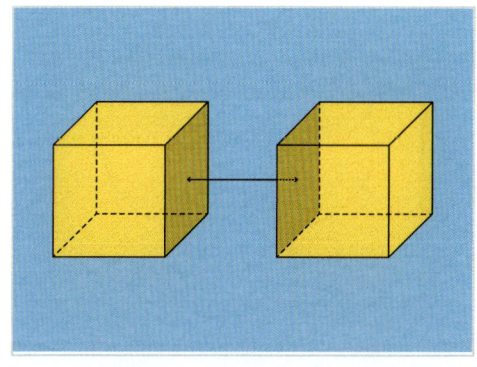

23. Jerry says, "A hexagon has six sides, so a pyramid with a hexagonal base must have six faces." Is Jerry correct? Explain your answer.

a
b
c

- Open question, no set answer

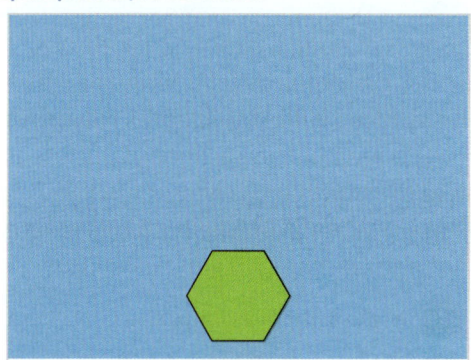

24. Which net folds into the open cube-shaped box shown in the image?

☐
☒
☐

1/3

■ net A ■ net B ■ **net C**

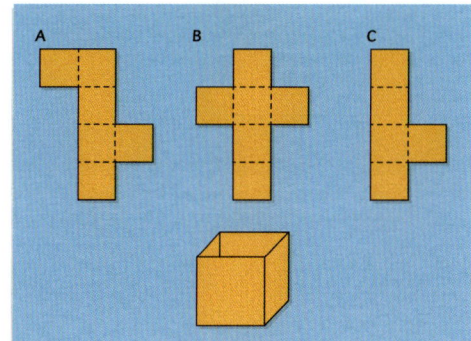

Level 3 *continued*

25. The nets need to be sorted into a Venn
a
b
c
diagram. What titles would you use for
group A and group B? Explain your answer.

- Open question, no set answer

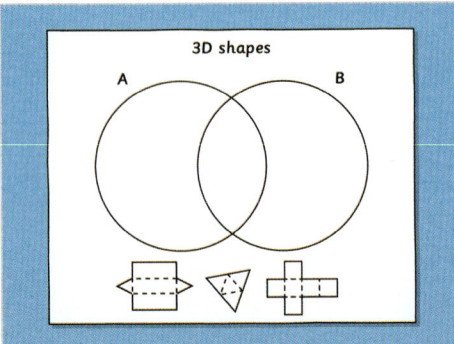

Level 4: Problem solving in greater depth: Solve multi-
step problems involving 3D shapes and their
2D representations.

✹ **Required:** 5/5   ✹ **Pupil Navigation:** on
✹ **Randomised:** off

26. Danny is making a cuboid-shaped box. The
1
2
3
box is 8 centimetres (cm) long and has two
square faces with side lengths measuring 5
cm. The area of the cardboard net he uses to
make the box is _____ cm².
Fill in the blank with the area of the net.
Don't include the units in your answer.

■ 210 ■ 40 ■ 50 ■ 25 ■ 160

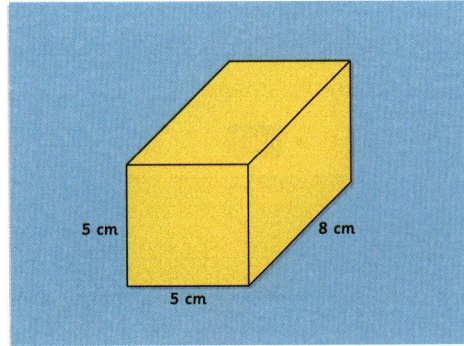

27. Ben has two identical tetrahedrons with
1
2
3
equilateral faces. If Ben joins the
tetrahedrons by matching two faces
together, how many **faces** does Ben's new
3D shape have?

■ 6 ■ 8

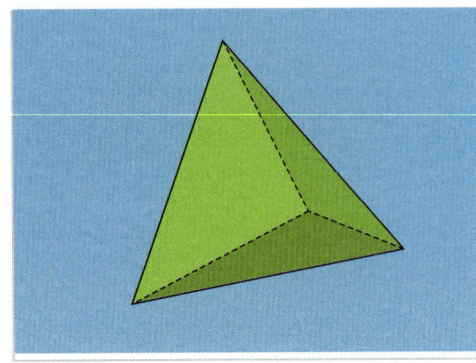

28. The dice has a different symbol on each face.
The diagram shows the dice from three
different angles. The symbol you can't see is
1/5
a heart.
Which symbol is the heart opposite?
Use the blank net to help you work out the
answer.

■ circle ■ diamond ■ star ■ arrow ■ cross

29. The 3D model is made from stacks of cubes.
1
2
3
How many cubes make up the 3D model in
total?

■ 20 ■ 10

Level 4 *continued*

30.  Peter puts a cube on the table and places a square-based pyramid on top of it, matching the two square faces. He walks around the table and counts the number of faces of the composite shape that he can see. How many faces does Peter count?

■ 9   ■ **8**   ■ 11

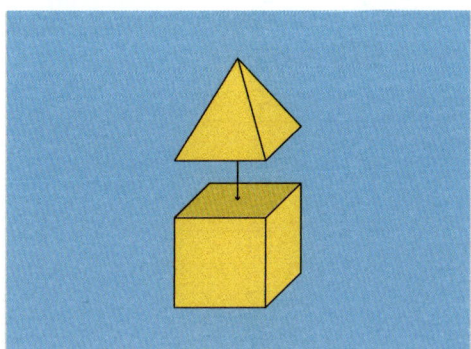

Calculate Angles at Different Points

Objective: I can calculate angles at a point, full turns, half turns and other multiples of 90 degrees.

Quick Search Ref: 10169

Level 1: Understanding: Know the sum of the angles on a straight line and around a point.

✱ **Required:** 7/10   ✱ **Pupil Navigation:** on   ✱ **Randomised:** off

---

1. There are 90° in a quarter turn. How many degrees are there in one **full turn**?

1/4   ■ 180°   ■ 270°   ■ **360°**   ■ 90°

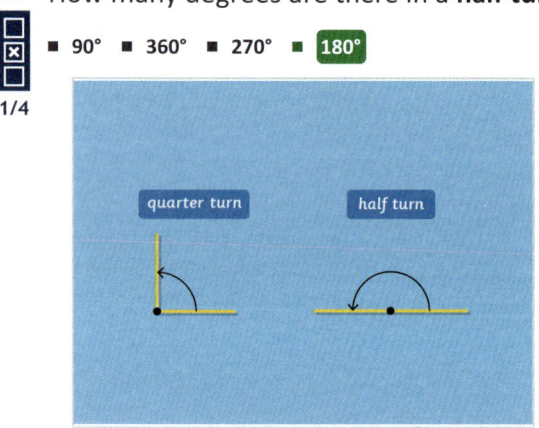

2. How many degrees are there in a **half turn**?

■ 90°   ■ 360°   ■ 270°   ■ **180°**

1/4

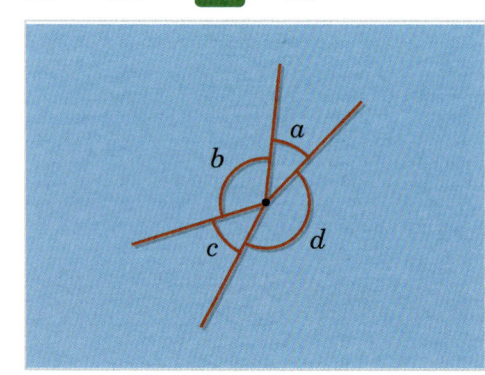

3. The sum of the angles around a point is the same as the number of degrees in one full turn. What is the sum of angles a, b, c and d?

1/4   ■ 180°   ■ 270°   ■ **360°**   ■ 720°

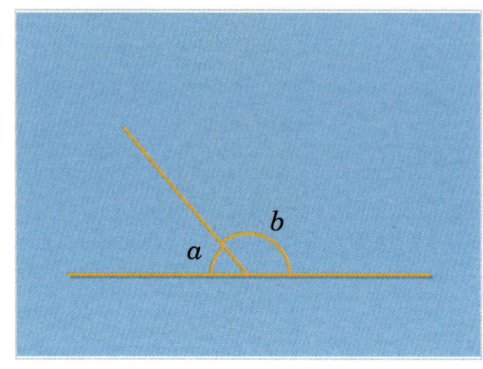

4. The sum of the angles on a straight line is the same as the number of degrees in a half turn. What is the sum of angles a and b?

1/3   ■ 360°   ■ **180°**   ■ 90°

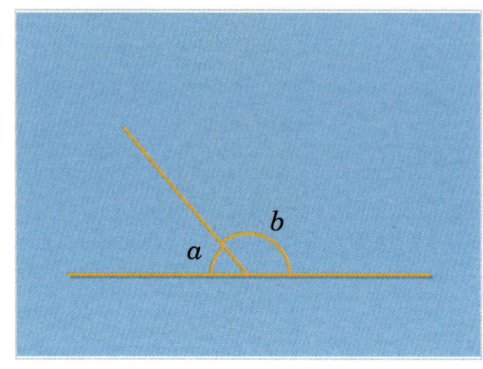

5. Select the calculation that you would use to find the size of angle a.

■ **180° − 80°**   ■ 360° − 80°   ■ 90° − 80°

1/3

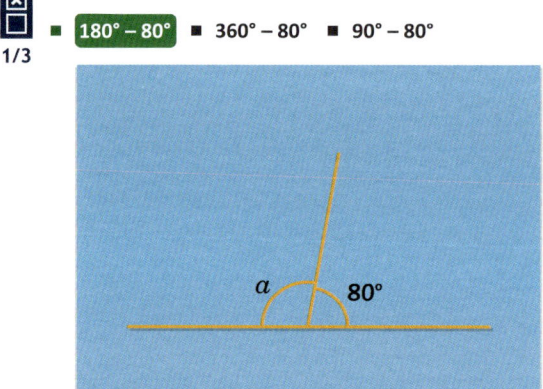

6. Select the number sentence that shows how to calculate the size of angle c.

■ 360° − a = c   ■ 180° − a − b = c   ■ 180° − b = c

1/4   ■ **360° − a − b = c**

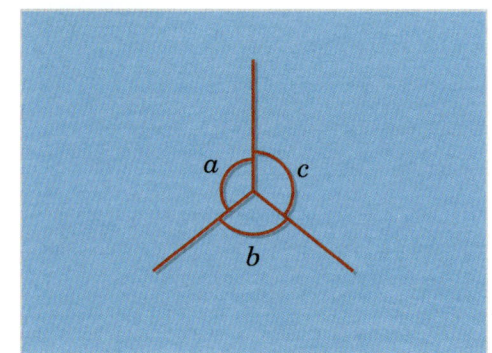

Level 1 *continued*

7. Select the number sentence that shows how to calculate the size of angle *a*.

1/3

■ 90° + 30° = a   ■ 180° − 90° − 30° = a
■ 360° − 90° − 30° = a

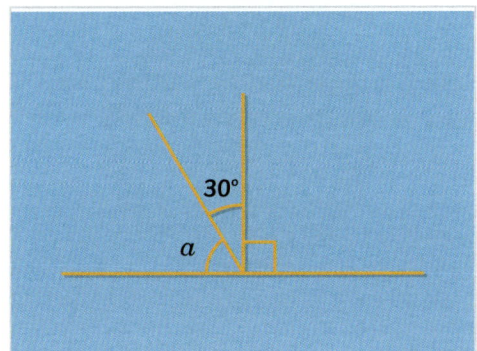

8. To find the size of angle *b,* what angle would you subtract 140° from?

1/3

■ 90°   ■ 180°   ■ 360°

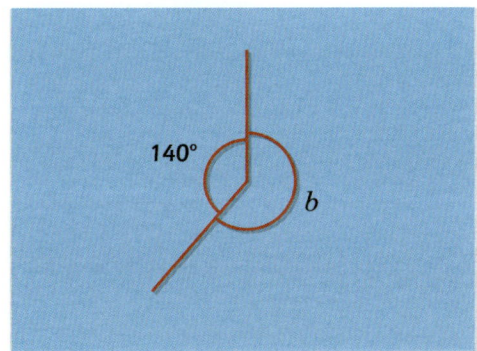

9. What is the sum of angles *a*, *b,* and *c*?

1/3

■ 180°   ■ 90°   ■ 360°

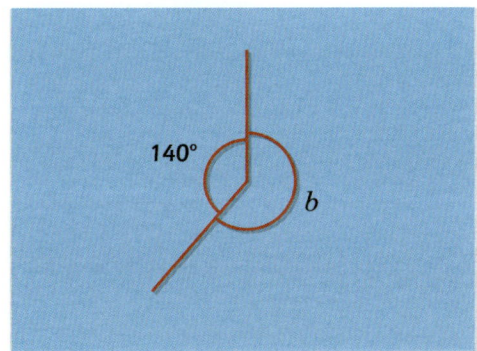

10. What is the sum of angles *a*, *b* and *c*?

1/4

■ 360°   ■ 180°   ■ 270°   ■ 90°

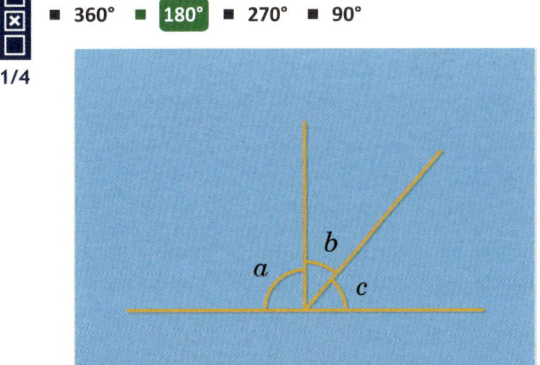

Level 2: Fluency: Calculate angles on a straight line and around a point.

✱ **Required:** 7/10   ✱ **Pupil Navigation:** on
✱ **Randomised:** off

11. What is the size of angle *a* in degrees?

1
2
3

■ 120   ■ 300   ■ 30

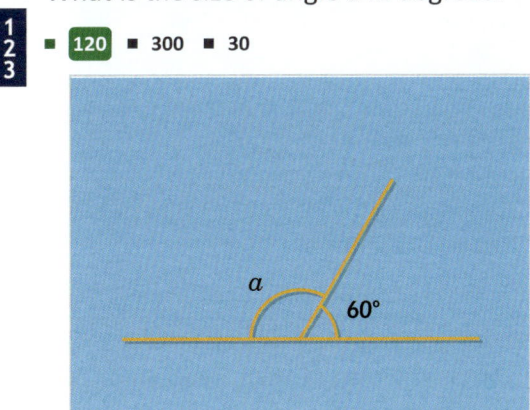

12. Calculate the size of angle *b* in degrees.

1
2
3

■ 244   ■ 64

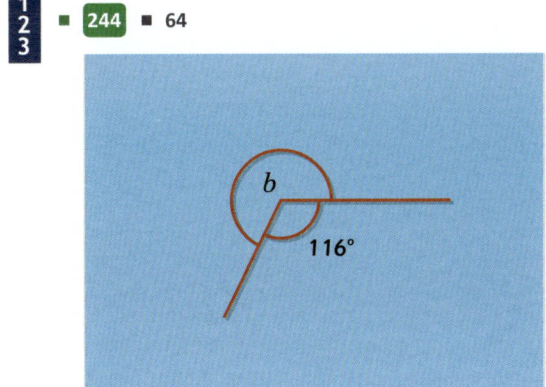

Level 2 continued

13. What is the size of angle *a* in degrees?

1
2
3
■ 230 ■ 120 ■ 240 ■ 250

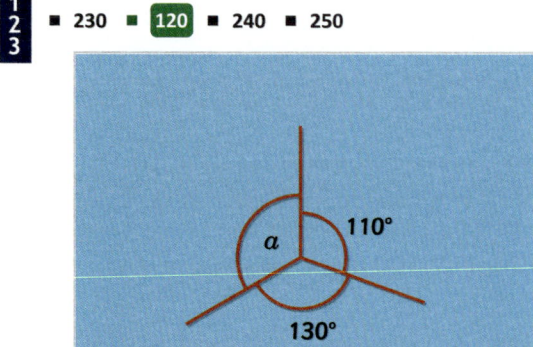

14. Calculate the size of angle *b* in degrees.

1
2
3
■ 108 ■ 165 ■ 72 ■ 123

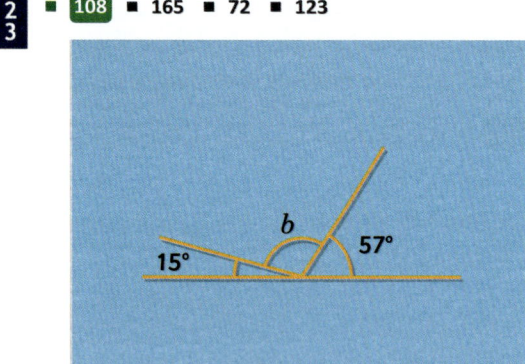

15. The image shows the diagonals of a rectangle crossing at their midpoints. What is the size of angle *a* in degrees?

1
2
3
■ 113 ■ 293

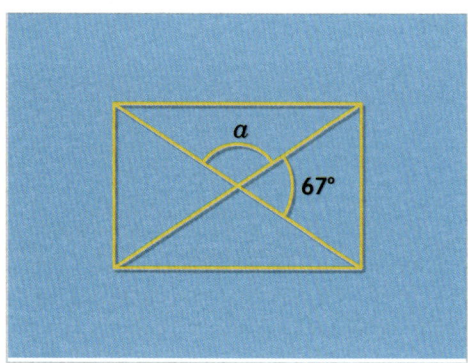

16. Adam cuts a cake into nine equally sized pieces. In degrees, what is the size of angle *a*?

1
2
3
■ 40 ■ 20

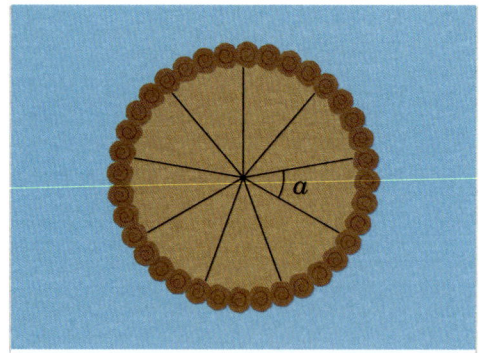

17. Jodie draws a parallelogram and a triangle on a straight line. What is the size of angle *a* in degrees?

1
2
3
■ 148 ■ 58 ■ 90

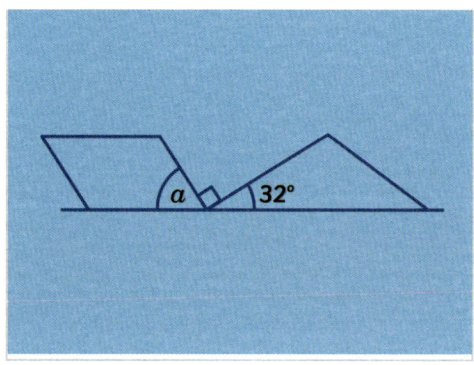

18. Two equilateral triangles are placed next to each other at the centre point of a square. In degrees, what is the size of angle *b*?

1
2
3
■ 240 ■ 300

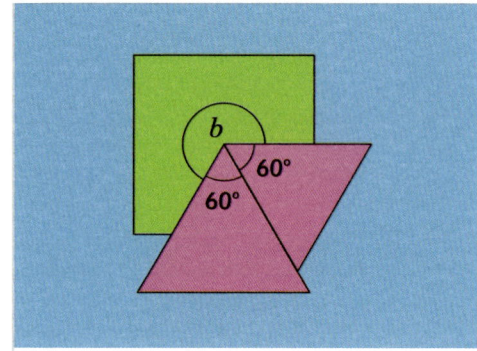

Level 2 continued

19. Calculate the size of angle *a* in degrees.

1
2
3

- 100 ▪ 280

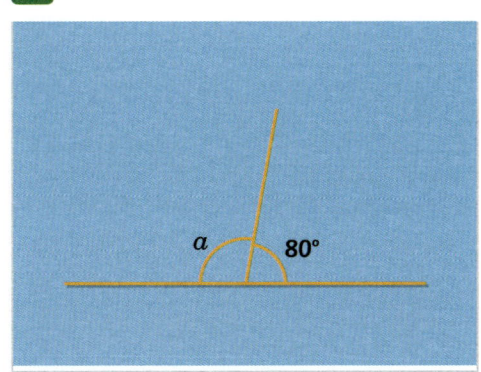

20. In degrees, what is the size of angle *b*?

1
2
3

- 220 ▪ 40

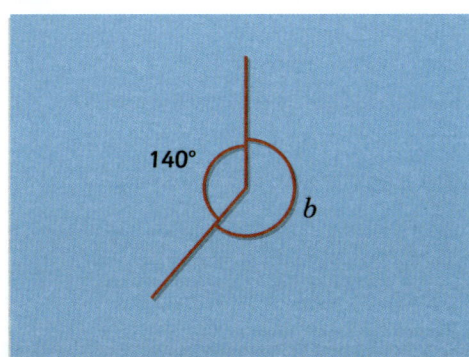

Level 3: Reasoning: Reason about angles at different points.

❋ **Required:** 5/5   ❋ **Pupil Navigation:** on
❋ **Randomised:** off

21. Bobby measures angles *a* and *b*. He says that
a
b
c
a is 63 degrees and *b* is 127 degrees. Explain how you know that he has made a mistake in his measurement.

- Open question, no set answer

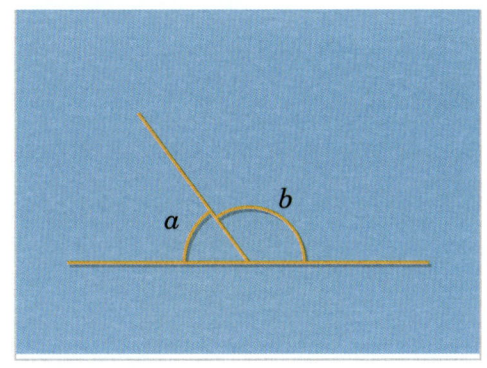

22. Prove that if angle *a* and angle *b* are both
a
b
c
acute, then angle *c* must be a reflex angle.

- Open question, no set answer

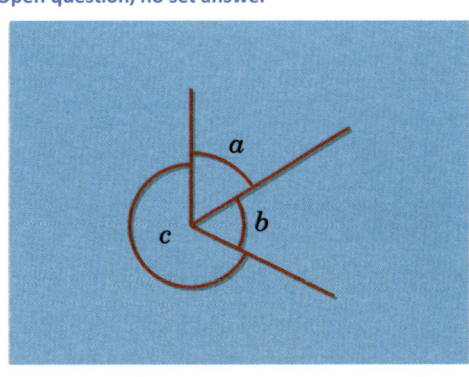

23. Look at the image and select the number sentence that is **incorrect**.

1/4

- $a + b + 30° + 100° = 360°$ ▪ $a = 180° - 30°$
- b + 30° = 180° ▪ $a = 360° - 100° - 30° - b$

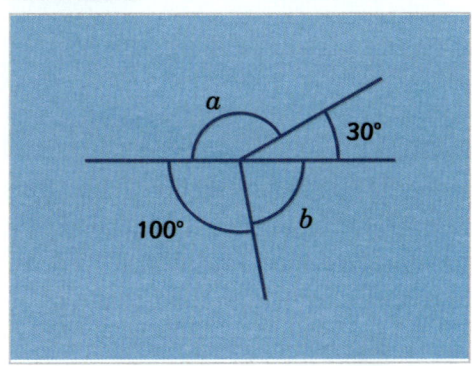

24. Kelly says, "If I know the size of angle *a*, I can
a
b
c
find the size of angle *b* and the size of angle *c* without measuring them."
Is Kelly correct? Explain your answer.

- Open question, no set answer

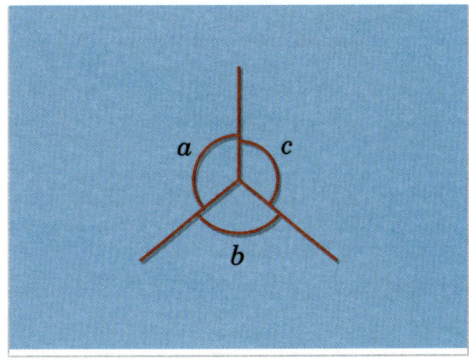

Level 3 *continued*

25. Tia says that if you join any three equilateral
a
b triangles at one vertex, you can make a
c straight line. Is Tia correct? Explain your
answer.

- Open question, no set answer

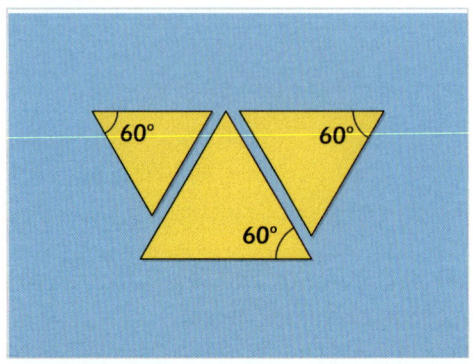

Level 4: Problem solving with greater depth: Solve
multi-step problems involving angles at
different points..

✷ **Required:** 5/5   ✷ **Pupil Navigation:** on
✷ **Randomised:** off

26. Peter cuts a cake exactly in half and puts one
1
2 half onto his plate. If Peter eats a 55° slice
3 from his plate, how big is the remaining
piece of cake on his plate in degrees?

■ 305   ■ **125**   ■ 180

27. The composite shape is made up of three
1
2 regular pentagons. What is the size of angle
3 *a* in degrees?

■ **36**   ■ 324

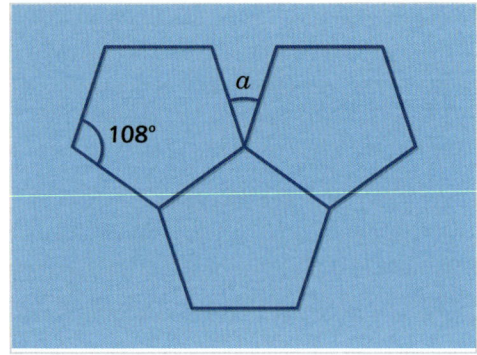

28. The image shows four lines meeting at a
1
2 point.
3
• Angle *a* is 52°.
• Angle *a* and angle *d* are on a straight line.
• Angle *b* is half the size of angle *d*.
What is the size of **angle c** in degrees?

■ **116**   ■ 64

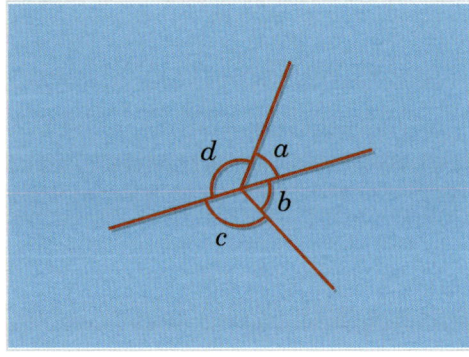

29. The pie chart shows the favourite pizza
1
2 toppings for children in class 6H. One-third
3 of the class like pepperoni and one-quarter
of the class like ham and pineapple. What is
the angle for the section of the class who like
cheese?

■ **150**   ■ 210

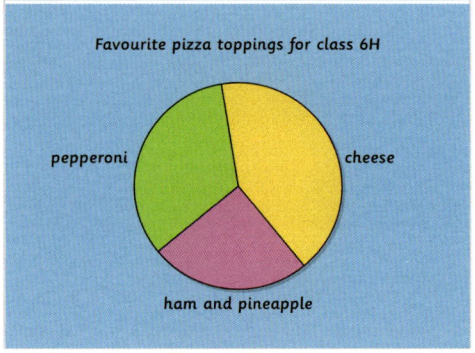

Level 4 *continued*

30. If the size of angle *a* is half the size of angle
b, how many degrees is angle *b*?

1
2
3

■ 78 ■ 117 ■ 39

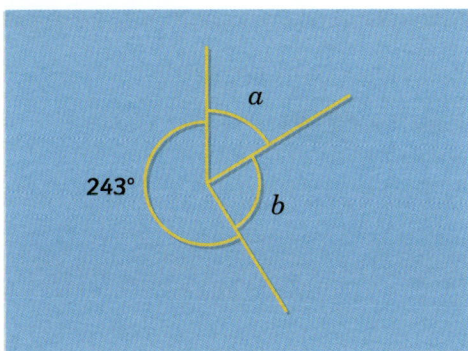

Estimate and Compare Acute, Obtuse and Reflex Angles

Objective: I can estimate and compare acute, obtuse and reflex angles.

Quick Search Ref: 10076

Level 1: Understanding: Compare acute, obtuse and reflex angles.

✿ **Required:** 7/10   ✿ **Pupil Navigation:** on   ✿ **Randomised:** off

1. Sort the following labels into order, starting with the type of angle in position 1.

- reflex angle - acute angle - right angle
- obtuse angle

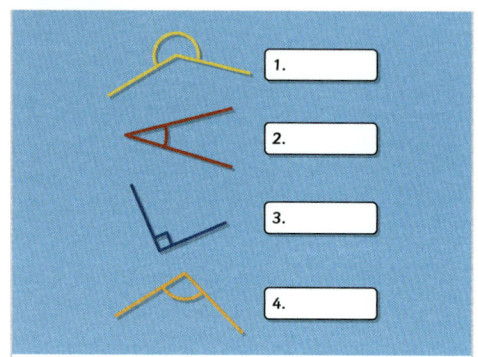

2. Select the **correct** statement about obtuse and reflex angles.

1/2

- A reflex angle is greater than an obtuse angle
- An obtuse angle is greater than a reflex angle.

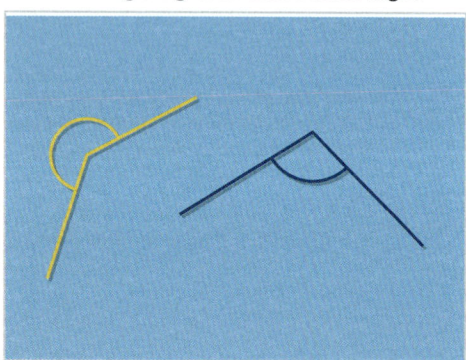

3. Sort the types of angle from smallest to largest.

- acute angle - right angle - obtuse angle
- reflex angle

type of angle	size
reflex	>180°
acute	<90°
obtuse	>90° and <180°

4. Which angle measures 20°?

1/4

- angle a - angle b - angle c - angle d

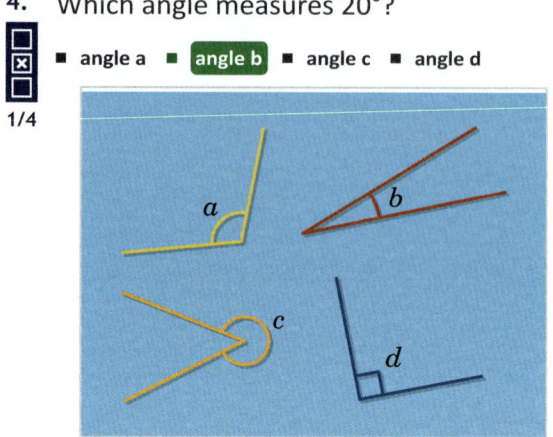

5. Which angle measures 250°?

1/4

- angle a - angle b - angle c - angle d

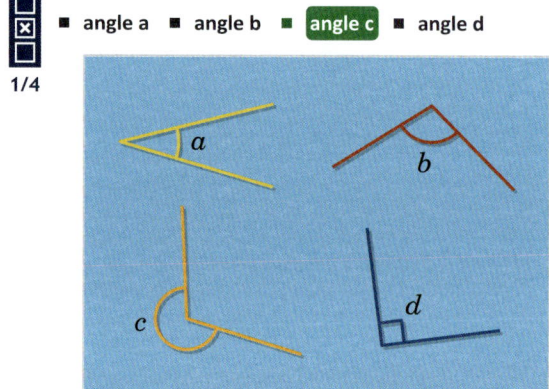

6. An angle measuring 359° is a _____ angle.
Enter the missing type of angle.

- reflex - right - obtuse - acute

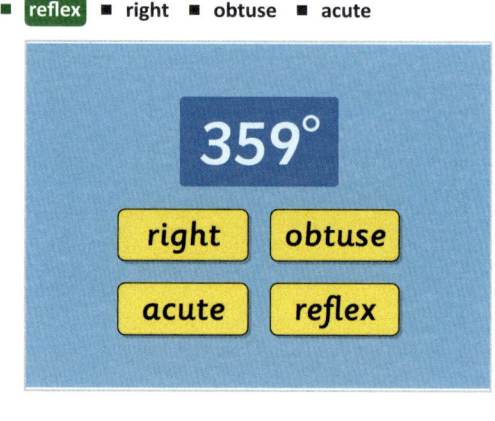

359°

| right | obtuse |
| acute | reflex |

Level 1 *continued*

7. Which is the reflex angle that measures 270°?

    1/5

    ■ angle a ■ angle b ■ angle c ■ angle d ■ angle e

    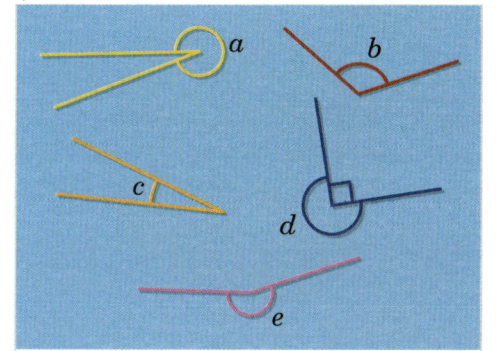

8. Sort the types of angles from largest to smallest.

    ■ reflex angle ■ obtuse angle ■ right angle
    ■ acute angle

    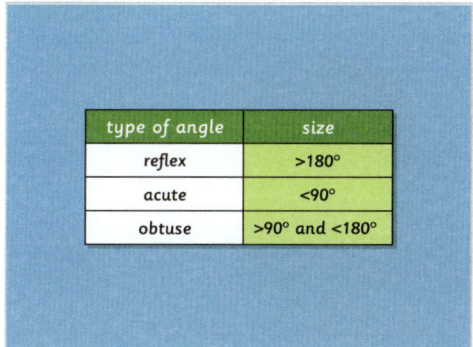

type of angle	size
reflex	>180°
acute	<90°
obtuse	>90° and <180°

9. Which angle measures 180 degrees (°)?

    1/4

    ■ angle a ■ angle b ■ angle c ■ angle d

    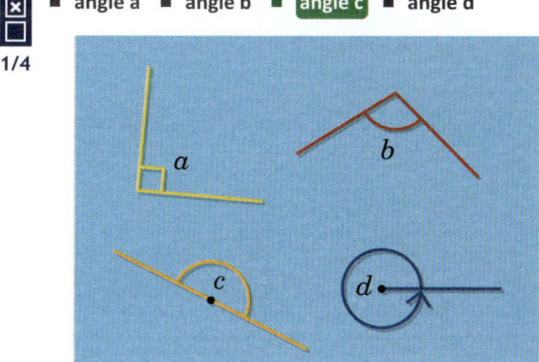

10. Which angle measures 340°?

    1/4

    ■ angle a ■ angle b ■ angle c ■ angle d

    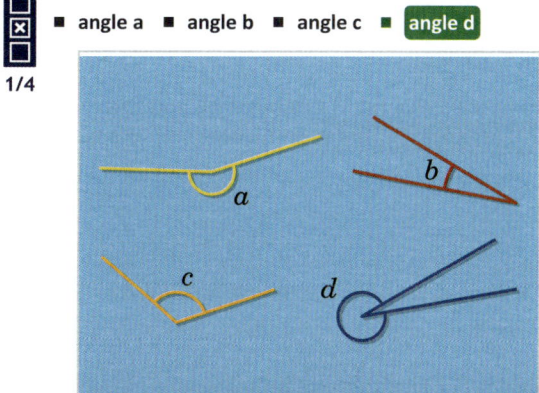

Level 2: Fluency: Compare and estimate acute, obtuse and reflex angles.

    ✿ **Required:** 7/10   ✿ **Pupil Navigation:** on
    ✿ **Randomised:** off

11. Half of a right angle is 45° and one-third of a right angle is 30°. Which of these acute angles is 30°?

    1/4

    ■ angle a ■ angle b ■ angle c ■ angle d

    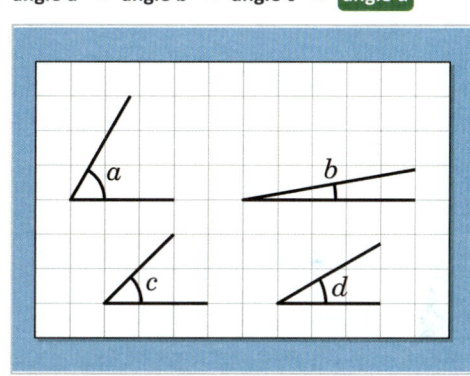

12. **Estimate** the size of angle *a* in degrees.

    1
    2
    3

    ■ 46 ■ 43 ■ 45 ■ 47 ■ 44

    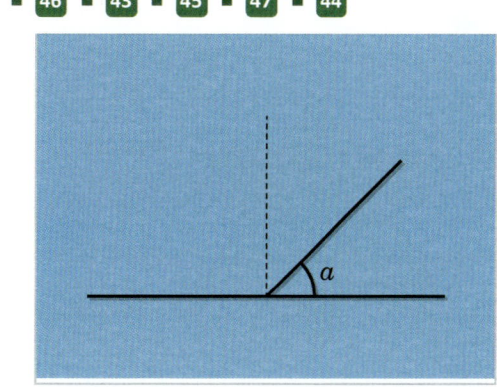

Level 2 *continued*

13. What is an approximate value for the size of angle *b*?

1/4

- between 45° and 50°   - between 90° and 100°
- between 120° and 140°   - between 170° and 180°

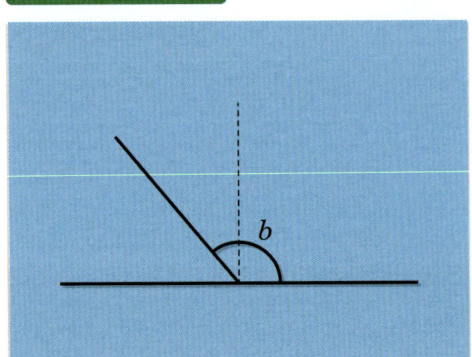

14. Which three angles can you add together to estimate the size of angle *x*?

3/5

- 90°   - 45°   - 10°   - 60°   - 180°

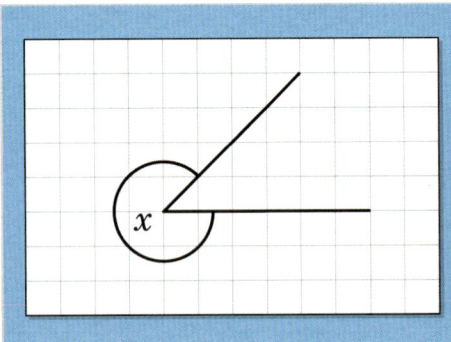

15. Cara uses one angle in the equilateral triangle to estimate the size of angle *x*. To the nearest 10°, what is the approximate size of angle *x*?

1
2
3

- 120   - 60

16. Jodie cuts a piece out of the pizza. To the nearest 10°, what is the approximate size of the reflex angle of the remaining pizza?

1
2
3

- 300   - 310   - 315   - 60   - 320

17. Estimate the size of the angle of the piece of cake. How many pieces will fit into the circular cake tin?

1
2
3

- 3   - 4

18. Angle *a* is approximately half the size of angle *b*. To the nearest 10°, what is the approximate size of angle *b*?

1
2
3

- 60   - 30

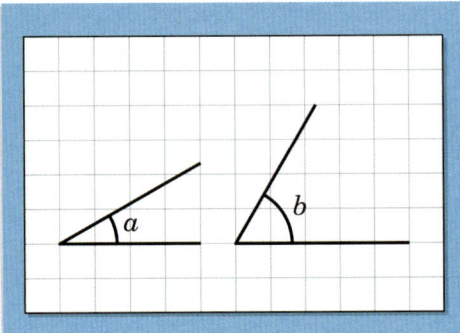

Level 2 *continued*

19. In the parallelogram, angle *b* is three times the size of angle *a*. How many degrees is angle *b*?

1 2 3

- **135** ■ 45

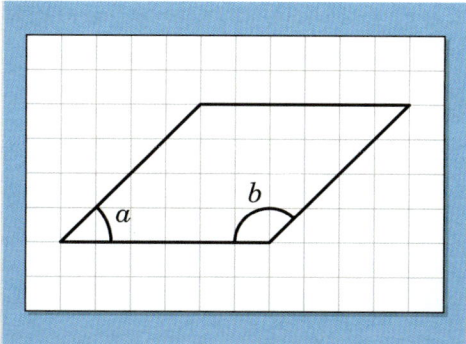

20. What is an approximate value for the size of angle *a*?

1/4

- ■ between 170° and 180° ■ between 90° and 100°
- **between 10° and 20°** ■ between 120° and 140°

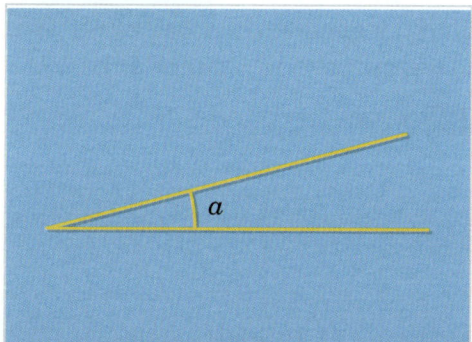

Level 3:   Reasoning: Compare and estimate measurements of angles.

✷ **Required:** 5/5   ✷ **Pupil Navigation:** on
✷ **Randomised:** off

21. Which angle is the **odd one out**?

1/5

- ■ 195° ■ 231° ■ **175°** ■ 335° ■ 206°

angle size	type of angle
195°	
231°	
175°	
335°	
206°	

22. One of the angles in the triangle is hidden by a rectangle. Bailey says that the hidden angle is an acute angle. Is Bailey correct? Explain your answer.

a b c

- **Open question, no set answer**

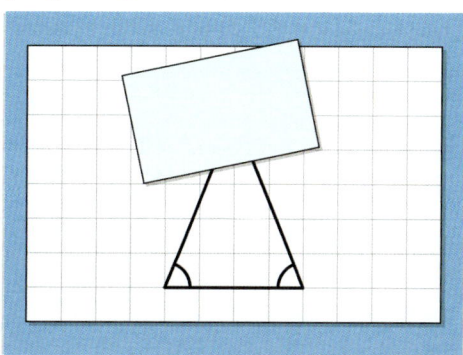

23. Winston estimates that the size of angle *a* is 100°. Is Winston's estimate sensible? Explain your answer and make your own estimate of the size of angle *a*.

a b c

- **Open question, no set answer**

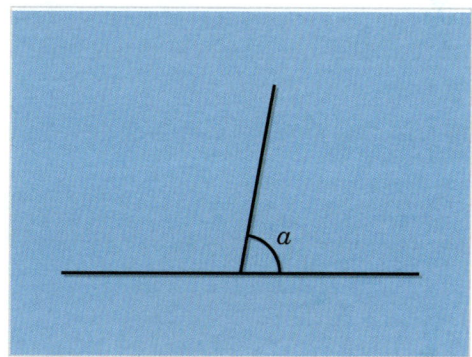

24. The shapes are all **regular polygons**. Estimate the size of each of the marked angles. Angle _____ is 120°.

1/4

- ■ a ■ d ■ b ■ **c**

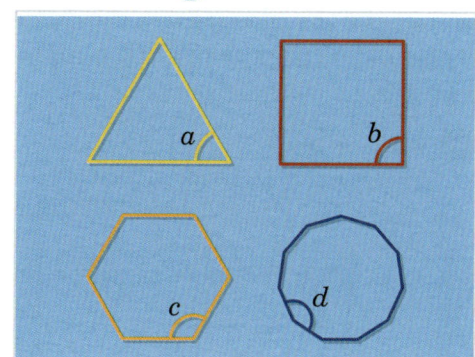

Level 3 *continued*

25. Liam says that angle *a* is the odd one out of
a
b   these three reflex angles. Explain why Liam
c   might say this.

- Open question, no set answer

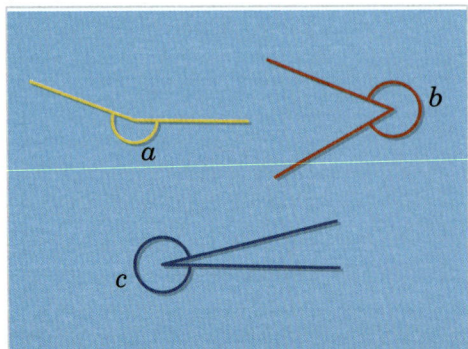

Level 4:   Problem Solving: Compare and estimate
angles of shapes.

✿ **Required:** 5/5   ✿ **Pupil Navigation:** on
✿ **Randomised:** off

26. Millie has three-quarters of a cake. If she
1
2   cuts the cake into identical slices, how many
3   slices can she cut in total?

■ **9** ■ **6**

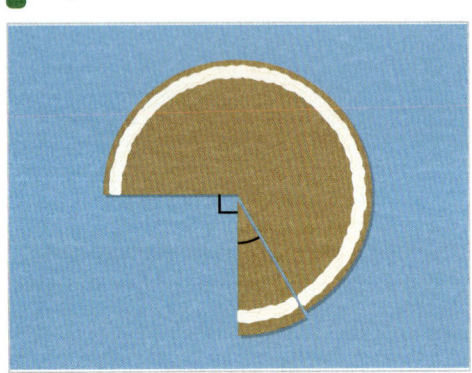

27. Estimate the size of angle *a* in the isosceles
1
2   triangle to find the size of angle *b* to the
3   nearest 10°.

■ **40** ■ **100** ■ **80**

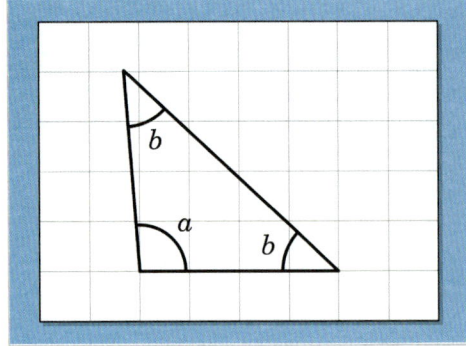

28. Use the following clues to find the size of the
1
2   angle shown in the image:
3   • Subtracting the angle from 360° creates a
   reflex angle greater than 270°.
   • The angle is a multiple of 10.
   • The angle is greater than half a right angle.
   • You can fit more than six of the angles in
   one whole turn.

■ **50**

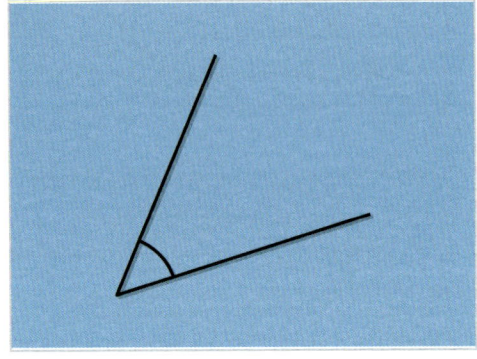

29. Ellie estimates that angle *a* is approximately
1
2   35°. If her estimate is wrong by 5°, what is
3   the largest possible value of angle *b*?

■ **150** ■ **140**

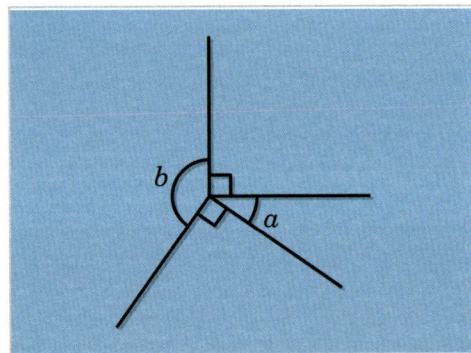

30. Angle *a* is half the size of angle *b* and angle *c*
1
2   is twice the size of angle *b*. Estimate the size
3   of angle *c* in degrees.

■ **120** ■ **30** ■ **60**

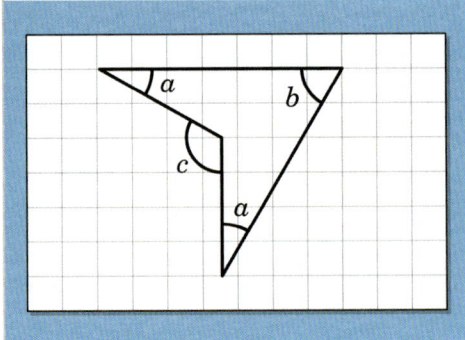

Measuring Angles

Objective: I know angles are measured in degrees and can measure angles.

Quick Search Ref: 10203

Level 1: Understanding: Understand angle vocabulary and recognise types of angles from their sizes.

✿ **Required:** 7/10   ✿ **Pupil Navigation:** on   ✿ **Randomised:** off

---

1. Which **two** statements about angles are true?

2/5

- ■ An angle measures the amount of turn about a point.
- ■ Angles are measured in millimetres (mm).
- ■ An angle is the distance between two points.
- ■ Angles are measured in degrees (°).
- ■ An angle is the distance around a circle.

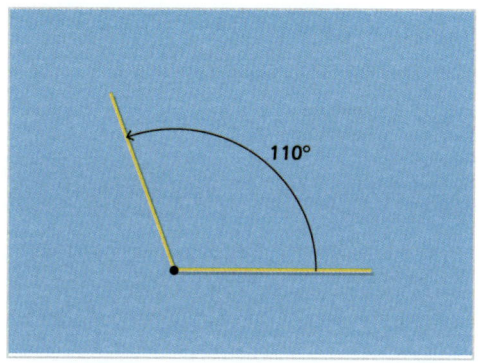

2. Select **two incorrect** statements about the number of degrees in different turns.

2/6

- ■ There are 25° in a quarter turn.
- ■ There are 90° in a quarter turn.
- ■ There are 50° in a half turn.
- ■ There are 180° in a half turn.
- ■ There are 270° in a three-quarter turn.
- ■ There are 360° in one full turn.

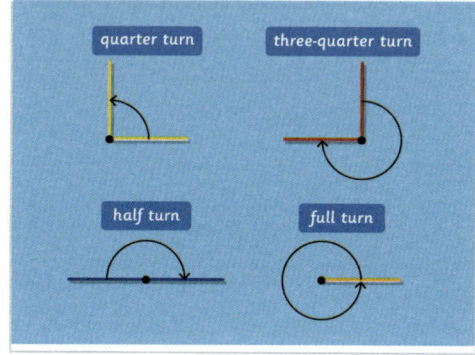

3. You use a _____ to measure an angle.
Enter the missing word.

- ■ protractor ■ ruler ■ compass ■ calculator

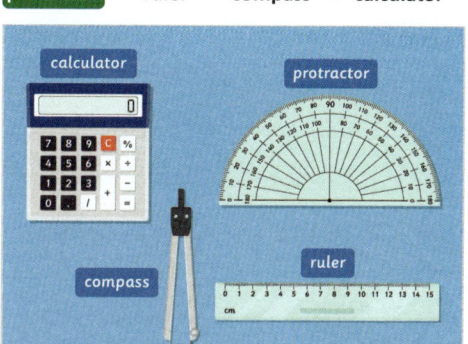

4. Sort the labels for the parts of the protractor into order, starting with the label in position 1.

- ■ outer scale ■ inner scale ■ centre point
- ■ base line

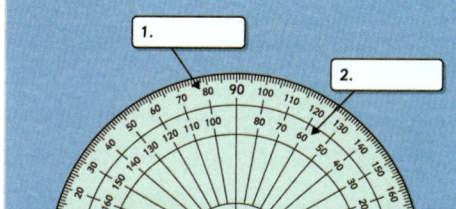

5. An angle of 240° is a _____ angle.
Enter the missing type of angle.

- ■ obtuse ■ reflex ■ acute ■ right

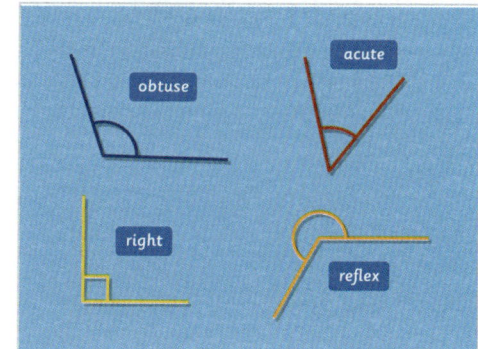

Level 1 *continued*

6. The measurement on the protractor shows an angle of 120°. What type of angle is this?

■ acute ■ right ■ **obtuse** ■ reflex

1/4

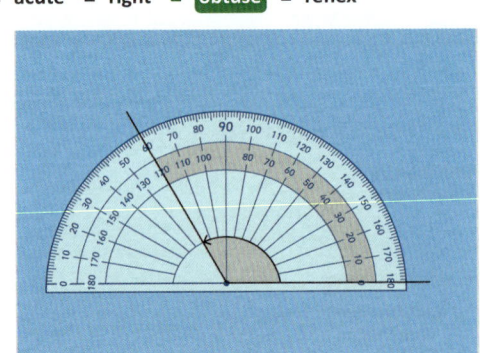

7. What type of angle is shown on the protractor?

■ right ■ **acute** ■ obtuse ■ reflex

1/4

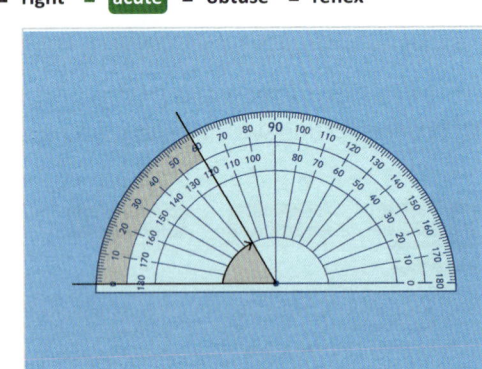

8. The measurement on the protractor shows an angle of 170°. What type of angle is this?

■ right ■ acute ■ reflex ■ **obtuse**

1/4

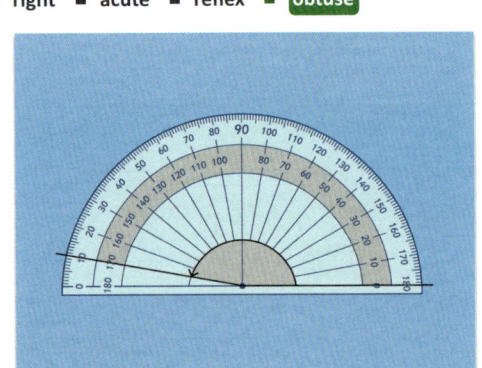

9. What type of angle is shown on the protractor?

■ **acute** ■ reflex ■ obtuse ■ right

1/4

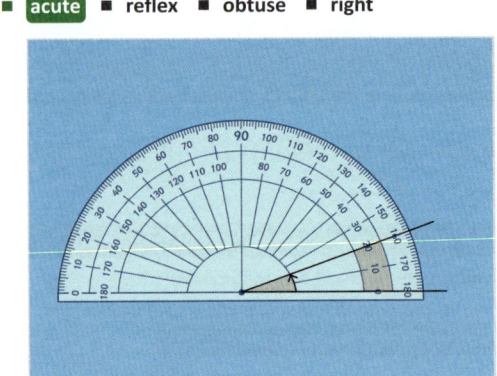

10. An angle of 100° is an _____ angle.
Enter the missing type of angle.

■ reflex ■ **obtuse** ■ acute ■ right

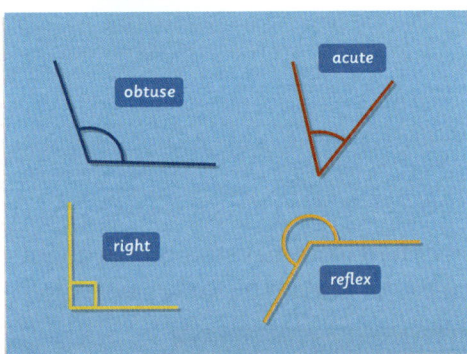

Level 2: Fluency: Read angle sizes on a protractor and calculate the size of a reflex angle.

❋ **Required:** 7/10   ❋ **Pupil Navigation:** on
❋ **Randomised:** off

11. The size of the angle shown on the protractor is _____ degrees (°).
Enter the missing number.

■ **30** ■ 150

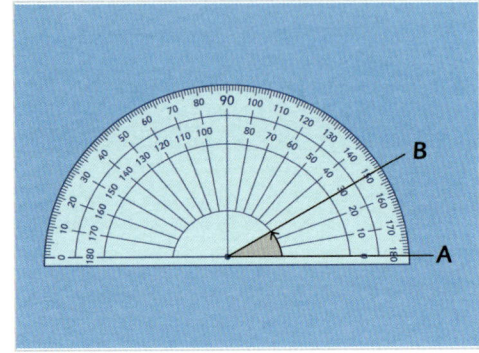

12. The size of the angle shown on the protractor is _____ degrees (°).
Enter the missing number.

■ **45** ■ 135 ■ 55

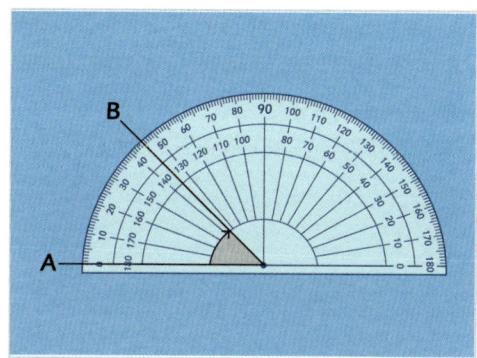

13. Which position shows an angle of 110° measured from line A?

1/3

■ position A ■ position B ■ **position C**

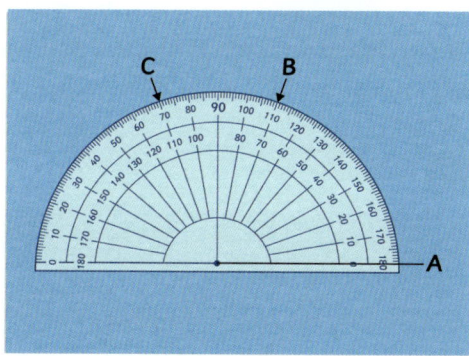

14. To calculate the size of the reflex angle *x*, you subtract the size of angle *y* from 360°. In degrees, what is the size of angle *x*?

■ **230** ■ 130 ■ 310

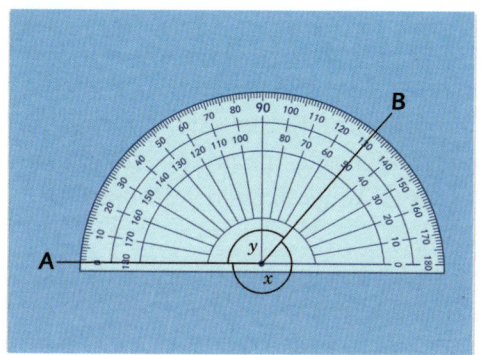

15. The positions marked by letters on the protractor show different angles of turn from line A. In degrees, what is the size of the obtuse angle?

■ **146** ■ **148** ■ 32 ■ **147** ■ 153 ■ 90

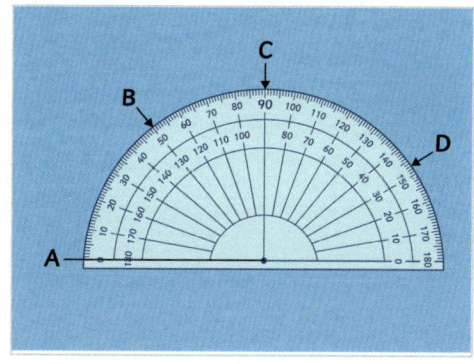

16. Emma has drawn line A. Which position does Emma need to use to draw another line at an angle of 75° from line A?

1/4

■ position A ■ position B ■ **position C** ■ position D

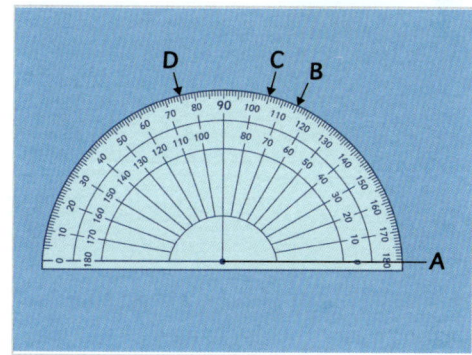

17. What is the size of angle *y* in triangle ABC?

■ **27** ■ **28** ■ **29** ■ 152 ■ 32

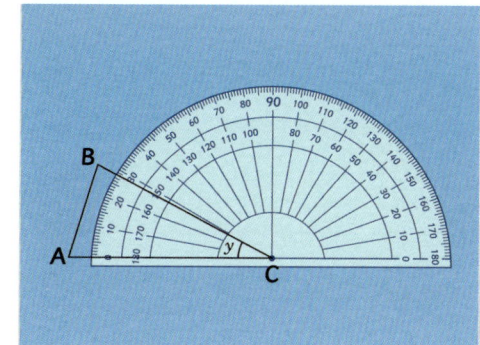

Level 2 *continued*

18. What is the size of angle *x* in triangle ABC?

1 2 3 ▪ 48 ▪ 46 ▪ 47 ▪ 53 ▪ 133

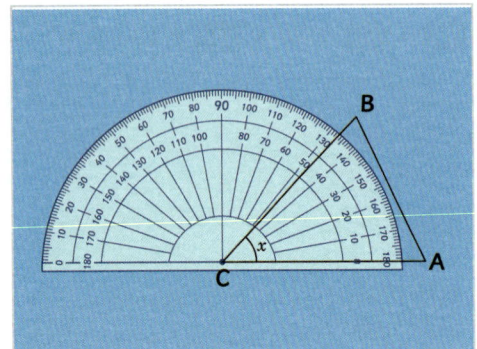

19. Which position shows an angle of 25° measured from line A.

1/3 ▪ position A ▪ position B ▪ position C

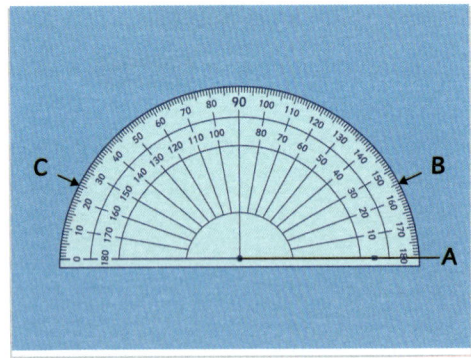

20. The size of the angle shown on the protractor is _____ degrees (°).
Enter the missing number.

1 2 3 ▪ 14 ▪ 12 ▪ 13 ▪ 167

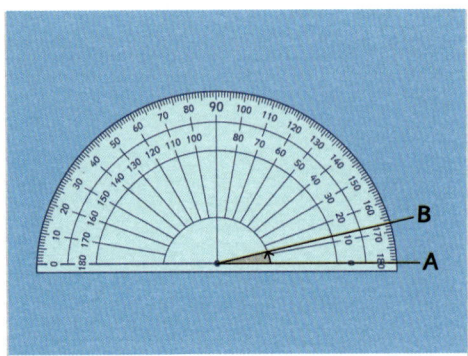

Level 3: Reasoning: Reason about measuring angles.

✱ **Required:** 5/5   ✱ **Pupil Navigation:** on
✱ **Randomised:** off

21. The image shows three protractors being used incorrectly to measure angle *x* between line A and line B. Explain how to correct the position of the protractor for each image.

a b c

- Open question, no set answer

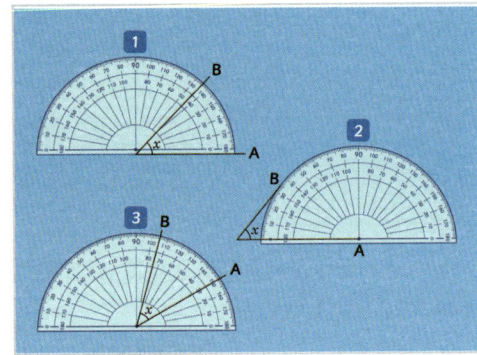

22. Rory has measured an angle using the wrong scale on his protractor. If he says that the angle measures 60°, what is the correct size of Rory's angle?

1 2 3 ▪ 120

23. Sophie says she can make line A and line B longer so that she can measure angle *x* with the protractor. Daniel says that making the lines longer will change the size of angle *x*. Who is correct? Explain your answer.

a b c

- Open question, no set answer

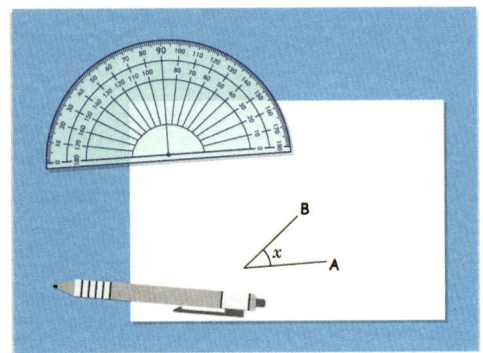

Level 3 continued

24. The scale on the protractor stops at 180°. How could you use this protractor to find the size of reflex angle *x*?

1/3
- ■ Measure angle y and add your answer to 180°
- ■ Measure angle y and subtract your answer from 180°
- ■ Measure angle y and subtract your answer from 360°

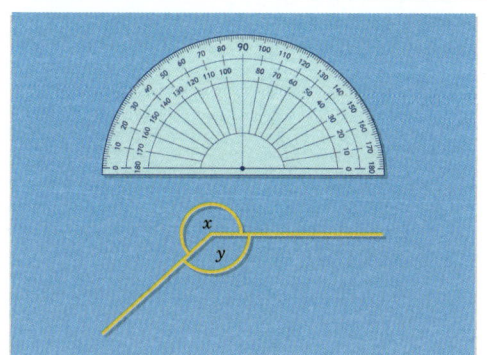

25. Noah has measured angle *x* and says that it's 35°. Explain how you know, without measuring, that Noah has made a mistake. What mistake might he have made?

- Open question, no set answer

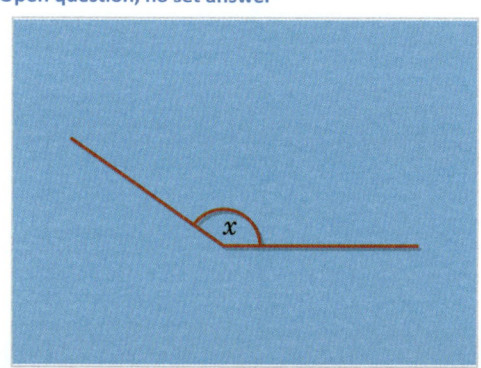

Level 4: Problem solving in greater depth: Solve problems involving measuring angles.

❋ **Required:** 5/5 ❋ **Pupil Navigation:** on
❋ **Randomised:** off

26. In degrees, what is the sum of all the angles in this regular polygon?

1 2 3
- ■ 540 ■ 108

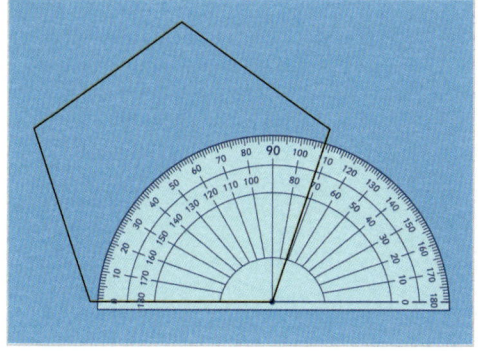

27. The size of the angle *y* is _____ degrees (°).
Enter the missing number.

1 2 3
- ■ 25 ■ 55 ■ 30

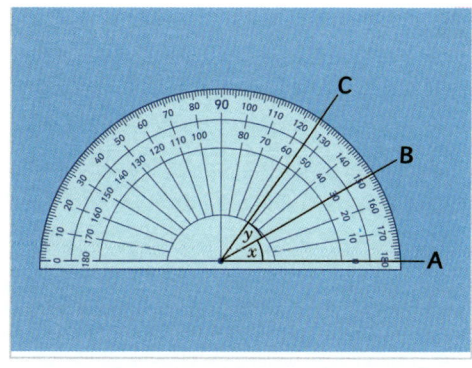

28. The heptagon has two identical reflex angles. What is the sum of the two reflex angles inside the heptagon in degrees?

1 2 3
- ■ 480 ■ 240

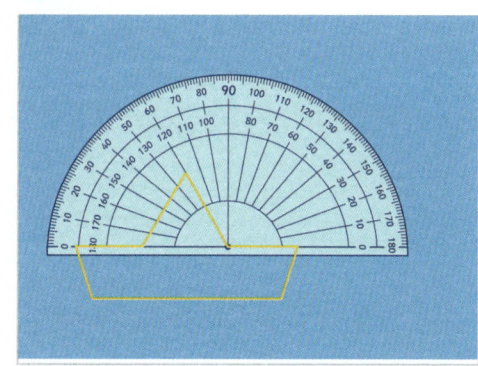

29. One corner of the quadrilateral has been torn off. What was the size of the angle in the missing corner in degrees?

1 2 3
- ■ 95 ■ 275 ■ 85

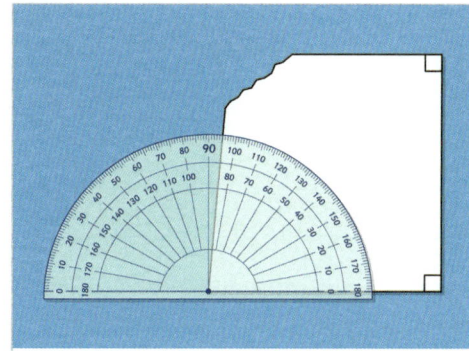

Level 4 *continued*

30. Angle *y* is how many degrees greater than angle *z*?

1 2 3

■ 60 ■ 155 ■ 25 ■ 85

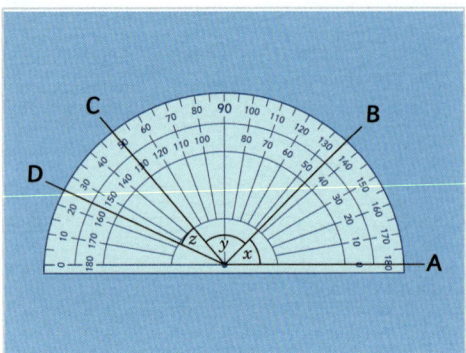

Identify, Describe and Represent the Reflection of a Shape

Objective: I can identify and describe the position of a shape following a reflection.

Quick Search Ref: 10327

Level 1: Understanding: Identify and describe the position of a shape following a reflection.

✸ Required: 7/10   ✸ Pupil Navigation: on   ✸ Randomised: off

1. Which image shows the reflection of triangle Z over the mirror line?

1/3

▪ image A ▪ **image B** ▪ image C

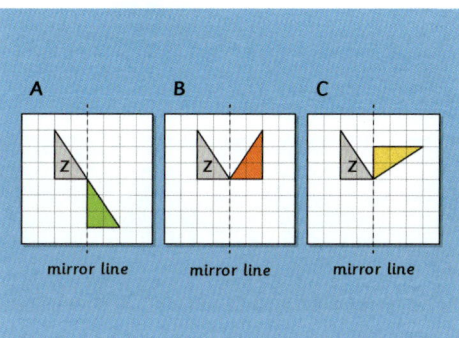

2. When you reflect a shape, the size of the shape _____ changes.
Select the missing word.

1/3

▪ always ▪ sometimes ▪ **never**

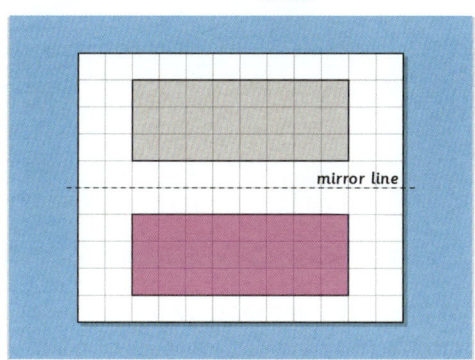

3. Which shape is a reflection of shape A?

1/4

▪ shape B ▪ shape C ▪ shape D ▪ **shape E**

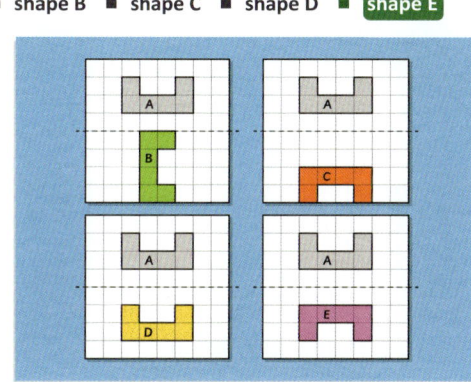

4. Which image shows a reflection of the original shape?

1/3

▪ **image A** ▪ image B ▪ image C

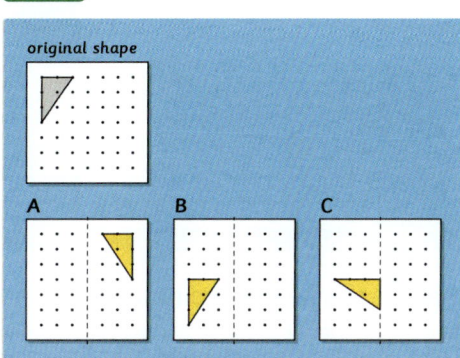

5. What are the coordinates of the square?
Give your answer in the form (x, y).

a b c

▪ (1, 3) ▪ 3, 1 ▪ **(3, 1)** ▪ (3 1) ▪ 3 1

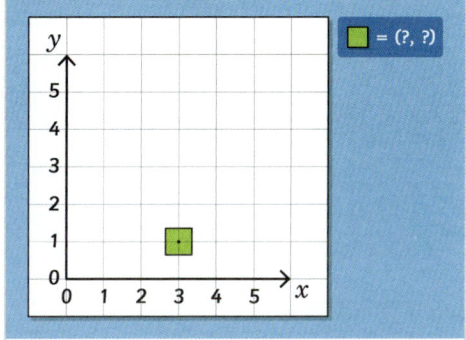

6. What are the new coordinates of the square after it is reflected over the mirror line?
Give your answer in the form (x, y).

a b c

▪ (1, 5) ▪ 5 1 ▪ **(5, 1)** ▪ (5 1) ▪ 5, 1 ▪ (3, 1)

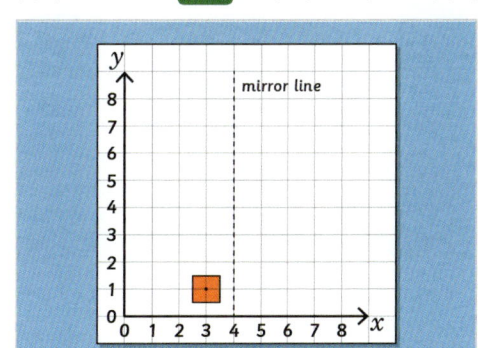

Level 1 *continued*

7. Reflect the circle over the mirror line. What are the new coordinates of the reflected circle?

a b c

Give your answer in the form (x, y).

■ **(2, 6)** ■ (2 6) ■ (6, 2) ■ 2 6 ■ (2, 2) ■ 2, 6

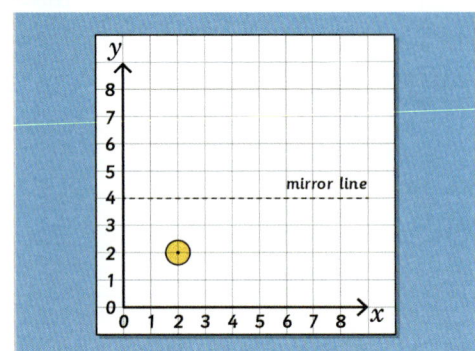

8. What are the new coordinates of the heart after it is reflected over the mirror line?

a b c

Give your answer in the form (x, y).

■ **(8, 2)** ■ (8 2) ■ (2, 8) ■ (2, 2) ■ 8, 2 ■ 8 2

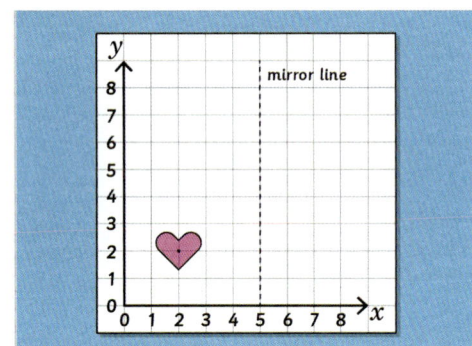

9. Which shape is a reflection of shape A over the mirror line?

1/4

■ shape B ■ **shape C** ■ shape D ■ shape E

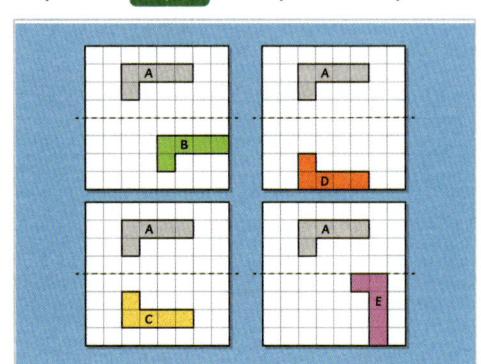

10. Which image shows the reflection of the pentagon over the mirror line?

1/3

■ **image A** ■ image B ■ image C

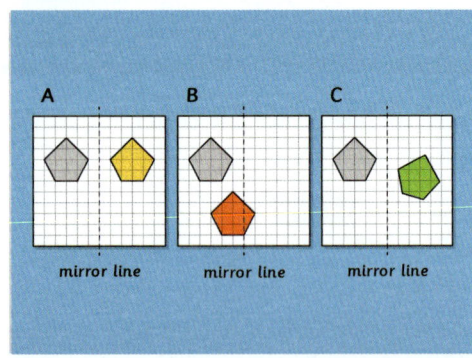

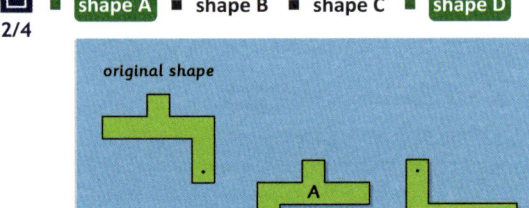

Level 2: Fluency: Find the coordinates of specific points after reflection.

✱ **Required:** 7/10   ✱ **Pupil Navigation:** on
✱ **Randomised:** off

11. Which **two** shapes are reflections of the original shape?

2/4

■ **shape A** ■ shape B ■ shape C ■ **shape D**

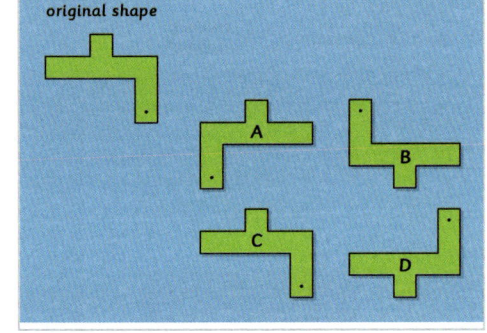

12. If you reflect the shape over the mirror line, what are the new coordinates of point *A*?

a b c

Give your answer in the form (x, y).

■ **(6, 6)** ■ (9, 6) ■ (4, 6)

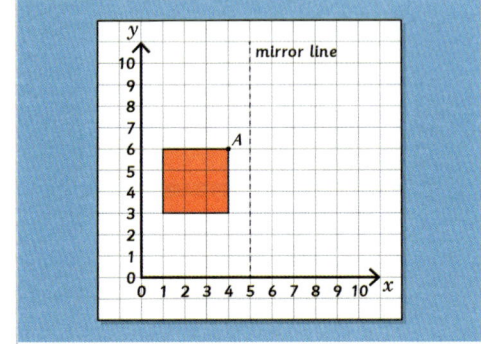

Level 2 continued

13. If you reflect the shape over the mirror line, what are the new coordinates of point *A*? *Give your answer in the form (x, y).*

- **(4, 8)** ■ (4, 6) ■ (4, 2)

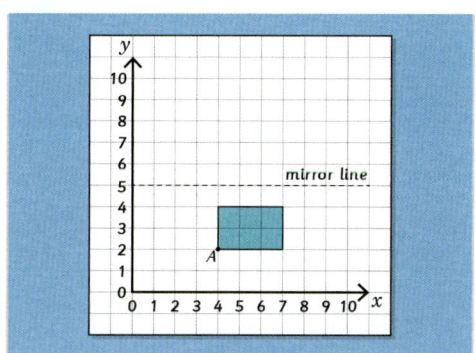

14. After this shape is reflected over the mirror line, the new coordinates of point *B* are _____. *Give your answer in the form (x, y).*

- **(7, 8)**

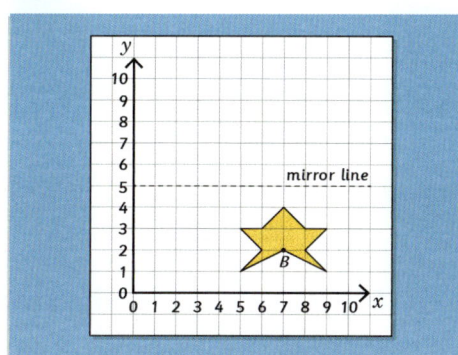

15. Reflect the shape over the mirror line. What are the new coordinates of point *C*? *Give your answer in the form (x, y).*

- **(9, 1)**

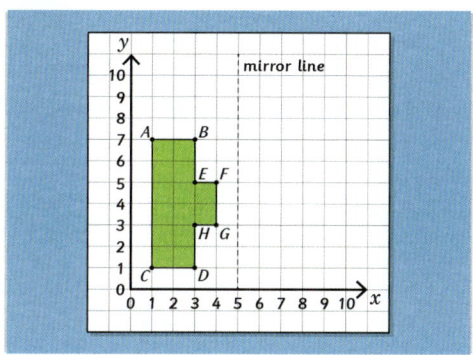

16. Samira notices that you can often see the reflection of buildings if they are next to water. Which image shows the correct reflection of the buildings in the water?

- ■ image A ■ image B ■ **image C**

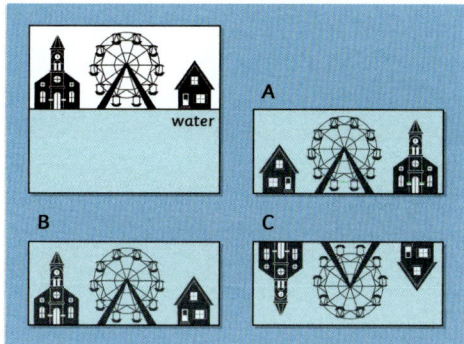

17. Cody writes his name and looks at its reflection in a mirror. Which image shows how his name looks in the mirror?

- ■ image A ■ **image B** ■ image C

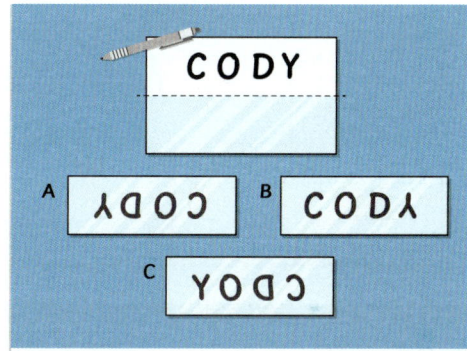

18. If you reflect the shape over the mirror line, what are the new coordinates of point *A*? *Give your answer in the form (x, y).*

- **(5, 2)** ■ (3, 2) ■ (7, 2)

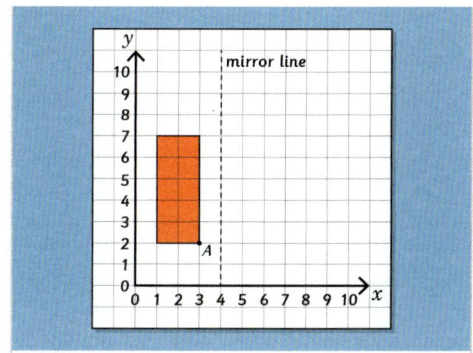

Level 2 continued

19. Which shape is a reflection of the original shape?

1/3

- **shape A** ▪ shape B ▪ shape C

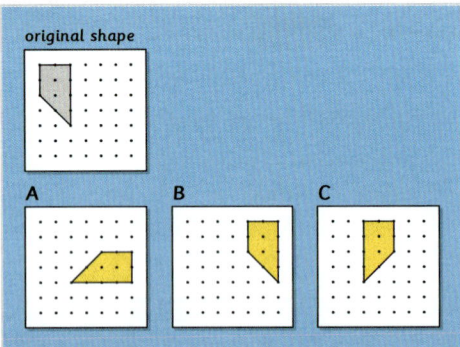

20. Which image shows a reflection of the original shape?

1/3

- image A ▪ image B ▪ **image C**

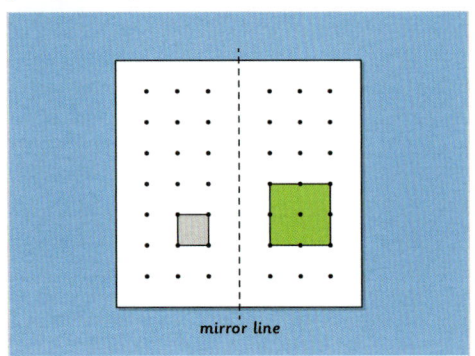

Level 3: Reasoning: Reason about reflecting shapes.

✿ **Required:** 5/5 ✿ **Pupil Navigation:** on
✿ **Randomised:** off

21. Harry reflects the shape over the mirror line. Explain how you know that Harry has made a mistake.

a
b
c

- Open question, no set answer

22. Shape B is a reflection of shape A over the mirror line. What are the coordinates of point *Z*?
Give your answer in the form (x, y).

a
b
c

- **(9, 1)**

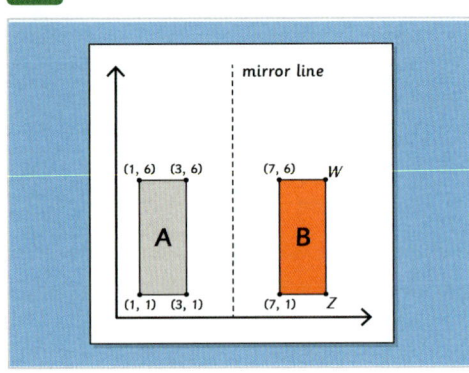

23. Triangle B is a reflection of triangle A. What are the coordinates of point *Z*?
Give your answer in the form (x, y).

a
b
c

- **(4, 7)**

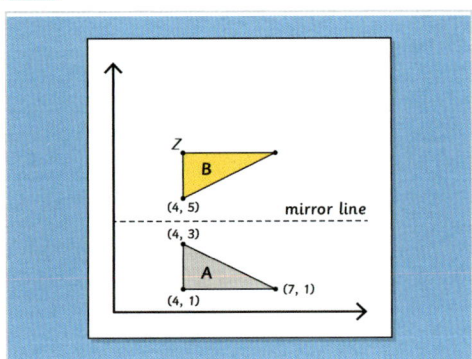

24. Mischa says, "When you reflect a shape on a grid, all the coordinates of the points on the reflected shape are different to the original points."

a
b
c

Is this true for **all** reflections? Explain your answer.

- Open question, no set answer

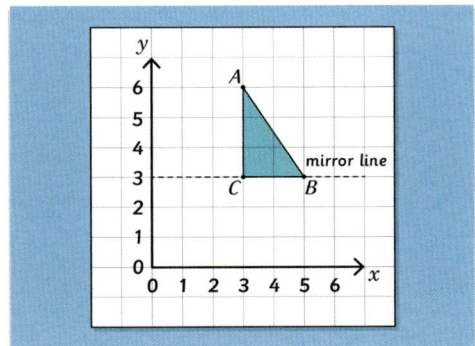

Level 3 *continued*

25. The shape in box A is reflected into the next box. This is repeated until each box contains a shape. Which **three** boxes will be exactly the same as box A?

3/6

■ B ■ C ■ D ■ E ■ F ■ G

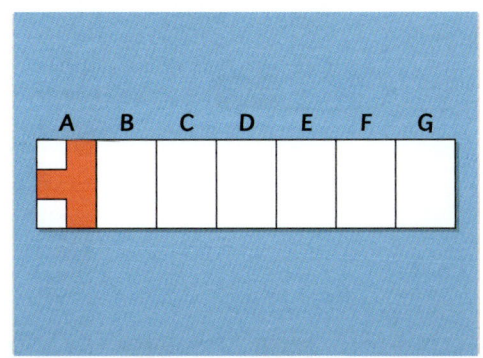

| **Level 4:** | Problem solving with greater depth: Solve multi-step problems involving reflection. |

✸ **Required:** 5/5   ✸ **Pupil Navigation:** on
✸ **Randomised:** off

26. Which **three** squares must be shaded to make a pattern that is symmetrical in both mirror lines?

3/6   ■ D7 ■ D5 ■ F5 ■ F6 ■ G5 ■ G6

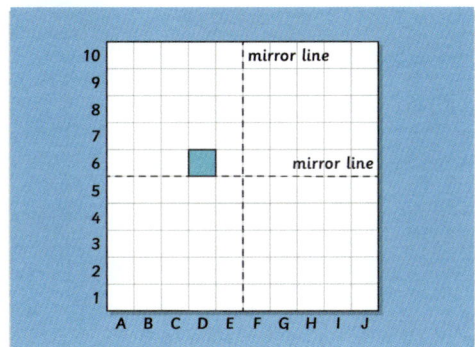

27. Reflect the pattern in the mirror line. Which **three** of these grid squares will **not** be shaded?

3/7   ■ A5 ■ B5 ■ D1 ■ F5 ■ H2 ■ H3 ■ J3

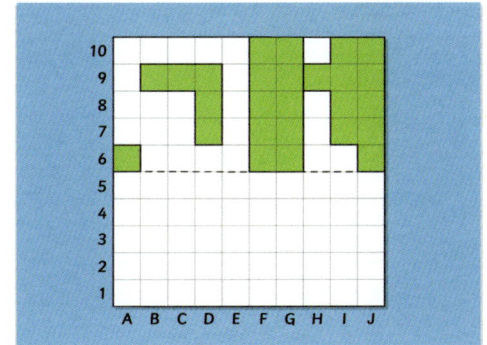

28. The coordinates of point *A* after it has been reflected are (9, 5). What are the coordinates of point *B* after it has been reflected?

a
b
c

■ (13, 9)   ■ (1, 9)

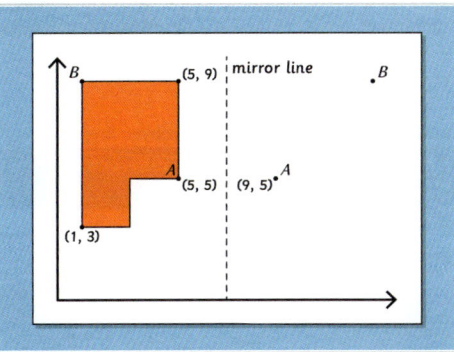

29. Lucia reflects the pattern in the mirror line. In total, what fraction of the squares are **not** shaded after Lucia reflects the pattern? *Put a forward slash (/) between the numerator and the denominator.*

a
b
c

■ 24/50   ■ 12/25   ■ 48/100   ■ 48

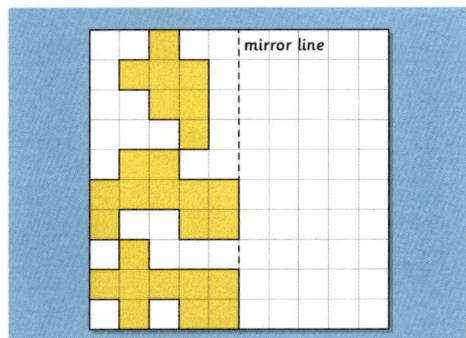

30. The shape has been reflected across the mirror line. The new coordinates of point *A* are (14, 3). What are the **new** coordinates for the point (5, 7)?

a
b
c

■ 3   ■ (11, 7)   ■ 6   ■ (2, 3)

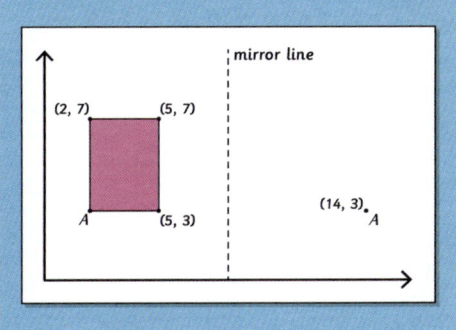

Identify, Describe and Represent Translation of a Shape

Objective: I can identify and describe the position of a shape following a translation.

Quick Search Ref: 10330

Level 1: Understanding: Translate and describe the translations of simple shapes.

✿ Required: 7/10   ✿ Pupil Navigation: on   ✿ Randomised: off

1. Translation is . . .

1/4
- when you 'flip' a shape over a mirror line.
- when you move a shape into a different position without 'flipping' it or rotating it.
- when you rotate a shape.
- when you make a shape bigger or smaller.

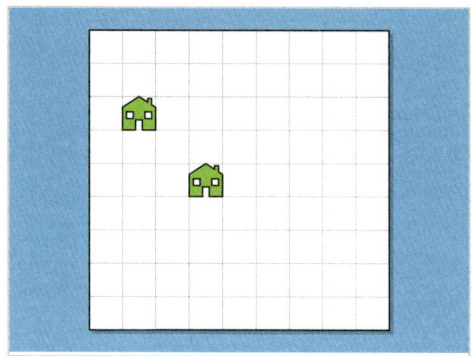

2. Which image shows a translation of the triangle?

1/3
- image A ■ image B ■ image C

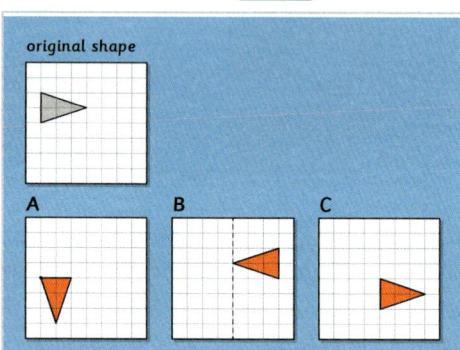

3. Which shape is a translation of shape A?

1/4
- shape B ■ shape C ■ shape D ■ shape E

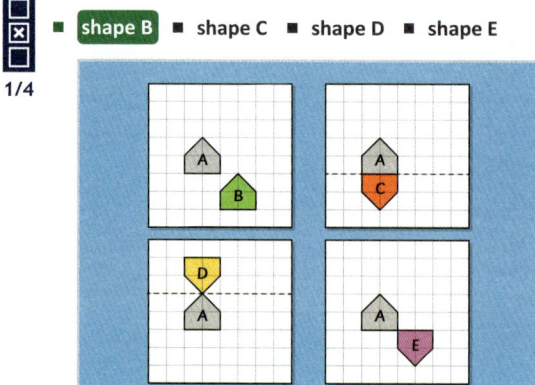

4. What is the translation from shape A to shape B?

1/5
- 3 right ■ 5 left ■ 5 right ■ 5 right, 2 up ■ 6 right

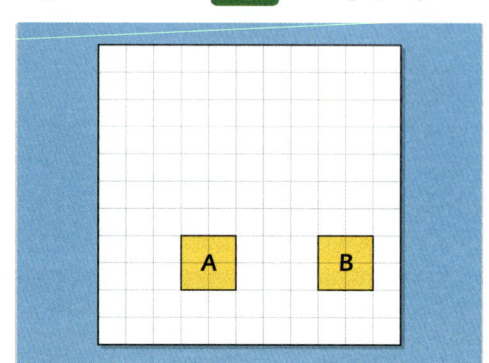

5. What is the translation from shape X to shape Y?

1/5
- 1 right, 5 up ■ 5 down ■ 1 left, 5 down ■ 7 up
- 1 right

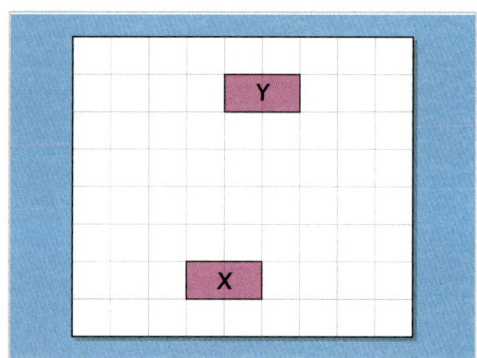

6. What are the coordinates of the star after a translation of 2 right and 1 down?
Give your answer in the form (x, y).

- (4 2) ■ (4, 4) ■ (4, 2) ■ (4, 3) ■ 4, 2 ■ (2, 4)
- (2, 2) ■ 4 2 ■ (2, 3)

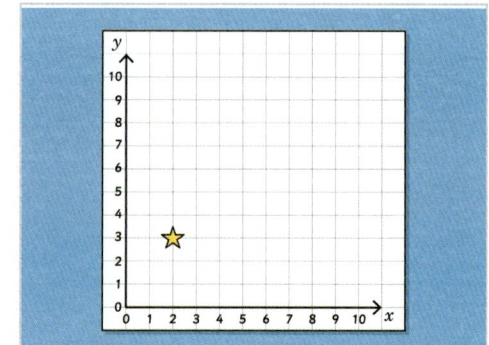

Level 1 *continued*

7. If you translate the shape 5 left and 1 down, what are the new coordinates of the square?
Give your answer in the form (x, y).

- 3, 4   ■ **(3, 4)**   ■ (3, 6)   ■ (8, 5)   ■ (3, 5)   ■ 3 4
- (4, 3)   ■ (3 4)

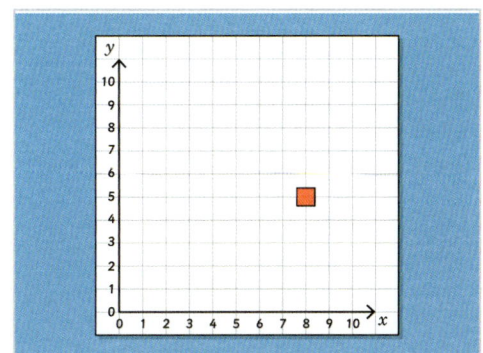

8. What are the coordinates of the heart after a translation of 5 left and 1 up?
Give your answer in the form (x, y).

- (7, 2)   ■ (3, 2)   ■ **(2, 3)**   ■ (2, 2)   ■ (2 3)   ■ (2, 1)
- 2 3   ■ 2, 3   ■ (7, 3)

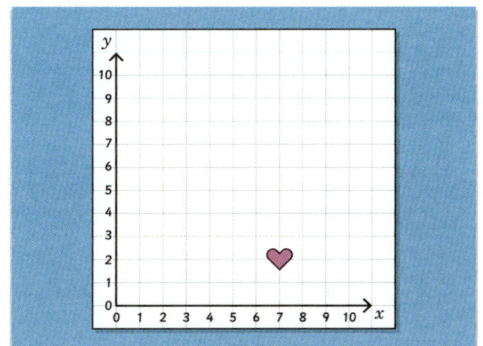

9. Which shape is a translation of shape V?

1/4

- shape W   ■ **shape X**   ■ shape Y   ■ shape Z

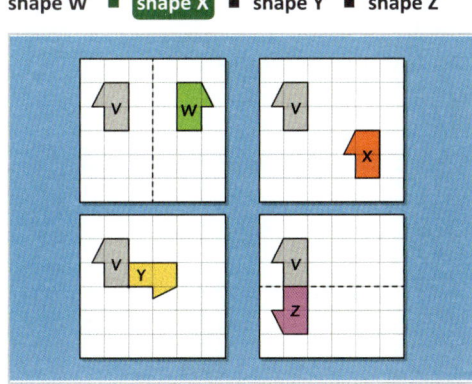

10. Which shape is a translation of shape E?

1/3

- shape F   ■ **shape G**   ■ shape H

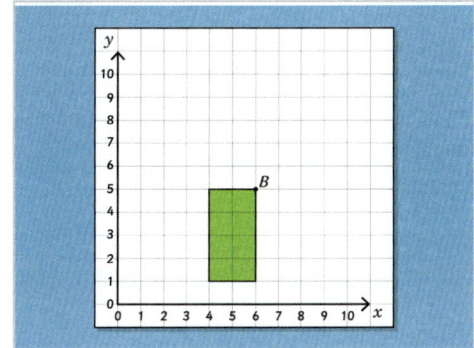

Level 2: Fluency: Translate more complex shapes.

✱ **Required:** 7/10   ✱ **Pupil Navigation:** on
✱ **Randomised:** off

11. What is the translation of this shape?
Select the two correct options.

2/6

- 7 down   ■ 4 left   ■ **4 right**   ■ 5 right   ■ **4 down**
- 4 up

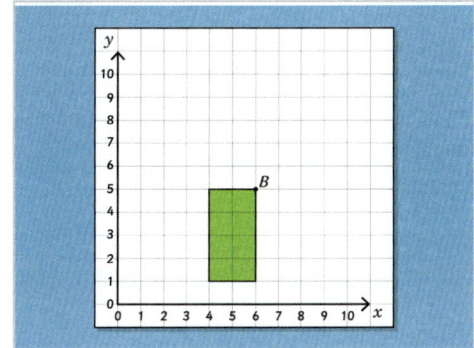

12. If you translate the shape 3 left and 2 up, what will be the new coordinates of point *B*?
Give your answer in the form (x, y).

- **(3, 7)**   ■ (3, 5)   ■ (6, 7)

13. If you translate the shape 2 left and 6 down, what will be the new coordinates of point *X*?

1/3

- (1, 7) ▪ (1, 1) ▪ (3, 1)

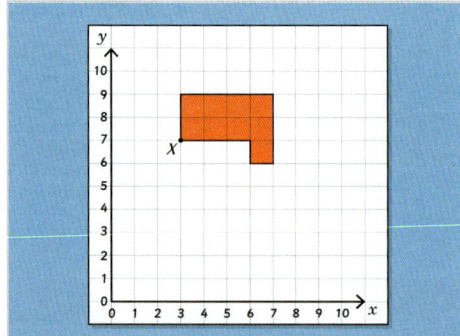

14. What are the coordinates of point B after a translation of 4 right and 2 up?
Give your answer in the form (x, y).

- (7, 6) ▪ (7, 8) ▪ (3, 8)

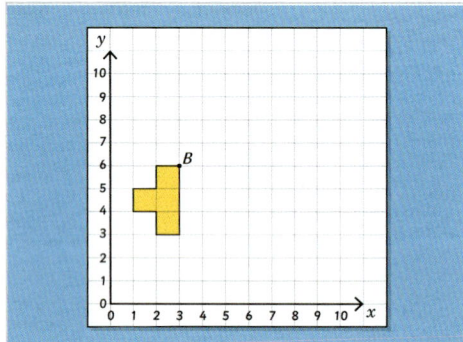

15. Translate the shape 3 left and 6 up. What are the new coordinates of point *A*?
Give your answer in the form (x, y).

- (4, 9) ▪ (4, 3) ▪ (7, 9)

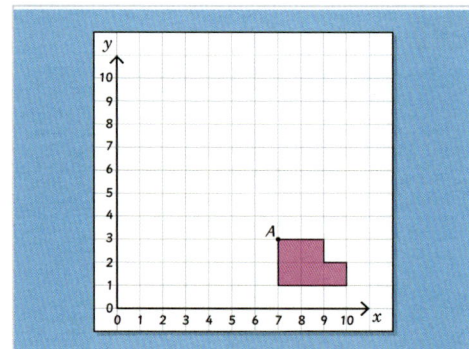

16. Jade draws a plan of the items in her playroom. She moves the toy box 6 right and 1 down. What are the new coordinates of the toy box?
Give your answer in the form (x, y).

- (8, 6) ▪ (7, 4) ▪ (7, 2)

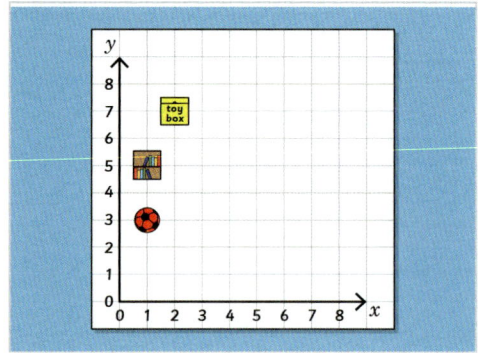

17. Pirate Big Ears moves the treasure chest 2 squares north and 6 squares west. What are the new coordinates of the treasure chest?
Give your answer in the form (x, y).

- (1, 4)

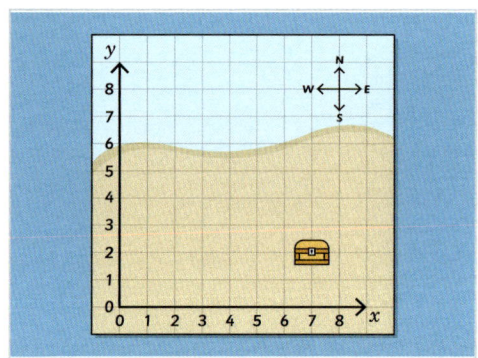

18. Danny moves the pool table in his games room. He translates the pool table _____ right and 2 down.
Enter the missing number.

- 6

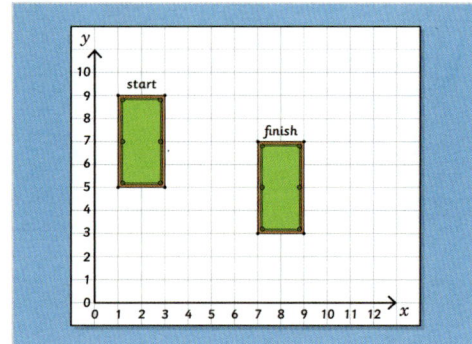

19. Which image shows a translation of the original shape?

■ image A   ■ image B   ■ **image C**

1/3

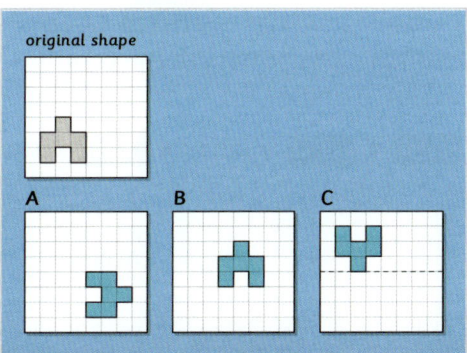

20. Which image shows a translation of the original shape?

■ image A   ■ **image B**   ■ image C

1/3

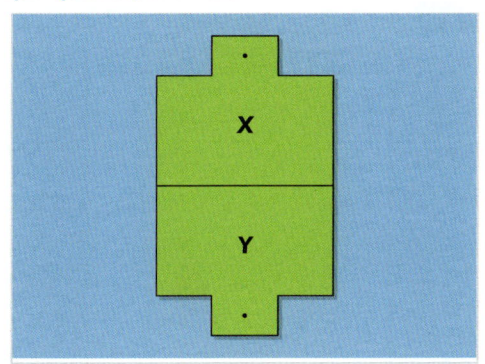

Level 3:   Reasoning: Reason about translating shapes.

✹ **Required:** 5/5   ✹ **Pupil Navigation:** on
✹ **Randomised:** off

21. Billy says that shape Y is a translation of shape X because it is the same size. Do you agree with Billy? Explain your answer.

a
b
c

- Open question, no set answer

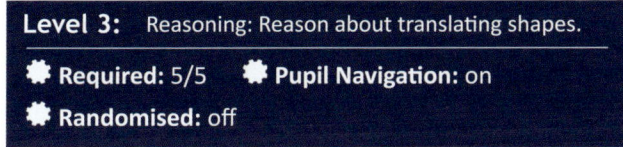

22. If you translate the shape so point *D* is at (0, 0), what will the new coordinates of point *B* be?

a
b
c

■ **(4, 2)**   ■ (4, 0)

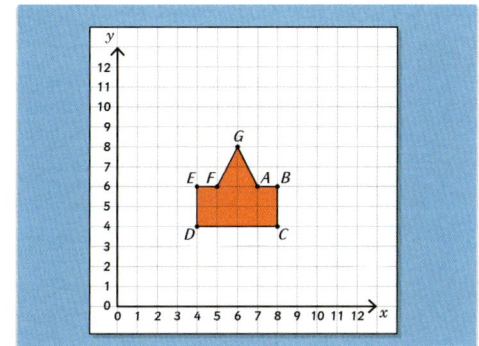

23. Tim says, "The shape has been translated 1 right and 4 down."
Is Tim correct? Explain your answer.

a
b
c

- Open question, no set answer

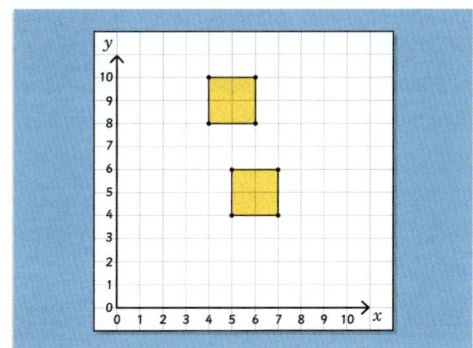

24. Ella knows the four coordinate points of a square are (3, 1), (3, 3), (5, 1) and (5, 3). Explain how she can find the new coordinates of each point after a translation of 2 right and 1 up, **without drawing the shapes**.

a
b
c

- Open question, no set answer

Level 3 *continued*

25. A triangle is translated on a coordinate grid.
What are the translated coordinates of point C?

a b c

Give your answer in the form (x, y).

- (9, 4)

point	original coordinates	translated coordinates
A	(3, 2)	(5, 3)
B	(3, 5)	(5, 6)
C	(7, 3)	?

Level 4: Problem solving: Solve multi-step problems related to translation.

❋ **Required:** 5/5   ❋ **Pupil Navigation:** on
❋ **Randomised:** off

26. A triangle is translated four squares down and two squares right. The coordinates of the translated triangle are (5, 8), (7, 9) and (9, 6). What are the coordinates of the **original** triangle?

3/5

- (5, 13) ▪ (6, 13) ▪ (7, 10) ▪ (8, 10) ▪ (3, 12)

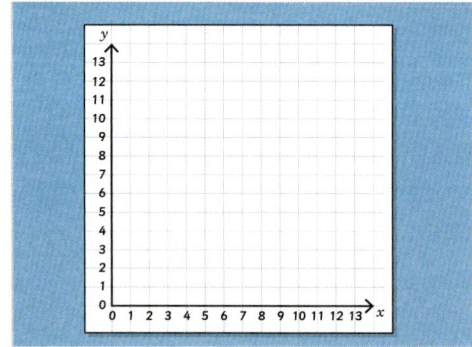

27. The scale on a map is 1 square = 1 mile. A windsurfer is at coordinates (3, 4) on the map. The wind blows her 4 miles east, but then changes direction and blows her 2 miles south. A storm then blows her 7 miles west and 4 miles north. What are the **new** coordinates of the windsurfer?

a b c

- (0, 6) ▪ (7, 2)

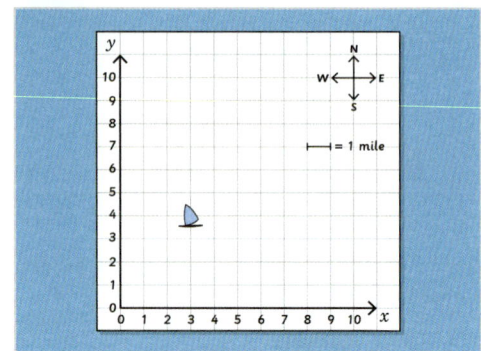

28. A parallelogram has been translated on a coordinate grid. What are the coordinates of points *A*, *B* and *C*?

3/5

- (14, 2) ▪ (19, 5) ▪ (21, 2) ▪ (20, 2) ▪ (20, 5)

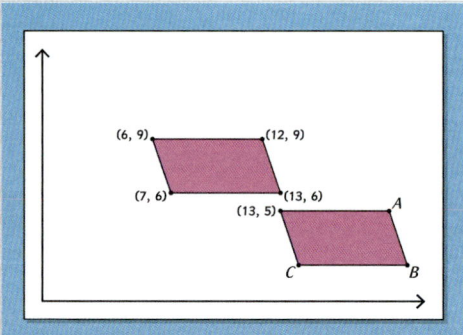

29. A pentagon has been translated four squares up and three squares to the right. What were the **original** coordinates of point *D*?

a b c

- (6, 1) ▪ (12, 9)

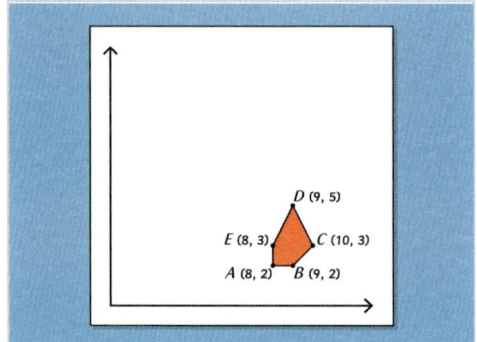

Level 4 *continued*

30. A square has been translated on a
a coordinate grid. What are the coordinates of
b point *A*?
c

Give your answer in the form (x, y).

- (8, 5)   ▪ (6, 3)   ▪ (4, 5)

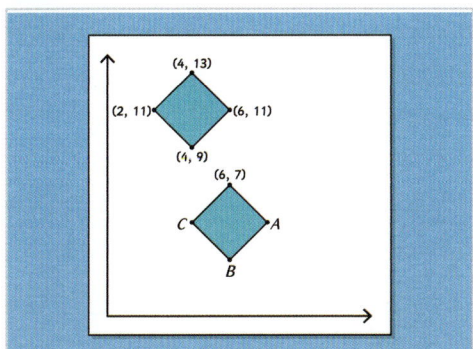

Reflection, Rotation and Translation of Shapes

Objective: I can identify and describe the position of a shape following rotation, reflection and translation.

Quick Search Ref: 10019

Level 1: Understanding: Definitions and recognising transformations.

✿ **Required:** 7/10   ✿ **Pupil Navigation:** on   ✿ **Randomised:** off

1. When **translating** a shape you:

- **move the shape into a different position without flipping or rotating it.**
- flip the shape over a mirror line to create a mirror image.
- turn the shape around a central point.

1/3

2. When **rotating** a shape you:

- move the shape into a different position without flipping or rotating it.
- **turn the shape around a central point.**
- flip the shape over a mirror line to create a mirror image.

1/3

3. When **reflecting** a shape you:

- move the shape into a different position without flipping or rotating it.
- turn the shape around a central point.
- **flip the shape over a mirror line to create a mirror image.**

1/3

4. When translating, rotating or reflecting a shape, the dimensions of the shape:

- **never change** - always change - sometimes change

1/3

5. Does the dotted line represent a line of reflective symmetry?

- Yes - **No**

1/2

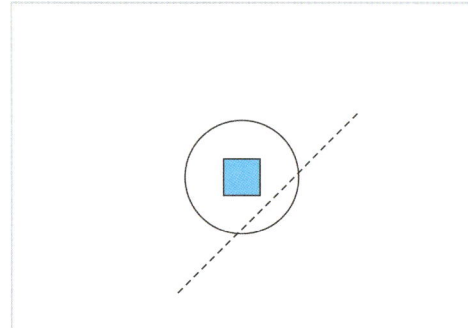

6. What transformation is shown in the diagram?
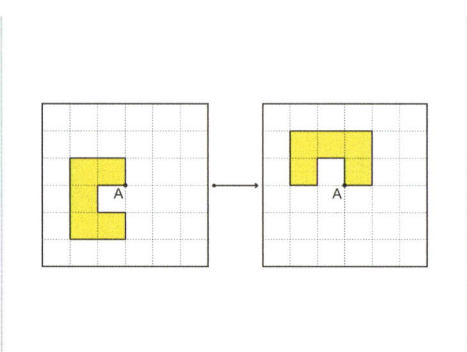
- reflection - **rotation** - translation

1/3

7. What transformation is shown in the diagram?
- **reflection** - rotation - translation

1/3

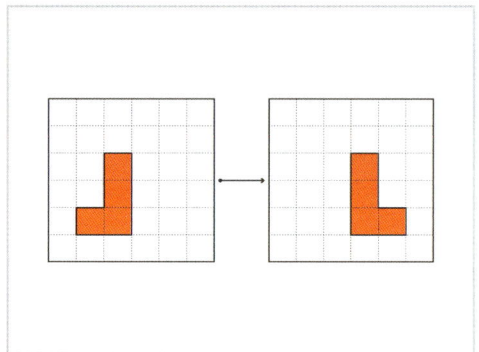

8. What transformation is shown in the diagram?
- reflection - rotation - **translation**

1/3

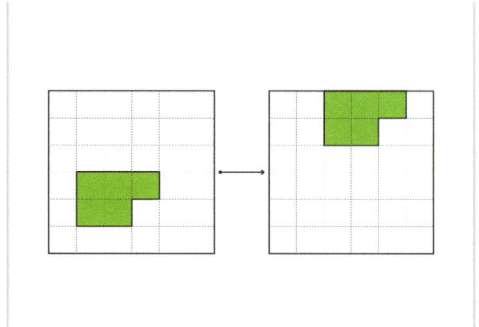

9. What transformation is shown in the diagram?

 ■ reflection ■ rotation ■ translation

1/3

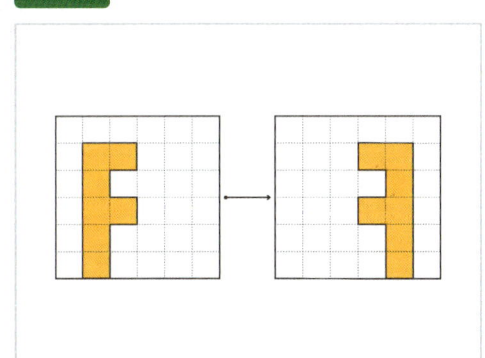

10. What transformation is shown in the diagram?

■ reflection ■ rotation ■ translation

1/3

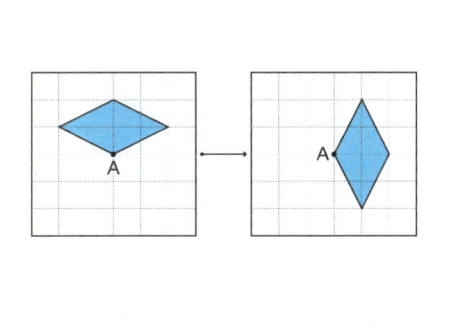

Level 2: Fluency: Recognising transformations and finding coordinates of shapes after transformation.

✱ **Required:** 7/10   ✱ **Pupil Navigation:** on
✱ **Randomised:** off

11. Which diagram shows the original diagram **reflected**?

■ diagram (a) ■ diagram (b) ■ diagram (c)

1/3

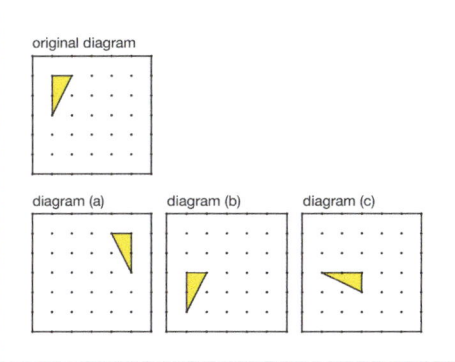

12. Describe the translation of shape A to shape B.

■ 4 right ■ 5 right ■ 5 right, 2 up ■ 6 right
■ 3 right

1/5

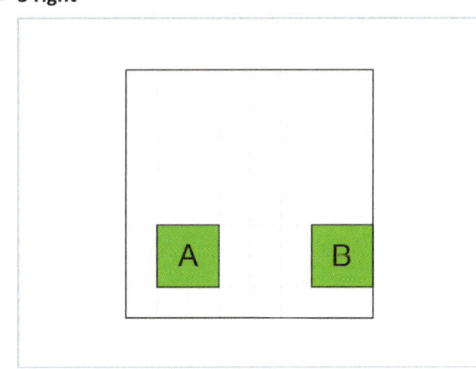

13. Which diagram shows **rotation**?

■ diagram (a) ■ diagram (b) ■ diagram (c)

1/3

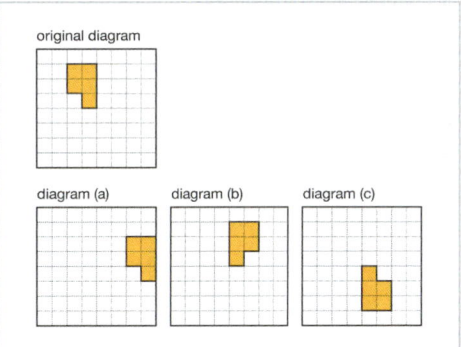

14. If you reflect the shape over the mirror line, what are the new coordinates of point A?

■ (4, 6) ■ (6, 6) ■ (9, 6) ■ (9, 3)

1/4

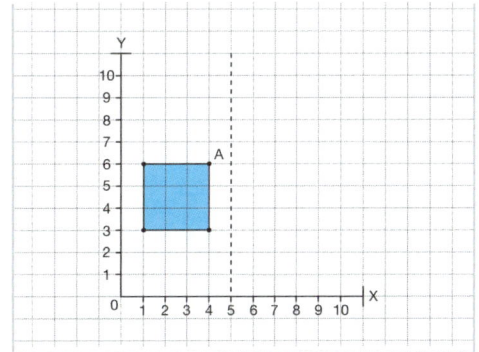

Level 2 *continued*

15. The triangle has been rotated around point C by how many degrees clockwise?

1/4

- 90° ■ **180°** ■ 270° ■ 360°

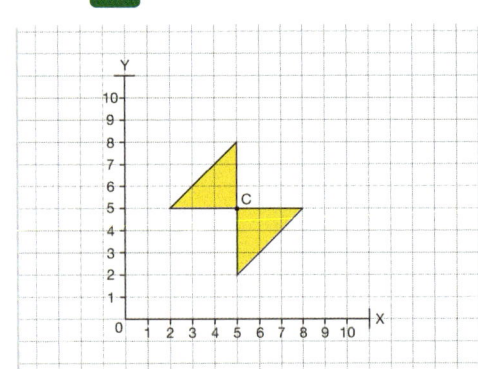

16. Which diagram shows translation?

1/3

- diagram (a) ■ **diagram (b)** ■ diagram (c)

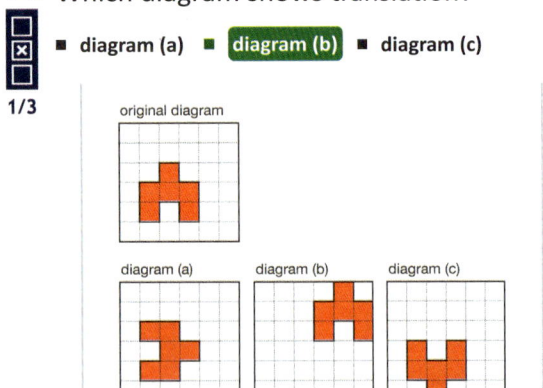

17. Which diagram shows a reflection of the original shape?

1/3

- **diagram (a)** ■ diagram (b) ■ diagram (c)

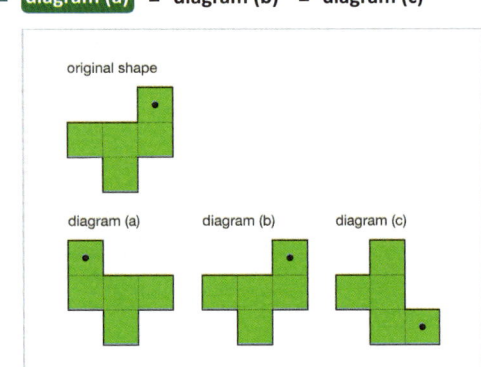

18. If you translate the shape 3 left and 2 up, what are the new coordinates of point B?

1/4

- (3, 5) ■ **(3, 7)** ■ (7, 3) ■ (9, 7)

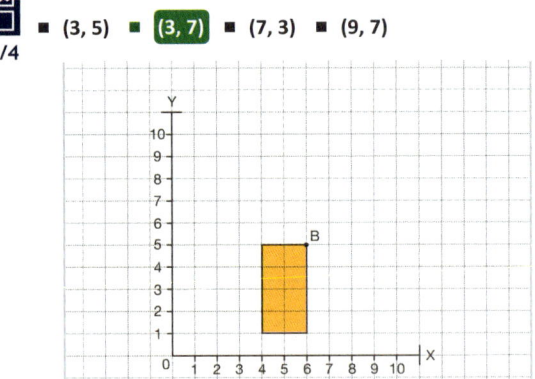

19. How would you describe the translation of the shape?

1/3

- 11 right, 2 down ■ **8 right, 2 down** ■ 9 right, 3 down

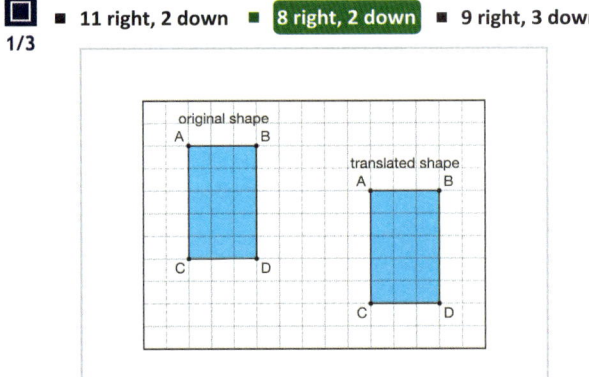

20. If you rotate the shape 90° clockwise around point B, what are the new coordinates of point C?

1/4

- (1, 5) ■ **(1, 9)** ■ (9, 1) ■ (9, 9)

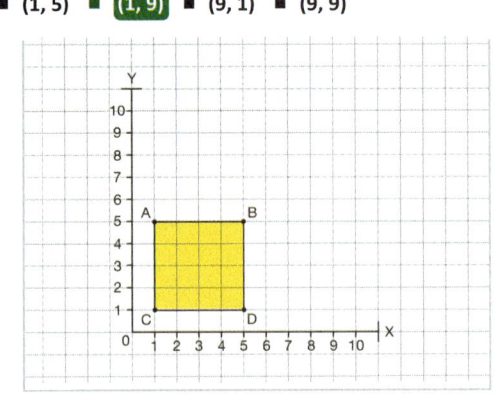

21. What sequence of transformation is shown from shape 1 to shape 2 to shape 3 in the diagram?

1/4

- rotation then reflection   ■ reflection then translation
- translation then reflection   ■ rotation then translation

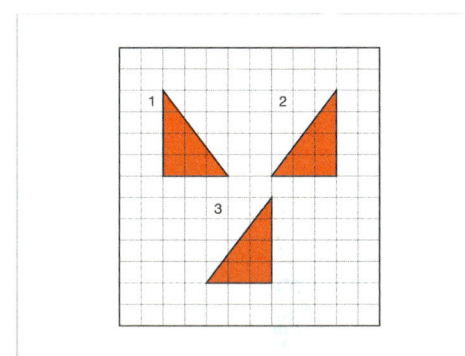

22. The blue rectangle is a reflection of the pink rectangle over the mirror line. What are the coordinates of point A?

- (9, 1)   ■ (1, 9)   ■ 9,1   ■ 91

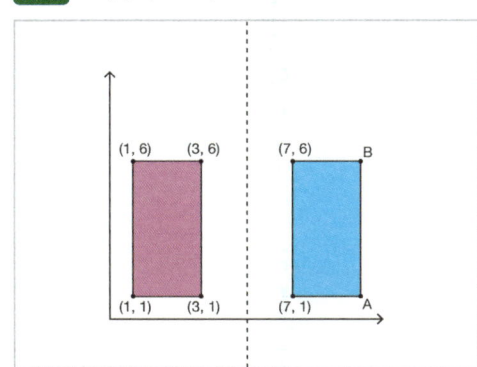

23. Jerome says, "If the square is translated three down and one left, the coordinates of the translated square's point C will be (2, 6)." Is he correct? Explain your answer.

- Open question, no set answer

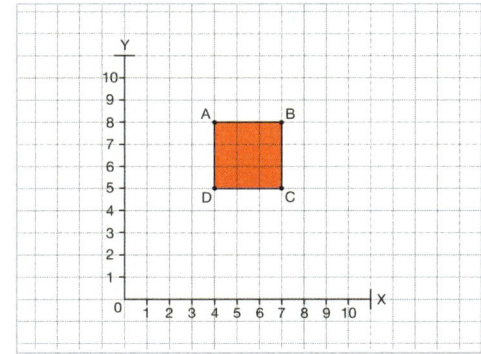

24. If you rotate the shape 270° clockwise around point C, what will the new coordinates of point G be?

1/5

- (2, 4)   ■ (4, 2)   ■ (8, 4)   ■ (10, 0)   ■ (12, 6)

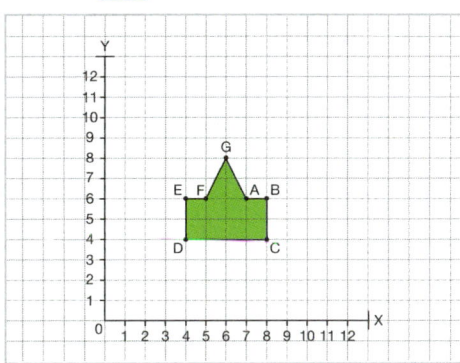

25. The green triangle is a reflection of the yellow triangle. What are the coordinates of the point A?

- (9, 3)   ■ 9,3   ■ 9 3   ■ (3, 9)

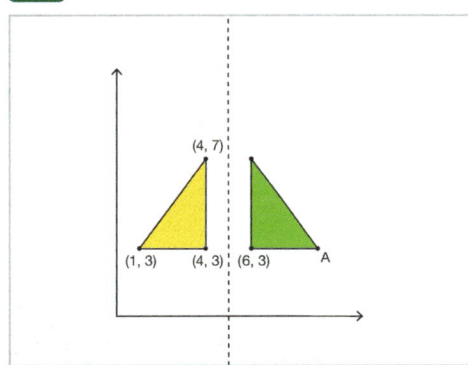

26. A rectangle is translated six squares up and three squares to the right. The coordinates of the translated rectangle are: (5, 7), (5, 10), (11, 10), (11, 7). What are the four coordinates of the original rectangle?

4/6

- (2, 1)   ■ (7, 5)   ■ (2, 4)   ■ (8, 4)   ■ (11, 9)
- (8, 1)

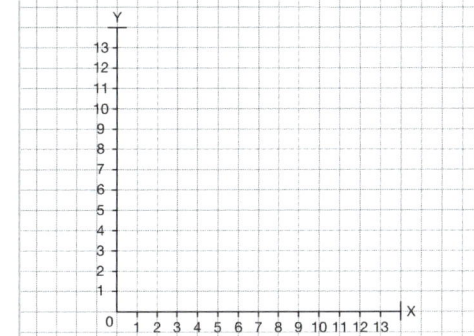

Level 4 *continued*

27. A parallelogram has been translated on a coordinate grid. What are the coordinates of point A, B and C?

3/5

▪ **(14, 2)** ▪ **(19, 5)** ▪ (21, 2) ▪ **(20, 2)** ▪ (20, 5)

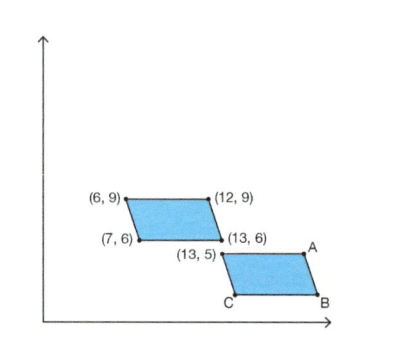

28. A triangle is translated four squares down and two squares right. The coordinates of the translated triangle are: (5, 8), (7, 9), (9, 6). What are the coordinates of the original triangle?

3/5

▪ **(5, 13)** ▪ (6, 13) ▪ **(7, 10)** ▪ (8, 10) ▪ **(3, 12)**

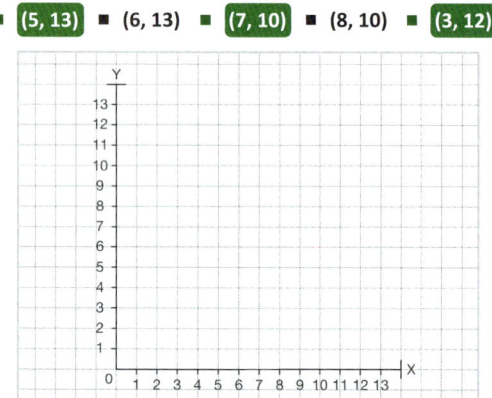

29. Which three numbers need to be shaded to make the pattern symmetrical in both mirror lines?

3/6

▪ **1** ▪ 2 ▪ 3 ▪ **4** ▪ 5 ▪ **6**

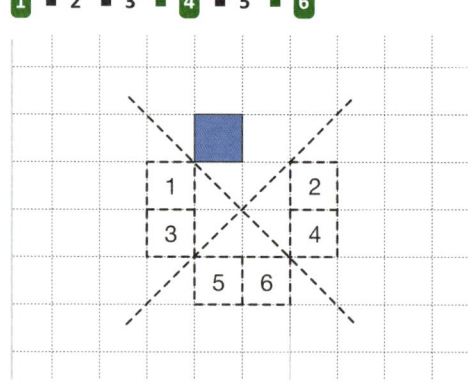

30. The square is rotated 90°, 180° and 270° clockwise around point a. What three boxes will the letter X appear in?

3/6

▪ **1** ▪ 2 ▪ 3 ▪ **4** ▪ 5 ▪ **6**

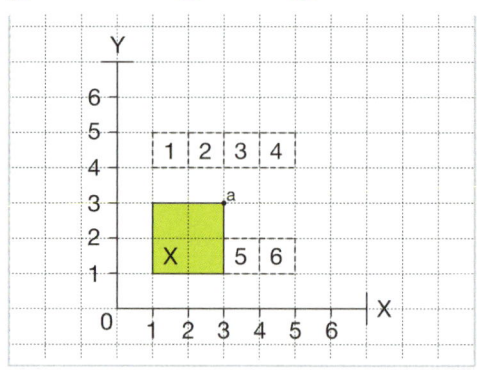

Translate and Reflect Shapes

Objective: I can identify and describe the position of a shape following a reflection and translation.

Quick Search Ref: 11680

Level 1: Understanding: Translate and reflect simple shapes and describe the position of a simple shape after a transformation.

✿ **Required:** 7/10 ✿ **Pupil Navigation:** on ✿ **Randomised:** off

1. Translation is . . .

1/4
- when you 'flip' a shape over a mirror line.
- when you move a shape into a different position without 'flipping' it or rotating it.
- when you rotate a shape.
- when you make a shape bigger or smaller.

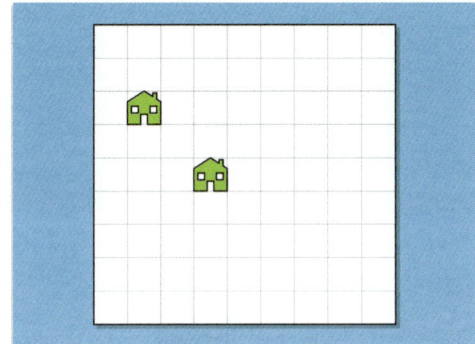

2. When you reflect a shape, the size of the shape _____ changes.
Select the missing word.

1/3
- always ■ sometimes ■ never

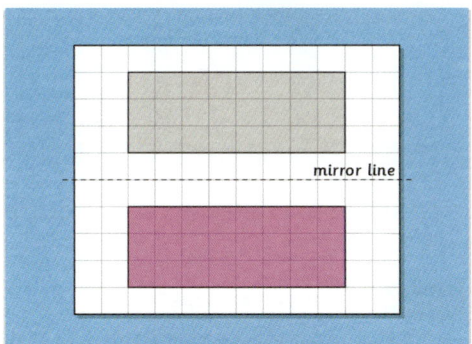

3. Which shape is a translation of shape V?
- shape W ■ shape X ■ shape Y ■ shape Z
1/4

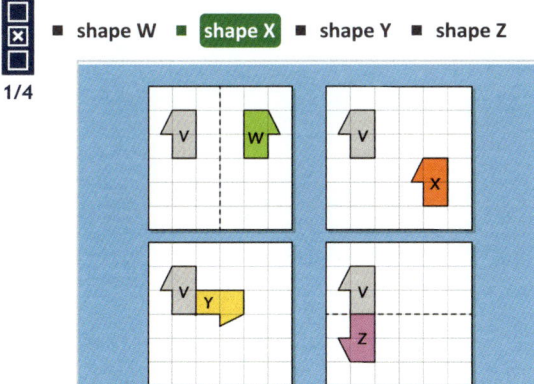

4. Which image shows the reflection of the pentagon over the mirror line?
1/3
- image A ■ image B ■ image C

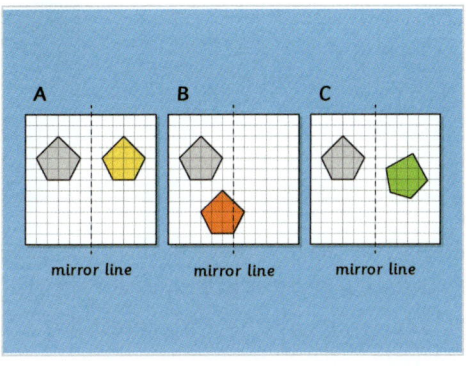

5. What is the translation from shape X to shape Y?
1/5
- 1 right, 5 up ■ 5 down ■ 1 left, 5 down ■ 7 up
- 1 right

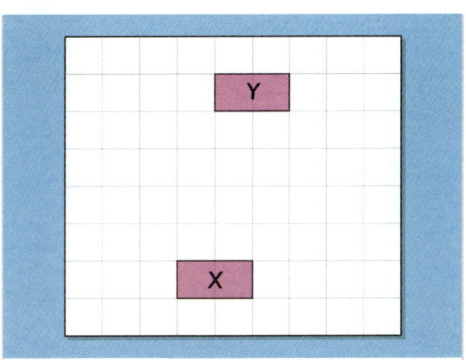

6. What are the new coordinates of the heart after it is reflected over the mirror line?
Give your answer in the form (x, y).

- (8, 2) ■ (2, 8) ■ 8, 2 ■ (8 2) ■ (2, 2) ■ 8 2

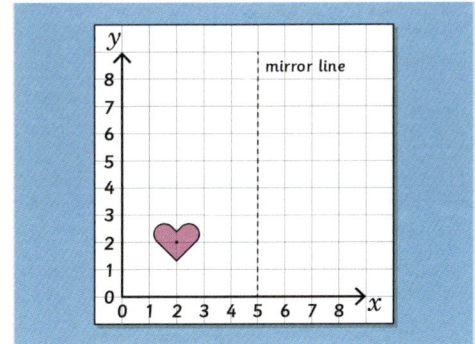

Level 1 *continued*

7. What are the coordinates of the heart after a translation of 5 left and 1 up?

a
b
c

Give your answer in the form (x, y).

- (3, 2) ■ **(2, 3)** ■ (2 3) ■ 2 3 ■ (7, 3) ■ (7, 2)
- (2, 2) ■ (2, 1) ■ 2, 3

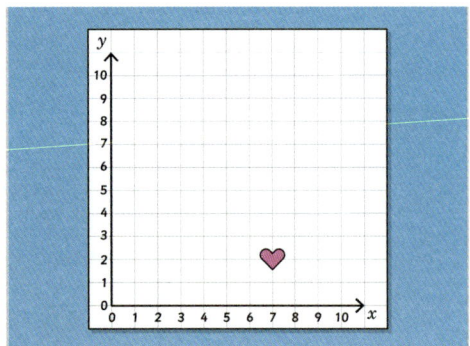

8. Reflect the circle over the mirror line. What are the new coordinates of the reflected circle?

a
b
c

Give your answer in the form (x, y).

- **(2, 6)** ■ (6, 2) ■ (2, 2) ■ (2 6) ■ 2 6 ■ 2, 6

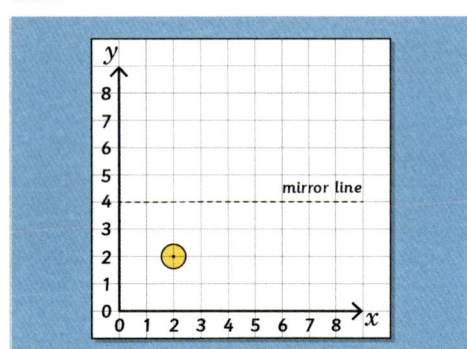

9. What is the translation from shape A to shape B?

1/5

- 3 right ■ 5 left ■ **5 right** ■ 5 right, 2 up ■ 6 right

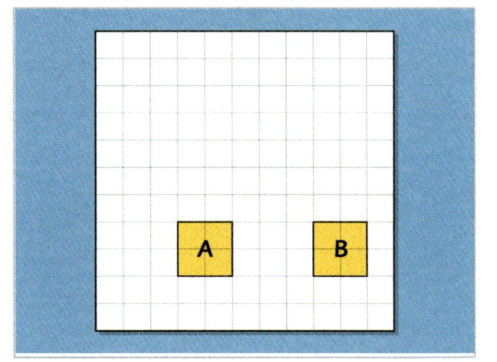

10. Which shape is a translation of shape E?

1/3

- shape F ■ **shape G** ■ shape H

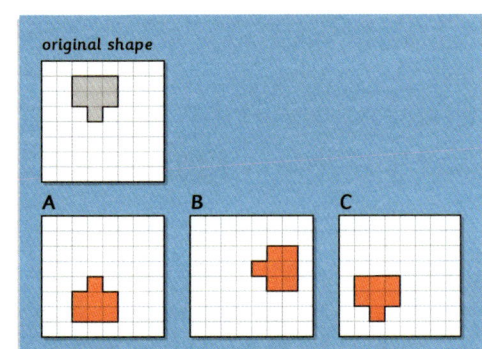

Level 2: Fluency: Translate and reflect shapes and describe the position of a shape after a transformation, including transformations in context.

✿ **Required: 7/10**   ✿ **Pupil Navigation:** on
✿ **Randomised:** off

11. Which image shows a translation of the original shape?

1/3

- image A ■ image B ■ **image C**

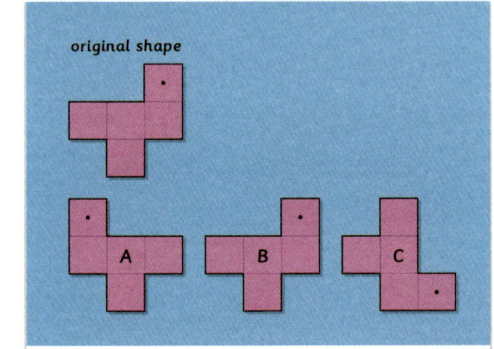

12. Which shape is a reflection of the original shape?

1/3

- **shape A** ■ shape B ■ shape C

Level 2 continued

13. Translate the shape 3 left and 6 up. What are
a the new coordinates of point *A*?
b
c *Give your answer in the form (x, y).*

- **(4, 9)** ■ (7, 9) ■ (4, 3)

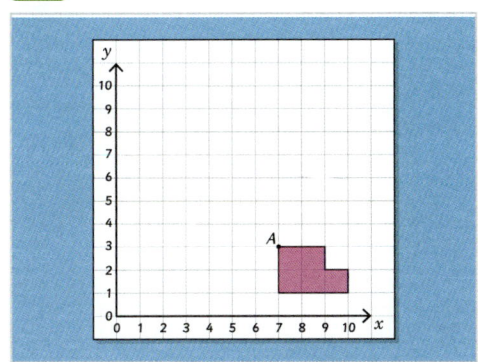

14. If you reflect the shape over the mirror line,
a what are the new coordinates of point *A*?
b
c *Give your answer in the form (x, y).*

- **(5, 2)** ■ (7, 2) ■ (3, 2)

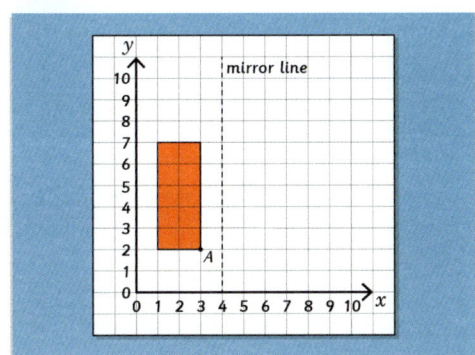

15. Reflect the shape over the mirror line. What
a are the new coordinates of point *C*?
b
c *Give your answer in the form (x, y).*

- **(9, 1)**

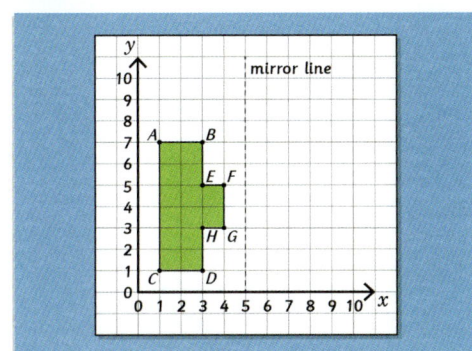

16. Danny moves the pool table in his games
1 room. He translates the pool table ____ right
2
3 and 2 down.
Enter the missing number.

- **6**

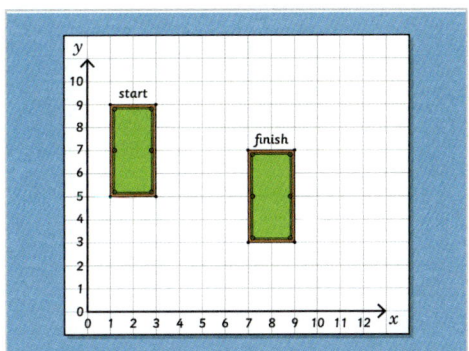

17. Cody writes his name and looks at its
reflection in a mirror. Which image shows
how his name looks in the mirror?

1/3
- ■ image A ■ **image B** ■ image C

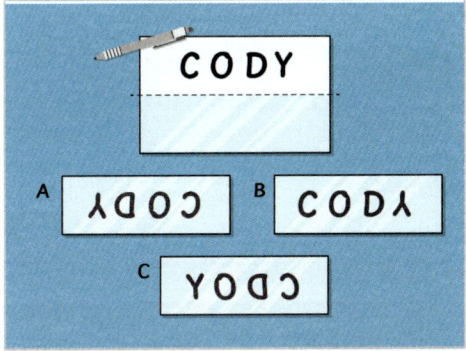

18. Pirate Big Ears moves the treasure chest 2
a squares north and 6 squares west. What are
b the new coordinates of the treasure chest?
c *Give your answer in the form (x, y).*

- **(1, 4)**

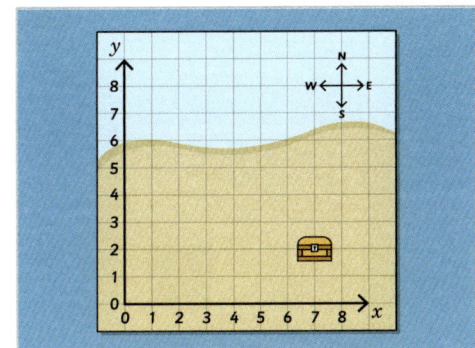

Level 2 *continued*

19. Which image shows a reflection of the original shape?

☐☒☐ 1/3

■ image A   ■ image B   ■ **image C**

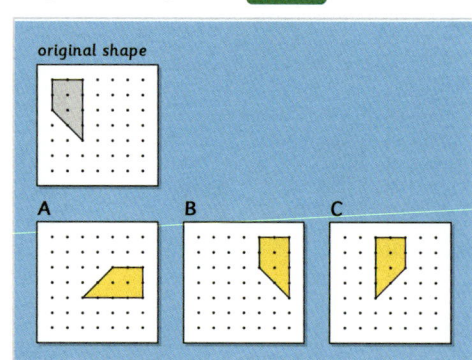

20. Which image shows a translation of the original shape?

☐☒☐ 1/3

■ image A   ■ **image B**   ■ image C

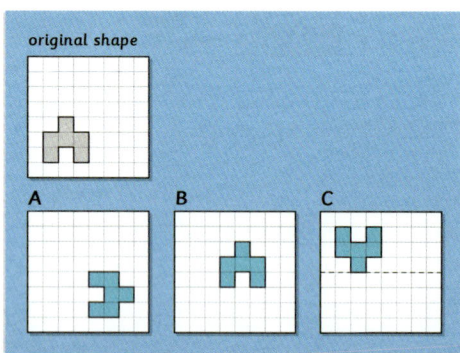

Level 3: Reasoning: Reason about translating and reflecting shapes.

✿ **Required:** 5/5   ✿ **Pupil Navigation:** on
✿ **Randomised:** off

21. Harry reflects the shape over the mirror line.
a b c Explain how you know that Harry has made a mistake.

- Open question, no set answer

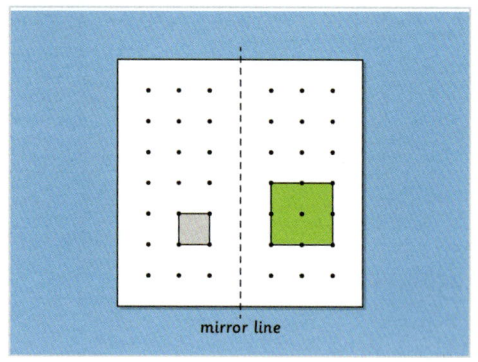

22. A triangle is translated on a coordinate grid.
a b c What are the translated coordinates of point C?
Give your answer in the form (x, y).

■ **(9, 4)**

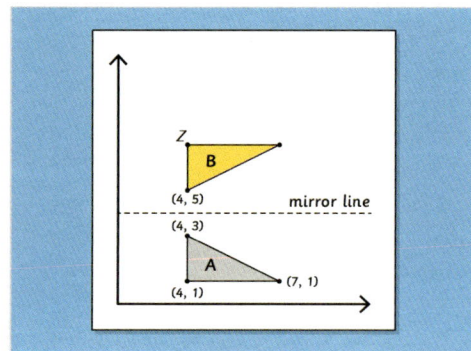

point	original coordinates	translated coordinates
A	(3, 2)	(5, 3)
B	(3, 5)	(5, 6)
C	(7, 3)	?

23. Triangle B is a reflection of triangle A. What
a b c are the coordinates of point Z?
Give your answer in the form (x, y).

■ **(4, 7)**

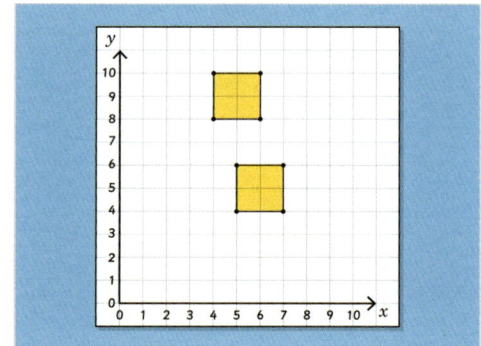

24. Tim says, "The shape has been translated 1
a b c right and 4 down."
Is Tim correct? Explain your answer.

- Open question, no set answer

Level 3 continued

25. Mischa says, "When you reflect a shape on a grid, all the coordinates of the points on the reflected shape are different to the original points."

a b c

Is this true for **all** reflections? Explain your answer.

- Open question, no set answer

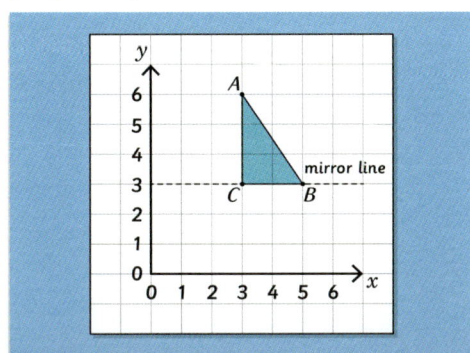

Level 4:	Problem solving with greater depth: Solve multi-step problems involving translating and reflecting shapes.

✳ **Required:** 5/5   ✳ **Pupil Navigation:** on
✳ **Randomised:** off

26. A triangle is translated four squares down and two squares right. The coordinates of the translated triangle are (5, 8), (7, 9) and (9, 6). What are the coordinates of the **original** triangle?

3/5

■ (5, 13) ■ (6, 13) ■ (7, 10) ■ (8, 10) ■ (3, 12)

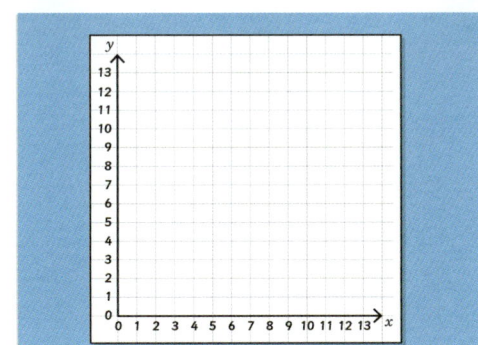

27. Which **three** squares must be shaded to make a pattern that is symmetrical in both mirror lines?

3/6

■ D7 ■ D5 ■ F5 ■ F6 ■ G5 ■ G6

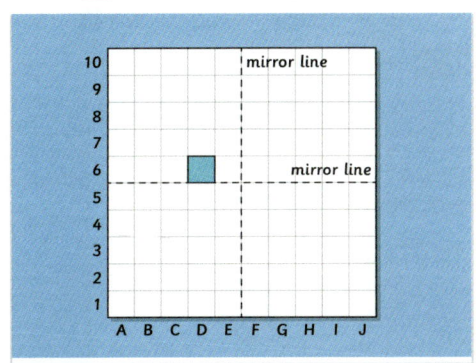

28. A parallelogram has been translated on a coordinate grid. What are the coordinates of points *A*, *B* and *C*?

3/5

■ (14, 2) ■ (19, 5) ■ (21, 2) ■ (20, 2) ■ (20, 5)

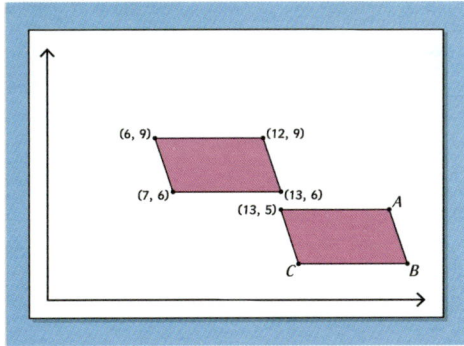

29. Lucia reflects the pattern in the mirror line. In total, what fraction of the squares are **not** shaded after Lucia reflects the pattern?
Put a forward slash (/) between the numerator and the denominator.

a b c

■ 24/50 ■ 12/25 ■ 48/100 ■ 48

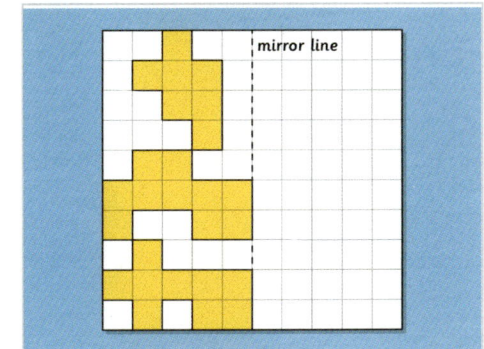

Level 4 *continued*

30. A square has been translated on a
a coordinate grid. What are the coordinates of
b point *A*?
c
Give your answer in the form (x, y).

- **(8, 5)** ▪ **(4, 5)** ▪ **(6, 3)**

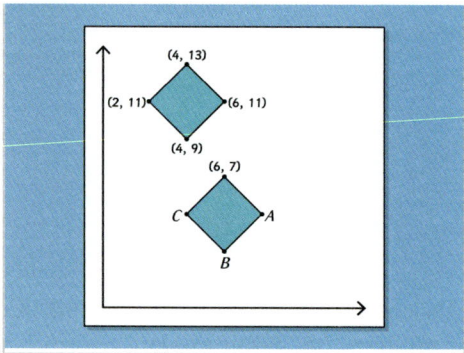

Geometry Topic Review

Objective: I can answer questions involving shapes, transformations and angles from the Year 5 curriculum.

Quick Search Ref: 10740

Level 1: Understanding

✻ **Required:** 7/10   ✻ **Pupil Navigation:** off   ✻ **Randomised:** off

1. Select the **two** 3D shapes.

2/5 ▪ circle ▪ **cube** ▪ triangle ▪ **cuboid** ▪ rectangle

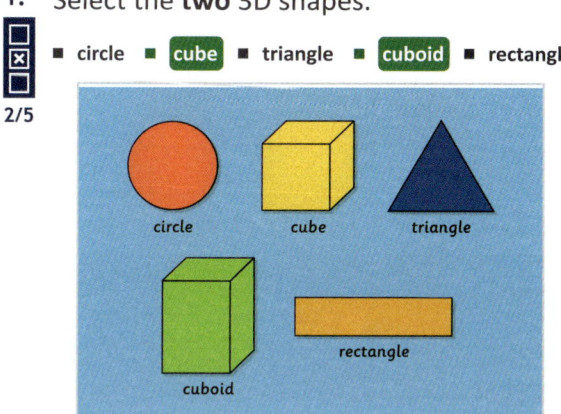

2. Select the **three** shapes that are not regular polygons.

3/5 ▪ **shape A** ▪ **shape B** ▪ shape C ▪ shape D
▪ **shape E**

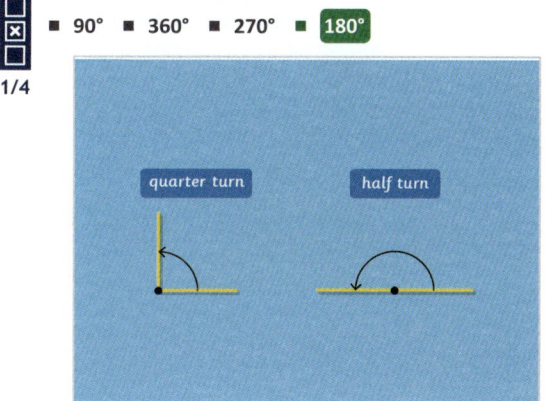

3. How many degrees are there in a **half turn**?

1/4 ▪ 90° ▪ 360° ▪ 270° ▪ **180°**

4. Which angle measures 340°?

1/4 ▪ angle a ▪ angle b ▪ angle c ▪ **angle d**

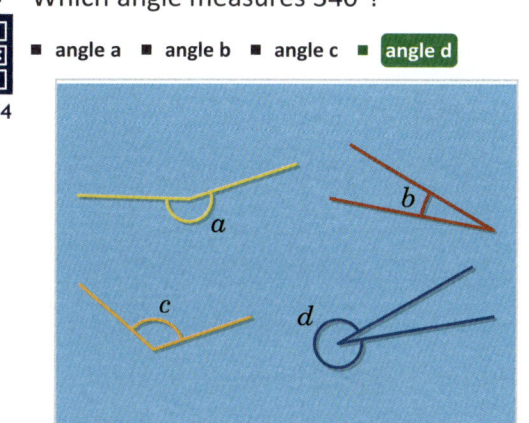

5. The measurement on the protractor shows an angle of 120°. What type of angle is this?

1/4 ▪ acute ▪ right ▪ **obtuse** ▪ reflex

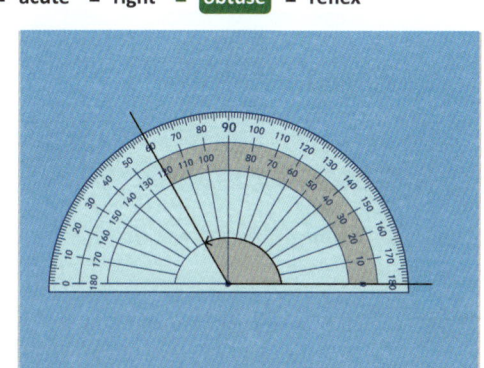

6. What is the size of angle *x* in the square?

1/3 ▪ 60° ▪ 90° ▪ **45°**

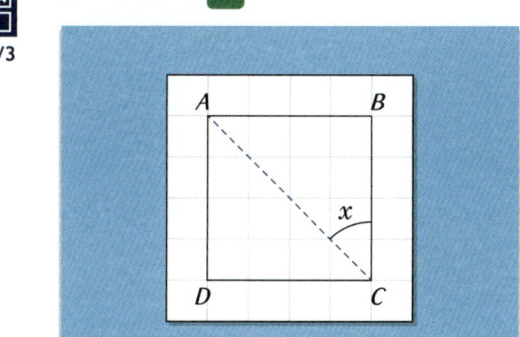

Level 1 *continued*

7. What are the new coordinates of the heart
a b c after it is reflected over the mirror line?
Give your answer in the form (x, y).

- **(8, 2)** ▪ 8, 2 ▪ (2, 2) ▪ (2, 8) ▪ (8 2) ▪ 8 2

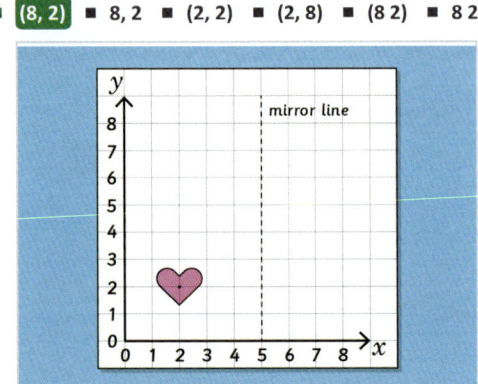

8. Which shape is a translation of shape V?

▪ shape W ▪ **shape X** ▪ shape Y ▪ shape Z

1/4

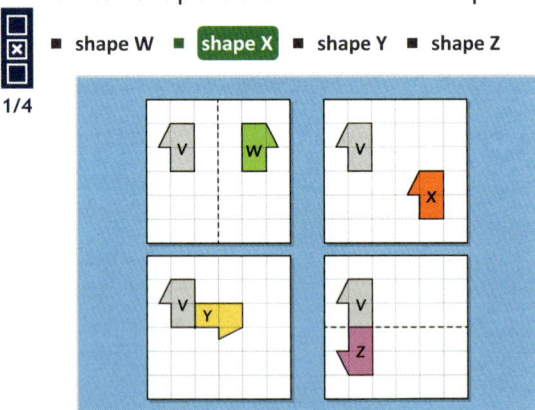

9. What type of angle is shown on the
protractor?

▪ **acute** ▪ reflex ▪ obtuse ▪ right

1/4

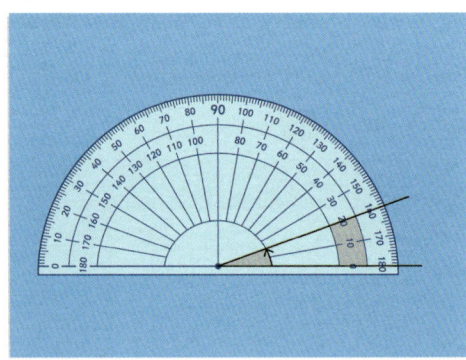

10. Shape *ABCD* is a rectangle. How long is side
a b c *DC*?
Include the units m (metres) in your answer.

▪ **8 metres** ▪ **8 m** ▪ 8

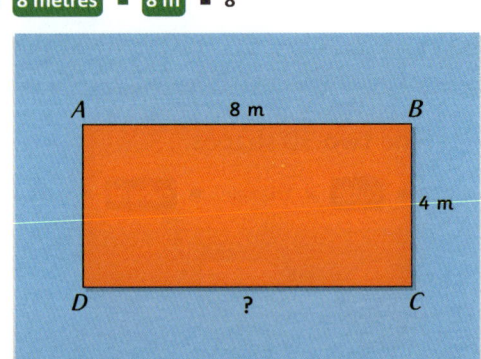

Level 2: Fluency

✿ **Required:** 7/10   ✿ **Pupil Navigation:** off
✿ **Randomised:** off

11. The net folds into which 3D shape?

▪ cube ▪ **cuboid** ▪ square-based pyramid
▪ triangular prism

1/4

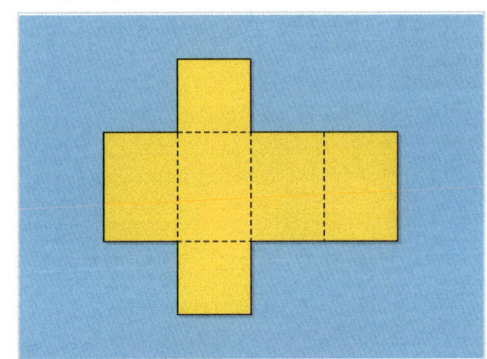

12. Select **two** options that describe the shape in
the image.

▪ a quadrilateral ▪ **a regular polygon** ▪ a hexagon
1/5 ▪ an irregular polygon ▪ **an equilateral triangle**

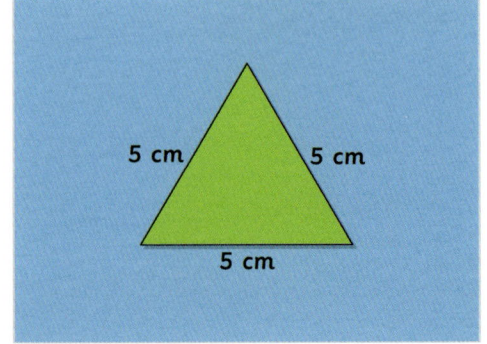

Level 2 *continued*

13. The size of the angle shown on the protractor is _____ degrees (°).
Enter the missing number.
123

- 14 - 12 - 13 - 167

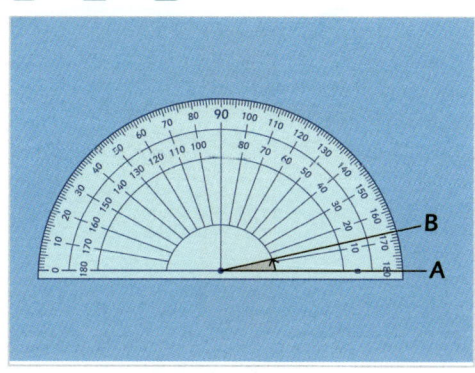

14. *ABCD* is a rectangle. Angle *x* is _____ degrees (°).
Enter the missing angle.
123

- 65

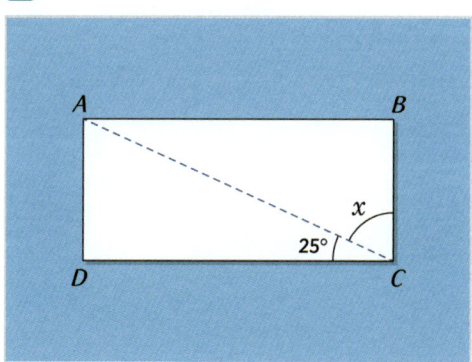

15. The net forms a 3D shape. How many edges does the 3D shape have?
123

- 9 - 5 - 10 - 6

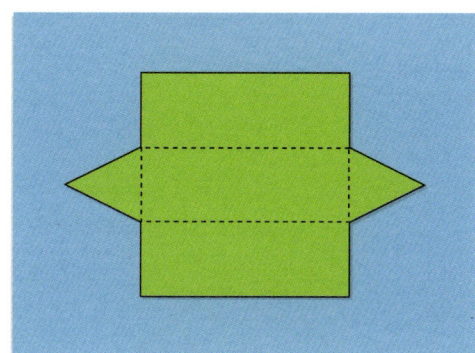

16. Angle *a* is approximately half the size of angle *b*. To the nearest 10°, what is the approximate size of angle *b*?
123

- 60 - 30

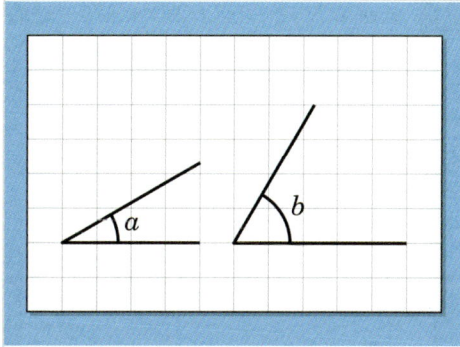

17. Danny moves the pool table in his games room. He translates the pool table _____ right and 2 down.
Enter the missing number.
123

- 6

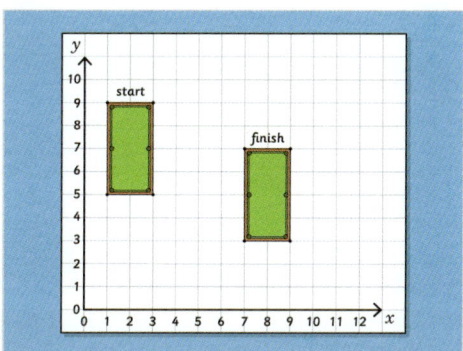

18. If you reflect the shape over the mirror line, what are the new coordinates of point *A*?
Give your answer in the form (x, y).
abc

- (5, 2) - (3, 2) - (7, 2)

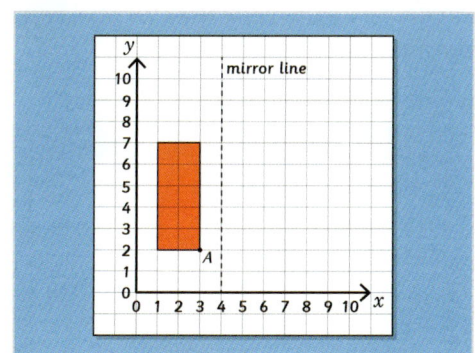

Level 2 *continued*

19. In the parallelogram, angle *b* is three times the size of angle *a*. How many degrees is angle *b*?

1 2 3

- **135** ▪ 45

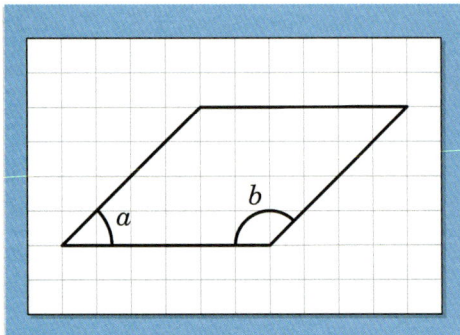

20. In degrees, what is the size of angle *b*?

1 2 3

▪ **220** ▪ 40

Level 3: Reasoning

❋ **Required:** 5/5   ❋ **Pupil Navigation:** off
❋ **Randomised:** off

21. One of the angles in the triangle is hidden by a rectangle. Bailey says that the hidden angle is an acute angle. Is Bailey correct? Explain your answer.

a b c

- Open question, no set answer

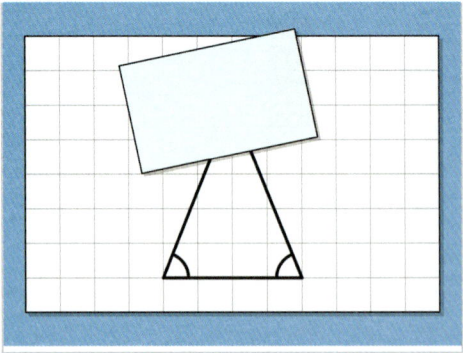

22. The image shows the net of a dice. Jen folds the net to make a 3D shape. When she rolls the dice, the letter B is showing on top. Which letter is on the **bottom** of the dice?

1/5

▪ A ▪ C ▪ D ▪ E ▪ F

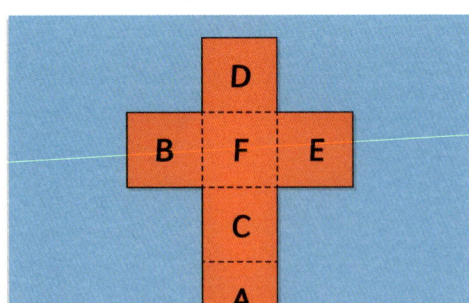

23. Noah has measured angle *x* and says that it's 35°. Explain how you know, without measuring, that Noah has made a mistake. What mistake might he have made?

a b c

- Open question, no set answer

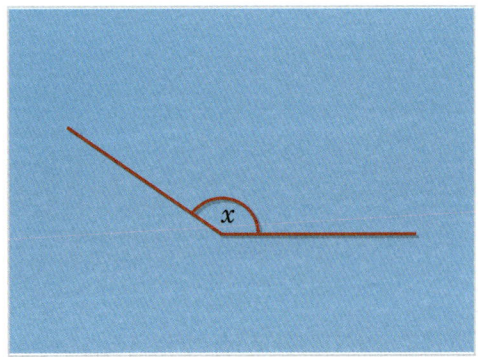

24. Kelly says, "If I know the size of angle *a*, I can find the size of angle *b* and the size of angle *c* without measuring them."
Is Kelly correct? Explain your answer.

a b c

- Open question, no set answer

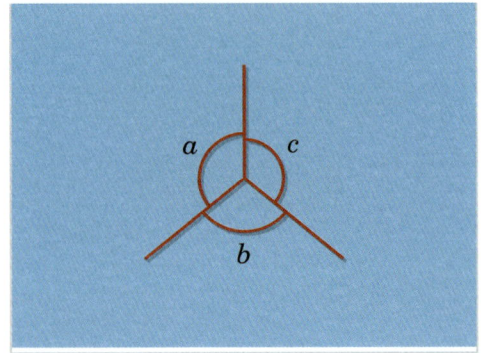

Level 3 *continued*

25. Shape B is a reflection of shape A over the
a b c mirror line. What are the coordinates of
point *Z*?
Give your answer in the form (x, y).

- **(9, 1)**

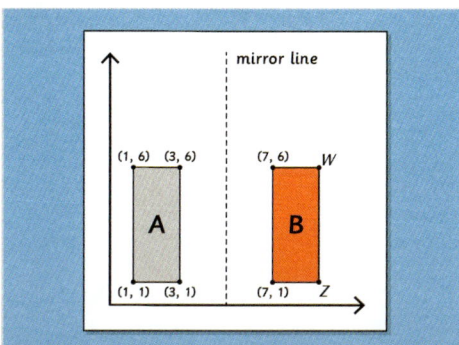

Level 4: Problem solving with greater depth

✴ **Required:** 5/5   ✴ **Pupil Navigation:** off
✴ **Randomised:** off

26. How many irregular polygons are there
1 2 3 inside the large square?

- **9** ▪ 2 ▪ 6 ▪ 4 ▪ 11

27. The scale on a map is 1 square = 1 mile. A
a b c windsurfer is at coordinates (3, 4) on the
map. The wind blows her 4 miles east, but
then changes direction and blows her 2 miles
south. A storm then blows her 7 miles west
and 4 miles north. What are the **new**
coordinates of the windsurfer?

- **(0, 6)** ▪ (7, 2)

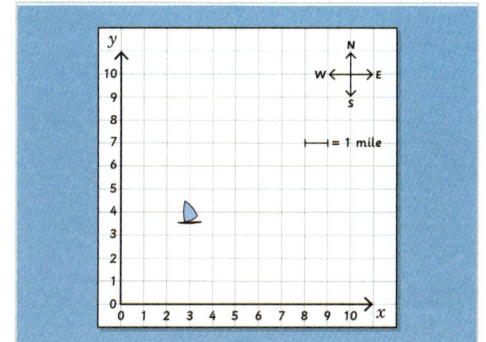

28. A rectangle has an area of 84 cm² and a
a b c perimeter of 38 cm. The lengths of the sides
of the rectangle are all whole numbers.
What is the length of the longer side of the
rectangle?
*Include the units cm (centimetres) in your
answer.*

- **12 centimetres** ▪ **12 cm** ▪ 7 centimetres ▪ 7
- 7 cm ▪ 12

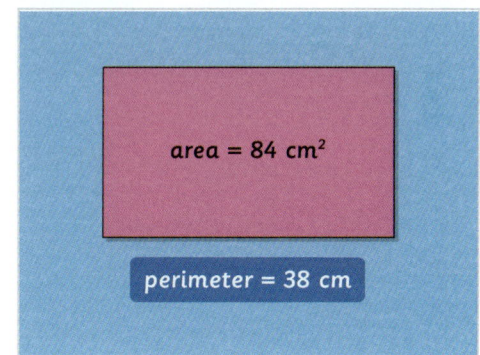

Level 4 *continued*

29. The composite shape is made up of three
regular pentagons. What is the size of angle
a in degrees?

1
2
3

- **36** ■ **324**

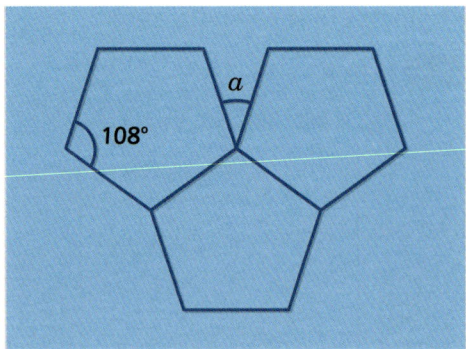

30. Ellie estimates that angle *a* is approximately
35°. If her estimate is wrong by 5°, what is
the largest possible value of angle *b*?

1
2
3

- **150** ■ **140**

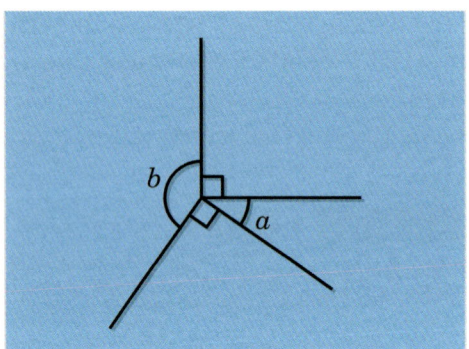

Mathematics Y5

Statistics

Tables
Line Graphs

Complete, Read and Interpret Information in Tables

Objective: I can complete, read and interpret information in tables, including timetables.

Quick Search Ref: 10056

1. Which **two** images show data presented in a table?

2/4

- ■ Image A ■ **Image B** ■ **Image C** ■ Image D

A

B

bus timetable		
Stopton	09:10	10:30
Finetown	09:35	10:55
Yayville	10:05	11:25
Canon	10:20	11:40

C

colour	tally	frequency
blue	||||| |||	8
yellow	||||	4
green	||||| |||||	10
red	||||| ||||	9

D

2. The table shows the number of goals scored in two World Cup tournaments. How many goals did Germany score at the World Cup in 2014?

- ■ **18** ■ 16

Goals scored in two World Cup tournaments

	Argentina	England	Germany
2010	10	3	16
2014	8	2	18

3. In Year 5, how many pupils have packed lunches?

- ■ 54 ■ **35** ■ 89

Children's dinners at Greenbank Primary

	packed lunches	school dinners	total
Year 3	23	67	90
Year 4	17	66	83
Year 5	35	54	89
Year 6	42	39	81
total	117	226	343

4. What is the total number of pupils in the school that have school dinners?

- ■ **226** ■ 343 ■ 117

Children's dinners at Greenbank Primary

	packed lunches	school dinners	total
Year 3	23	67	90
Year 4	17	66	83
Year 5	35	54	89
Year 6	42	39	81
total	117	226	343

5. What is the **total** number of pupils in years 3 to 6?

- ■ 226 ■ **343** ■ 117

Children's dinners at Greenbank Primary

	packed lunches	school dinners	total
Year 3	23	67	90
Year 4	17	66	83
Year 5	35	54	89
Year 6	42	39	81
total	117	226	343

6. The timetable shows where and when a bus stops on its route from Stopton to Canon. Select the **two** times when a bus stops at Yayville.

2/6

- ■ 10:20 ■ 11:40 ■ **10:05** ■ 09:10 ■ **11:25**
- ■ 10:30

bus timetable			
Stopton	09:10	10:30	
Finetown	09:35	10:55	
Yayville	10:05	11:25	
Canon	10:20	11:40	

Level 1 continued

7. The timetable shows where and when a bus stops on its route from Stopton to Canon. Where does the bus stop at 10:55?

1/4

- Stopton ■ Finetown ■ Yayville ■ Canon

bus timetable

Stopton	09:10	10:30
Finetown	09:35	10:55
Yayville	10:05	11:25
Canon	10:20	11:40

8. The timetable shows where and when a bus stops on its route from Stopton to Canon. Select the **two** times when a bus stops at Finetown.

2/6

- 10:55 ■ 10:20 ■ 11:40 ■ 09:10 ■ 09:35
- 10:30

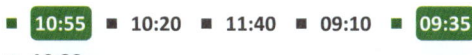

bus timetable

Stopton	09:10	10:30
Finetown	09:35	10:55
Yayville	10:05	11:25
Canon	10:20	11:40

9. In Year 4, how many pupils have school dinners?

- 66 ■ 83 ■ 17

Children's dinners at Greenbank Primary

	packed lunches	school dinners	total
Year 3	23	67	90
Year 4	17	66	83
Year 5	35	54	89
Year 6	42	39	81
total	117	226	343

10. The table shows the number of pieces of fruit that the pupils at Appleby Infant School ate during two weeks. In which **class** did the pupils eat 125 pieces of fruit in week 2?

1/3

- Reception ■ Year 1 ■ Year 2

Fruit eaten at Appleby Infant School

	Reception	Year 1	Year 2
week 1	121	132	127
week 2	113	145	125

Level 2: Fluency: Calculate and use data from a table.

❋ **Required:** 7/10 ❋ **Pupil Navigation:** on
❋ **Randomised:** off

11. The pupils at Court Lane primary have taken part in a survey on their favourite sports. Which sport is most popular in Year 6?

1/4

- football ■ rugby ■ cricket ■ tennis

Favourite sports of pupils at Court Lane Primary

sport	Year 4	Year 5	Year 6
football	10	12	7
rugby	7	12	16
cricket	5	7	1
tennis	3	1	5

12. The pupils at Court Lane Primary have taken part in a survey on their favourite sports. How many **more** pupils prefer rugby in Year 6 than in Year 4?

- 4 ■ 9 ■ 5

Favourite sports of pupils at Court Lane Primary

sport	Year 4	Year 5	Year 6
football	10	12	7
rugby	7	12	16
cricket	5	7	1
tennis	3	1	5

Level 2 continued

13. Three schools are competing in a hockey competition. What is the **total** number of games that Irwell Primary has played so far?

■ **5** ■ 14 ■ 6 ■ 4

Hockey competition results

	Dee Primary	Thames School	Irwell Primary	total
won	3	1	1	
drawn	2	2	2	
lost	1	1	2	
total				

14. Three schools are competing in a hockey competition. How many **more** games have been drawn than won so far?

■ **1** ■ 6 ■ 5

Hockey competition results

	Dee Primary	Thames School	Irwell Primary	total
won	3	1	1	
drawn	2	2	2	
lost	1	1	2	
total				

15. How many minutes does train B take to travel from Faceside to Vertiseas?

■ **33** ■ 48 ■ 17 ■ 36 ■ 31

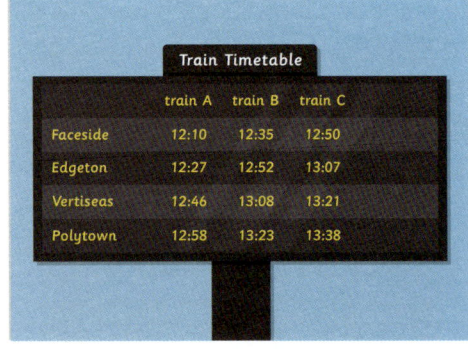

Train Timetable

	train A	train B	train C
Faceside	12:10	12:35	12:50
Edgeton	12:27	12:52	13:07
Vertiseas	12:46	13:08	13:21
Polytown	12:58	13:23	13:38

16. Ryan arrives at the train station in Brazley at 08:30. If he catches the next train, what time will Ryan arrive in Coloton?

1/4

■ 09:05 ■ 09:25 ■ **09:35** ■ 09:47

Train Timetable

Argtown	08:10	08:25	08:45	09:02
Brazley	08:15	08:28	08:47	09:04
Uruaton	08:20	08:35	08:55	09:12
Perville	08:49	09:11	09:21	09:37
Coloton	09:05	09:25	09:35	09:47

17. Katie is meeting a friend at Perville station at 09:15. What is the **latest** time that she can leave Argtown to reach Perville in time to meet her friend?

1/4

■ 08:10 ■ **08:25** ■ 08:45 ■ 09:02

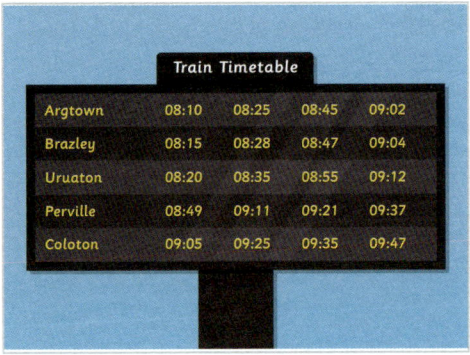

Train Timetable

Argtown	08:10	08:25	08:45	09:02
Brazley	08:15	08:28	08:47	09:04
Uruaton	08:20	08:35	08:55	09:12
Perville	08:49	09:11	09:21	09:37
Coloton	09:05	09:25	09:35	09:47

18. How many minutes does **train A** take to travel from Faceside to Polytown?

■ **48**

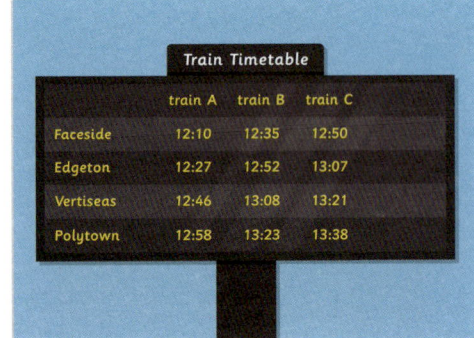

Train Timetable

	train A	train B	train C
Faceside	12:10	12:35	12:50
Edgeton	12:27	12:52	13:07
Vertiseas	12:46	13:08	13:21
Polytown	12:58	13:23	13:38

Level 2 *continued*

19. Three schools are competing in a hockey competition. What is the **total** number of games that have been **lost** in the competition so far?

■ 4 ■ 6 ■ 5

Hockey competition results

	Dee Primary	Thames School	Irwell Primary	total
won	3	1	1	
drawn	2	2	2	
lost	1	1	2	
total				

20. The pupils at Court Lane primary have taken part in a survey on their favourite sports. Which sport is least popular in Year 4?

1/4 ■ football ■ rugby ■ cricket ■ tennis

Favourite sports of pupils at Court Lane Primary

sport	Year 4	Year 5	Year 6
football	10	12	7
rubgy	7	12	16
cricket	5	7	1
tennis	3	1	5

Level 3: Reasoning: Interpret data in tables.

✿ **Required:** 5/5   ✿ **Pupil Navigation:** on
✿ **Randomised:** off

21. Ellen says that her investigation results show that the marble travels fastest through washing up liquid. Is Ellen correct? Explain your answer.

- Open question, no set answer

Time taken for a marble to move through different liquids

liquid	time taken (s)
oil	5.7
water	3.4
washing up liquid	6.2
vinegar	3.1

22. The 08:10 from Argtown is delayed by 20 minutes at Brazley. Janie says she would arrive at Uruaton earlier by catching the 08:25 train instead of the 08:10 train. Is Janie correct? Explain your answer.

- Open question, no set answer

Train Timetable

Argtown	08:10	08:25	08:45	09:02
Brazley	08:15	08:28	08:47	09:04
Uruaton	08:20	08:35	08:55	09:12
Perville	08:49	09:11	09:21	09:37
Coloton	09:05	09:25	09:35	09:47

23. The table shows the number of fiction and non-fiction books borrowed from the school library during October. What is the missing value for non-fiction books borrowed by pupils in Year 5?

■ 31

Books borrowed from school during October

	fiction	non-fiction	total
Year 3	29	16	45
Year 4	34	14	48
Year 5	18	?	
Year 6	23	29	52
total	104	90	194

24. Freya says that Junior Masterbrain is on for longer than Sports Special. Is Freya correct? Explain your answer.

- Open question, no set answer

TV GUIDE

time	programme
16:00	Monkey Business
16:15	Supermouse
16:25	Junior Masterbrain
17:00	News Update
17:20	Sports Special
18:00	Nature is Nice
19:00	News

Level 3 continued

25. The timetable shows a typical school day. Freddie has a 30 minute drum lesson which starts at one o'clock. Select **two correct** statements about Freddie's school day.5

2/5

- ■ Freddie misses 15 minutes of the history lesson.
- ■ Freddie misses 30 minutes of the history lesson.
- ■ Freddie spends 6.5 hours at school.
- ■ Freddie misses half an hour of the sport lesson.
- ■ Freddie spends more time in history than in sport.

School timeable

09:00	register
09:15	assembly
09:30	maths
10:30	playtime
11:00	English
12:00	dinner
13:15	topic
14:00	sport
15:30	home

Level 4: Problem solving in greater depth: Solve multi-step problems involving data presented in tables.

� **Required:** 5/5   ✿ **Pupil Navigation:** on
✿ **Randomised:** off

26. How many minutes **longer** is the film than the wildlife programme?

1
2
3

- ■ **90** ■ 165 ■ 75

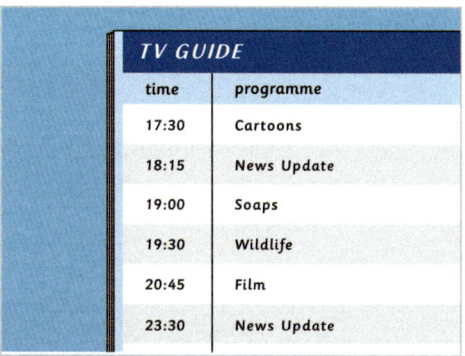

TV GUIDE	
time	**programme**
17:30	Cartoons
18:15	News Update
19:00	Soaps
19:30	Wildlife
20:45	Film
23:30	News Update

27. Mrs Halstead plans four activities for each of her art classes:

1
2
3

sketching: 10 minutes
clay modelling: 15 minutes
pastel colouring: 15 minutes
painting: 20 minutes

How many **more** minutes do the pupils in **Year 6** need to be able to complete all four activities during their lesson?

- ■ **5** ■ 55

Art class timetable

class	time
Year 3	09:00 – 09:45
Year 4	10:15 – 11:00
Year 5	12:30 – 13:30
Year 6	14:20 – 15:15

28. Mark arrives at Paperton bus stop at 08:48. How many **minutes** will he have to wait for the next bus to Pen Village?

1
2
3

- ■ **53** ■ 7

bus timetable

	bus 1	bus 2	bus 3	bus 4	bus 5
Paperton	07:31	08:17	08:55	09:41	10:27
Stapleville	07:49	08:39			10:51
Pen Village		08:54		10:03	11:13
Markermont	08:11	09:09	09:32	10:22	11:25

Level 4 *continued*

29. Four runners compare their times for completing the same race in May and June. How many minutes is the biggest time improvement between May and June?

▪ -8 ▪ **8** ▪ 10 ▪ 3 ▪ -3

Race results

	May (minutes)	June (minutes)
Alfie	51	45
Dean	37	34
Lucy	43	53
Becky	62	54

30. The table shows the results of a survey into where people went on holiday last summer. How many **more** people had a holiday abroad in August than had a holiday in the UK in July?

▪ 20 ▪ **21** ▪ 41

Holiday survey

	UK	abroad	total
June	7		9
July		14	34
August	12		
total		57	96

Interpret Information Presented in a Line Graph

Objective: I can read, compare and calculate data presented in a line graph.

Quick Search Ref: 10300

1. Which image shows data presented in a line graph?

☐
☒
☐

1/3

■ image A   ■ image B   ■ **image C**

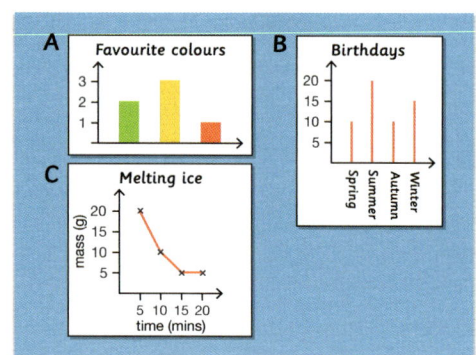

2. Which table has values that can be represented on a line graph?

☐
☒
☐

1/3

■ table A   ■ **table B**   ■ table C

3. The image shows a line graph of how far a snail crawls over time. How many centimetres (cm) are represented by one square on the *y*-axis?

1
2
3

■ **2.5**   ■ 5   ■ 1

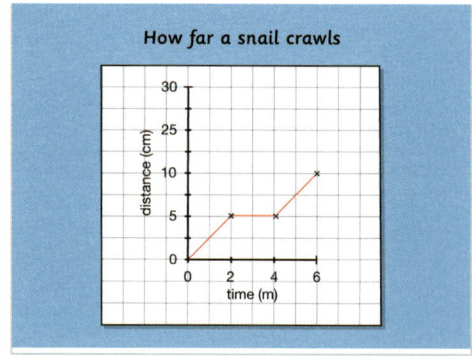

4. The graph shows the average temperature in a city during one year. In degrees Celsius (°C), what is the average temperature in April?

1
2
3

■ **12**

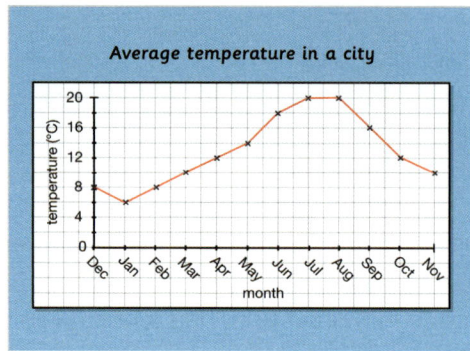

5. Each year on Chloe's birthday, her mum records Chloe's height in centimetres (cm). In which year was Chloe 117 cm tall?

1
2
3

■ **2016**

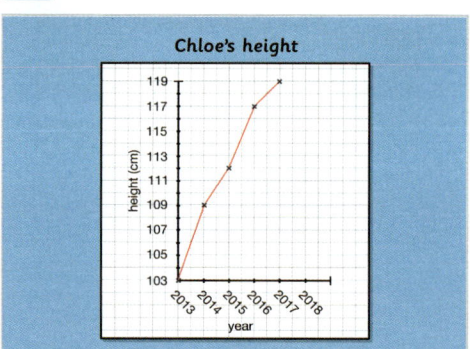

6. How many dollars ($) are the same as £60 on the conversion graph?

1
2
3

■ **40**   ■ 90

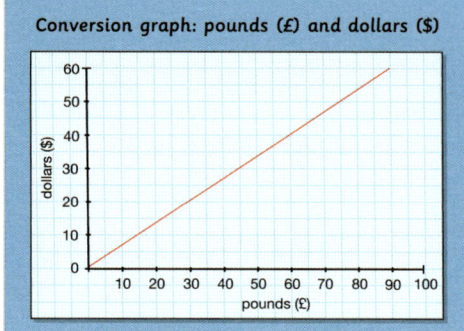

Level 1 *continued*

7. How many pounds (£) are the same as $30 on the conversion graph?

1 2 3

- ▪ 45 ▪ 20

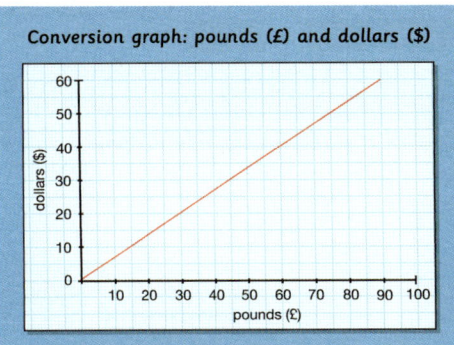

8. In which **two** months was the average temperature in Manchester 12°C?

☐☒☐ 2/7

- ▪ January ▪ March ▪ April ▪ May ▪ June
- ▪ September ▪ October

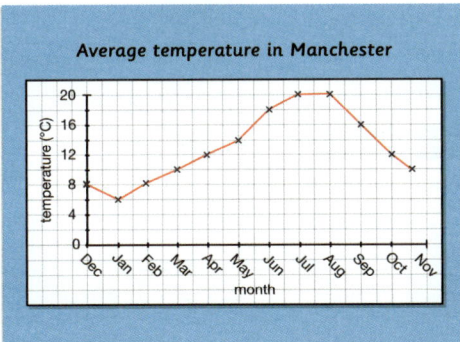

9. The image shows a line graph of how far a snail crawls over time. How many minutes (m) are represented by one square on the *x*-axis?

1 2 3

- ▪ 1 ▪ 2.5 ▪ 2

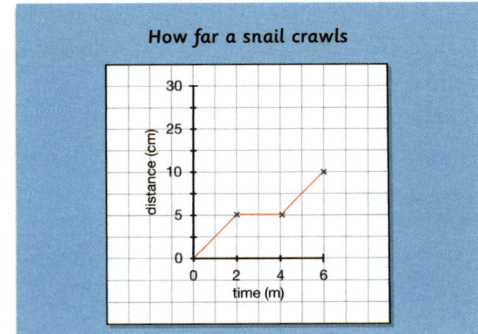

10. Each year on Chloe's birthday, her mum records Chloe's height in centimetres (cm). How many centimetres tall was Chloe on her birthday in 2014?
Don't include the units in your answer.

1 2 3

- ▪ 109

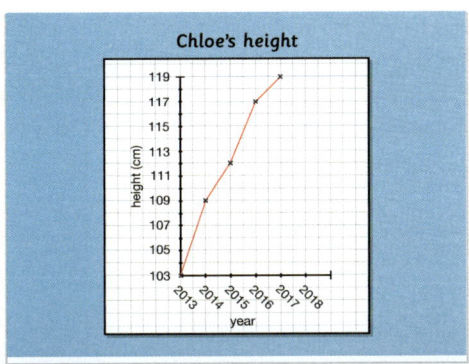

Level 2: Fluency: Compare and calculate data in a line graph.

✸ **Required:** 7/10   ✸ **Pupil Navigation:** on
✸ **Randomised:** off

11. In millions, what was the population of the UK in 1965?

a b c

- ▪ 55 million ▪ 55 ▪ 55000000 ▪ 55,000,000

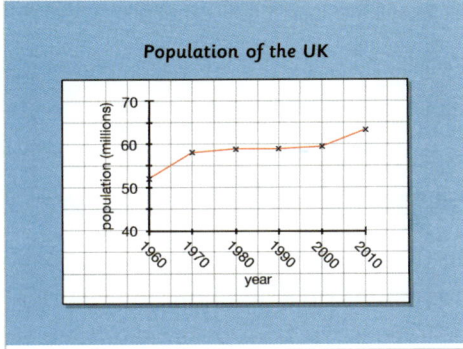

12. Jack's heart rate is measured every five minutes during an exercise session. How many minutes of exercise has he completed when his heart rate starts to decrease for the first time?

1 2 3

- ▪ 15 ▪ 30

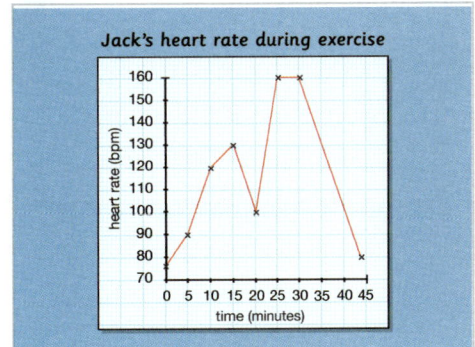

Level 2 continued

13. The noise level at a football match is
measured in decibels every ten minutes. In
decibels, what is the **difference** between the
loudest and the quietest noise levels?

■ **40** ■ 60 ■ 100

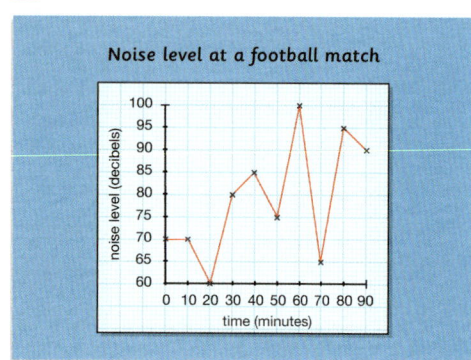

14. The temperature in Freetown is recorded
every hour from 21:00 to 09:00. For how
many hours is the temperature **below** 0°C?

■ **7** ■ 8

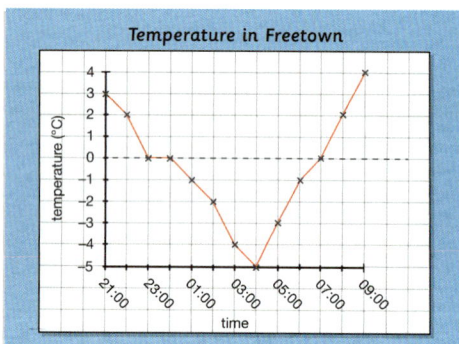

15. The speed of a steam train is recorded every
ten minutes on a two-hour journey. By how
many miles per hour (mph) did the train's
speed decrease between 50 and 70 minutes
into the journey?

■ **30** ■ 50 ■ 80

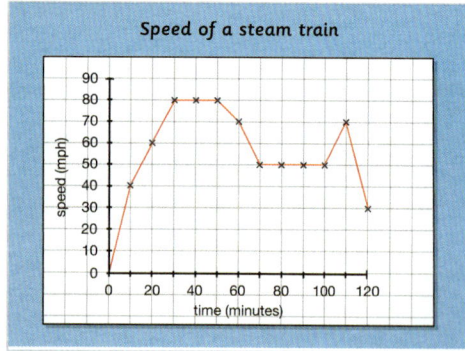

16. What is the difference between the highest
amount of rainfall in Brazil and the highest
amount of rainfall in the UK?
*Include the units mm (millimetres) in your
answer.*

■ **160 millimetres** ■ **160 mm** ■ 160

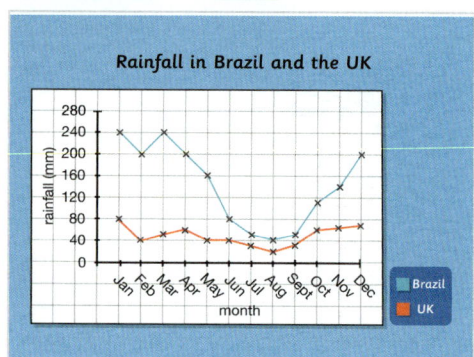

17. The length and weight of a baby boy and girl
are plotted on the same graph. How long are
the babies when their lengths and weights
are the same?
*Include the units cm (centimetres) in your
answer.*

■ **60 centimetres** ■ 6 ■ **60 cm** ■ 6 cm
■ 6 centimetres ■ 60

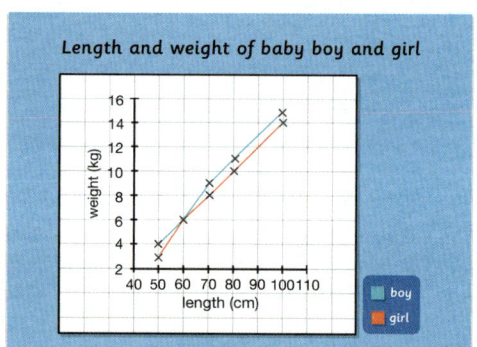

18. What is the difference between the lowest
amount of rainfall in Brazil and the lowest
amount of rainfall in the UK?
*Include the units mm (millimetres) in your
answer.*

■ **20 millimetres** ■ **20 mm** ■ 20

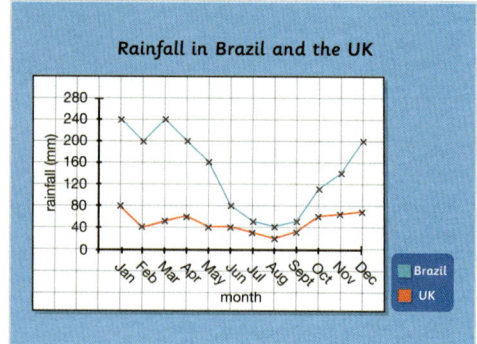

Level 2 continued

19. Jack's heart rate is measured every five minutes during an exercise session. How many minutes does it take for his heart rate to increase from 90 beats per minute (bpm) to 130 beats per minute?

- ■ 10

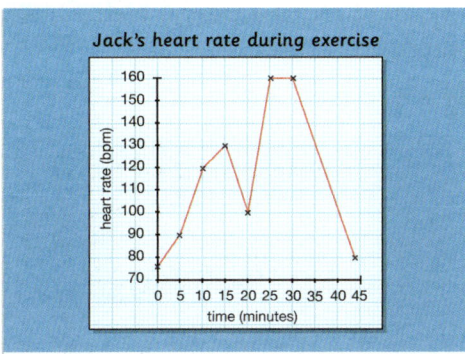

20. The noise level at a football match is measured in decibels every ten minutes. At what time **during the first 50 minutes** of the football match is the noise level the loudest?

- ■ 40 ■ 85 ■ 60

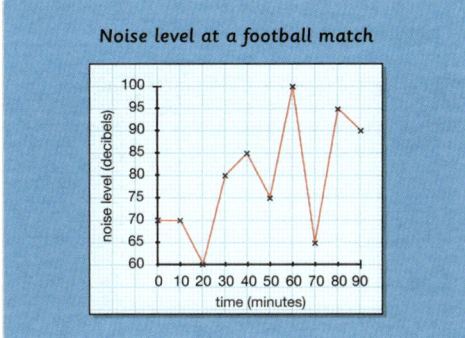

Level 3: Reasoning: Reason about the interpretation and presentation of data in a line graph.

✻ **Required:** 5/5   ✻ **Pupil Navigation:** on
✻ **Randomised:** off

21. Jacob says that the temperature of the water is 40°C after 5.5 minutes. Is Jacob correct? Explain your answer.

- Open question, no set answer

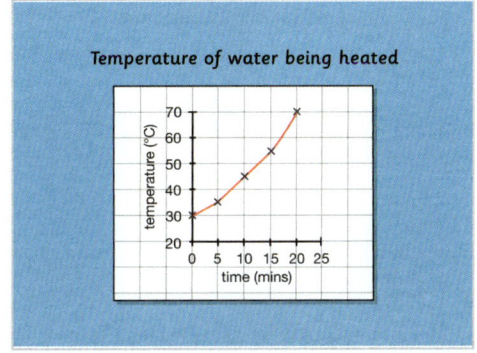

22. Mark carries out a traffic survey and records the colour of each car that he sees. Explain why Mark can't use a line graph to present his data.

- Open question, no set answer

23. Use the conversion graph to calculate how many pounds (£) are the same as $80.

- ■ 120

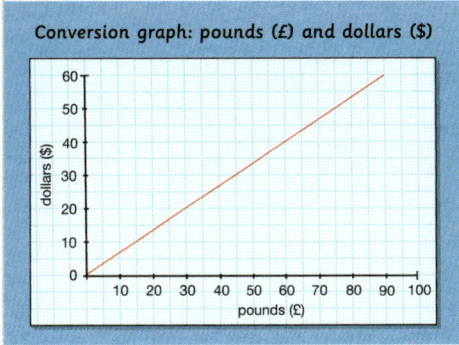

Level 3 *continued*

24. Alfie drives away from home to go shopping. He parks his car while he shops and then goes back home. If Alfie plots his shopping trip on a line graph, which diagram shows what his graph would look like?

1/3

- ■ diagram A ■ diagram B ■ **diagram C**

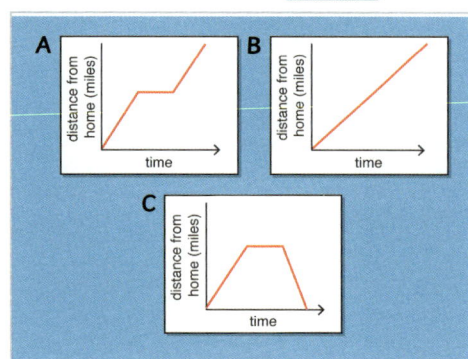

25. Select one **incorrect** statement about the data on the graph.

1/4

- ■ The rainfall in Brazil is always greater than the rainfall in the UK.
- ■ The rainfall in August is less than in December in both Brazil and the UK.
- ■ **The biggest difference between the rainfall in Brazil and in the UK is in February.**
- ■ Brazil never has less than 4 cm of rainfall.

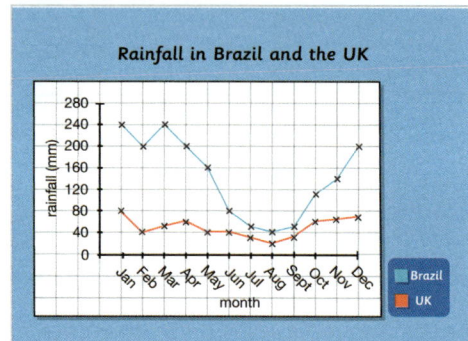

Level 4: Problem solving in greater depth: Solve multi-step problems using data from a line graph.

✷ **Required:** 5/5   ✷ **Pupil Navigation:** on
✷ **Randomised:** off

26. The graph shows the depth of water in a tank over a period of time. If the tank is full when the depth is 0.5 metres (m), how many more minutes did it take to fill the tank than to empty it?

1
2
3

- ■ **5** ■ 10

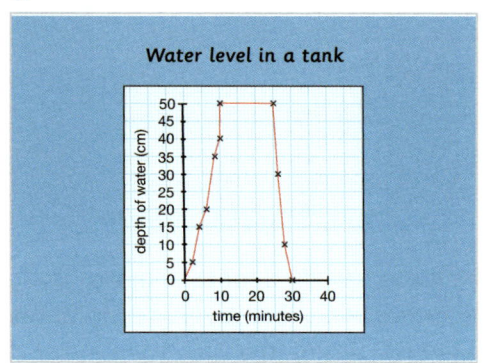

27. Jack's heart rate is measured every five minutes during an exercise session. How many beats per minute (bpm) has his heart rate **risen** after 22 minutes?

1
2
3

- ■ **44** ■ 76 ■ 120

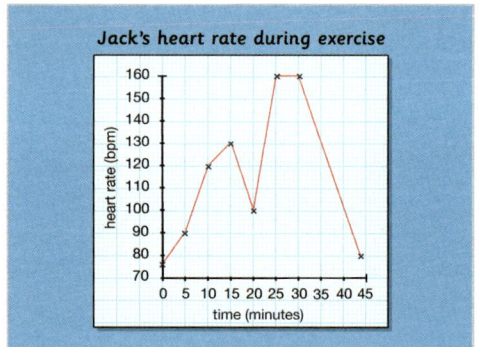

Level 4 *continued*

28. Alice runs 100 metres (m) from the start line
to the finish line on a running track. She has
a rest, then runs back to the start line. How
many seconds does it take Alice to complete
150 metres of the track?

1
2
3

- ■ 50 ■ 30

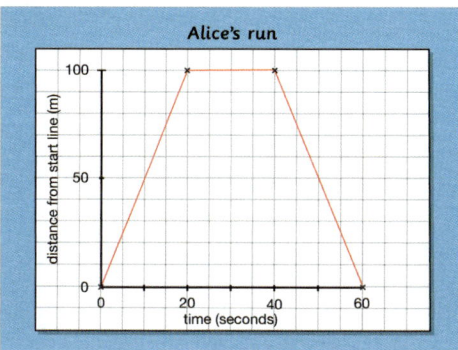

29. Lucy wants to change £45 into euros (€). Use
the conversion charts to find out how many
euros she will get for £45.

1
2
3

- ■ 25 ■ 30

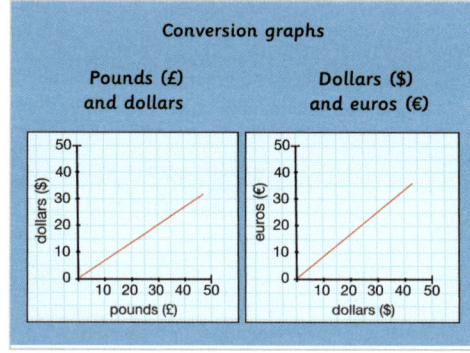

30. The populations of the towns Headly and
Footon are recorded every year and plotted
on a line graph for a five year period. How
many months is the population of Footon
higher than Headly?

1
2
3

- ■ 6 ■ 12 ■ 1

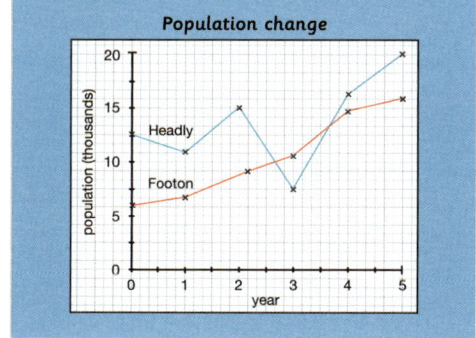

Ref:10300   Interpret Information Presented in a Line Graph

LbQ Super Deal
Class set of tablets and charging cabinet

Class charging & storage cabinet

+ 32 x
Pupil 8" tablets with protective cover

+ 1 x
Teacher 10" tablet with protective cover

*Special Offer Price

£1,100 per year on 3 years compliant operating lease

Subject to a £150 initial documentation fee

LbQ Question Set subscription required to be eligible

Min 1 LbQ subscription per set £200/year or £500/3 years

Learning by Questions app pre-loaded

Includes 3 years advanced replacement warranty on tablets (damage not covered)

Prices exclude VAT and delivery

Option to renew equipment or purchase at end of agreement

Available in United Kingdom and Republic of Ireland only

Price subject to change at any time

£1,100* per year for 3 years including warranty

bett AWARDS 2019 WINNER INNOVATOR OF THE YEAR

era 2019 WINNER

Place your orders with our sales partner LEB who will organise the paperwork for you:

Email: orders@lbq.org
Tel: 01254 688060

Learning by Questions

Specifications

Charging & Storage Cabinet

- 33-bay up to 10" tablet charging cabinet
- 2 easy access sliding shelves
- 4 efficient fans for ventilation
- locking doors with keys
- 4 castors / 2 handling bars overload, leakage and lightning surge protection
- unladen weight 96kg
- CE / ROHS / FCC compliancy
- 3 years warranty

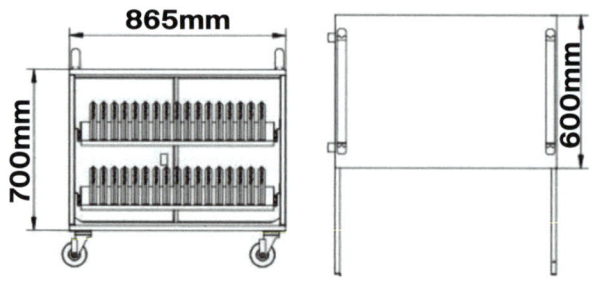

Student and Teacher Tablets Configuration

All tablets with LbQ Tasks app pre-loaded and tablets installed in cabinet including charging cables for quick and easy deployment in classrooms.

	8" Android Tablet with Protective Cover	10" Android Tablet with Protective Cover
Display		
Resolution	1280 x 800	1920 x 1200
16:10 display ratio	✓	✓
Capacitive 5-touch	Capacitive 5-touch	Capacitive 10-touch
System		
Cortex 64bit Quad Core 1.5GHz CPU	✓	✓
2GB of RAM	✓	✓
16Gb of storage	✓	✓
Android 7.0	✓	✓
Front and rear Camera	✓	✓
Input / Output Ports		
1 x Micro SD Slot	✓	✓
1 x Micro USB (PC / device / charger)	✓	✓
Micro-HDMI output	✓	✓
1 x Earphone, 1 x Speaker, 1 x Mic	1 x Earphone, 1 x Speaker, 1 x mic	1 x Earphone, 2 x Speaker, 1 x mic
Communication		
Wifi – 802.11a/b/g/n	2.4 GHz / 5 GHz	2.4 GHz
GPS module	✓	✓
Bluetooth	✓	✓
Power		
5V 2A	✓	✓
Battery	3500 mAh battery	7000 mAh battery
Physical		
Colour: metal black	✓	✓
Weight	300g	560g
Dimensions	205 x 114 x 8mm (approx.)	263 x 164 x 9mm (approx.)
Warranty		
3 years advanced replacement for faulty tablets - does not cover damage CE / ROHS / FCC compliant	✓	✓